Digital VLSI Design with Verilog

John Michael Williams

Digital VLSI Design with Verilog

A Textbook from Silicon Valley Polytechnic Institute

Second Edition

 Springer

John Michael Williams
Wilsonville, OR
USA

Additional material to this book can be downloaded from http://extras.springer.com.

ISBN 978-3-319-04788-1   ISBN 978-3-319-04789-8 (eBook)
DOI 10.1007/978-3-319-04789-8
Springer Cham Heidelberg New York Dordrecht London

Library of Congress Control Number: 2014938203

© Springer International Publishing Switzerland 2014

This work is subject to copyright. All rights are reserved by the Publisher, whether the whole or part of the material is concerned, specifically the rights of translation, reprinting, reuse of illustrations, recitation, broadcasting, reproduction on microfilms or in any other physical way, and transmission or information storage and retrieval, electronic adaptation, computer software, or by similar or dissimilar methodology now known or hereafter developed. Exempted from this legal reservation are brief excerpts in connection with reviews or scholarly analysis or material supplied specifically for the purpose of being entered and executed on a computer system, for exclusive use by the purchaser of the work. Duplication of this publication or parts thereof is permitted only under the provisions of the Copyright Law of the Publisher's location, in its current version, and permission for use must always be obtained from Springer. Permissions for use may be obtained through RightsLink at the Copyright Clearance Center. Violations are liable to prosecution under the respective Copyright Law.

The use of general descriptive names, registered names, trademarks, service marks, etc. in this publication does not imply, even in the absence of a specific statement, that such names are exempt from the relevant protective laws and regulations and therefore free for general use.

While the advice and information in this book are believed to be true and accurate at the date of publication, neither the authors nor the editors nor the publisher can accept any legal responsibility for any errors or omissions that may be made. The publisher makes no warranty, express or implied, with respect to the material contained herein.

Printed on acid-free paper

Springer is part of Springer Science+Business Media (www.springer.com)

To my loving grandparents,
William Joseph Young (*ne* Jung) and
Mary Elizabeth Young (*nee* Egan)
who cared for my brother Kevin and me
when they didn't have to.

Preface to the Second Edition

Like the first edition, this book is based on the lab exercises and order of presentation of a course developed and given by the author over a period of years at what is now Silicon Valley Polytechnic Institute, San Jose, California.

To the author's best knowledge, this course was—and still is—the only one ever given which (*a*) presented the entire verilog language; (*b*) involved implementation of a full-duplex serdes simulation model; or (*c*) included design of a synthesizable digital PLL.

The author wishes to thank the owner and CEO of *Silicon Valley Polytechnic Institute*, Dr. Ali Iranmanesh, for his patience and encouragement during the course development and in the preparation of this book.

In the second edition, many minor typographical errors have been corrected, as have been several other errors newly discovered in the text and figures.

Major upgrades in the second edition are:
- Modified Day 1 presentation making it more useful to verilog beginners
- Dozens of new figures
- Expansion or clarification of explanations on almost every page
- Upgrade of the simulation figures to be in color
- New coverage of the features of *SystemVerilog* and *VerilogA/MS*
- A new summary introduction to each chapter and lab exercise
- IEEE Stds references including SystemVerilog as well as verilog
- A new, optional lab checklist for recording learning progress

As was done for the first edition, corrections, changes, and teaching information will be posted online at *Scribd*.

Table of Contents

Chapter 1: Introductory Material 1
  1 Course Description 1
  2 Using this book 1
      2.1 Textbook Extras 2
      2.2 Performing the lab exercises 3
      2.3 Proprietary Information and Licensing Limitations............ 3
  3 Textbook References 4
      3.1 Supplementary Textbooks 4
      3.2 Interactive Language Tutorial 4
      3.3 Recommended Free Verilog Simulator 5
      3.4 Reading References... 5
  4 Lab Checklist... 9

Chapter 2: Week 1 Class 1 13
  2 Today's Agenda:.. 13
      2.1 Introductory Lab 1 14
      2.2 Lab 1 Postmortem and Lecture 25
      2.3 Verilog vectors 29
      2.4 Logical (Boolean) Operators 32
      2.5 Bitwise Operators: Vectors and Reduction......................... 33
      2.6 Operator Lab 2... 34
            2.6.1 Lab postmortem 35
      2.7 First-Day Wrapup 35
            2.7.1 VCD File Dump 35
            2.7.2 SDF File Dump 36
            2.7.3 The Importance of Synthesis 37
      2.8 Additional Study 37

Chapter 3: Week 1 Class 2 39
  3 Today's Agenda:.. 39
      3.1 More Language Constructs 40
            3.1.1 Traditional module header format............................ 40
            3.1.2 Modern module header format................................... 40
            3.1.3 Header formats contrasted 41
            3.1.4 Verilog comments 43
            3.1.5 always blocks 44
            3.1.6 Initial blocks 44
            3.1.7 Continuous assignments 45
            3.1.8 Vectors and vector values 45
            3.1.9 Parameters.. 48
            3.1.10 Commenting with verilog macroes........................... 48
      3.2 Parameter and Conversion Lab 3 50
            3.2.1 Lab postmortem 52
      3.3 Procedural control... 53
            3.3.1 Procedural Control Constructs 53

	3.3.2 Conditional Expression Operator	53
	3.3.3 Combinational and Sequential Logic	54
	3.3.4 Verilog Strings and Messages	56
	3.3.5 Shift Registers	58
	3.3.6 Reconvergence Design Note	59
3.4	Nonblocking Control Lab 4	60
	3.4.1 Lab postmortem	65
3.5	Additional Study	66

Chapter 4: Week 2 Class 1 67

4	Today's Agenda:	67
4.1	Net Types, Simulation, and Scan	67
	4.1.1 Variables and Constants	69
4.2	Identifiers	69
4.3	Concurrent *vs* Procedural Blocks	70
4.4	Miscellaneous Other Verilog Features	70
4.5	Backus-Naur Format (BNF)	70
4.6	Verilog Semantics	70
4.7	Modelling Sequential Logic	73
4.8	Design for Test (DFT): Scan Lab Introduction	76
4.9	Simple Scan Lab 5	79
	4.9.1 Lab postmortem	88
4.10	Additional Study	88

Chapter 5: Week 2 Class 2 89

5	Today's Agenda:	89
5.1	PLLs and the SerDes Project	89
	5.1.1 Phase-Locked Loops	89
	5.1.2 A 1x Digital PLL	90
	5.1.3 Introduction to SerDes and PCI Express	96
	5.1.4 The SerDes of this course	98
	5.1.5 A 32 x Digital PLL	100
5.2	PLL Clock Lab 6	101
	5.2.1 Note on Synthesis don't_touch	108
	5.2.2 Lab postmortem	112
5.3	Additional Study	113

Chapter 6: Week 3 Class 1 115

6	Today's Agenda:	115
6.1	Data Storage and Verilog Arrays	115
	6.1.1 Memory: Hardware and Software Description	115
	6.1.2 Definitions of Memory Size	116
	6.1.3 Verilog Arrays	116
6.2	A Simple RAM Model	119
	6.2.1 Verilog Concatenation	120
6.3	Memory Data Integrity	120
	6.3.1 Error Checking and Correcting (ECC)	123
	6.3.2 ECC from parity	123
	6.3.3 Parity for SerDes Frame Boundaries	127

Table of Contents xi

    6.4 Memory Lab 7 128
              6.4.1 Lab postmortem 132
    6.5 Additional Study... 132

Chapter 7: Week 3 Class 2... 133
    7   Today's Agenda:... 133
    7.1 Counter Types and Structures.. 133
              7.1.1 Introduction to Counters 133
              7.1.2 Terminology: Behavioral, Procedural, RTL, Structural 135
              7.1.3 Adder Expression vs Counter Statement 137
              7.1.4 Counter Structures 137
              7.1.5 Ripple Counter 138
              7.1.6 Carry Look-Ahead (Synchronous) Counter...................... 139
              7.1.7 One-Hot and Ring Counters... 140
              7.1.8 Gray Code Counter 140
    7.2 Counter Lab 8 141
              7.2.1 Lab postmortem 146
    7.3 Additional Study... 146

Chapter 8: Week 4 Class 1... 147
    8   Today's Agenda:... 147
    8.1 Contention and Operator Precedence.. 147
              8.1.1 Verilog Net Types and Strengths...................................... 148
              8.1.2 Verilog Strength Usage... 149
              8.1.3 Race Conditions, Again 151
              8.1.4 Unknowns in Relational Expressions 155
              8.1.5 Verilog Operators and Precedence................................... 156
    8.2 Digital Basics: Decoder and Three-State Buffer 157
    8.3 Strength and Contention Lab 9 157
              8.3.1 Strength Lab postmortem 165
    8.4 Back to the PLL and the SerDes 165
              8.4.1 Named Blocks 165
              8.4.2 The PLL in a SerDes.. 166
              8.4.3 The SerDes Packet Format Revisited 168
              8.4.4 Behavioral PLL Synchronization (language digression) 168
              8.4.5 Unsynthesizability of the Behavioral PLL Code 177
              8.4.6 Synthesizable, Pattern-Based PLL Synchronization....................... 177
    8.5 PLL Behavioral Lock-In Lab 10 178
              8.5.1 Lock-in Lab postmortem 181
    8.6 Additional Study... 181

Chapter 9: Week 4 Class 2... 183
    9   Today's Agenda:... 183
    9.1 State Machine and FIFO design.. 183
              9.1.1 Verilog Tasks and Functions 183
              9.1.2 A Function For Synthesizable PLL Synchronization 186
              9.1.3 Concurrency by fork-join... 187
              9.1.4 Verilog State Machines 189
              9.1.5 FIFO Functionality 190
              9.1.6 FIFO Operational Details... 192

        9.1.7 A Verilog FIFO 197
   9.2 FIFO Lab 11 205
        9.2.1 Lab postmortem 209
   9.3 Additional Study 209

Chapter 10: Week 5 Class 1 211
   10 Today's Agenda: 211
   10.1 Rise-Fall Delays and Event Scheduling 211
        10.1.1 Types of Delay Expression 211
        10.1.2 Verilog Simulation Event Queue 215
        10.1.3 Simple Stratified Queue Example 217
        10.1.4 Event Controls 220
        10.1.5 Event Queue Summary 221
   10.2 Scheduling Lab 12 222
        10.2.1 Lab postmortem 228
   10.3 Additional Study 229

Chapter 11: Week 5 Class 2 231
   11 Today's Agenda: 231
   11.1 Built-in Gates and Net Types 231
        11.1.1 Verilog Built-in Gates 231
        11.1.2 Implied Wire Names 232
        11.1.3 Net Types and Their Default 233
        11.1.4 Structural Use of Wire vs Reg 234
        11.1.5 Port and Parameter Syntax Note 235
        11.1.6 A D Flip-flop from SR Latches 236
   11.2 Netlist Lab 13 239
        11.2.1 Lab postmortem 242
   11.3 Additional Study 242

Chapter 12: Week 6 Class 1 245
   12 Today's Agenda: 245
   12.1 Verilog Procedural Control Statements 245
        12.1.1 Verilog case Variants 249
        12.1.2 Procedural Concurrency 253
        12.1.3 Verilog Name Space 257
   12.2 Concurrency Lab 14 258
        12.2.1 Lab postmortem 261
   12.3 Additional Study 261

Chapter 13: Week 6 Class 2 263
   13 Today's Agenda: 263
   13.1 Hierarchical Name Access 263
        13.1.1 Verilog Arrayed Instances 266
        13.1.2 generate Statements 267
        13.1.3 Conditional Macroes and Conditional generates 267
        13.1.4 Looping generate Statements 269
        13.1.5 generate Blocks and Instance Names 269
        13.1.6 A Decoding Tree with generate 274

Table of Contents   xiii

    13.2 Generate Lab 15 278
        13.2.1 Lab postmortem 284
    13.3 Additional Study 285

Chapter 14: Week 7 Class 1 287
  14 Today's Agenda: 287
    14.1 Serial-Parallel Conversion 287
        14.1.1 Simple Serial-Parallel Converter 287
        14.1.2 Deserialization by function and task 289
    14.2 Lab Preface: The Deserialization Decoder 291
        14.2.1 Some Deserializer Redesign—An Early ECO 293
        14.2.2 A Partitioning Question 294
    14.3 Serial-Parallel Lab 16 295
        14.3.1 Lab postmortem 300
    14.4 Additional Study 300

Chapter 15: Week 7 Class 2 301
  15 Today's Agenda: 301
    15.1 UDP's, Timing Triplets, and Switch-level Models 301
        15.1.1 User-Defined Primitives (UDP's) 301
        15.1.2 Delay Pessimism 305
        15.1.3 Gate-Level Timing Triplets 305
        15.1.4 Switch-Level Components 307
        15.1.5 Switch-Level Net: The trireg 312
    15.2 Component Lab 17 313
        15.2.1 Lab postmortem 321
    15.3 Additional Study 321

Chapter 16: Week 8 Class 1 323
  16 Today's Agenda: 323
    16.1 Parameter Types and Module Connection 324
        16.1.1 Summary of Parameter Characteristics 324
        16.1.2 ANSI Header Declaration Format 324
        16.1.3 Traditional Header Declaration Format 324
        16.1.4 Instantiation Formats 325
        16.1.5 Parameter Format Values 325
        16.1.6 ANSI Port and Parameter Options 326
        16.1.7 Traditional Module Header Format and Options 326
        16.1.8 defparam 327
    16.2 Connection Lab 18 327
        16.2.1 Connection Lab postmortem 333
    16.3 Hierarchical Names and Design Partitions 333
        16.3.1 Hierarchical Name References 333
        16.3.2 Scope of Declarations 334
        16.3.3 Design Partitioning 335
        16.3.4 Synchronization Across Clock Domains 337
    16.4 Hierarchy Lab 19 340
        16.4.1 Lab postmortem 344
    16.5 Additional Study 344

Chapter 17: Week 8 Class 2 345
  17 Today's Agenda: 345
    17.1 Verilog configurations 345
      17.1.1 Libraries 345
      17.1.2 Verilog Configuration 346
    17.2 Timing Arcs and specify Delays 347
      17.2.1 Arcs and Paths 347
      17.2.2 Distributed and Lumped Delays 348
      17.2.3 specify Blocks 351
      17.2.4 specparams 352
      17.2.5 Parallel vs. Full Path Delays 353
      17.2.6 Conditional and Edge-Dependent Delays 354
      17.2.7 Conflicts of specify with Other Delays 356
      17.2.8 Conflicts Among specify Delays 356
      17.2.9 Conflicts with SDF Delays 357
    17.3 Timing Lab 20 357
      17.3.1 Lab postmortem 361
    17.4 Additional Study 361

Chapter 18: Week 9 Class 1 363
  18 Today's Agenda: 363
    18.1 Timing Checks 363
      18.1.1 Timing Checks and Assertions 363
      18.1.2 Timing Check Rationale 364
      18.1.3 The Twelve Verilog Timing Checks 365
      18.1.4 Negative Time Limits 369
      18.1.5 Timing Check Conditioned Events 371
      18.1.6 Timing Check Notifiers 371
    18.2 Pulse Filtering 372
      18.2.1 PATHPULSE Syntax 373
      18.2.2 specparam Improved Pessimism 374
    18.3 Miscellaneous time-Related Types 375
    18.4 Timing Check Lab 21 376
    18.5 Additional Study 380

Chapter 19: Week 9 Class 2 381
  19 Today's Agenda: 381
    19.1 The Sequential Deserializer 381
    19.2 PLL Redesign 382
      19.2.1 Improved VFO Clock Sampler 383
      19.2.2 Synthesizable Variable-Frequency Oscillator 384
      19.2.3 Synthesizable Frequency Comparator 387
      19.2.4 Modifications for a 400 MHz 1x PLL 389
      19.2.5 Wrapper Modules for Portability 393
    19.3 Sequential Deserializer I Lab 22 393
      19.3.1 Lab postmortem 413
    19.4 Additional Study 413

Chapter 20: Week 10 Class 1 415
20 Today's Agenda: 415
20.1 The Concurrent Deserializer 415
20.1.1 Dual-porting the Memory 416
20.1.2 Dual-clocking the FIFO State Machine 416
20.1.3 Upgrading the FIFO for Synthesis 417
20.1.4 Upgrading the Deserialization Decoder for Synthesis 417
20.2 Concurrent Deserializer II Lab 23 418
20.2.1 Lab postmortem 444
20.3 Additional Study 444

Chapter 21: Week 10 Class 2 445
21 Today's Agenda: 445
21.1 The Serializer and The SerDes 445
21.1.1 The SerEncoder Module 446
21.1.2 The SerialTx Module 447
21.1.3 The SerDes 447
21.2 SerDes Lab 24 447
21.2.1 Lab postmortem 467
21.3 Additional Study 467

Chapter 22: Week 11 Class 1 469
22 Today's Agenda: 469
22.1 Design for Test (DFT) 469
22.1.1 Design for Test Introduction 469
22.1.2 Assertions and Constraints 470
22.1.3 Observability 470
22.1.4 Coverage 471
22.1.5 Corner-Case vs. Exhaustive Testing 472
22.1.6 Boundary Scan 473
22.1.7 Internal Scan 475
22.1.8 BIST 476
22.1.9 Design For Test Summary 479
22.2 Scan and BIST Lab 25 479
22.2.1 Lab postmortem 489
22.3 DFT for a Full-Duplex SerDes 490
22.3.1 Full-Duplex SerDes 490
22.3.2 Test Logic Questions 491
22.4 Tested SerDes Lab 26 491
22.4.1 Lab postmortem 505
22.5 Additional Study 505

Chapter 23: Week 11 Class 2 507
23 Today's Agenda: 507
23.1 SDF Back-Annotation 507
23.1.1 Back-Annotation 507
23.1.2 SDF Files in Verilog Design Flow 508
23.1.3 Verilog Simulation Back-Annotation 509

- 23.2 SDF Lab 27 509
  - 23.2.1 Lab postmortem 515
- 23.3 Additional Study 515

Chapter 24: Week 12 Class 1 517
- 24 Today's Agenda: 517
  - 24.1 Wrap-up: The Verilog Language 517
    - 24.1.1 Verilog-1995 vs 2001 (or 2005) Differences 517
    - 24.1.2 Verilog Synthesizable Subset Review 517
    - 24.1.3 Constructs Not Exercised in this Course 518
    - 24.1.4 List of all verilog system tasks and functions 520
    - 24.1.5 List of all verilog compiler directives 521
    - 24.1.6 Verilog PLI 521
  - 24.2 Continued Lab Work (Lab 23 or later) 523
  - 24.3 Additional Study 523

Chapter 25: Week 12 Class 2 525
- 25 Today's Agenda: 525
  - 25.1 Deep-Submicron Problems and Verification 525
  - 25.2 Deep Submicron Design Problems 525
  - 25.3 The Bigger Problem 527
    - 25.3.1 Modern Verification 528
    - 25.3.2 Formal Verification 529
    - 25.3.3 Nonlogical Factors On The Chip 529
  - 25.4 System Verilog 531
    - 25.4.1 Some Features of SystemVerilog 531
    - 25.4.2 SystemVerilog Conclusion 538
  - 25.5 Verilog-AMS 538
    - 25.5.1 Introduction 538
    - 25.5.2 Relationship to Other Languages 539
    - 25.5.3 Analogue Functionality Overview 540
    - 25.5.4 Analogue and Digital Interaction 540
    - 25.5.5 Example: VAMS DFF 541
    - 25.5.6 Benefits of VAMS 541
  - 25.6 Continued Lab Work (Lab 23 or later) 542
  - 25.7 Additional Study 542

Index 543

Chapter 1
Introductory Material

1 Course Description

This book may be used as a combined textbook and workbook for a 12-week, 2-day/week interactive course of study. The book is divided into chapters, each chapter being named for the week and day in the anticipated course schedule.

The course was developed for attendees with a bachelor's degree in electrical engineering, or the equivalent, and with digital design experience. In addition to this kind of background, an attendee was assumed familiar with software programming in a modern language such as C or C++.

Someone fulfilling the course requirements would expect to spend approximately 12 hours per week for 12 weeks to learn the content and do all the labs. Of course, now that it is a book, a reader can expect to be able to proceed more at his or her own pace; therefore, the required preparation can be more flexible than for the programmed classroom presentation.

> **Topic List (partial):**
>
> **Discussion:** Modules and hierarchy; Blocking/nonblocking assignment; Combinational logic; Sequential logic; Behavioral modelling; RTL modelling; Gate-level modelling; Hardware timing and delays; Verilog parameters; Basic system tasks; Timing checks; Generate statement; Simulation event scheduling; Race conditions; Synthesizer operation; Synthesizable constructs; Netlist optimization; Synthesis control directives; Verilog influence on optimization; Use of SDF files; Test structures; Error correction basics; SystemVerilog, VerilogA/MS.
>
> **Lab Projects:** Shift and scan registers; counters; memory and FIFO models; digital phase-locked loop (PLL); serial-parallel (and v-v) converter; serializer-deserializer (serdes); primitive gates; switch-level design; netlist back-annotation.

2 Using this book

The reader is encouraged to read the chapters in order but always to assume that topics may be covered in a multiple-pass approach: Often, a new idea or language feature will be mentioned briefly and explained only incompletely; later, perhaps many pages later, the presentation will return to the idea or feature to fill in details.

Each chapter ends with supplementary readings from two highly recommended works, textbooks by Thomas and Moorby and by Palnitkar. When a concept remains difficult, after discussion and even a lab exercise, it is possible that these, or other publications listed in the References, may help.

2.1 Textbook Extras

The *Springer **Extras*** zip file contains problem files and complete solutions to all the lab exercises in the book. The contents have been updated for the second edition, but older solution elements, such as default.cfg and *.spj files, have been retained for readers using older versions of the software. A redundant backup of everything is stored with the Extras in a tar file, for easy copying to disc in a Linux or Unix working environment.

Be sure to read the **ReadMe.txt** file in the Extras before using the stored material.

The **misc** directory in the Extras contains an include file required for Lab 1, plus some other side files. It contains PDF instructions for basic operation of the VCS or QuestaSim simulators.

The misc directory also contains nonproprietary verilog library files, written by the author, which allow approximately correct simulation of a verilog netlist. These netlist models are not correct for design work in TSMC libraries, but they will permit simulation for training purposes. **DO NOT USE THESE VERILOG LIBRARIES FOR DESIGN WORK.** If you are designing for a TSMC tape-out, use the TSMC libraries provided by Synopsys, with properly back-annotated timing information.

As a suggestion, a good working environment in which to organize the contents of the Extras which accompanies this Textbook would be set up as shown in the following figure:

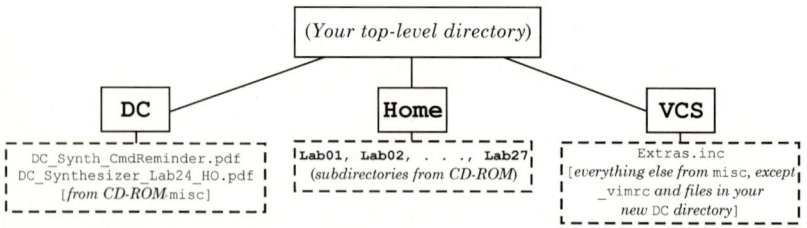

If you create the DC and VCS directories as suggested, you will have used everything in the Extras misc directory, so you will not need a misc directory in your working environment. None of the misc contents are TSMC or Synopsys proprietary.

The majority of the exercise answer subdirectories which depend upon the tcbn90ghp_v2001.v simulation library contain a zero-length copy of that .v file. This was done to reduce the space occupied by the answers in the Extras material. Whenever you encounter a zero-length file named tcbn90ghp_v2001.v, you should replace it with a full-length version from your VCS directory (see above) or from the misc data directory in your copy of the Extras. The contents of tcbn90ghp_v2001.v are copyrighted but are not TSMC or Synopsys proprietary.

The _vimrc file in the Extras misc directory is a convenient startup file for the ***vim*** text editor, which is the recommended text editor for verilog. The _vimrc file should be

Digital VLSI Design with Verilog

renamed to .vimrc and copied to your home directory ("~"), if running Linux or Unix. If you are running Windows, _vimrc should be copied to the startup directory which you have configured for *vim*.

2.2 Performing the lab exercises

The book contains step-by-step lab instructions. Start by setting up a working environment in a new directory, as above, and copying everything in the Extras into it, preserving directory structure.

The reader must have access to simulation software to simulate any of the verilog source provided in the book or the lab solutions. Readers without access to EDA tools can do almost all the source verilog lab simulations (except the serdes project) with the demo version of the Silos simulator delivered on CD-ROM with the Thomas and Moorby or Palnitkar hardcopy texts. A more functional simulator is available from *Aldec* (student version) as described below. Verilog simulators also often are supplied free with FPGA hardware programming kits. Netlist simulations of designs based on ASIC libraries generally will require high-capacity tools such as VCS or QuestaSim.

The Synopsys Design Compiler synthesizer and VCS simulator should be used for best performance of the labs. Almost all verilog simulator waveform displays were created using the Synopsys VCS simulator, which is designed for large, ASIC-oriented designs.

Keep in mind that all verilog simulators are incomplete works: Different simulators emphasize different features, but all have some verilog language limitations. Attempts have been made in the text to guide the user through the most important of these limitations.

For the professional reader's interest, the Extras labs in this edition were performed using Synopsys Design Compiler and VCS versions dated from 2007 through 2010 on a 1 GHz x86 machine with 384 megs of RAM, running Red Hat Enterprise Linux 3. TSMC 90-nm front-end libraries (*typical* PVT) from Synopsys were used for synthesis.

2.3 Proprietary Information and Licensing Limitations

Publishing of **operational performance details** of VCS, Design Compiler, or QuestaSim may require written permission from Synopsys or Mentor, and readers using the recommended EDA tools to perform the labs in this book are advised not to copy, duplicate, post, or publish any of their work showing specific tool properties without verifying first with the manufacturer that trade secrets and other proprietary information have been removed. This is a licensing issue unrelated to copyright, fair use, or patent ownership. Know your tools' license provisions!

Also, the same applies to the **TSMC library files** which may be available to licensed readers. The front-end libraries are designed for synthesis, floorplanning, and timing verification only, but they may contain trade secrets of TSMC. Do not make copies of anything from the TSMC libraries, including TSMC documentation, without special permission from TSMC and Synopsys.

The verilog simulation library files delivered with the accompanying Extras resemble those in tool releases; however, they are not produced by Synopsys or TSMC. These files should be considered copyrighted but not otherwise proprietary; they are available for copying, study, or modification by the purchaser of this book, for individual learning purposes.

Verilog netlists produced by the synthesizer are not proprietary, although the *Liberty* models used by the synthesizer are proprietary and owned by Synopsys. Specific details of synthesized netlist quality may be considered proprietary and should not be published or distributed without special permission from Synopsys.

3 Textbook References

At present, all references in the Textbook for verilog are to IEEE Std 1364-2005 and to the later *SystemVerilog* Std 1800-2012 document, unless otherwise stated. Except (*a*) for inclusion of existing *C* or *C++* code, or (*b*) for complex *assertion* composition, *SystemVerilog* and verilog 2005 differ very little. In some of the SystemVerilog citations, the reader will find additional features not present in the verilog 1364 references. SystemVerilog is discussed in some detail in the final lecture of the course.

3.1 Supplementary Textbooks

The Thomas and Moorby and Palnitkar textbooks are referenced many times in our Textbook for optional, supplementary exercises and readings. These may be useful to clear up understanding of specific topics presented during the course. Each of these books (full hardcopy version) includes a **Simucad** Silos® Verilog-2001 (MS Windows) simulator on CD-ROM, with Silos simulation projects. This is a capacity-limited demo version of the Silos simulator, but it will work with many of the smaller lab exercises in this course.

Purchasers of newer full-text versions of the Thomas and Moorby or Palnitkar books may not have obtained a copy including Silos. If so, a copy may be obtained from Springer at http://extras.springer.com/2002/978-1-4020-7089-1, by downloading the "Entire Contents".

References to the Thomas and Moorby textbook are for the 2002 (5th) edition. This edition uses verilog 2001 headers throughout.

3.2 Interactive Language Tutorial

The *Evita* Verilog Tutorial is available for use in an MS Windows environment. This is a free download for individual use. Sponsored by *Aldec* at http://www.aldec.com. The language level is verilog-1995 only. When downloading, mention the course unless you want sales followup.

This interactive tutorial includes animations, true-false quizzes, language reference, search utility, and a few minor bugs and content errors. It is very easy to run and may be useful especially for understanding course content from a different perspective.

3.3 Recommended Free Verilog Simulator

In addition to the Evita free tutorial, which is somewhat out-of-date, *Aldec* provides a free, downloadable Verilog-2001 simulator which is fully documented and runs in MS Windows; this is a version with generous but limited capacity and is highly recommended.

However, for more than a 20-day unlicensed trial, it is necessary to have the Windows machine connected to the internet in order for the Aldec site to inspect the machine and create the license.

3.4 Reading References

Note: Active-HDL, Design Compiler, Design Vision, DVE, HSPICE, Liberty, ModelSim, NanoSim, QuestaSim, Silos, VCS, and Verilog (capitalized) are trademarks of their respective owners.

Anonymous. (2013). *IEEE Std 1800–2012: Standard for SystemVerilog*. Verilog is a proper subset of SystemVerilog and ceases to be an independent standard with the Std 1364–2005 edition. Beginning in 2010, all updates to the LRM for verilog have been in IEEE Std 1800, only. Happily, this means little further change in verilog per se. The complete SystemVerilog manual currently is available free at online. Retrieved August, 2013 from http://standards.ieee.org/getieee/1800/download/1800-2012.pdf.

Anonymous. (2010). Design for Test (DFT). Chapter 3 of *The NASA ASIC Guide: Assuring ASICs for Space*. Retrieved February 3, 2010 from http://klabs.org/DEI/References/design_guidelines/content/guides/nasa_asic_guide/Sect.3.3.html.

Anonymous. (2013). *SystemVerilog*. Background and basic extensions of verilog, with examples. Retrieved August 10, 2013 from http://en.wikipedia.org/wiki/SystemVerilog.

Anonymous. (2013). *SystemVerilog introduction*. An overall introduction and complete details of the language are posted online. Retrieved August 11, 2013 from http://www.systemverilog.in/systemverilog_introduction.php.

Anonymous. (2014). *SerDes transceivers*. Freescale Semiconductor, Inc. For example, "MC92600: Quad 1.25 Gbaud SERDES", 2003 Manual. Retrieved January 3, 2014 from http://cache.freescale.com/files/timing_interconnect_access/doc/user_guide/MC92600UM.pdf.

Anonymous. (2009). *Verilog-AMS language reference manual*, Draft v. 2.3.1 of June, 2009. Accellera. Available for download (free for individuals) at the Accellera web site. This is the current language standard. This document is the one which should be used to resolve disputes or clarify understanding of the language; it contains many examples and detailed explanations. Presumably, IEEE eventually will produce a Std document more or less along the same lines.

Bansal, S., Krishna, N. S., & Mishra, A. (2009) *Circuit design hint: Calculating corner-independent timing closure*. Retrieved January 1, 2014 from http://www.eetimes.com/document.asp?doc_id=1276946.

Barrett, C. (Ed.). (1999). *Fractional/Integer-N PLL basics*. Texas Instruments Technical Brief SWRA029, August, 1999. Retrieved January 3, 2014 from http://focus.ti.com/lit/an/swra029/swra029.pdf.

Bertrand, R. (2014). The basics of PLL frequency synthesis. In: *Online radio and electronics course* (2002–04). Retrieved January 3, 2014 from http://www.nsarc.ca/hf/pll.pdf.

Bhasker, J. (2005). *A Verilog HDL primer* (3rd ed.). Allentown, Pennsylvania: Star Galaxy Publishing.

Cipra, B. A. (1993). The ubiquitous Reed_Solomon codes. *SIAM News*, *26*(1), 1993. Retrieved January 3, 2014 from http://www.eccpage.com/reed_solomon_codes.html.

Cummings, C. E. (2002). Simulation and synthesis techniques for asynchronous FIFO design (rev. 1.1). Originally presented at San Jose, California: The *Synopsys Users Group Conference*, 2002. Retrieved January 3, 2014 from http://www.sunburst-design.com/papers/CummingsSNUG2002SJ_FIFO1_rev1_1.pdf.

Cummings, C. E., & Alfke, P. (2002) Simulation and synthesis techniques for asynchronous FIFO design with asynchronous pointer comparisons (rev. 1.1). Originally presented at San Jose, California: The *Synopsys Users Group Conference*, 2002. Retrieved January 3, 2014 from http://www.sunburst-design.com/papers/CummingsSNUG2002SJ_FIFO2_rev1_1.pdf.

IEEE Std 1364-2005. (2005). *Verilog hardware description language*. Piscataway, New Jersey: The IEEE Computer Society, 2005. Revised in 2001 and reaffirmed, with some additional SystemVerilog compatibility, in 2005. See the preceding reference to IEEE Std 1800, SystemVerilog.

Keating, M., et al. (2007). *Low power methodology manual for system-on-chip design*. Springer Science and Business Solutions. Available from Synopsys as a free PDF for personal use. Retrieved November 6, 2007 from http://www.synopsys.com/lpmm.

Knowlton, S. (2007). *Understanding the fundamentals of PCI express*. Synopsys White Paper, 2007. Available at the Technical Bulletin, Technical Papers page. Retrieved October 17, 2007 from http://www.synopsys.com/products/designware/dwtb/dwtb.php. Free registration and login.

Koeter, J. (1996). *What's an LFSR?*. Retrieved January 3, 2014 from http://focus.ti.com/lit/an/scta036a/scta036a.pdf.

Kundert, K. S., & Zinke, O. (2004). *The designer's guide to Verilog AMS*. Boston: Kluwer Academic. Kluwer now is a subsidiary of Springer. This is the only textbook available on the subject, as of March 2009. It includes an excellent chapter on top-down design and its strength is in the mixed and analogue features. It contains a VAMS language specification which now is somewhat out of date; for example, the syntax is verilog-1995, only. There are a few small mistakes (many corrected by errata posted at the authors' web site).

Mead, C., & Conway, L. (1980). *Introduction to VLSI systems*. Menlo Park, CA: Addison-Wesley. Excellent but old introduction to switch-level reality and digital transistor design and fabrication.

Palnitkar, S. (2003). *Verilog HDL* (3rd ed.). Palo Alto, California: Sun Microsystems Press. A good basic textbook useful for supplementary perspective. Also includes a demo version of the *Silos* simulator on CD-ROM. Our daily *Additional Study* recommendations include many optional readings and exercises from this book.

Plank, J. S. (1997). A tutorial on Reed-Solomon coding for fault-tolerance in RAID-like systems. *Software: Practice & Experience, 27*(9),995–1012. Available online as *Technical Report CS-96-332*, Department of Computer Science, University of Tennessee. Retrieved January 3, 2014 from http://web.eecs.utk.edu/~plank/plank/papers/CS-96-332.html.

Seat, S. (2002). Gearing up Serdes for high-speed operation. Retrieved January 3, 2014 from http://www.eetimes.com/document.asp?doc_id=1205307.

Shah, A. (2013). *Whats new in Systemverilog 2009?* New constructs and examples for the 2005 and 2009 updates. Retrieved August 10, 2013 from http://testbench.in/SystemVerilog_2009_enhancements.html.

Suckow, E. H. (2003). *Basics of high-performance SerDes design: Part I* (2003-04-13). Retrieved January 3, 2014 from http://www.datasheet-thriftstore.com/basics-of-high-performance-serdes-design-part-i/; and, Part II (2003-04-26). Retrieved January 3, 2014 from http://www.datasheet-thriftstore.com/basics-of-high-performance-serdes-design-part-ii/.

Sutherland, S. (2014). The IEEE Verilog 1364–2001 standard: What's new, and why you need it. Based on an *HDLCon 2000* presentation. Retrieved January 3, 2014 from http://www.sutherland-hdl.com/papers/2000-HDLCon-paper_Verilog-2000.pdf.

Thomas, D. E., & Moorby, P. R. (2002). *The Verilog hardware description language* (5th ed.). New York: Springer (reprinted in 2008). A very good textbook which was used in the past as the textbook for this course. The hardcopy version includes a demo version of the *Silos* simulator on CD-ROM. Our *Additional Study* recommendations include many optional readings and exercises from this book.

Thompson, D. (2011). The benefits of FireWire on PCI Express. *EETimes* online. Retrieved January 3, 2014 from `http://www.eetimes.com/design/microcontroller-mcu/4219628/The-benefits-of-FireWire-on-PCI-Express`.

Turpin, M. (2014). The dangers of living with an X (bugs hidden in your Verilog). Retrieved January 6, 2014 from `http://www.arm.com/files/pdf/Verilog_X_Bugs.pdf`.

Wallace, H. (2001) *Error detection and correction using the BCH code*. Retrieved January 3, 2014 from `http://www.aqdi.com/bch.pdf`.

Weste, N., & Eshraghian, K. (1985). *Principles of CMOS VLSI design: A systems perspective*. Menlo Park, CA: Addison-Wesley. Old, but overlaps and picks up where Mead and Conway leave off, especially on the CMOS technology per se.

Wood, B. (2009). Backplane tutorial: RapidIO, PCIe, and Ethernet. *DSP Designline*, January 14, 2009 issue. Retrieved January 3, 2014 from `http://www.eetimes.com/document.asp?doc_id=1275655`.

Zarrineh, K., Upadhyaya, S. J., & Chickermane, V. (2001). System-on-Chip Testability Using LSSD Scan Structures. *IEEE Design & Test of Computers*, May–June,83–97.

Ziegler, J. F., & Puchner, H. (2004). *SER—History, Trends, and Challenges*. San Jose, Cypress Semiconductor Corporation. (stock number `1-0704SERMAN`). Contact: `serquestions@cypress.com`.

Zimmer, P. (2014). Working with PLLs in PrimeTime—avoiding the 'phase locked *oops*'. Drafted for a *Synopsys User Group presentation in 2005*. Retrieved January 3, 2014 from `http://www.zimmerdesignservices.com/zimmer_pll_update_051405.pdf`.

4 Lab Checklist

This check-off list is provided for your convenience in recording your progress through the course. If you skip a Lab or Lab Step, or have to interrupt your study, this list will help keep you on track to come back later.

Lesson	Lab	Step	Progress							
Week n Class n	nn	n	Read notes	Thinking about it	10%	25%	50%	90%	100%, still questions	Done
Week 1 Class 1 Introductory	01	1								
		2								
		3								
		4								
		5								
Operator	02	1								
		2								
		3								
Week 1 Class 2 Conversion	03	1								
		2								
		3								
Nonbl'k Control	04	1								
		2								
		3								
		4								
		5								
		6								
Week 2 Class 1 Simple scan	05	1								
		2								
		3								
		4								
		5								
		6								
		7								
		8								
		9								
Week 2 Class 2 PLL clock	06	1								
		2								
		3								
		4								
		5								
		6								
		7								
		8								
Week 3 Class 1 Memory	07	1								
		2								
		3								
		4								
		5								
		6								
Week 3 Class 2 Counter	08	1								
		2								
		3								
		4								
		5								

Lesson	Lab	Step	Progress								
Week *n* Class *n*	*nn*	*n*	Read notes	Thinking about it	10%	25%	50%	90%	100%, still questions	Done
Week 4 Class 1 `Wire strength`	9	1								
		2								
		3								
		4								
		5								
		6								
		7								
`PLL lock-in`	10	1								
		2								
		3								
Week 4 Class 2 `FIFO`	11	1								
		2								
		3								
		4								
		5								
Week 5 Class 1 `Sim. scheduling`	12	1								
		2								
		3								
		4								
		5								
Week 5 Class 2 `Netlist`	13	1								
		2								
		3								
Week 6 Class 1 `Concurrent`	14	1								
		2								
		3								
		4								
		5								
Week6 Class 2 `Generate`	15	1								
		2								
		3								
		4								
		3								
Week 7 Class 1 `Ser-Par convert`	16	1								
		2								
		3								
Week 7 Class 2 `Component`	17	1								
		2								
		3								
		4								
		5								
		6								
		7								
Week 8 Class 1 `Connection`	18	1								
		2								
		3								
		4								
		5								
		6								

Lesson	Lab	Step	Progress							
Week n Class n	nn	n	Read notes	Thinking about it	10%	25%	50%	90%	100%, still questions	Done
Hierarchy	19	1								
		2								
		3								
		4								
		5								
		6								
Week 8 Class 2	20	1								
Timing		2								
		3								
		4								
		5								
Week 9 Class 1	21	1								
Timing check		2								
		3								
		4								
		5								
		6								
		7								
		8								
		9								
		10								
Week 9 Class 2	22	1								
Sequential Des I		2								
		3								
		4								
		5								
		6								
		7								
		8								
		9								
		10								
		11								
		12								
		13								
Week 10 Class 1	23	1								
Concurrent Des II		2								
		3								
		4								
		5								
		6								
		7								
		8								
		9								
Week 10 Class 2	24	1								
SerDes		2								
		3								
		4								
		5								
		6								
		7								
		8								

Lesson	Lab	Step	Progress							
Week n Class n	nn	n	Read notes	Thinking about it	10%	25%	50%	90%	100%, still questions	Done
Week 11 Class 1	25	1								
Scan & BIST		2								
		3								
		4								
		5								
		6								
		7								
		8								
		9								
Tested SerDes	26	1								
		2								
		3								
		4								
		5								
		6								
		7								
		8								
		9								
Week 11 Class 2	27	1								
SDF		2								
		3								
		4								

Chapter 2
Week 1 Class 1

2 Today's Agenda:

Introductory Material
Topics: Course content and organization.
Summary: This is a verilog language course. It is heavily lab-oriented and its goal is netlist synthesis, more than simulation, as a content-related skill. The labs and the homework reading assignments are of utmost importance.

Introductory Lab 1
We start work immediately with a brief Lab session which introduces tool operation of the synthesizer and simulator. It is meant to be executed by rote; explanations will follow.

Lab 1 Postmortem
After having seen a verilog design through the tool flow, the enrollee is explained the basic language constructs which were involved.

Reality-Check Self-Exam
This is a quiz to help in-class enrollees understand how well prepared they are for the course. It is not collected or graded or distributed with the Textbook.

Lecture: Verilog vectors and numerical expressions
Topics: Vector notation and numerical constants.
Summary: Verilog vector declarations and assignment statements are presented. Numerical expressions in hexadecimal, binary and other bases are described. Simulation delay expressions and verilog timescale specifications are introduced.

Verilog Operator Lab 2
This exercises basic vector assignments and associated simulation delays. It also exercises simple boolean operators and shows how a verilog module header may be written either in ANSI standard format or in the older, verilog-1995, "K&R C" format.

Lab 2 Postmortem
After lab Q&A, vector extensions and reduction operators are explained. Advantages of the ANSI module header format are given. Module timescale and time resolution are discussed.

Lecture: VCD and SDF files
Summary: Simulator VCD files and synthesizer SDF file formats are described briefly and their purposes explained.

2.1 Introductory Lab 1

Topic and Location: This is a small design intended to be looked at, simulated, and synthesized to a netlist <u>without</u> much understanding of the verilog.

Begin this lab by logging in and changing to the `Lab01` directory.

Preview: We simply glance over the verilog source files used to define the design, named `Intro_Top`. We also look at the verilog testbench file which has been provided and which instantiates the the top of the design; this testbench is what stimulates the design during simulation. The testbench file is not used for synthesis. We then synthesize a netlist from the verilog design, using a synthesizer script file provided. We end this lab by examining a schematic of the synthesized netlist in a netlist viewer.

Deliverables: This lab is a rote exercise and requires no deliverables other than the rote-synthesized netlist.

Lab Procedure: This is a stereotyped, automated lab. Just do everything by rote; the lab will be explained in a postmortem lecture which will follow it immediately. Verilog source code files have a `.v` extension. The files in the design have been listed in the text file, `Intro_Top.vcs`, for your convenience in using the VCS simulator.

An include file, `Extras.inc`, provided in the Extras `misc` directory, will have to be referenced properly in the `TestBench.v` file or copied to a location usable by the simulator of your choice. To do this, you will have to edit `TestBench.v`, changing the path in `` `include "../../VCS/Extras.inc" `` to match your setup. Notice that there is no semicolon at the end of a `` `include ``. If you comment out the `` `include ``, the simulation will work, but no VCD file will be created during simulation.

NOTE: There is a script named "`Clean`" in your `~/bin` directory. Because VCS and DC create multitudinous temporary side files, this script will be very useful to run after every invocation of these tools.

However, be careful of two things:

1. **NEVER** name a file or directory ***ending*** in "`Synth`" (*e. g.*, `MySynth`). `Clean` will delete such things without warning.

   Naming a file something like "`MySynth.txt`" will be safe, because it ends in an extension (not in `Synth`).

2. `Clean` prompts individually to delete certain side files which may be useful.

   In general, answer 'y' to these prompts; however, be careful not to delete files named "*something*.`log`" unless you are sure you don't want them.

Step 1. In your `Lab01` directory, use a text editor to look at the top level of the design in `Intro_Top.v`.

Digital VLSI Design with Verilog   15

The top level structure may be represented schematically this way:

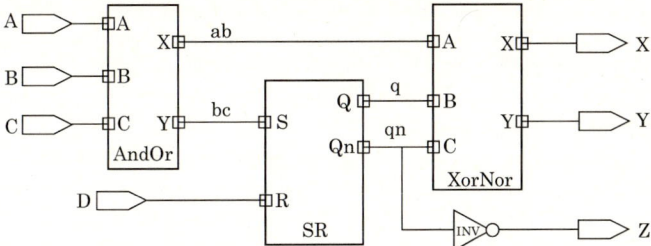

Fig. 2-1. Schematic representation of Intro_Top.v. The blocks are labelled with their module names. Connecting wire names also are shown.

In the verilog file Intro_Top.v, documentation and code comments begin with the verilog comment token, "//". A copy of the contents is reproduced below for your convenience.

```
// ==============================================================
// Intro_Top:  Top level of a simple design using verilog
// continuous assignment statements.
//
// This module contains the top structure of the design, which
// is made up of three lower-level modules and one inverter gate.
// The structure is represented by module instances.
//
// All ports are wire types, because this is the default; also,
// there is no storage of state in combinational statements.
//
// ANSI module header.
// --------------------------------------------------------------
// 2004-11-25 jmw: v. 1.0 implemented.
// ==============================================================
module Intro_Top (output X, Y, Z, input A, B, C, D);
  //
  wire ab, bc, q, qn;    // Wires for internal connectivity.
  //
  // Implied wires may be assumed in this combinational
  // design, when connecting declared ports to instance ports.
  // The #1 is a delay time, in `timescale units:
  //
  assign #1 Z = ~qn;  // Inverter by continuous assignment statement.
  //
   AndOr  InputCombo01   (.X(ab), .Y(bc), .A(A), .B(B), .C(C));
     SR   SRLatch01      (.Q(q), .Qn(qn), .S(bc), .R(D));
  XorNor  OutputCombo01  (.X(X),  .Y(Y),  .A(ab), .B(q), .C(qn));
  //
endmodule // Intro_Top
```

Notice that everything is contained in a "module". The top module, which is shown in Intro_Top.v, includes output and input port declarations (I/O's), wire declarations, one assignment statement (a *continuous assignment* wire-connection statement), and three component instance declarations. The components are declared in other modules. This module, combined with the component instances, represents the structure of the design.

We have defined the component instances in Intro_Top ourselves, with our own names, for this lab, using verilog module declarations, as will be shown below. Those component instances are an "AndOr" named InputCombo01, a "SR" named SRLatch01, and an "XorNor" named OutputCombo01. The instance names are entirely arbitrary. The "01" endings are not required but they do make large numbers of similar instances more easily distinguishable.

Step 2. Look at the verilog which will be used to run logic simulation on this design: It is in a file named TestBench.v, which includes the instance of Intro_Top which is to be exercised. The contents of this file are reproduced below.

An equivalent schematic of TestBench.v is as follows:

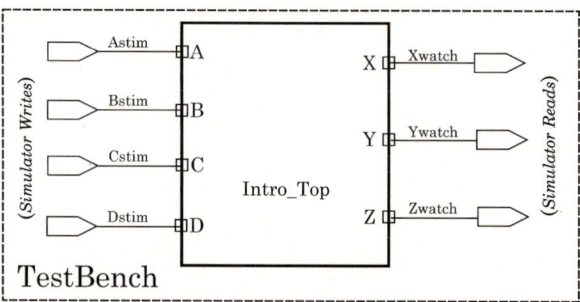

Fig. 2-2. Schematic view of the Lab01 TestBench module.

The verilog simulator will use the testbench (cleverly named "TestBench") to apply stimuli to the inputs of Intro_Top on the left, and it will read results from the outputs on the right. The testbench module in TestBench.v will be omitted when synthesizing the design, because it is not part of the design.

```
// ==============================================================
// TestBench:  Simulation driver module (stimulus block)
// for the top-level block instance of Intro_Top.
//
// This module includes an initial block which assigns various
// values to top-level inputs for simulation.   initial blocks
// are ignored in logic synthesis.
//
// No module port declaration.
// --------------------------------------------------------------
// 2004-11-25 jmw: v. 1.0 implemented.
// ==============================================================
//
`timescale 1 ns/100ps    // No semicolon after `anything.
//
module TestBench;        // Stimulus blocks have no port.
//
wire Xwatch, Ywatch, Zwatch;      // To connect to design instance.
reg  Astim, Bstim, Cstim, Dstim;  // To accept initialization.
//
initial
  begin
  //
  // Each '#' precedes a delay time increment, here in 1 ns units:
  //
  #1  Astim = 1'b0;   // For Astim, 1 bit, representing a binary 0.
  #1  Bstim = 1'b0;   // This occurs at time 1 + 1 = 2.
  #1  Cstim = 1'b0;
 (other stimuli omitted here)
  #50 Dstim = 1'b1;
  #50 Astim = 1'b0;
  #50 Cstim = 1'b0;
  #50 Dstim = 1'b0;
  #50 $finish;        // Terminates simulation 50 ns after the last stimulus.
  end // No semicolon after end.
//
// The instance of the design is named Topper01, and its
// ports are associated by name with stimulus input and simulation
// output wires:
//
Intro_Top Topper01 ( .X(Xwatch),  .Y(Ywatch),   .Z(Zwatch)
                   , .A(Astim),   .B(Bstim),    .C(Cstim), .D(Dstim)
                   );
//
endmodule // TestBench.
```

Notice the use of two different keyboard characters which may appear very similar:

- In the <u>timescale specifier</u>, **`timescale**, the '`' is a backquote. This character also is used in macroes such as `define, `include, `ifdef, `else, `endif, etc.
- In the <u>width specifier</u> for literal constants, **1'b1**, etc., the "'" is a single-quote. It mostly is used to separate the number of characters in a number from the numerical value, as in 1'b1, 2'b01, 16'h7a2c, etc.

Step 3. Look briefly at the verilog for the three other `modules` in the design; these `modules` are in files `AndOr.v`, `SR.v`, and `XorNor.v`.

Here is the structure of `AndOr.v`:

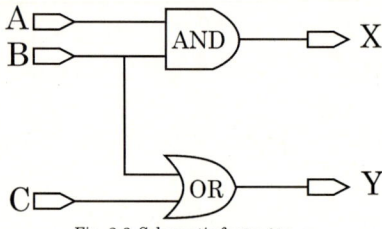

Fig. 2-3. Schematic for `AndOr.v`.

The verilog in `AndOr.v` is this:

```
// ==============================================================
// AndOr:    Combinational logic using & (and) and | (or).
// This module represents simple combinational logic including
// an AND and an OR expression.
//
// ANSI module header.
// --------------------------------------------------------------
// 2004-11-25 jmw:  v. 1.0 implemented.
// ==============================================================
module AndOr (output X, Y, input A, B, C);
   //
   assign #10 X = A & B;
   assign #10 Y = B | C;
   //
endmodule // AndOr.
```

The only assignment statements here are *continuous assignments*, recognizable by the **assign** verilog keyword. These are `wire` connection statements. The '`=`' sign merely connects the left side to the right side. This connection is permanent and can not be modified during simulation, although changes in value on the right will cause reevaluation of the left-hand target of the **assign**.

> **assign** means: "connect *something* to this wire on the left"

The *something* is on the right. For example, "`assign #10 X = A & B;`" in the `AndOr` module above means to (*a*) take notice of any change in value on the right; then to (*b*) wait 10 time units; then to (*c*) evaluate (the value of the `A` input) **and**ed with (the value of the `B` input); and, finally, to (*d*) use the result to replace whatever previously was the value of the `X` output on the left.

The "`#10`" literals above are simulator programmed delays, in nanoseconds.

Here is the structure of SR.v:

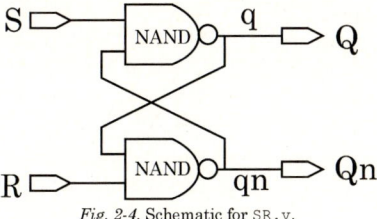

Fig. 2-4. Schematic for SR.v.

The verilog in SR.v is this:

```
// ================================================================
// SR:   An S-R Latch using ~ and &.
// "SR" means Set-Reset.
// This module represents the functionality of a simple latch,
// which is a sequential logic device, using combinational
// ~AND expressions connected to feed back on each other.
//
// ANSI module header.
// ----------------------------------------------------------------
// 2005-04-09 jmw: v. 1.1 modified comment on wire declarations.
// 2004-11-25 jmw: v. 1.0 implemented.
// ================================================================
module SR (output Q, Qn, input S, R);
   wire q, qn;   // For internal wiring.
   //
   assign #1 Q  = q;
   assign #1 Qn = qn;
   //
   assign #10 q  = ~(S & qn);
   assign #10 qn = ~(R & q );
   //
endmodule // SR.
```

Here we have four more <u>continuous assignments</u> which again map to wiring in the schematic.

Here is the structure of `XorNor.v`:

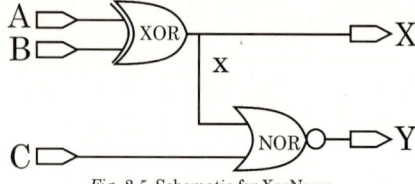

Fig. 2-5. Schematic for XorNor.v.

The verilog in `XorNor.v` is this:

```
// =============================================================
// XorNor:    Combinational logic using ^ and ~|.
// This module represents simple combinational logic including
// an XOR and a NOR expression.
//
// ANSI module header.
// -------------------------------------------------------------
// 2005-04-09 jmw: v. 1.1 modified comment on wire declarations.
// 2004-11-25 jmw: v. 1.0 implemented.
// =============================================================
module XorNor (output X, Y, input A, B, C);
  wire x; // To illustrate use of internal wiring.
  //
  assign #1 X = x;  // Verilog is case-sensitive; 'X' and 'x' are different.
  //
  assign #10 x = A ^ B;
  assign #10 Y = ~(x | C);
  //
endmodule // XorNor.
```

Here we see a `wire` declaration and three more <u>continuous assignments</u>.

Step 4. Load `TestBench.v` into the simulator and simulate it, if necessary using the handout sheet (***VCS Simulator Summary*** or ***QuestaSim Simulator Summary***) provided in PDF format with the Extras materials. You probably have copied this file during setup to your VCS directory.

Don't ponder the result; just be sure that the simulator runs without issuing an error message, and that you can see the resulting waveforms displayed, as in the figures below.

Since the first edition of this book, Synopsys has distributed a fully-functional, new version of VCS. A realistic comparison of the old vs. new VCS is provided immediately below, as well as in several future lab exercises. There are differences in functionality, but none of them matters much for purposes of this course.

Digital VLSI Design with Verilog   21

Illustrations of simulation waveforms in the old VCS gui look like the following:

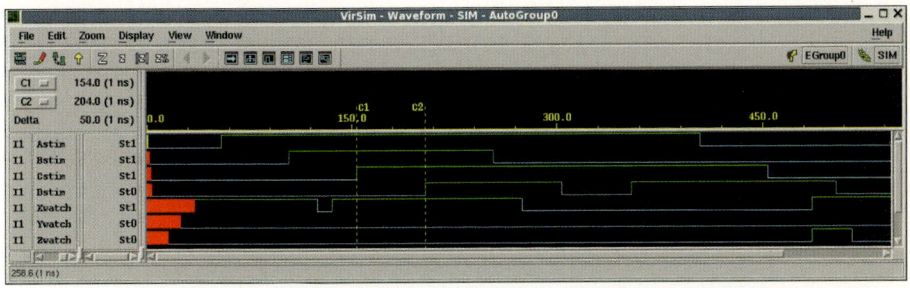

Fig. 2-6: `Intro_Top` simulation waveforms in the Synopsys VCS simulator (old gui).

Almost all simulation figures in the Textbook use the old gui, which was invoked by using the startup command, `vcs -RI` and omitting `-gui`.

The new gui, now recommended for all design work, is called **DVE**. Basic usage of DVE is described in your *VCS Simulator Summary* handout sheet, and it looks like this when first invoked:

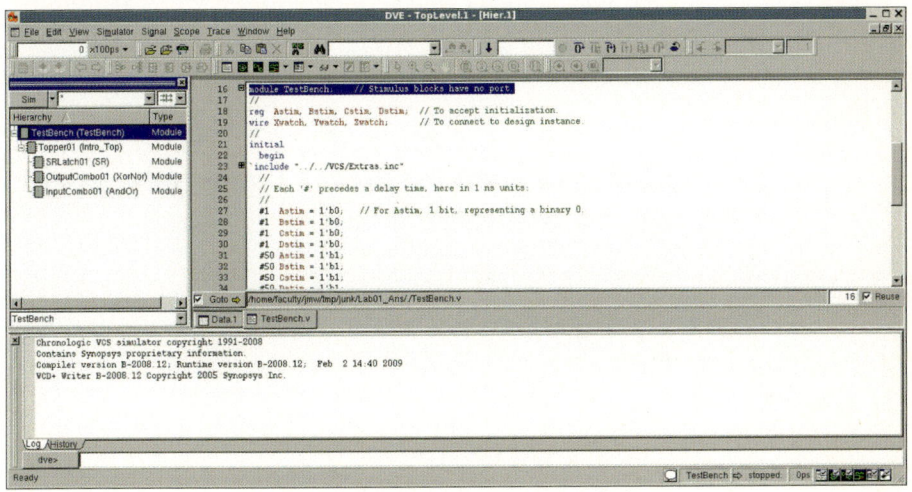

Fig. 2-7: New VCS initial simulator startup display.

After running the simulation in this Step, your new wave window will look about as follows:

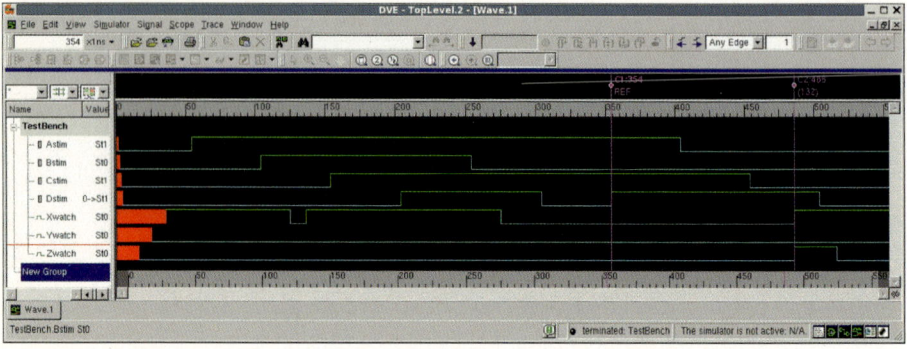

Fig. 2-8: `Intro_Top` simulation waveforms in the Synopsys VCS simulator (new gui).

The DVE still uses VCS as its simulator, but the graphic displays are improved and are more portable to native Linux (or Unix) than was the old gui. However, for our purposes, there is almost no difference between the two.

Other simulators present about the same waveforms:

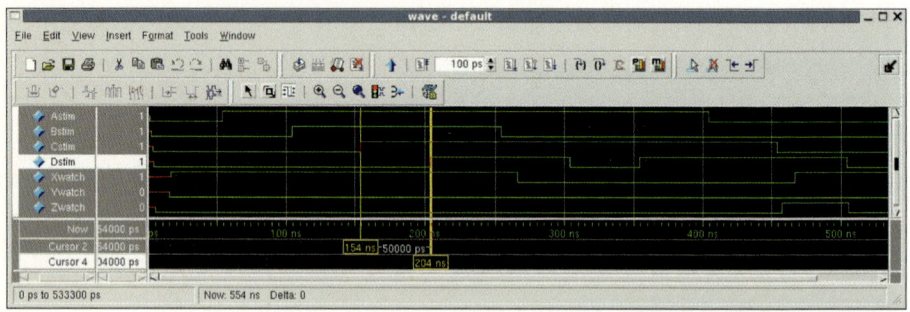

Fig. 2-9: `Intro_Top` simulation waveforms in the Mentor QuestaSim simulator.

Digital VLSI Design with Verilog

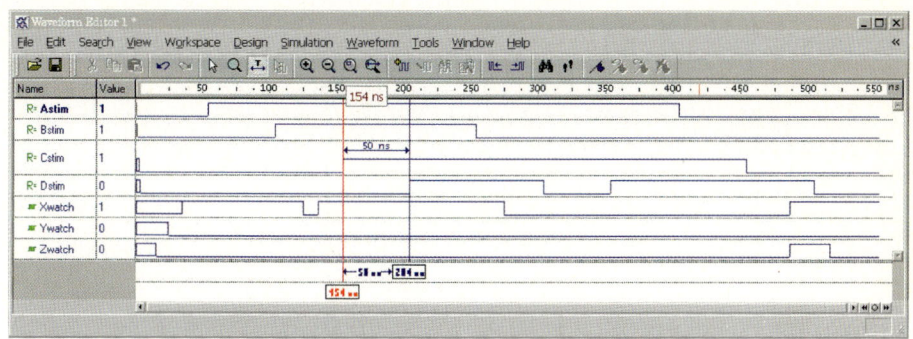

Fig. 2-10: `Intro_Top` simulation waveforms in the Aldec Active-HDL simulator.

When you are satisfied with your simulator waveforms, exit VCS.

Step 5. The Design Compiler (DC) text interface, `dc_shell`, is the standard interface used by designers to synthesize, and especially optimize, digital logic. It includes a schematic display tool named `design_vision`. Typically, various `dc_shell` options, and directives embedded in the code, are tweaked repeatedly before the resulting netlist is satisfactory, meeting all timing and area requirements. These are text-editing exercises. There is a great advantage of text scripting in these activities: Graphical manipulations of synthesizer commands can be inefficient, undocumented, irreproducible, and error-prone when tweaking a synthesis netlist.

We shall use `design_vision` very occasionally in this course, just to display schematics. We use the *TcL* (*T*ool *C*ontrol *L*anguage) scripting interface for almost everything. In the past, *TcL* was indicated by "-t", for `dc_shell-t` or `design_vision-t`. However, in late 2008, Synopsys completed its transition to *Tcl* syntax. Currently, invoking `design_vision` or `dc_shell` automatically selects the *Tcl* interface, and the "-t" invocations no longer are used.

There isn't much to synthesize in `Intro_Top`, it's already described almost on a gate level. But, continuous assignment statements aren't gates; so, during synthesis, there will be a noticeable difference when their expressions are replaced visibly by gates.

Load the design into the logic synthesizer and synthesize it to a gate-level netlist, using the ***DC Synthesizer Lab 1 Summary*** provided. The gates we shall be using for synthesis in this course are provided in a 90-nm library from the TSMC fab house. **DO NOT** perform the schematic steps in the Lab 1 ***Summary*** unless time permits the running of optional Step 6 below.

Step 6 (*optional*). As already mentioned, our graphical interface to DC is named `design_vision`. But, instead of DC, a better way to see a schematic representing a verilog design is to use new VCS. VCS can be faster and neater than DC in displaying a schematic, and it can create schematics for unsynthesizable designs.

So, use the optional instructions in your *DC Synthesizer Lab 1 Summary* sheet to invoke VCS and create a schematic of the source verilog for the `Intro_Top` design:

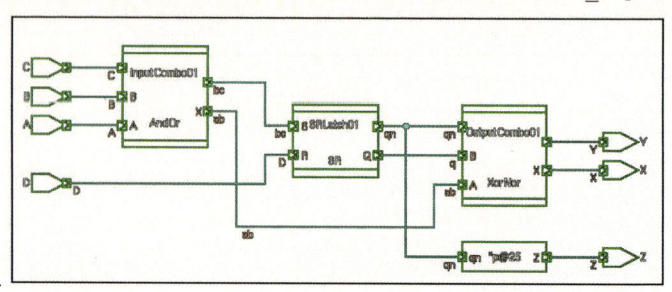

Fig. 2-11: The new VCS can create a schematic showing the contents of `Intro_Top`.

The `TestBench` level of the design can be displayed, but more interesting is the `Intro_Top` schematic, which, as seen above, shows the three submodules and the inverter on the `Z` output.

We shall not use verilog-generated schematics in this course; however, you are free to create them out of curiosity, if you wish. A schematic view almost always is useless in understanding `modules` of any size, over, say, about 20 blocks or gates. However, some utilities (not covered in this course) can trace timing-critical logic schematically through a netlist, and the graphical view of the single path provided can be very helpful in discovering errors or making small changes in a design.

Digital VLSI Design with Verilog

If you decide to resynthesize the design using `design_vision`, the corresponding top-level schematic will appear more or less as shown:

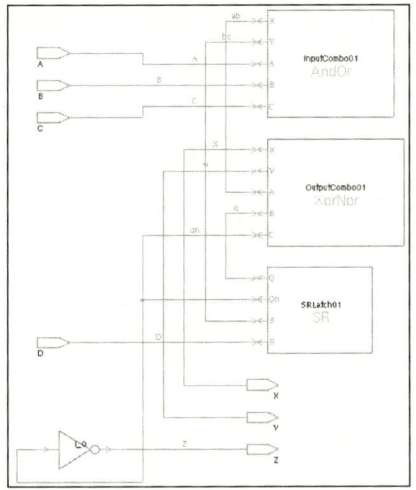

Fig. 2-12: A `design_vision` hierarchical schematic of the synthesized `Intro_top` netlist.

In addition to the rectangles of VCS, `design_vision` can be configured to represent the shapes of elementary logic gates.

End of Lab 1.

A Note on Lab Postmorta:

When we hold a lab during scheduled class time, it may be followed, if time permits, by a **postmortem** meeting, even if the lab work is not yet finished.

The postmortem is intended to give an opportunity to expand on the course content and explain or emphasize new features introduced in the lab.

2.2 Lab 1 Postmortem and Lecture

Keywords. There were several verilog keywords used in the Lab 1 files: `module`, `endmodule`, `assign`, `wire`, `reg`. All were lower case.

All verilog keywords are lower case; but, <u>the language is case-sensitive</u>. Thus, one can avoid any possible keyword name conflict by declaring ones own names always with at least one upper-case character. So declaring, `module Module ... endmodule`, like this,

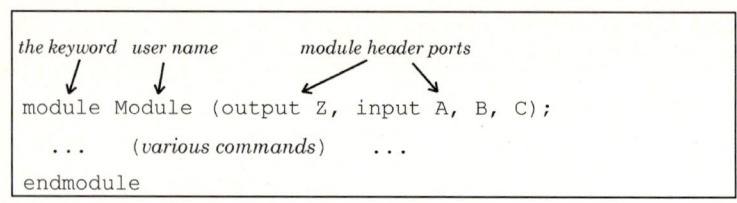

```
         the keyword   user name        module header ports
               ↓           ↓              ↙         ↘
         module      Module   (output Z,  input  A, B, C);
             ...    (various commands)   ...
         endmodule
```

is perfectly legal, although unusual and a bit vague in practice.

Comments. In verilog, the comment tokens are the same as in C++: "//" for a line comment; "/*" and "*/" for a block comment. Block comments are allowed to include multiple lines or to be located in the middle of a line.

Modules. The only user-declared object type in verilog is the ***module***; everything else is predeclared and predetermined. Designers just decide how to name and use the other data types available. This makes verilog a simple language, and it permits a verilog compiler to be implemented bug-free.

A module is delimited by the keywords module and endmodule. Everything between these keywords defines the functionality of the module. A module is the smallest verilog object which can be compiled or simulated. It corresponds to a design block on a schematic. In synthesizer terminology, a verilog module is referred to as a "design".

For example, here is a simple module which *and*s two inputs with an output delay of one time unit:

```
module And2 (output Z, input A, B);
   assign #1 Z = A & B;
endmodule
```

The assign vector, introducing a *continuous assignment*, is explained below.

Initial Blocks. Our TestBench module contained an ***initial*** block. An initial block includes one or more procedural statements which are read and executed by the simulator, beginning just before simulation time 0. An initial block is executed just once during simulation. Statements in an initial block are delayed by various amounts and are used to define the sequence of test-vector assignments in a testbench.

Our own Lab 1 testbench initial block included many statements. Blocks of statements in verilog are grouped by the keywords, begin and end. Only statements end with a semicolon; blocks do not require a semicolon.

For example, from Lab01,

```
initial
  begin
  #1   Astim = 1'b0;
  #1   Bstim = 1'b0;
  #1   Cstim = 1'b0;
  ...  (other stuff)  ...
  #50  Dstim = 1'b0;
  #50  $finish;
  end
```

To terminate a simulation, always add a `$finish` command at the end of the testbench `initial` block. Otherwise, the simulation may hang (depending on simulator configuration) and may never stop. The `$stop` command may be used in an `initial` block to create a breakpoint which doesn't permanently `$finish` a simulation.

Module Header. A module begins with a module header. The header starts with the declared module name and ends with the first semicolon in the module. Except testbenches, which usually have no I/O, module headers are used to declare the I/O ports of the module. There are two different formats used for module headers in verilog: The older, traditional *verilog-1995* format, and the newer, ANSI-C, *verilog-2001* format. We shall use only the newer format, because it is less repetitive that the older one and is less prone to user error.

In the module header, ANSI **port** declarations each begin with a direction keyword, `output`, `input`, or `inout`; the keyword may be followed with bus indices when a port is more than 1 bit wide. Ports may be declared in any order, but usually it is best to declare `output` ports first (we'll see why later). Each direction keyword declaration is followed by one or more names of the I/O ports (wires). To declare additional port names, one uses a comma (','), another direction keyword, and another list of one or more names.

For example, here is a header of a `module` named `ALU` which is declared with one 32-bit output, two 16-bit inputs, and two 1-bit inputs:

```
module ALU (output[31:0] Z, input[15:0] A, B, input Clock, Ena);
```

Assignment Statements. In the `TestBench` `initial` block of our Lab 1 exercise, there were two different kinds of verilog statement, ***continuous assignment*** statements to `wires`, and ***blocking assignment*** statements to `regs`. A blocking assignment is one kind of a procedural assignment. In verilog, `wire` types only can be connected and can not be assigned procedurally. One declares a different kind of data object, a **reg**, to assign values procedurally. While procedural statements are being executed, `regs` can be changed multiple times, in different ways. To get a procedural result out of its containing `module` and into the design, one uses continuous assignment to assign the `reg` to a `wire` or port.

For example,
```
module And2 (output Z, input A, B);
reg Zreg;
// A continuous assignment connection statement:
assign #1 Z = Zreg;     // Puts the value of Zreg on the output port.
// Procedural statements:
initial
  begin
     Zreg = 1'b0;
  #2 Zreg = A && B;
  #5 $finish;
  end
endmodule
```

This *and* gate model is rather useless, (the `initial` block terminates the simulation at time 2 + 5 = 7); but, it shows how a `reg` and a `wire` may be used together. `Zreg` is set to 0 as the simulation starts; the delay on the assignment of `Zreg` to `Z` causes `Z` to go from unknown to 0 at time 1. At time 2, `Zreg` is updated from the module inputs; and, because of that, `Z` may change for the last time at time 3.

The blocking assignments in the `TestBench` and in the model above are indicated by '`=`'. There is another kind of procedural assignment, the nonblocking assignment, which is indicated by "`<=`". We shall study nonblocking assignments later.

Simple Boolean operators. Verilog has about the same operator set as C or C++. In particular, the ***logical*** operators, '`!`', "`&&`" and "`||`", return a '1' or a '0', representing *true* or *false*, respectively. The ***bitwise*** operators, '`&`', '`|`', and '`~`' are the same as the logical "`&&`", "`||`", and '`!`', respectively, when the operands are all just one bit wide. For multiple-bit operations, the bitwise operators return multiple-bit results. The bitwise exclusive-*or*, '`^`', has no boolean correspondent. Any of these operators may return a logic level of '`x`' if a bit value of '1' or '0' can't be determined.

Four Logic Levels. In addition to the logically defined levels '**1**' and '**0**', verilog has two special levels to accommodate simulation states: '**x**' means either '1' or '0', but that the simulator can not determine which one, creating effectively an unknown state; '**z**' means "turned off" (three-state off) and usually means the same as '**x**' except when multiple drivers on a single wire are in contention.

It usually is good practice to specify the width of every verilog literal expression, even when the width is just 1. So, we usually try to write "`1'b0`" instead of just '0'. In this kind of expression, the width <u>in bits</u> precedes the quote (`'`); then comes the number base ('b', 'h', or 'd'—binary, hex, or decimal); and, finally, the literal value.

Thus, `4'b000z` means a 4-bit value with least-significant bit turned off (z); `4'b0x00` means a 4-bit value with bit 2 unknown (we count from bit 0 on the right). `12'b101` and `12'h5` have the same value, 5, and both are 12 bits wide.

Digital VLSI Design with Verilog

Delay. A delay value is preceded by '#' and always is in decimal base. The meaning of a delay value is determined by a ***timescale*** macro, usually at the beginning of the first file compiled for simulation. For example, `` `timescale 1ns/100ps `` means that #1 equals a 1 ns delay, and that the simulator should calculate all delays to precision no better than 100 ps (= 0.1 ns).

Component Instantiation. Module instances are the structural basis of all big designs. A module must be declared before it can be instantiated. A module instance also may be referred to as a component instance, especially when the instance is from a library of relatively small gates. Instantiations are statements, and, like other statements in verilog, they end with a semicolon.

The basic syntax of instantiation is: *module_name instance_name port_map*; for example, from Lab 1, the device under test was instantiated in `TestBench` this way:

```
Intro_Top Topper01 ( .X(Xwatch), .Y(Ywatch), .Z(Zwatch)
                   , .A(Astim),  .B(Bstim),  .C(Cstim), .D(Dstim)
                   );
```

The declared name of the instantiated `module` was `Intro_Top`; this instance was named `Topper01`, and the port map followed in the outer parentheses.

Port Map. The preceding instantiation shows how to map the port names of the module, `Intro_Top`, to the wires in the `TestBench` module: Each port name is preceded by a period (.), and the mapped wire, connected to that port, is immediately after it in parentheses.

Note that verilog is a free-form language; the way the three lines of the above `Intro_Top` port map was written with vertically aligned '**(**', '**,**', and '**)**' is just the present author's preference and is not required.

2.3 Verilog vectors

Verilog vector notation is used for ordered sets—busses—of logical bits.

Verilog has similar *vector* and *array* constructs. We'll look at arrays later.

A declared ***vector*** can be assigned to another one without explicit indexing; or, alternatively, a bit or part select may be made. A ***select*** means that part of the vector is being named ("<u>select</u>ed") explicitly.

All the following are legal (controlling statements omitted):
```
module VectorExamples ( output[7:0] OutHigh, OutLow1, OutLow2
                      , output OutBit, input ClockIn, ... (other inputs) ...
                      );
   reg[7:0]   HighByte, LowByte1, LowByte2;
   reg Bit;
   ...        // The below continuous assigns go near here.
   ...        // Other stuff.
   HighByte      = LowByte1;       // Assigns one vector to another.
   Bit           = LowByte2[7];    // This is called a "bit select".
   LowByte2[3:0] = LowByte2[7:4];  // Two "part selects".
   ...        // Other stuff.
endmodule
```
The actual, omitted controlling statements of the above might be as follows:
```
   assign OutHigh = HighByte;      // Four continuous assignments shown here.
   assign OutLow1 = LowByte1;
   assign OutLow2 = LowByte2;
   assign OutBit  = Bit;
   ... (various assignments, etc.) ...
   always @(posedge ClockIn)       // LowByte1 is not assigned here.
     begin
       HighByte      = LowByte1;
       Bit           = LowByte2[7];
       LowByte2[3:0] = LowByte2[7:4];
     end
```
All the `always` assignments above are procedural, because a `reg` must be assigned procedurally.

Notice the difference between verilog and *C* language declarations: The index range in verilog appears after the type name (`reg` or `wire`), not after the name being declared:

Verilog	C or C++
`reg[7:0] HighByte, LowByte;`	`char HighByte[8], LowByte[8];`

In verilog, "register" means "regular line-up". Registered data is the defining characteristic of <u>R</u>egister-<u>T</u>ransfer <u>L</u>ogic (RTL), the level of abstraction most frequently used in digital simulation and synthesis.

Don't read verilog "`reg`" as *register*! Perhaps *storage register* would be better, but it still would be a little wrong. The "`reg`" declaration represents localization of storage of information, even if of just one bit. In verilog, a `reg` localizes a logic state. The `wire` types never are considered `reg`s, even when regularly lined up in busses. Verilog `wire`s are just plain wires; they just connect logic from one place to another. Because they have multiple endpoints, `wire`s are unlocalized. Both nets (`wire`s) and `reg`s may vary in value during simulation, but only `reg`s are called *variables* in verilog.

The value of a `wire` can be changed only if the logic driving it changes value. The value of a `reg` can be changed by assigning it a different value to it in an `initial` or `always`

Digital VLSI Design with Verilog   31

block, often using several different drivers. An assignment to a `reg` is called a *procedural* assignment, because it can be done by a procedure of individual statements inside an `initial` or an `always` block.

Continuous assignments are done by continuous assignment statements ("`assign`" keyword) and represent single, permanent connections of something to a net. For example, the net here is a `wire` type:

```
module ModuleName(...);
...
wire x;
assign x = a & b | c;   // LHS is wire; RHS may be wire or reg(s).
...
endmodule
```

All the assignments in the first lab's `Intro_Top` design were continuous assignments to `wire`s; there were no `reg`'s in that design (excluding the testbench); therefore, there were no procedural assignments.

Procedural assignments are done in procedural blocks ("`always`" or "`initial`" keyword) and represent logical changes in a particular order. For example,

```
module ModuleName(..., input a, b, c);
reg   x;
wire y;
//
assign y = a & b;
assign y = y | c; // y driven in two different assigns probably is an error!
...
always @(a,b,c)
  begin         // LHS must be reg; RHS may be wire or reg.
  x = a & b;
  x = x | c;    // x gets (a & b) | c.   No problem.
  end
endmodule
```

Unlike `initial` blocks, `always` blocks may be run many times during a simulation; in fact, they ***always*** run when something on their RHS changes.

In any kind of assignment statement, binary, hexadecimal, decimal, and octal expressions are allowed for constant values (*literal*s):

1'b1, 1'b0, 1'bx, 1'bz.   8'b00zz_xx11. 64'h33b4_1223_1112_af01.

Here, `8'b1010_0010` is the same value as `8'ha2` or `8'hA2` or `8'd162` or `8'o242`.

During logic simulation, limitations on displayed expressions tend to fail safely, in the sense that ambiguity is propagated throughout the values represented by a hex, decimal, or octal numeral. This is just a display change, not a value change. If we have a four-bit binary expression which contains just one 'z' or 'x', then writing that expression in hex will cause the result to be expressed in hex the same as though all four bits had been found to be 'z' or 'x', respectively. The same holds for decimal or octal numeral representations. Thus, assigning a variable 8'b000z_xx11 means that it will be displayed in hex as 8'hzx. The 'x' is stronger than the 'z', so 4'b0z0x is displayed as 4'hx.

The significance of the bits in a vector is unalterable. If 8'ha2 (= 8'b10100010) is assigned to an 8-bit variable declared reg[7:0], then bit [7] is the MSB, which gets a '1'. If it is assigned to a reg[0:7], bit [0] is the MSB, and it still gets a '1'. The MSB always is on the left of a verilog vector.

```
// Some example vector operations:
`timescale 1ns/100ps
module Vector;
   reg [0:15] MyBus;    // A vector of 16 bits of storage.
                        // The verilog MSB always is on the left.
   wire[15:0] mybus;    // Another vector, again of 16 bits.
   ...
   MyBus[0:2] = mybus[10:8]; // This is called a "part select".
   MyBus = mybus;
   MyBus[0:15] = mybus[15:0];
   ...
endmodule
```

`timescale detail: It allows use of fractional delay expressions such as #0.1. For example,

```
`timescale 1ns/10ps
...
#0.05 X = Y & Z; // The delay is 50 ps.
```

Timing expressions such as #0.05 or #10 are ignored by the logic <u>synthesizer</u>; for the synthesizer, unknowns are treated as known, but in special ways. Neither 'x' nor 'z' is meaningful as an actual value for synthesis: 'x' means that the simulator can't decide whether to assign a '1' or '0'; 'z' refers to three-state gate functionality, not to a logic state. But, in synthesis, the synthesizer has to apply its own interpretation to assignments of 'x' or 'z', in order to make the synthesized gates simulate between '1' and '0' the same way (except timing) as would the presynthesis design in hardware.

2.4 Logical (Boolean) Operators

There are three of them:

        &&   Binary logical *and*

        || Binary logical *or*

        !   Unary *not* (negation)

They perform the usual Boolean operations. They always express the scalar value of a single bit, which must be '1' for *true*; '0' for *false*; or, 'x' for either '1' or '0' but cannot determine which.

Operands are allowed to be vectors. However, a vector operand can be *false* only if all its bits are '0'. Also, any 'x' or 'z' bit in a vector operand makes the operand's value the same as that of a one-bit 'x'.

2.5 Bitwise Operators: Vectors and Reduction

The following four are **bitwise operators**:

&	Binary bitwise **and**
|	Binary bitwise **or**
^	Binary bitwise **exclusive-or** (*xor*)
~	Unary bitwise **negation**

In a statement, they return a vector matching the width of the destination.

To apply the binary bitwise operators to a pair of vectors, align the two vectors on their LSBs; and, if the vectors are not the same width, fill the shorter one with '0'.

To apply the unary operator (~), invert every bit in the vector. In the vector, 'x' or 'z' bits invert to 'x'.

Examples:

```
reg[3:0] Nybble; reg[1:0] Bits, NewBits; reg Bit;
Nybble = 4'bx111;
  Bits = 2'b01;
   Bit = 1'bx;
...
Nybble = Nybble & Bits & Bit;  // --> 4'bx111 & 4'b0001 & 4'b000x --> 4'b000x
NewBits = Bit | Nybble;        // --> 2'b0x | 2'b11 --> 2'b11
```

The bitwise operators &, |, and ~ are equivalent to the boolean operators &&, ||, and !, respectively, when they are applied to 1-bit wide operands.

The binary bitwise operators (&, |, and ^) may be used as **reduction operators**. A reduction operator reduces a vector to a single bit by applying the logical operation to all the bits in the vector, as though the vector was a set of inputs to a single logic gate.

More examples of bitwise operations. The setup initializations in the left block are used in the two sets of statements in the middle and the right:

```
reg[3:0] a, b, z;   z = a & b;   --> z = 4'b0100   y = a & b;   --> y = 1'b0
reg y;              z = a | b;   --> z = 4'b1110   y = a | b;   --> y = 1'b0
a = 4'b0110;        z = a ^ b;   --> z = 4'b1010   y = a ^ b;   --> y = 1'b0
b = 4'b1100;        z = ~a;      --> z = 4'b1001   y = ~a;      --> y = 1'b1
```

Contrast the above bitwise statements with the corresponding boolean ones, in which any vector operand not all-0 means *true* (or, '1'):

```
z = a && b;   --> z = 4'b0001     y = a && b;   --> y = 1'b1
z = a || b;   --> z = 4'b0001     y = a || b;   --> y = 1'b1
z = !a;       --> z = 4'b0000     y = !a;       --> y = 1'b0
```

Here are four examples of the use of bitwise operators as **reduction operators**, assuming the initializations above:

```
z = &b;   // --> 4'b000(1 & 1 & 0 & 0) --> 4'b0000
z = |b;   // --> 4'b000(1 | 1 | 0 | 0) --> 4'b0001
z = ^b;   // --> 4'b000(1 ^ 1 ^ 0 ^ 0) --> 4'b0000
y = &b;   // -->            (1 & 1 & 0 & 0) --> 1'b0
```

2.6 Operator Lab 2

Topic and Location: Some simple vector-related and miscellaneous exercises.

Log in and change to the `Lab02` directory; do all your work here for this lab.

Preview: We begin with some elementary vector assignment exercises to see what is, and is not, legal. We then move on to some elementary boolean operations on vectors and selects. We end by reformatting our standard ANSI module header to a traditional format, just be be acquainted with the latter (which still is used in legacy code).

Deliverables: 1. A correctly simulating `Vector.v` file as in Step 1A. 2. A copy of the Step 1A `Vector.v` file including `My48bits` assignments. 3. Another file with a `module` (OK to use `Vector.v` again) including the boolean assignments of Step 2.

Lab Procedure:

Step 1: Vector exercise:

A. Create a file named `Vector.v`, and copy the `module Vector` content into it from the above lecture example which is about two pages up. Add output and input port declarations, and fill in the missing parts ("...") of this `module` any way you want.

But, use continuous assignment to drive some value onto `mybus`, and use blocking assignments in an `initial` block to schedule the assignments shown in the lecture example. Supply your own time delays. Check your work by simulating in VCS.

B. Now try to reverse bus directions by adding `MyBus[0:2] = mybus[8:10]` or `MyBus[15:0] = mybus[15:0]`. Compile in the simulator. What happens?

C. If you tried `mybus = MyBus` in your `initial` block, what kind of problem might occur?

D. Declare a 48-bit reg named `My48bits` and assign it in an `initial` block as follows:

```
#5 My48bits = 'bz;
#5 My48bits = 'bx;
#5 My48bits = 'b0;
#5 My48bits = 'b1;
```

Notice that no width specification is given in these literal values. Simulate after adding `#5 $finish`. What happens? Suppose you used `1'bz`, `1'bx`, etc.?

Digital VLSI Design with Verilog

Step 2: Boolean operator exercise:

A. Declare a reg[4:0] named X and two others named A and B. Try simulating after assigning A and B to X as follows: Initialize A = 5'b01010; B = 5'b11100; then, run these in order: X = A & B; X = A | B; X = A ^ B; X = ~A | ~B.

B. After that, try X[0] = &A; X[1] = |A; X[2] = &B; X[3] = ^A & ~^B; X[4] = A[4] ^ B[4]; X = ^X.

C. Parentheses may be used for grouping: Try X = ((~A & ^B) | (A & B))^A;

These operators also work with unequal-sized vectors; we'll look at this later in the course.

2.6.1 Lab Postmortem

Things to have noticed:

    Vector extension for logic 0, x, or z, but not 1.

    Binary vs. reduction operators.

What happens with an assigned delay of #0.1 and `timescale 1ns/1ns?

2.7 First-Day Wrapup

2.7.1 VCD File Dump

The logic simulator can create a VCD (Value-Change Dump) file. The VCD file format is specified in IEEE Std 1364 section 18 or IEEE Std 1800 section 21.7. Briefly, a VCD file provides an ASCII file record of a simulation in an extremely compact format. Because it is plain ASCII text, such a file is accessible to homemade scripts or other utilities for various purposes—for example, for estimation of test coverage or detection of timing violations of rules not in force when the simulation was run.

The VCD file consists of a header with information such as a design module name, and a table of abbreviations of variable names. Ports can be dumped optionally, as though they were variables in the module. Variable names are represented by one-character ASCII symbols such as '!', '#', or '('. The body of the file is a list of simulation times written as #*time* (absolute; not a delay), followed immediately by a list of variable values which were new at that time. For this reason, simulator output files of this kind often are called "time-value" files.

VCD file example:

```
...
$scope module TestBench $end
...
$var reg  1  #  Cstim $end
$var reg  1  $  Dstim $end
$var wire 1  %  Xwatch $end
$var wire 1  *  C $end
...
#30
0#
#40
0$
0*
...
#350
1%
...
```

The values dumped can be selected as level-only (4-state) or level plus strength.

In Lab 1, you generated a VCD file named `VCS_SimRun.VCD` when you ran the `Intro_Top` simulation; you may examine it later, if you want. The above example was taken from one version of it.

Because the VCD file format is an IEEE standard, it is portable across all tools reading or writing this format. One important use for it in a design flow is as a sample of simulation activity for an estimate of design dynamic power dissipation. Power estimation is beyond the present course.

2.7.2 SDF File Dump

The <u>S</u>tandard <u>D</u>elay <u>F</u>ile format, also an IEEE standard, resembles that of lisp or EDIF, with punctuation supplied almost entirely by parentheses.

For example, the SDF delays for a 0 drive-strength, noninverting buffer instance in a netlist, named U3 and of type `BUFFD0`, might be:

```
(CELL
  (CELLTYPE "BUFFD0")
  (INSTANCE U3)
  (DELAY
    (ABSOLUTE
      (IOPATH I Z (0.064:0.064:0.064) (0.074:0.074:0.074))
    )
  )
)...
  (INTERCONNECT U3/Z U4/A (0.002:0.002:0.002) (0.003:0.003:0.003)
```

An SDF file can be used to back-annotate delays into a netlist. The logic synthesizer can create this kind of file, but usually it is most useful when created by a tool with access to a chip floorplan or layout.

An SDF file contains a complete timing representation of a design netlist, with delay information based on a library and calculated during a static timing analysis or other delay extraction process. We shall look at this file again toward the end of the course, when we discuss back-annotation of layout timing into a simulation or layout netlist.

2.7.3 The Importance of Synthesis

Although simulation might be considered enough for a front-end designer, this is quite incorrect. True, the simulator may be used to validate functionally the original design (source verilog) as well as any synthesized netlist; but, the timing used by the simulator is entirely artificial—it is entered manually by the designer, based on estimates and expectations. Only after a netlist has been synthesized, can the propagation delays of the gates, the setup, hold, and other clocking constraints, and the (back-annotated) trace delays be estimated with any accuracy.

More importantly, a netlist can be fabricated; it is a product of value in the marketplace. Simulator waveforms can't be marketted or sold, so they are strictly for the benefit of the designer. Thus, a good netlist is a designer's contribution to the success of the company, and the logic synthesizer is the tool which almost always is the means of creating that netlist.

2.8 Additional Study

Read Thomas and Moorby (2002) chapter 1 on verilog basics.

Read Thomas and Moorby chapter 3 on synthesizable verilog through section 3.4. Ignore 3.4.4 (`casez` and `casex`).

Do Thomas and Moorby 1.6 Ex. 1.2 (different adders).

Find the `.VCD` file in the `Lab01` directory and open it in a text editor. When you ran your `Lab01` simulation, this `.VCD` file was created because of a directive, `` `include "../../VCS/Extras.inc" ``, which was present in the `Lab01` testbench file (`TestBench.v`). Examine the VCD file, just to see what it looks like. We won't be making use of VCD in this course, but it is good to know about. Detailed discussion is in Bhasker (2005), section 10.10.

Optional readings in Palnitkar (2003):

Read through the exercises in chapters 1–4, and try a few if you want.

Study the S-R latch model in section 4.1; there is a runnable example on the Palnitkar CD as `Chap04/Chap_EG/SR_Latch.v`. We shall do further work with the S-R latch in a later lab.

Study section 2.6, a ripple-carry counter example. The code is available on the Palnitkar CD as `Chap02/Chap_EG/ripple.v`. We shall study various counter designs later in the course.

Read section 9.5.6 for some general information on VCD files and section 10.4 on SDF files. We'll do a lab on SDF format in one of our last chapters.

Chapter 3
Week 1 Class 2

3 Today's Agenda:

Lecture: More language constructs
Topics: Traditional module header format. Comments, procedural blocks, integer and logical types, constant expressions, implicit type conversions and truncations, parameter declarations, and macro (`` `define``) definitions.

Summary: We expand on the previously introduced, basic verilog constructs allowed within a module. The emphasis is on procedural blocks and vector value conversions. Verilog parameters are introduced as the preferred way of propagating constants through the modules of a design; the main alternative, globally `` `defined`` macro compilation, also is presented.

Type Conversion Lab 3
A trivial design illustrates the use of `` `define`` to control parameters, and the use of comments. We exercise constraints to synthesize for area and for speed. We also do exercises in signed and unsigned vectors, and in vector assignments of different width.

Lab postmortem
This is mostly lab Q&A; also, the use of vector negative indices is discussed.

Lecture: Control constructs; strings and messages
Topics: The `if`, `for`, and `case`; the conditional expression; event controls; the distinct contexts for blocking and nonblocking assignments. The `forever` block and edge event expressions. Messaging system tasks. Shift registers.

Summary: Procedural control and the basic `if`, `for`, and `case` statements are introduced. The relationship of `if` to the conditional ("`?:`") is described, with warning about 'x' evaluations. Blocking assignments for unclocked logic; nonblocking assignments for clocked logic. The `forever` block is presented but its use is not encouraged. `posedge` and `negedge` for sequential logic are presented. Assertions based on `$display`, `$strobe`, and `$monitor` are advocated as good design practice. Finally, a shift register is presented briefly to prepare for the next lab session.

Nonblocking Control Lab 4
After flip-flop and latch models, verilog for a serial-load and parallel-load shift register is coded, simulated, and synthesized. A simple procedural model of a shift-register is given.

Lab postmortem
In addition to lab Q&A, some thought-provoking questions about synthesis and assertions are asked.

3.1 More Language Constructs

Today we'll present enough of the verilog language to be able to design something meaningful. Also, we'll be cementing our understanding of certain basic constructs which we shall use again and again in the rest of the course. After an initial lab on conversion basics, we'll look into the essential concept of a shift register and then design one in lab.

3.1.1 Traditional Module Header Format

We never shall use the traditional 1995 format for our lab designs, and it is deprecated for all new design entry; however, many tools (and older textbooks) still write headers in the older format, so the 1995 format should be understood.

The traditional format follows that of K&R *C*, whereas the newer format follows that of ANSI *C*. In both verilog formats, a port name automatically is associated with an implied net of the same name, making all ports `wire` types by default.

Traditional format:

```
module module_name
       ( names of outputN, inputN, inoutN );  // Port names alone.
output1  width_if_not_1 port_name;  // Type, width, and name of port.
output2  width_if_not_1 port_name;  //    "      "      "   "   "
...
input1   width_if_not_1 port_name;  //    "      "      "   "   "
input2   width_if_not_1 port_name;  //    "      "      "   "   "
...
inout1   width_if_not_1 port_name;  //    "      "      "   "   "
inout2   width_if_not_1 port_name;  //    "      "      "   "   "
...
reg      width_if_not_1 reg_name;   // Could be anywhere in the module.
...
endmodule
```

We'll see an example of this shortly.

3.1.2 Modern Module Header Format

The main difference is that in the 2001 format, header declarations are entirely contained in the header, which occupies the first statement in the module. The first semicolon in a 2001 module always terminates the header.

The 2001 format has all N names and widths in the header:

```
module  module_name
        ( name of outputN width_if_not_1 port_name, ...,
          name of inputN width_if_not_1 port_name, ...,
          name of inoutN width_if_not_1 port_name, ... );
... nothing more to declare about ports; but, for each output port with width > 1:
reg  width new_reg_nameN;
assign outputN = new_reg_nameN;  // All work below will be with new_reg_nameN.
...
endmodule
```

3.1.3 Header Formats Contrasted

In the 1995 format, only the names of the ports are in the first statement (first parenthesis); declarations of width or directionality follow and may be located anywhere in the body of the module.

This implies that, in the traditional format, it is necessary to enter the name of every port at least twice: Once in the first parentheses and once in a subsequent directionality and width declaration. In a small module, this is only a small burden; however, in a module with dozens or hundreds of I/O's, it greatly increases the risk of error.

In the traditional format, because the name of every `output` port had to be typed twice anyway, it was allowed, and actually was common practice, to type it yet a third time, declaring the port to be a `reg` type if the design intent was to assign it procedurally. This saved the (trivial) effort of declaring a separate internal `reg` and using a continuous assignment statement to wire the internal `reg` to the `output` port, as is almost always done in modern design.

However, this apparently saved effort caused three problems: First, we now had three declarations of the same name, a condition which was confusing to designers who wanted their declarations each to reserve a different name for every different object. Second, it increased the likelihood of a maintenance error: The three different naming statements always had to be coordinated whenever any I/O was changed in the design. Third, it required different `output` ports to be treated fundamentally differently—with no evident difference in the initial parentheses naming those ports: Ports later named a `reg` had to be assigned procedurally, only in `always` blocks, and ports not later named a `reg` had to be assigned only in wiring statements such as continuous assignments.

Finally the traditional "header" never had a well-defined location in the `module`; in principle, the header went on and on interminably, until finally, somewhere inside the module, the last I/O named in the initial parentheses was given a width and direction.

All these problems were overcome by the 2001 format, in which all I/O declarations were confined only to the header, and only were allowed to occur once.

Two design management problems with the traditional format also arose:

First, because `module` contents could modify the header's meaning, especially the I/O widths, at the start of a project, 1995 headers could not be sketched usefully and distributed to the design team. In 2001 format, it is easy to distribute complete headers and require that no designer modify a header without management approval.

Second, a `parameter` (= a named configuration constant; see below) declared and defaulted after the header in the 1995 format was fundamentally ambiguous: It could not be made clear which `parameters` should be overridden in a `module` instantiation, and which were meant to be strictly internal to the `module`. In the 2001 format, although `parameters` not declared in the header still can be overridden in an instantiation, design-team procedures easily can be stated so that instance overrides are restricted to `parameters` in the header.

Here are two alternative `module` declarations specifically illustrating the header differences. The module functionality marked "..." has been omitted.

In 1995 header format:

```
module MyCounter
        (CountOut, CountReady, StartCount, Load, Clock, Reset);
output[15:0] CountOut;
output       CountReady;
input[15:0]  StartCount;
input        Load, Clock, Reset;
reg[15:0]    CountOut;  // Could be anywhere in the module.
...
endmodule
```

The 1995-format header never really ends.

In 2001 header format:

```
module MyCounter
        (output[15:0] CountOut, output CountReady
        , input[15:0] StartCount, input Load, Clock, Reset);
// Header has ended; module itself now begins:
reg[15:0] CountOutR;
assign CountOut = CountOutR;
...
endmodule
```

One additional aspect of the difference is that the above 2001 continuous assignment to the `CountOut` net allows for specification of different delay values for rise and fall. This is a minor advantage but a real one. Procedural delays can have only one delay value, implying that 1995-style direct declaration of `CountOut` as a `reg` would mean that it could be assigned only one value for rise and fall. The different ways of assigning verilog delays will be presented later.

In 2001 format, it still is possible to declare an `output` port to be a `reg`,

```
module My2001Module (output reg[15:0] CountOut, ...);
```

Digital VLSI Design with Verilog

however, this should be avoided in a modern, manually-entered design. Although `reg` output port declarations in the 2001 format create only a minor inconvenience, we shall not allow them in our coursework.

Side note: An `input` or `inout` port never can be a `reg`, because a `reg` on an `input` only could be changed procedurally, and there are no header procedures. Therefore, a `reg` input always would be uninitialized and would contend with anything wired to that `input` when the module was instantiated, creating a perpetual (and illegal) 'x' on that `input`.

3.1.4 Verilog Comments

There are two different formats, the same as for C++:

- Line comment: "//" Starts anywhere on a line.
- Block comment: "/*" begins a comment which doesn't end until "*/".

Comment examples:

```
reg a, b;
...
assign X = a & b; // Put the AND of a and b on X.
//
assign Y = a /*left operand*/ & b /*right operand*/;
//
assign Z = (a>5)? A & b : a | b;
/* The last assignment puts the AND of a and b on Z
   if a is greater than 5; otherwise, it puts the OR
   of a and b on Z. */
```

Comments also may be effected in nonexistent-macro regions: To do this, refer to an undefined macro variable name in "`ifdef *undefined_name*"; this will cause everything between that `ifdef and the next `endif to be ignored by a verilog-conforming compiler (simulation or synthesis). Thus, an undefined macro name may be used effectively to commented out code. Later, the designer may `define the *undefined_name* and thereby uncomment the code and put it back into the verilog. This is discussed more fully below. Note that using the macroes makes it easy to differentiate commented running code from code comments not intended to be executed.

Comments can be used by tools to impose constraints not part of the verilog language: Synthesis directives or other attributes referring locally to verilog code constructs can be embedded in line comments if the comment containing them includes certain tool-specific keywords. A tool looking for keywords will read at least part of every verilog comment in the code. A typical such comment would be in the form of "//rtl_synthesis ...", with a command or constraint following the keyword "rtl_synthesis" on the same line.

3.1.5 `always` Blocks

A change of value in the sensitivity list (event control expression) of an `always` block during simulation can cause the statements in the block to be reread and reexecuted. A logic synthesizer usually will synthesize logic for every `always` block. Omission of a variable from the `always` block sensitivity list means that the state of the logic does not change when that variable does, possibly implying latched data of some sort. We shall discuss this idea in detail later in the course.

Thomas and Moorby (2002: section 7.2 and elsewhere), includes examples of `always` blocks without sensitivity lists. There is nothing wrong with this usage; but, in this course, we avoid it for reasons of style. A synthesis tool may flag it as an error.

Omission of an entire event control (sensitivity list) is not common in design practice, because location of an event control at the top of a block makes the functionality of the block more easily understood. In this course, we recommend never to use "always" except when immediately followed by an event control ("always @(*variable_list*)").

```
...
always                          // Avoid this style.
  #10 ClockIn <= !ClockIn;
...
always@(ClockIn)                // Recommended style.
  #10 ClockIn <= !ClockIn;
```

3.1.6 `initial` Blocks

An `initial` block has no sensitivity list and is read only once, beginning before simulation time 0, a time when no event control can be expressed.

Example (from the `Lab01` testbench):

```
module TestBench;
reg Astim, Bstim, Cstim, Dstim;   // To accept initialization.
. . .
initial
  begin
`include "../../VCS/Extras.inc"
  #1   Astim = 1'b0;
  #1   Bstim = 1'b0;
  #1   Cstim = 1'b0;
  #1   Dstim = 1'b0;
  #50  Astim = 1'b1;
  ...
  end
. . .
```

All `initial` blocks are ignored by synthesis tools. Because `initial` blocks can not be activated and disabled concurrently but only can schedule events after various delays from time 0, it usually makes good sense only to use one `initial` block in a `module` intended for simulation; more than one would just make it hard to determine the order in time of the events to be scheduled.

Exceptions to limiting the design to one `initial` block may be useful when a second `initial` block is reserved to implement a simulation clock (using `forever`) or is reserved for actions unrelated to design functionality, for example to define an SDF file to load, or to set up a `$monitor` task.

3.1.7 Continuous Assignments

We should mention these statements again here, for completeness. The three most common concurrent blocks in verilog are `always` blocks, `initial` blocks, and continuous assignment blocks—in addition to structural (hierarchical) design instances. These three blocks, and hierarchical instances, are the designer's main work when entering a design in verilog. A continuous assignment is sensitive to a change of anything on its right-hand side.

A few examples:

A continuous assignment is a `wire` assignment sensitive to the expression on the RHS:

```
assign #2 X = A[0] && B[0] || !Clock;
```

A procedural assignment is sensitive to changes in its event control list:

Xreg is not updated on changes in `Clock`:

```
always@(A[0], B[0]) Xreg = A[0] && B[0] || !Clock;
```

Yreg is updated only when `Clock` goes to `1'b1`:

```
always@(posedge Clock) Yreg = a && b;
```

3.1.8 Vectors and Vector Values

Vectors are one-dimensional lists used to describe the contents of multivalued `reg` or other variables. All vectors are read as numerical values. Most verilog vector types (`reg` or net) are read as unsigned by default. Exceptions are `integer` and `real` types, which are read as signed. A `real` type generally is not synthesizable. The width of an `integer` is 32 (bits), if not declared less.

Assuming no overflow, the bit-pattern representing a specific number is identical, whether the number's storage is signed or unsigned; the difference is when the value is being used in a comparison of some sort. When the value is used, the MSB of a signed number is used to determine whether the number is positive or negative. An MSB value of `'1'` means *negative*, as is usual in two's complement arithmetic. The MSB of an unsigned number simply is the most significant bit of its value; an unsigned number can not be negative, no matter what its bit pattern.

Example of the meaning of the sign bit, using the ">" relational operator:
```
reg[31:0] A;
integer    I;
...
A = -1;   // The default is to treat "-1" as a decimal integer.
I = -1;   // Now, both A and I hold 32'hffff_ffff.
//
if ( I > 32'h0 )
     $display("I is positive.");
else $display("I is not positive.");   //Prints "I is not positive."

if ( A > 32'h0 )
     $display("A is positive.");
else $display("A is not positive.");   //Prints "A is positive."
...
```

Some details on **verilog operators**, which will be covered later in more detail:

A. Relational operators:

>	returns true if the left value is greater than the right.
>=	returns true if the left value is greater than or equal to the right.
<	returns true if the right value is the greater.
<=	returns true if the right value is greater than or equal to the left.

As in the first sign-bit "`if (I > 32'h0)`" statement above, all these operators return `false` otherwise. They also return `false` if either value is nonnumerical.

B. Logical operators `!`, `&&`, and `||` treat operands as `true` when any bit in a vector operand is nonzero; they return `false` for all zero-bits, only.

C. Binary bitwise operators `&`, `|`, and `^` operate bit-by-bit. The unary `~` operator inverts each bit in a vector. Every binary bitwise operator can be used as a **reduction operator**: For example, `&A` expresses the *and* of all bits in vector `A`.

Unnamed **constant** values are called ***literals***. A literal expression consists of a width specification, a delimiting tick symbol (`'`), a numerical base specifier, and finally a number expressed in that base: The entire literal thus is *width*`'`*base*(*value*).

For example, `5'h1d` defines a 5-bit wide vector literal with hex value `1d`. The value may be found as the sum of its products: We have `1d` = 16*1 + 1*d = 16 + 13 = decimal 29. The expression, `16'h1d`, also evaluates to decimal 29, but it is much wider, being expressable as `16'h001d`. Notice that decimal 29 also is equal to binary 16 + binary 13 = binary 10000 + binary 01101 = binary 11101, which, as a 16-bit quantity, becomes `16'b0000_0000_0001_1101`.

If width and base are omitted, a literal number such as `1d` is assumed in verilog to be in decimal format; so, `1d` = 1x10 + d x 1 = 10 + 14 = 24. This is something usually to be avoided. If a whole number such as `48` is written without width or base, it is assumed to

Digital VLSI Design with Verilog

be a 32-bit integer (signed). In writing verilog, it is best never to omit the width or base of any literal numerical value. There are a few exceptions which will be explained later.

Vector **type conversions** are performed consistently and simply, avoiding the need for explicit conversion operators. Verilog actually has no user-defined type, so the rules can be quite simple:

1. The significance of vector bits is predefined; the MSB is always on the left.
2. All values in an expression are aligned to the rightmost bit (LSB).
3. All expressions on the right-hand side of an assignment statement are adjusted to the width of the destination before operators are applied.
4. Width adjustment is by simple truncation of excess bits, or by filling with 0 bits. When an operand is signed, the widening fill is by the value of the original MSB (sign bit).

These four rules hold for most of the binary operators. However, according to IEEE Std 1364 section 5.4.1, or IEEE Std 1800 sections 11.4.7 and 11.4.9, <u>logical</u> and <u>reduction</u> operators are treated specially: Such expressions are evaluated on the RHS first; then the width is adjusted to that of the destination (LHS).

As an example of a logical expression, we have,

```
reg[3:0] A, B;
reg Z; ...
Z = A && B;
```

Here, A and B each are evaluated true or false as 4-bit vectors; then, the logical *and* is taken. The resulting single bit LSB is assigned to Z.

Likewise, suppose we have,

```
reg[3:0] Z;
reg A, B; ...
Z = A && B;
```

Here, the *and* of A and B is evaluated; then, the resulting LSB is lengthened to 4 bits for assignment to Z.

Other examples of type conversions for vector operations:

```
reg X;
reg[3:0] A, B, Z;
...
X = A && B;  // X gets 1'b0 only if either A or B is 4'b0.
Z = A && B;  // Z gets 4'b0 only if either A or B is 4'b0.
             // Otherwise, Z gets 4'b0001
Z = A & B;   // Each bit of Z get the and of the
             // corresponding bits of A and B.
X = !Z;      // X gets 1'b1 only if all bits of Z are 0.
X = ~Z;      // Z narrows to its LSB and is inverted; so X gets !Z[0].
```

Truncation and widening examples: We look at unsigned expressions, only, for now. Notice how a variable can be initialized when it is declared; this is for simulation, only:

```
reg        A = 1'b1, B = 1'b0;
reg[3:0]   C = 4'b1001, D = 4'b1100;
reg[15:0]  E = 16'hfffe;
...
C = A | B;  // C gets A=4'b0001 | B=4'b0000 --> 4'b0001.
C = E & D;  // C gets E=4'b1110 & D=4'b1100 --> 4'b1100.
E = A + C;  // E gets A=16'h1 + C=16'h9 --> 16'ha.
```

3.1.9 Parameters

Verilog includes **parameter declaration**s. Parameters are named constants which must be assigned a value when declared. If declared in a `module` header, the value assigned is a default which is applied whenever that `module` is instantiated. This default value may be overridden at that point in the `module` where it is instantiated.

Inside a module,

```
module ModuleName ( ... I/O's ... );  // <-- semicolon ends header.
...
parameter Awid = 32, Bwid = 16;
...
reg[Awid-1:0] Abus, Zbus;
wire[Bwid-1:0] Bwire;
...
endmodule
```

In a module header,

```
module ModuleName
       #(parameter Awid = 32, Bwid = 16)  // <-- no comma!
        ( ... I/O's ... );  // <-- semicolon ends header.
...
reg[Awid-1:0] Abus, Zbus;
wire[Bwid-1:0] Bwire;
...
endmodule
```

Parameters are unsigned by default and are as wide as required to hold the value assigned.

3.1.10 Commenting with Verilog Macroes

Conditional commenting with *macro*es. To comment out a block of statements or other design data, one way is to use the verilog block comment tokens, /* and */.

This can be useful when unsynthesizable code would cause the synthesizer to issue an error. But, often, a better way is to know that when (our) synthesizer reads a verilog file, it defines a global macro DC before it starts. This macro is empty, but its name is

Digital VLSI Design with Verilog

always declared. This means that a designer can comment out code for the synthesizer but not for the simulator by surrounding the offending lines with `ifndef DC ("<u>if</u> DC <u>n</u>ot <u>def</u>ined") and `endif.

For example, the synthesizer rejects verilog system tasks. So, use the DC macro to make verilog system-task assertions invisible to the synthesizer but visible to a simulation compiler (which actually can use them).

For example,

```
always@(posedge Clk)
  begin
  Xreg = NewValue;
  `ifndef DC // = "if DC not defined"
  $display("time=%05d:  Xreg=[%h]", $time, Xreg);
  `endif
  Yreg = ...
  end
```

If you leave a verilog system task visible to the synthesizer, the least that will happen is that you get a warning—which, like all warnings, must be studied and understood to determine whether it should be ignored. This may be more of a waste of time than just adding the `ifndef.

The Silos demo simulator included with the Thomas & Moorby and Palnitkar books named in the *References* does not accept `ifndef, but it does accept `ifdef and `else; so, a more portable way of writing the example above would be to add an `else line.

This is shown here:

```
always@(posedge Clk)
  begin
  Xreg = NewValue;
  `ifdef DC
  `else
  $display("time=%05d:  Xreg=[%h]", $time, Xreg);
  `endif
  ...
```

The $display system task is like printf in *C* or *C++*: It formats a message to be printed to the terminal window of the simulator. It is, or at least should be, ignored during netlist synthesis. We shall discuss details of $display later in today's lecture.

A good coding rule is to use `define and other macro definitions sparingly: They are global, and more than one of the same name may exist inadvertently in a large design. Then, one value during compilation may replace another one, possibly causing design errors.

To avoid errors because of compilation order, a defined macro should be undefined as soon as it has served its purpose, preferably in the same file as the one in which it was defined. Exception: `timescale.

Example:

```
`define BLOCKING
`define Awid 32
module ModuleName ( ... I/O's ... );
  ...
  reg[`Awid-1:0] Abus, Zbus;    // The ` selects the `Awid value.
  ...
  always@( ... )
    begin
    `ifdef BLOCKING
      Qreg = Areg | Breg;
    `else
      Qreg <= Areg | Breg;
    `endif
    end
endmodule
`undef BLOCKING   // Use `undefs unless you want these to
`undef Awid       // be seen in subsequently compiled files.
```

A *macro* also may be called a **compiler directive**. In verilog, macroes almost always merely are defined or assigned a text value to be substituted. In principle, a macro could be used, as in C, to execute code during simulation, but not all tools will accept executable substitutions.

3.2 Parameter and Conversion Lab 3

Topic and Location: Simple use of defined macro values and parameters in synthesis. Type conversions, truncations, and numerical interpretation of vector values.

Work in the Lab03 directory for these lab exercises.

Preview: In this lab we present a small verilog design consisting of a counter and a bit-converter; the latter represents hardware which reformats the counter's output by padding it with zeroes to a configurable width. This design is parameterized already, and the exercise merely requires changing the macro which controls the parameter values and synthesizes the design. Basic constraints for netlist optimization then are exercised. The lab then changes topic to elementary signed and unsigned arithmetic operations to be checked by simulation. Finally, some elementary exercises are simulated to show truncation and widening of verilog vectors.

Deliverables: 1. An area-optimized netlist and a speed-optimized netlist synthesized in Step 1. 2. A second set of Step 1 area and speed optimized netlists with the Converter padding value changed. 3. One or more new verilog modules in a file named Integers.v, with assignment statements as described in Step 2. 4. A variety of assignment statements in one or more modules in a file named Truncate.v completes Step 3.

Digital VLSI Design with Verilog

Lab Procedure:

Step 1: Synthesis of a design with parameterized widths. The design in the `Lab03` directory consists of a top-level verilog module in `ParamCounterTop.v` and two submodules in `Counter.v` and `Converter.v`. A schematic is given in figure 3-1, with only those pin names shown which entail a name change in the port mapping. For example, the port `OutEnable` is connected to an `Enable` pin; therefore, the `Enable` name is shown:

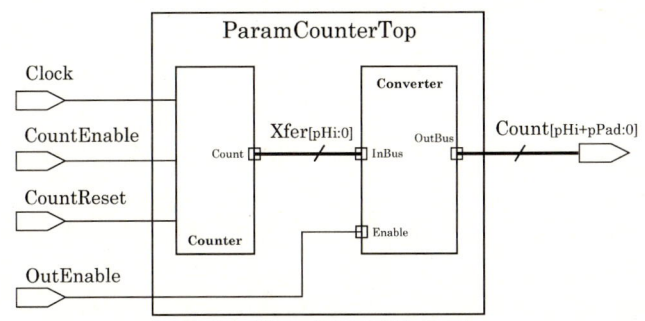

Fig. 3-1. Schematic for `Lab03`, Step 1.

The design is of a resettable up-counter which puts its result in the lower-order bits of an output bus of configurable width. The design is intended to show how parameters are used to configure a design. In this exercise, only bus widths are parameterized.

Don't bother simulating, unless you want to do it after finishing this lab.

In this and all subsequent labs, to run the synthesizer, use the lab-specific `.sct` synthesis script file, if any, provided for you. This particular synthesis `.sct` file begins with numerous *EDIF*-specific commands, just to show how a synthesizer might have to be configured to generate an *EDIF* netlist for fabrication purposes. Such commands will be omitted in all our subsequent labs.

When a `.sct` file is not provided, you should create your own by copying an existing `.sct` file and modifying it in a text editor. Such modification generally will include changes in the names of the verilog input files, the name of the synthesis library (working directory), and the names of log files and synthesized netlists. The `.sct` constraints will have to be modified with special attention to the intended design functionality.

In practice, hardly anyone ever writes down a synthesis script file except by starting with a copy of a preexisting one.

When you are ready, change `PADWIDTH` to 8, and synthesize the design; optimize it for area and then for speed. To synthesize for area, just comment out all speed-related constraints (but leave the design rules); to synthesize for speed, leave these constraints in, but keep the area goal at 0. Area and speed are somewhat correlated in practice.

After experimenting with `PADWIDTH` set to 8, increase the value of the `pPad` parameter and resynthesize to see what happens.

Step 2: Creation of a design to simulate arithmetic. Next, try some signed and unsigned arithmetic, using the simulator to view the results:

Create a verilog `module` in a new file `Integers.v` and check the results of the following (you should change the variable names if you wish to do them all in one `initial` block in one `module`, simultaneously):

A. `integer A = 16, B = -8;`
     `X = A + B;`
     `X = A - B;`
     `X = B - A;`

B. Same as A above with all declared as `reg[31:0]` (use `32'd` to assign to this `reg`).

C. (optional) Same as B above with the following successive changes: Just `X` declared as `integer`; just `A` declared `integer`, and just `B` declared `integer`.

Step 3: To see the effect of truncation and widening of various types, create a new verilog `module` in a file `Truncate.v` and try the following in the simulator: Declare data objects as follows: `integer Int; reg[7:0] Byte; reg[31:0] Word; reg[63:0] Long;` and `reg[127:0] Dlong`. Initialize each one with `'b0`. Then,

A. Assign `6'd1` and then later `-6'd1` to each of the above objects and notice what happens, both with hex and decimal radix display:

B. Assign `36'd1` and then later `-36'd1` to each as in A.

C. Assign 1 and then −1 to `Int`, and use `Int`, not a literal, to assign each of the others as in A and B above.

D. Assign `32'h7eee_777f` to `Word` and then use `Word` to assign each of the others as above. This value has a 0 MSB and so should represent a positive number.

E. Finally, assign `32'hf777_eee7` to `Word` and assign to each of the others. The 1 MSB should represent a negative number.

Note: In new *VCS*, vector display formats are associated with variables, not with displayed traces; so, it is not possible to view a variable in more than one radix at a time.

3.2.1 Lab Postmortem

Side note: Vector negative indices are allowed, although they are not used in this course. The width always is given by the difference, plus 1.

For example, a declaration of,

> `reg[3:-7] CoeffA;`

may be interpreted by a tool to represent an 11-bit decimal fraction with LSB equal to 2^{-7} = $1/2^7$ = $1/128$. Used in DSP modelling.

Digital VLSI Design with Verilog

3.3 Procedural Control

3.3.1 Procedural Control Constructs

We shall study three procedural control constructs at this point: `if`, `case`, and `for`. These only are allowed in a procedural block—`initial` or `always`.

Use **if** for explicit priority or for ranges of values. The `if` often is easier to use than the other control constructs when describing a short list of mutually exclusive events such as a clock and an asynchronous control.

```
if (expr) statement1;
else if (expr) statement2;
else statement3;
```

Use **case** when alternatives are specific values, are numerous, or are conceptually unprioritized, for example to implement a table lookup or small memory addressing scheme. The verilog `case` breaks automatically and does not "fall through" on a match the way the C language `case` (in a *switch* statement) does. Good practice is never to omit the `default` of a `case` statement; we'll return to this issue, and the `case` statement, later in the course.

```
case (expr)
    alt1:  statement1;
    alt2:  statement2;
    ...    ...
    default:  default_stmt;
endcase
```

The `case` alternatives usually are constant expressions but may be variables; the `case` expression usually is a variable.

Use **for** everywhere. The **for** is the preferred looping construct in verilog; it works about the same way as the C `for`. However, the C language unary increment and decrement expressions, `i++`, `++i`, `i--`, and `--i`, are not allowed, although they are allowed in SystemVerilog. In verilog, one must use `i = i + 1` or `i = i - 1`.

```
for ( loop var init;  loop reentry expression;  loop var update )
    statement(s);
```

3.3.2 Conditional Expression Operator

The conditional expression operator, *control_expr* ? *True_expr* : *False_expr*. This is used like a C function call returned value—in an expression. It is an expression, not a statement. The expression always is interpreted by the current synthesizer as combinational logic. If the control expression evaluates to *true* (non-0), the *True_expr* expression is its value; otherwise, it evaluates to the *False_expr*.

Example:
```
wire[31:0] X;
integer A, B;
...
// Put the greater of A or B into X; A if they are equal:
assign #2 X = (A>=B)? A : B;
```
This is very useful for a mux connection to a `wire`, because `if` is not allowed in a continuous assignment (`if` may be used only in a procedural block).

3.3.3 Combinational and Sequential Logic

- Simple `always@` block syntax: Use ',' in the event control (sensitivity list); the 1995 traditional but inconsistent "`or`" is deprecated. If any change causes the block to respond, the logic is combinational. When a `posedge` or `negedge` expression causes insensitivity to the opposite edge, the result is sequential logic:

  `always@(posedge Clk, posedge Reset)` means the same as the 1995 verilog `always@(posedge Clk or posedge Reset)`.

  It also is possible to embed event controls inside an `always` block:
  ```
  always@(posedge clk)
    begin
    xbus[1] <= 1'b1;
    @(posedge Enable)  // Execution stops here until Enable goes to '1'.
       begin
       Dbus       <= 8'haa;
       xbus[7:4]  <= 'b0;
       end
    xbus[2] <= 1'b0;
    ...
    end
  ```

- For complicated combinational constructs, consider using blocking assignments in an `always` block, with the result possibly put on the right side of a continuous assignment statement.

  Example:
  ```
  always@(Ain, Bin, Cin, Din, temp1, temp2)
    begin
    temp1 = Ain^Bin;
    temp2 = Cin^Din;
    ComboOut = (temp1 & temp2) | Ain^Din;
    end
  assign #5 OutBit = ComboOut | OtherComboOut;  // Collect the delays here.
  ```

- nonblocking assignments. These are statements within a procedural block. They differ from blocking assignments in two ways: (*a*) during simulation, at any specific

Digital VLSI Design with Verilog

time, nonblocking statements higher in a procedural block do not block reading of lower ones; and, (b) at any given simulation time, nonblocking statements are evaluated along with blocking statements, but they are executed only <u>after</u> all blocking assignments and net updates are complete.

- Use nonblocking assignments in an `always@` block for clocked sequential logic. Nonblocking evaluation occurs when blocking evaluations do, but assignment is scheduled after all blocking assignments on a given clock event, thus ensuring that combinational input values will not have been altered by nonblocking updates before being clocked in.

  Avoid: Blocking assignments for clocked sequential logic, nonblocking assignments for combinational logic.

  Avoid: Mixed blocking and nonblocking assignments in one procedural block.

  Avoid: Delay of `#0` to fix up race conditions caused by failure to avoid!

- Implementation of clocks and storage registers. Clocks imply sequential elements (storage), because the clocked values remain constant between edges. Let's ignore level-sensitive latches for now; if we do so, sequential elements are synthesized by use of the edge specifiers, `posedge` or `negedge`, in an event control. Sensitivity only to an edge implies storage on the opposite edge.

Here are examples of clocks paired with equivalent schematics:

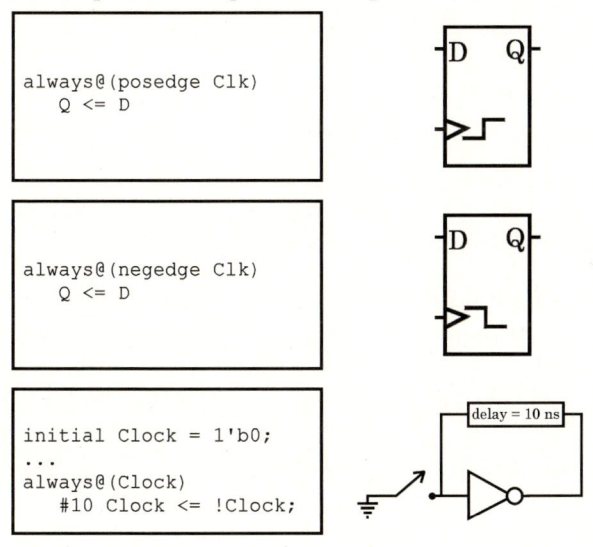

```
always@(posedge Clk)
    Q <= D
```

```
always@(negedge Clk)
    Q <= D
```

```
initial Clock = 1'b0;
...
always@(Clock)
    #10 Clock <= !Clock;
```

A clock generator usually appears only in a testbench and may be defined, as in the last clock example above, by means of a single, delayed nonblocking assignment in a level-sensitive `always` block:

```
reg Clock;
...
always@(Clock)
   #10 Clock <= !Clock; // ~Clock also OK.
```

This example makes use of the concurrent `always` statement. The assignment must be nonblocking, so that the `always` block event control will be sensitive to the inversion. The clock `reg` must be initialized somewhere else.

Another way to define a clock is to use the procedural `forever` statement in an `always` or (more usually) `initial` block. Being procedural, a `forever` must be enclosed in a concurrent block in the module in which it is to be run.

For example,

```
reg Clock;
...
initial // This one only for clock generation.
   begin
   Clock <= 1'b0;
   forever
     #10 Clock <= !Clock;
   end
```

This clock would work with blocking assignments, because of the "`forever #10 ...`". However, with blocking assignments, anything clocked by it would require special treatment to ensure proper data setup timing.

Be careful not to use an `initial` block to initialize things that should be synthesized or used to represent hardware initialization: This is a very important difference between coding of simulation software and coding of hardware.

3.3.4 Verilog Strings and Messages

The verilog string type is for literal strings, only. We shall not use strings in this course except in <u>system task</u> messages. String values may be stored in a `reg` vector, or in a memory object (array of bytes); but, be careful, especially with unicode text or other nonASCII character encodings. Messaging system tasks all print to a simulator (console) text screen; they resemble *C* language *printf* functions. See Thomas and Moorby (2002) sections B.4 and F.1, IEEE Std 1364 section 17.1, or IEEE Std 1800 chapter 21 for more information on messaging during simulation.

The three most useful messaging tasks:

$display(*format, args_for_display*) for a simulation result printout when it is encountered procedurally but <u>before</u> RHS evaluations (before nonblocking assignments) at that simulation time.

$strobe(*format, args_for_display*) for a simulation result printout when it is encountered procedurally but <u>after</u> RHS evaluations (after nonblocking assignments) at that simulation time.

$monitor(*format, args_for_display*) for print-on-change procedural simulation results <u>after</u> RHS evaluations and nonblocking assignments <u>whenever</u> one of its args changes. Usually invoked in an `initial` block, often under a condition or after a delay.

Assertions are routines embedded in the design, by the designer's foresight, to check that a condition holds and <u>to announce a warning or error during simulation</u>. Like simulation itself, they are a design activity as much as a verification activity.

Assertions *assert* that something should be true, remaining silent when it is; and, they report when it is false. Assertions bring attention to conditions which otherwise might be overlooked after a design change or under unusual or complex simulation conditions.

Whereas the primary purpose of an assertion is to warn the designer when the asserted condition has failed, assertions also may be used to generate simulation errors and cause a running simulation to pause or terminate. We shall deal only with assertion messages for now.

A simple, homemade assertion statement using the **$time** system task:

```
reg X, Y;
...
if (X!=Y)
  $strobe("\n***Time=%04d.   X=%1b == Y=%1b failed.\n", $time, X, Y);
//
// Typical screen output when Y unexpectedly goes unknown:
//       "***Time=0912.   X=0 == Y-x failed."
```

Messages not only contain text strings, but they almost always also should report values of design variables; so, they must provide readable and useful formats for those values.

The most useful format specifiers are these (* = most common):

```
*   %h    hex integer
*   %d    decimal integer
    %o    octal integer
*   %b    binary integer
    %v    strength level
    %c    single ASCII character
*   %s    character string
*   %t    simulation time in timescale units
    %u    2-value data of 0, 1
    %z    4-value data of 0, 1, x, z
*   %m    module instance name (see below)
```

Any of the integer formats except %t may be preceded by "n", in which n is an integer, to constrain the minimum number of unused leading places displayed. In addition, "0n" fills remaining leading displayed places with zeroes, as in the assertion example above.

For example,

```
integer x; ...; x = 5;
$display("%s = [%04b] = [%4b].", "The value", x, x);
```

prints to the simulator console, "The value = [0101] = [101].". Otherwise, using "%b" with no n, the integer would be printed with a default of 29 leading blanks, because integers are 32 bits wide.

There is a special string replacement option, %m, which appears to be a format specifier but which actually does not format anything; instead, it is replaced in the output with the full hierarchical name of the instance in which the system task was executed. Hierarchical names are explained in detail in *Week 6 Class 2*.

3.3.5 Shift Registers

A shift register is a register of bits which shifts the bit-pattern up or down in the register. Often, "shift right" is used to describe shifting down, and "shift left" is used for shifting up. Up *vs.* down refers to the unsigned numerical verilog value interpreted from the register contents. Counters count up or down by similar reasoning. The same terminology is used in software assembly-language programming, in which the contents of a register in the CPU or memory are shifted one way or the other.

Binary representation is the most useful way to understand what happens during a shift. For example, suppose an 8-bit register with this content: 8'b0001_1001. A shift up changes the contents to 8'b0011_0010. The leftmost bit (MSB) is lost, and a new '0' appears on the right (because a '0' would be shifted into the register by default). A shift down changes the original contents to, 8'b0000_1100. If this register, in hardware, was hooked up so that its MSB input connected to its LSB output during a shift down,

Digital VLSI Design with Verilog

then the original contents, shifted down, would be, `8'b1000_1100`: The LSB '1' would be shifted into the MSB of the register.

When a shift register is connected to other design elements, the new bit appearing on one end or the other depends on how the register has been connected. What happens to the bit shifted out also depends on the design.

3.3.6 Reconvergence Design Note

A shift register with programmable storage is a good way to introduce the problem of reconvergent fanout—in this case, of a clock.

To use a shift register as something other than a 1-bit FIFO or a latency control, the shifting should be capable of being disabled; this allows the shift register to store its current value for one or more clocks. A simple way to achieve this is to gate the shift clock, as shown in figure 3-2.

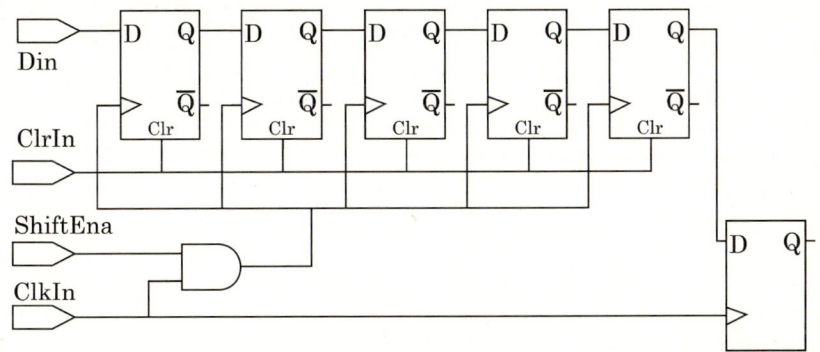

Fig. 3-2. Shift register which stores data when clock is gated off. The clock gate may cause a reconvergence error.

However, in the application shown, the clock which shifts the data also clocks an independent flip-flop as shown on the far right. This clock has reconvergent fanout, because its logical effect, after traversing the shift register, reconverges on the (D input of the) external flip-flop. From the schematic alone, the problem in this case is likely to be setup: The *and* gate on the shift clock causes it to arrive later than the clock on the external flip-flop. Thus, even with a design allowing for setup on the individual components, the external flip-flop may be clocked too soon to capture the current value on the last shift flop.

This is a general problem in any design with a gated clock. It is overcome by using specially designed shift flops, with an input-enable control pin, or by adding delay to the clock on all components which are independent of the shift register and which receive data from it.

The simple *and* gate shown above in figure 3-2 also illustrates a new need for a special shift flop: With the logic shown, an asynchronous `ShiftEna` could cause a glitch on the gated clock, with unpredictable results. A realistic clock gate would use a transparent latch which was enabled by the inactive edge of the clock to be gated.

Such a latch is shown in figure 3-3:

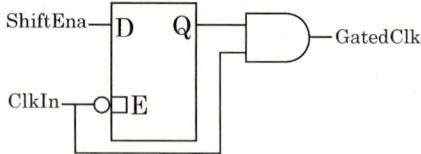

Fig. 3-3. A glitch-proof gated clock assumed active high.

Of course, the additional delay of the latch would make the reconvergence delay longer, possibly requiring more design effort to avoid it.

Possible reconvergence is something always to keep in mind; however, the synthesizer is designed to accommodate reconvergence and can be expected to create a netlist from the (gated-clock) design in figure 3-2 without the setup error described. Of course, if the designer wants the external logic to receive late data, then the synthesizer must be constrained to create it that way. We shall not discuss this problem further at this point in the course. In the next lab, we avoid reconvergence by adding shift-enable logic manually to each of our shift-register flip-flops.

3.4 Nonblocking Control Lab 4

Topic and Location: Models and assertions for D flip-flops, D latches; also for shift registers, serial- and parallel-load, built using the flip-flop models.

Work in the Lab04 directory. Create subdirectories under Lab04 if you want.

Preview: Nonblocking assignment statements are used for clocked sequential logic elements such as flip-flops. In this lab, we model single flip-flops with various features such as asynchronous preset or clear. We also model latches to see how this is done. We then use the flip-flop models to construct a shift register with serial load of data; this shift register is tested by simulating it. Next, we modify the shift register by adding multiplexers so that it can be loaded all bits in parallel; we synthesize this design for area and for speed. Finally, a vastly simpler procedural shift register is presented, and subtleties of its design are explored in simulation and synthesis.

Deliverables: Step 1: A verilog source module with three different D flip-flop models. Synthesized netlists of this module optimized for area and speed. Step 2: A simulatable verilog source module of a flip-flop with synchronous clear. Step 3: A source module with three different latch models and synthesized netlist; plus, assurance that latches were synthesized as expected. Step 4: A structural, serial-load shift register model in at least two different files which simulates correctly. Optional Step 5: A structural parallel-load shift register model in at least three different files; this should simulate correctly and should be synthesized to two different netlists, one optimized for area and the other for speed. Step 6: Two simulation models of a procedural shift register, one using nonblocking assignments and the other blocking.

Lab Procedure:

Step 1: Use the following examples to model three different D flip-flops in a single verilog `module`. Add an assertion that warns when preset and clear are asserted simultaneously; write the assertion so that it will be ignored by DC during synthesis. Check your design by simulating it; then, synthesize it, optimizing first for area and then for speed.

Simple D flip-flop:
```
always@(posedge clk1) Q <= D;
```

D flip-flop with asynchronous clear:
```
always@(negedge clk1, posedge clr)
   begin
   if (clr == 1'b1)
       Q <= 1'b0;
   else Q <= D;
   end
```

D flip-flop with asynchronous preset and clear:
```
always@(posedge clk2, negedge pre_n, negedge clr_n)
   begin
   if      (clr_n == 1'b0) Q <= 1'b0; // clear has priority over preset.
   else if (pre_n == 1'b0) Q <= 1'b1;
   else                    Q <= D;
   end
```

Because these three `always` blocks will coexist in the same `module`, their procedural variables (Q regs) will have to be declared with three different names. Also, to put the three regs into the (hardware simulation) design, each one will have to be wired to a module output port with its own continuous assignment statement.

Step 2: Write a verilog model of a D flip-flop with synchronous clear. Simulate it to check your design.

Step 3: Use the following examples to model three different D latches corresponding to the flip-flops in Step 1. You may wish to copy your design from Step 1 and modify the flip-flop code to represent latches. Decide which "simple D latch" to use for each one from the examples below. As in Step 1, add an assertion that warns when preset and clear are asserted simultaneously. Check your design by simulating it; then, synthesize it, optimizing first for area and then for speed. Check this netlist by inspection of the schematic to be sure that it indeed models three latches.

Simple D latch:
```
always@(D) if (ena1==1'b1) Q = D;
```

One assumes that the enal variable is declared somewhere and controlled externally; for example, enal might be a module input, as probably would be D.

What would happen if the sensitivity list was always@(enal)?

What kind of simple D latch would be the following?

```
always@(D, enal) if (enal==1'b1) Q = D;
```

D latch with asynchronous clear and with enable asserted low:

```
always@(D, clr, ena_n)
    begin
    if (clr == 1'b1)
        Q = 1'b0;
    else if (ena_n==1'b0) Q = D;
    end
```

D latch with asynchronous preset and clear asserted low:

```
always@(D, pre_n, clr_n, ena2)
    begin
    if      (clr_n == 1'b0) Q = 1'b0; // clear has priority over preset.
    else if (pre_n == 1'b0) Q = 1'b1;
    else if ( ena2 == 1'b1) Q = D;
    end
```

Step 4: Serial-load shift register. A shift register shifts on every clock if shifting is enabled; otherwise, it holds previous data. See the schematic in figure 3-4.

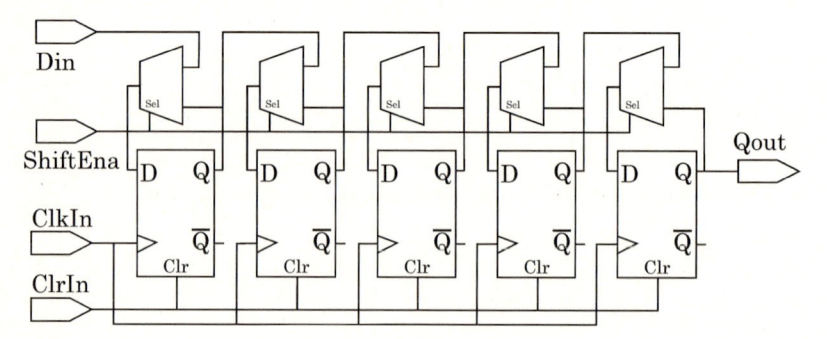

Fig. 3-4. Shift register with serial load.

Write a verilog model of a D flip-flop, with asynchronous clear, in its own file. Your flip-flop should have Q and Qn outputs. Use this flip-flop to construct, in another file, a 5-bit shift register with serial load. This serially loaded shift register should have one D input, as shown above. The contents of the shift register can be loaded by shifting data for 5 clocks; or, it can be cleared asynchronously.

Digital VLSI Design with Verilog

To allow the shift register to hold its data while the clock still is running, use a multiplexer ("*mux*") to supply each flip-flip D input with one of these two possible inputs: (*a*) With shift enabled, the mux should connect the previous Q to the next D; or, (*b*) with shift disabled, the mux should connect each flip-flop's D input with its own Q output.

The schematic representation of a 2-input mux is shown in figure 3-5, and their relationship to the rest of the design is in figure 3-4 above.

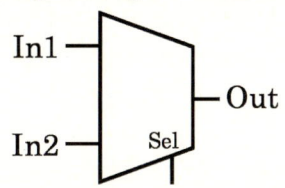

Fig. 3-5. Schematic mux symbol.

Your mux model may be written this way:

```
always@(sel, in1, in2)
   if (sel==1'b0)
      outbit = in1;
   else outbit = in2;
```

A mux is combinational, not sequential, logic, so blocking assignments are fine. Notice that if either of in1 or in2 had been omitted from the always sensitivity list, a latched state would exist, and we would not have a mux but rather a latch of some kind.

Another way to write a mux (*not in this Step*, please):

```
assign outbitWire = (sel==1'b0)? in1 : in2;
```

The D flip-flop model should be in a separate file from the shift-register, and so should be the mux model.

Simulate your design briefly to check it for functionality.

Optional: Synthesize your design twice, optimized for area first and then for speed.

Step 5 (optional): Parallel-load shift register. Modify the design from Step 4 so that all 5 bits of the shift register can be loaded with new data on one clock.

The easiest way to do this is to add a third selection to your Step 4 muxes: The third choice should connect each flip-flop D input to its respective bit on a new, 5-bit, parallel-load input data bus. See figure 3-6.

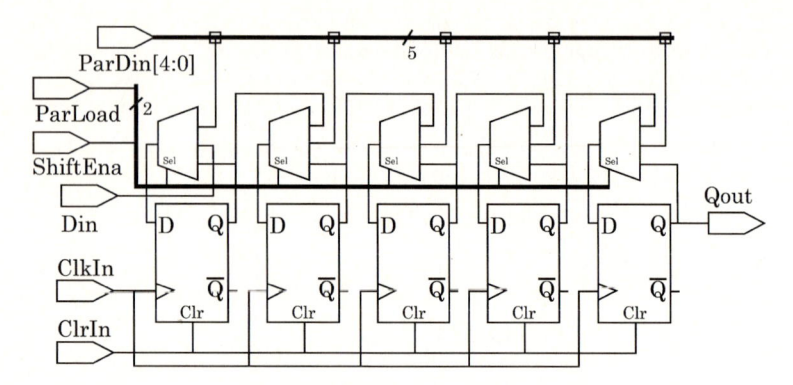

Fig. 3-6. Shift register with parallel load.

In this parallel-load design, when shift is not enabled, each flip-flop Q is connected to its own D; when shift is enabled, each flip-flop Q is connected to the D of the next flip-flop. When parallel-load is selected, each flip-flop Q is unconnected (except the last one), and each flip-flop D is connected to a bit on the parallel data input bus.

Be sure that your new muxes always select exactly one of the three intended connections no matter how they are operated.

Step 6: Rewrite the serial-load shift operation of Step 4 as a ***procedural model*** in a single `always` block like this:

```
always@(posedge ShiftClock)
  begin
  QReg[0] <= Din;
  QReg[1] <= QReg[0];
  QReg[2] <= QReg[1];
  QReg[3] <= QReg[2];
  QReg[4] <= QReg[3];
  end
```

This kind of design will not require a DFF model at all. To add the Step 4 muxes, you may use conditional expressions on the right instead of unconditional `Qreg` expressions:

```
always@(posedge ShiftClock)
  begin
  QReg[0] <= (ShiftEna==1'b1)?   Din   : QReg[0];
  QReg[1] <= (ShiftEna==1'b1)? QReg[0] : QReg[1];
  QReg[2] <= (ShiftEna==1'b1)? QReg[1] : QReg[2];
  QReg[3] <= (ShiftEna==1'b1)? QReg[2] : QReg[3];
  QReg[4] <= (ShiftEna==1'b1)? QReg[3] : QReg[4];
  end
```

This shift register model can be implemented using just one always block and no design hierarchy. Incidentally, putting the same delay, say, "#1", in front of each assignment statement in the code above should delay each shift by that delay but not otherwise affect the logic. However, some simulators will not simulate the delayed statements correctly—another good reason not to use delays in procedural blocks in your design work.

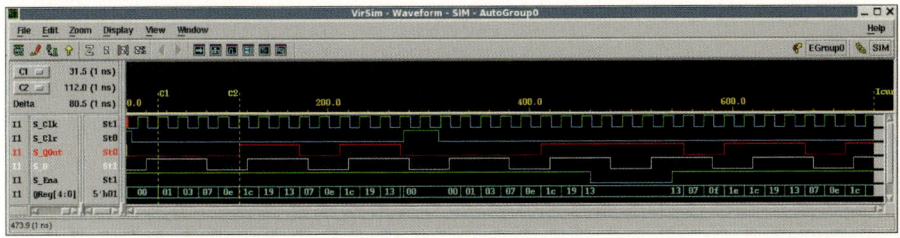

Fig. 3-7. Simulation of the behavioral shift register model above, with a 0.5 ns lumped output delay not shown in the verilog.

Simulate to verify (see figure 3-7); then, try to synthesize the procedural model for area and speed. As it happens, the current version of the synthesizer may not synthesize nonblocking assignments correctly if they are delayed, but you should try anyway, to see what happens.

Then, just remove all the delays above and resynthesize.

What happens in simulation if the nonblocking assignments are replaced by blocking assignments in the code fragment immediately above? In the code fragment above, how much total time does it take for one shift?

You can synthesize a procedural shift register by removing the delays (above) or by changing the assignments to blocking ones. However, if you use blocking assignments, the statements then have to be ordered in reverse of that shown. With nonblocking assignments and equal (or no) delays as above, order is irrelevant, because the old value on the right will not be updated before it has been assigned.

To study the special features of nonblocking assignments, simulate the `Scheduler.v` model which you will find in your `Lab04` directory.

3.4.1 Lab Postmortem

Latches are a problem for the synthesizer. This is not unintentional; guess why?

Do you think the synthesizer can create better logic with a structural model than with a procedural one? Why?

How might the flip-flop models be improved? What about 'x' or 'z' states?

In the serial-load shift-register design, how might an assertion be added to check to make sure that shift was not disabled before at least five valid data had been loaded?

3.5 Additional Study

As previously assigned, read Thomas and Moorby (2002) chapter 1 on basics and chapter 3 (on synthesizable verilog) through section 3.4.3. Do Thomas and Moorby 1.6 Ex. 1.2.

Read Thomas and Moorby chapter 2 on logic synthesis. We shall study finite-state machine design in verilog later, so read through the FSM sections lightly.

For inferred latches, study Thomas and Moorby Example 2.7 in section 2.3.2 and do 2.9 Ex. 2.2.

Optional: Read Thomas and Moorby section 5.2 on parameters, but ignore p. 149 on `defparam`.

Optional readings in Palnitkar (2003):

Section 9.2.2 on parameters.

Sections 3.2.9, 3.3.1, and 9.5.3 for details on strings and messages.

Sections 14.4–14.6 on logic optimization,

If you are puzzled by procedural controls, read through chapter 7 to see if explanations there might help. You may find the Palnitkar CD answers to exercises in section 7.11 beneficial; however, use of `forever` or `while` in the coursework, like use of `always` without an event control, generally is discouraged.

Read section 15.2 for an overview of assertion-based verification, which is very important in modern, large, complex designs.

Chapter 4
Week 2 Class 1

4 Today's Agenda:

Verilog Lecture
Topics: Variable `reg` and net types; constants; basic simulation and relation to synthesis; basic system tasks and PLI. Internal scan.
Summary: We study a variety of features of the verilog language: We begin by explaining `reg` and net variables again, and we distinguish several of the types of net available in verilog. We mention BNF format for clarifying syntax. We then return to sequential logic synthesis, using latch and mux similarities to illustrate how combinational logic should be modelled. We introduce the basics of clock simulation and point out some simple ways of avoiding race conditions. We wrap up with a brief presentation of internal and boundary scan, which increase observability of design state in the hardware.

Simple Scan Lab 5
Using the combinational `Intro_Top` design of the first lab of the course, we add a JTAG port, register all inputs and outputs using flip-flops, install muxes, and link the flip-flop muxes to form a scan chain. Finally, we add an assertion to be sure the scan mode operates correctly.

Lab Postmortem
We discuss glitches and scan-chain operational refinements.

4.1 Net Types, Simulation, and Scan

4.1.1 Variables and Constants

As previously mentioned, **variables** are of two different kinds, `reg` and net. However, the situation is more complicated.

A `reg`, an unsigned type, is about the same as either of the corresponding signed types, `integer` and `real`. Any of these `reg`-like types can be assigned a value procedurally and retains the value assigned until the value is changed by a subsequent assignment. Both `integer`s and `real`s are 32 bits wide; a `reg` may be any width, from one bit up to the limit acceptable to the tool in use. For flexibility, in *verilog-2001*, a `reg` may be declared *signed*, in effect allowing declaration of integers of any width. A `real` is not synthesizable in the digital design tools we use, and we shall ignore it in most of the rest of this course.

The word *net* does not refer to a specific type but rather a generic characteristic of connectivity. Unlike a `reg`, a net type has no storage capability; a net just connects or communicates.

The type of a net may be `wire`, `tri`, `wand`, or `wor`—or any of several other types to be studied later. A `wire` and a `tri` are functionally identical, and the different names are just mnemonics. Multiple drivers on a `wire` may be a design error; on a `tri`, they probably are three-state drivers. A `wand` almost always is multiply driven, and it drives elements to which it is attached with the ***wired and*** of its drivers; a `wor` drives with the ***wired or***. Thus, a `wand` or `wor` effects a logic operation in addition to a connection.

Multiple drivers on `wand` or `wor` are effected by using two or more concurrent assignments. This can be done (*a*) by wiring the net to two or more instance output pins, or (*b*) by two or more continuous assignment statements to the net.

For example,

```
wire ToSubDir, In01, In02;
wor  SelTwo;
 . . .
assign SelTwo   = In01;
assign SelTwo   = In02;  // Thus, SelTwo evaluates (In01 or In02).
assign ToSubDir = SelTwo;
 . . .
// Much more common, and more efficient, would be this:
assign ToSubDir = In01 | In02;  // = In01 || In02; also is OK.
```

It should be mentioned that neither `wand` nor `wor` is used often in modern cmos design; regardless, we'll be looking into details of all these expressions later in the course.

Constants may be literal numerical values, parameter values, or string values.

Because `integers` already are exactly 32 bits wide, they may be assigned constant values without specifying width. However, it usually is wise to provide a width for every literal constant; this means that operations on the literal will have unambiguous width, making the results well defined. Examples are the four fundamental 1-bit states,

1'b1, 1'b0, 1'bx, and 1'bz;

or, typical hexadecimal or binary expressions such as,

7'h15 (hex) and 16'b0001_0101_1100_1111 (binary).

A `parameter` value is by default unsigned. When initialized by a decimal integer literal, a `parameter` becomes 32 bits wide by default; however, a `parameter` may be declared of any width. Examples are

```
parameter    x = 5, y = 11;      // Both are unsigned & 32 bits wide.
parameter[4:0] z = 5'b11000;     // Unsigned and 5 bits wide.
parameter    z = 5'b11000;       // Unsigned and probably 5 bits wide.
```

There is no string type in verilog, but string literals enclosed in quotes may be used in simulation screen messages or assertions. A string of bytes also may be assigned to a `reg` of adequate width, but this usage is not common in digital design except when programming messages into a ROM or RAM.

Digital VLSI Design with Verilog

A typical simulation example is,

```
$display("Error at time=[%0d].", $time);
```

If the preceding `$display` is executed at simulator time 531, the displayed message will be,

```
"Error at time = [531]."
```

There are no "global" variables or constants in verilog; all must be declared somewhere within a `module`.

4.2 Identifiers

Identifiers in verilog, which is to say, (*a*) declared names of variables or constants, or (*b*) names of modules, blocks or other objects, all are:

- made of ASCII alphanumeric characters, '_', and '\'
- case-sensitive
- of any length allowed by the tool in use
- begin with a letter (alphabetical), underscore ('_'), or backslash ('\').

So, "Z", , "Z12", "z12", "_z23", and "_z23 " are legal identifiers; "2z12" is not.

Identifiers beginning with '\' are called *escaped identifiers* and, unlike identifiers in general, may contain any ASCII character except a blank. These identifiers are terminated by a blank (' '), which for them is a delimiter and not part of the name. Escaped identifiers are used by tools to avoid name conflicts during translation or for portability and usually are not written manually in verilog design source code.

4.3 Concurrent vs. Procedural Blocks

Concurrent blocks. These are blocks of code which simulate in no well-defined order relative to one another. Verilog source files, for example, may be compiled in a particular order, but the compilation order does not define the order of simulation. The most important concurrent block is the `module` instance. Likewise, continuous assignment statements, `initial` blocks, and `always` blocks are concurrent within a `module`. Other kinds of concurrent block, including `primitives` and the `specify` block, will be introduced later in the course.

Procedural blocks. These are blocks of code within a concurrent block which are read in order (procedural sequence) during simulation and, if executed, usually are executed in the order read. Procedural blocks may contain:

- blocking assignment statements
- nonblocking assignment statements
- procedural control statements (`if`; `for`; `case`; `forever`)
- `function` or `task` calls (later)

- event controls ('@')
- `fork-join` "parallel" statements (later)
- nested procedural blocks enclosed in `begin ... end`.

4.4 Miscellaneous Other Verilog Features

Macroes are like C preprocessor directives but begin with '`' instead of '#'. They are not strictly language elements, because they do not relate to other specific constructs and may appear anywhere on a new line in the code. Some macro examples are `` `define``, `` `timescale``, `` `ifdef``, `` `include``.

System tasks and system functions will be covered lightly in this course. They are simulation constructs and may appear in procedural blocks, only. For example, $strobe, $display, $sdf_annotate, $stop, $finish.

Timing checks will be covered in detail later. They are predefined assertions which execute concurrently. They may appear only in `specify` blocks. Examples are $width, $setup, and $hold.

The **PLI** (Programming-Language Interface) is a library of C routines which allow a user to define new system tasks or functions which extend the functionality of a verilog simulator. We shall discuss this toward the end of the course but shall not use it at all.

4.5 Backus-Naur Format (BNF)

This is a way of defining the syntax of a language and is used widely in verilog and other standards contexts. It gives an alternative view to text specifications of language syntax and helps to resolve ambiguity. We shall not use it in this course, but it is nice to know. The Thomas and Moorby (2002) appendix G treats it extensively.

The BNF rationale is extremely simple and essentially hierarchical: The allowed substructure of any primary element of syntax is provided in a list following "::="; subelements are broken down following ":=". For example, suppose for simplicity's sake the BNF for a logical operator was,

```
logic_op ::= boolean_op | bitwise_op
             boolean_op := && | || | !
```

Then, from this we may deduce that a logical operator always will be a bitwise operator or a boolean `&&`, `||`, or `!`. An attempt to use subtraction (-) as a logical operation then would be an error easily recognized from the BNF given in this example.

4.6 Verilog Semantics

Verilog is an HDL, and its meaning is hardware. So, in a VLSI context, verilog means logic gates and wires or traces. The logic gates correspond to `reg`s, module instances, `integer`s, and so forth; the wires correspond to *net*s of various types.

Like any programming language, verilog works by expressions and statements. An *expression* is just something that can be evaluated (represents a value). An example of an expression is a sum or logical product, for example "5+7" or "A&&B". An expression just evaluates to something; it doesn't change anything.

However, a *statement* assigns a value to something and thus changes it. The changed value either is stored locally, or it is routed somewhere else. For example, in a procedural block, the statement "Zab = A && B;" assigns the logical *and* of A and B (an expression) to a reg named Zab. The two equivalent concurrent statements, (*a*) continuous assignment "assign Z = A && B;" or (*b*) component instantiation "and and01(Z, A, B);" change the value of a net instance named Z.

Statements have to be controlled somehow, and the usual way is by simulation of a change in a variable. For example, "assign Z = A&B;" will be reexecuted every time the simulator simulates a change in A or B. An always block may contain numerous expressions and statements, and the designer usually must provide such a block with an event control ("sensitivity list"); the always block statements all will be reread only when a variable in the sensitivity list changes.

For example, a block controlled by "always@(A)" will be reread only when A changes, even if B changes and one of the block's statements is "Zreg = A&B;". However, a block controlled by "always@(*)" will be reread whenever any variable in an expression in the block is changed. This last is different from "always" with no event control; the plain always will be reexecuted when anything in it changes, variable or expression.

Latches and Muxes. Edge sensitivity in an always block is introduced by the always keywords posedge or negedge. Assuming only level sensitivity, the meaning of an incomplete sensitivity list in an always block is a <u>latch</u> of some kind; a complete sensitivity list, by contrast, usually means combinational logic such as a collection of combinational logic gates or a mux. However, if a control construct such as an if or case causes a value to be ignored, a latched state can be created in any kind of always block, even one with a complete sensitivity list.

For example, suppose that, in a module, we have declarations of "wire a, b, sel;" and "reg z;". Consider the inputs and outputs of the following four different always blocks:

always@(a, b, sel)
  if (sel==1'b1)
    z = a;
  else z = b;
	always@(*)
	  if (sel==1'b1)
	    z = a;
	  else z = b;
		always@(sel)
		  if (sel==1'b1)
		    z = a;
		  else z = b;
			always@(a, b)
			  if (sel==1'b1)
			    z = a;
			  else z = b;

The leftmost two always blocks then represent muxes, because all the input-related variables in those expressions are in the sensitivity list, and every alternative in the if is assigned to the output (z). The other two always blocks above represent nonstandard latches of some kind which are disabled when the variable(s) in the sensitivity list does not change. The rightmost one (always@(a, b)) is sensitive to every variable on the right-hand side of a statement; thus, it represents a mux which is latched by `sel` in an abstract sense: The selection is latched to the previous one; the output value itself is not latched.

Here is a nonstandard latch which contains no procedural control construct:

```
always@(a, v)     // Typo! always@(a, b) intended.
  z = a | b;
```

In this block, the value of z is latched against changes in b. This kind of latch often is created by a typographical error in the sensitivity list.

A simple transparent latch can be modelled correctly in an `always` block by omission of the "`sel==1'b0`" alternative in what otherwise would be a mux.

Here in figure 4-1 is an example of this:

```
// verilog simple transparent latch:
always@(D, Sel)
  if (Sel==1'b1)
    Q = D;
```

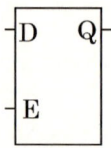

Fig. 4-1. Simple transparent latch by procedural omission. `Sel` renamed to E(nable).

The always block above is the preferred style for a synthesizable latch. The synthesizer will find a single component in the library for it, if the library has one.

Another, very different, way to write a synthesizable latch would be by continuous assignment:

```
assign Z = (Sel==1'b1)? D : Z;
```

Although the preceding `always`-block model generally will synthesize to a latch library component, a continuous assignment latch typically would synthesize to explicit combinational logic with feedback.

The continuous-assignment latch might be synthesized as in figure 4-2:

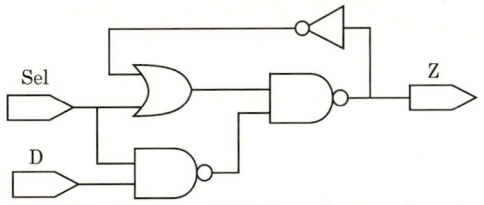

Fig. 4-2. Simple transparent latch by continuous assignment.

A continuous assignment statement is treated as combinational logic by the synthesizer, so it does not attempt to find a library sequential element for the netlist in figure 4-2. For this reason, it usually is not a good idea to use a combinational representation for a latch. This advice generally should be heeded, although we shall use combinational S-R latches in our course work for instructional reasons.

When a latch is described functionally in combinational code, the precise gate structure is up to the synthesizer, and this structure may be changed unpredictably during synthesis optimization. Thus, the timing (including possible glitching) of a combinationally-implemented latch is more uncertain than when it has been described more directly in a simple sequential construct such as by the `always` block above, as represented by figure 4-1. It is best to avoid the whole problem, wherever possible, by writing clocked constructs implying flip-flops instead of enabled constructs implying transparent latches.

4.7 Modelling Sequential Logic

We now turn to some aspects of model building in verilog which are intended to produce accurate synthesis results. For reasons to be expanded later, we avoid latches per se entirely here and discuss only clocked latching elements—which is to say, flip-flops.

Clocks. A clocked block in verilog is an `always` block with an event control containing an edge expression, `posedge` or `negedge`. A clocked block never includes more than one clock, and it should not include any level sensitivity. However, multiple edge expressions may appear if only one of them is applied to a clock.

For example,

```
always@(posedge clk) ...
always@(negedge clk) ...
always@(posedge clk, negedge clear) ...
```

The following is not recommended and will not be synthesized:

```
always@(posedge clk, clear, preset) ...
```

Asynchronous controls. These usually are a single preset or clear but may be both a preset and a clear. If verilog is written to represent both preset and clear, the `always`-block code can not avoid implying a priority for one of the controls.

For example,

```
always@(posedge Clk, negedge Preset_n, negedge Clear_n)
   if (Preset_n==1'b0)
              Q <= 1'b1;
   else if (Clear_n==1'b0)
              Q <= 1'b0;
        else  Q <= D;
```

In this model, priority is given to `Preset_n` over `Clear_n` if both are asserted at the same time. Thus, a correct netlist should include logic creating this priority, even though assertion of both probably would represent an operational error. However, the designer's intent usually will be to synthesize a single gate with (*a*) two pins for the controls, and (*b*) either no priority (random priority) or some sort of internal gate structure effecting a priority.

The possible confusion is illustrated here:

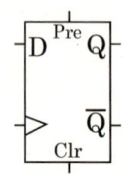

Fig. 4-3. A flip-flop with two asynchronous controls, `Clr` (sets Q to 0 and Qn to 1) and `Pre` (sets Q to 1 and Qn to 0). If `Clr` and `Pre` both are asserted, the result depends on the library; but, if it works, both Q and Qn might go to 0.

Avoid more than one asynchronous control if possible: The block may not be synthesizable if the library does not include a component with both a preset and a clear pin; and, if it does synthesize, the simulation may not match the synthesized netlist if both controls ever should be asserted at once. It may be possible to pass a synthesizer a constraint or other directive which would control its library access so that cells would be chosen with specific preset-clear priority for particular instances. In the absence of such a constraint, the synthesizer might insert logic to implement exactly the verilog simulation priority; or, it simply might ignore the verilog asynchronous control priority entirely.

Anyway, when one or more asynchronous controls is present in a clocked block, the clock always will have the lowest priority.

Race conditions. A race condition results from code which allows an ambiguous value (maybe '1'? or maybe '0'?) to be scheduled during simulation. This generally means that the corresponding hardware will not be functional.

For example, within one `always` block, delayed nonblocking assignments can allow concurrency and therefore a race:

```
always@(posedge Clk)
  begin
  #2 X <= 1'b1;
  #2 X <= 1'b0;
  #3 Y = X;   // Ambiguous value of X used.
  end
```

The same concurrency can occur between different always blocks:

```
always@(posedge Clk) #1 X = a;
always@(posedge Clk) #1 X = b;
```

To avoid the majority of race conditions, **never** mix nonblocking assignments with blocking assignments; and, **never** assign to the same variable from more than one always block. Incidentally, the synthesizer will refuse to synthesize these styles, although the simulator will permit them.

In addition, it is good design to latch all outputs on major design components (large blocks or `modules`). Do the latching with flip-flops: Flip-flops on all outputs guarantee that when the clock edge occurs, current values only, and not possibly conflicting ones, will be supplied to inputs of other components.

Finally, never schedule a delay of #0, except maybe in a testbench initial block. We shall study reasons for this later in the course. For now, just don't do this:

```
always@(posedge Clk)
  begin
  #0 Q1 <= a;   // Likely error!  Never use #0!
  #0 Q2 <= b;   // Likely error!
  ...
  end
```

Synthesizable language subset. Not all the language is synthesizable. To be sure that what you simulate also will synthesize, here some rules of thumb:

- Don't mix edge expressions and level-change expressions in a single sensitivity list.
- If you write a delay expression in an `always` block, use it with blocking assignments, only. The synthesizer will object to any delayed nonblocking assignment, either by reporting an error or by issuing a severe warning. We shall see why later.
- When you code delays for simulation, try to do it this way: Avoid delays in `always` blocks; instead, estimate the total delay(s) on the output of each such block, and move that estimated total to a continuous assignment statement.

For example, code the way shown here:

```
module MyModule (output X, Y, rest of sensitivity list);
... local declarations ...
assign #5 X = Xreg;   // estimated total delay = 5.
assign #7 Y = Yreg;   // estimate = 7.
...
always@(posedge Clock) // A very strange flip-flop!
  begin
  x1   = (a && b) ^ c;  // was delayed #3
  Xreg = x1 | x2;       // was delayed #2
  Yreg = &(y1 + y2);    // was delayed #2
  end
endmodule
```

- Don't assign to the same variable in more than one `always` block.
- Don't mix blocking and nonblocking assignments in a single `always` block.
- Generally, avoid coding for level-sensitive sequential constructs (latches); use flip-flops instead.

4.8 Design for Test (DFT): Scan Lab Introduction

Modern tools automatically will insert scan into a design; this usually is done late in the design cycle, after the unscanned design has been well simulated. Understanding the mechanics of scan insertion will help you to recognize and correct insertion errors or inefficiencies.

Scan is called "scan" because scan registers allow logic states at hardware test points to be ***scanned*** (shifted serially) in and out of a design.

The purpose of scan is to enable the observation of changes inside a design or a whole chip. Scan registers are hardware test points; they allow a designer or test engineer to see what is going on in the hardware. Scan registers are not just simulation devices; they stay in the design after it is taped out and implemented in silicon.

There are two major kinds of scan, ***internal*** scan and ***boundary*** scan.

In internal scan, the inputs and outputs of all, or some selected, blocks of combinational logic in the design are changed into scan elements. This is done by replacing every flip-flop or latch with a scan flip-flop or scan latch. Obviously, every I/O of any combinational block must be either a chip pin or a sequential element such as a flip-flop or latch.

A special test port is used for controlling scan operation; this port is called a *JTAG* port ("Joint Test Activity Group") after the standards group that defined it.

The new scan components functionally are the same as the ones they replace, except that they are muxed together into a serial scan chain, which is just a giant shift register distributed over all or part of the design. When the scan muxes are in the *operate* or *normal* mode, the design components operate as they were designed to do, the scan

Digital VLSI Design with Verilog

components substituting for the ones which they replaced. When the muxes are in the *scan* mode, the design doesn't operate any more, but all logic values stored in scanned registers can be shifted out of the design and inspected for correctness. Thus, by scanning in new inputs, operating, and then scanning out the result, every combinational block in the design can be tested for correct operation, revealing design errors or (random) hardware failures, if any.

Figures 4-4 through 4-6 illustrate how internal scan insertion is done in a design with preexisting sequential elements, in this case flip-flops.

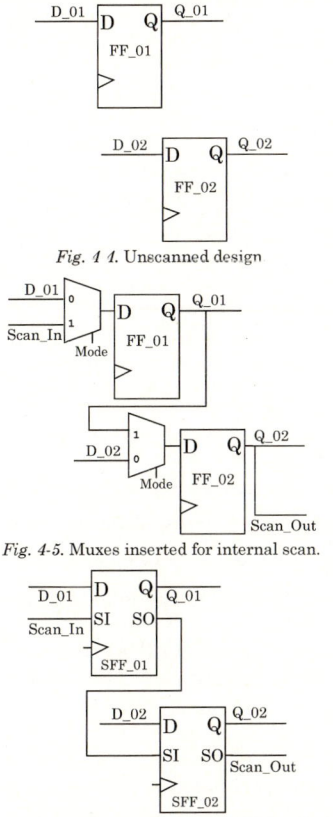

Fig. 4-4. Unscanned design

Fig. 4-5. Muxes inserted for internal scan.

Fig. 4-6. Library internal scan components. Mode select not shown.

In our lab exercise, we shall do this kind of scan insertion, but with some modification to allow for the fact that the original design we have chosen does not include any sequential element to replace.

Boundary scan mostly is used for board-level hardware testing; it monitors the boundary of a chip. In boundary scan, the pins on a chip are connected to additional scan components which in turn connect to the gates inside the chip. This allows chip test patterns to be applied by shifting in, and the results to be observed by shifting out, after the chip has been mounted on its intended circuit board. Boundary scan thus eliminates the need for a "bed of nails", or other external hardware probe, on every test point on the board. A modern ball-grid chip package may have well over 1,000 pins spaced apart by a fraction of a millimeter; this makes a bed-of-nails approach impractical and possibly harmful to the tiny, delicate contact points.

Boundary scan usually is combined with some sort of built-in self-test protocol. In addition to a *JTAG* port, boundary scan typically requires a Test Activity Port controller (TAP controller), which is a built-in state machine—a complexity we shall ignore for now.

In the next lab, we'll use what we have learned about shift registers to insert scan flip-flops into the `Intro_Top` design of the first lab of the course. It should be emphasized that we shall be using ***two*** muxes per flip-flop, instead of the one required for normal internal scan. This is because the original design does not have even one flip-flop; and, therefore, to get the original design to operate completely normally, we need the extra muxes to bypass the flip-flops entirely. So, in scan mode, flip-flops will exist to allow the scan programming to work, and this will require one mux per flip-flop, as in figure 4-5 above; in normal mode, there will be no flip-flop allowed, and this will require the second set of muxes to send input and output data around the flip-flops. Yes, it is true: We are using scan insertion as an excuse for a teaching exercise in verilog.

Our JTAG test port will consist of just five special one-bit ports: A scan-***in*** port, a scan-***out*** port, a scan ***clock*** port, a scan ***clear*** port, and a scan ***mode*** port.

We want to insert scan into our old Lab 1 design just to see how it is done. Issues of clocking and settling of logic will be addressed again later in the course.

Before we start this lab, here is a more detailed outline of what we shall do:

First, we'll add a *JTAG* port and install flip-flops around the combinational logic in `Intro_Top`. Because `Intro_Top` was purely combinational, this means we'll install a flip-flop on every I/O. After installing the flip-flops, we'll restore functionality to the `Intro_Top` design by clocking the normally combinational data through the flip-flops. The only clock available is the *JTAG* scan clock, so we'll use that one. The rest of the *JTAG* port will remain unused at this point. The `Intro_Top` design then will become a synchronous design, clocked by the scan clock and with its original functionality intact, except for slower operation because of having to wait for clock edges on the inputs and outputs.

Second, we'll install muxes, two for each flip-flop. Each mux at this point will be held in one *select* state (the *operate* or *normal* mode). For example, two `Intro_Top` outputs with muxes added to the new flip-flops would look as shown in figure 4-7.

Digital VLSI Design with Verilog

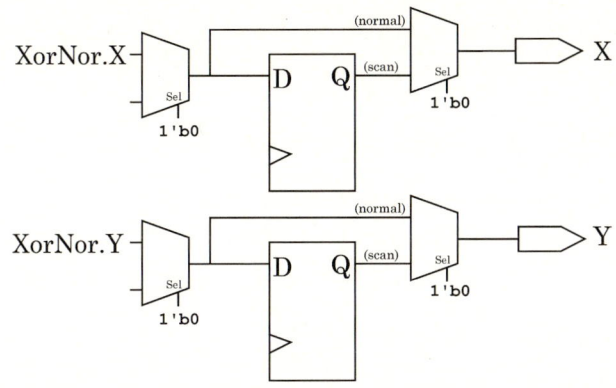

Fig. 4-7. Two outputs in `Intro_Top` after connection of muxes to the new flip-flops. Clock and clear logic omitted for clarity; '0' on *select* is assumed to select *normal* mode.

If we put the muxes in the *scan* mode, the design would not do anything, because the scan inputs to the input muxes are unconnected at this point. So, to allow immediate verification of our verilog by simulation, we shall install the muxes so that they bypass the flip-flops entirely. This means that the `Intro_Top` design again becomes purely combinational, with no synchronizing delays, but with some small propagation delay added by the mux logic.

Notice that the design fragment in figure 4-7, if held in its *normal* mode, completely ignores the states of the flip-flops and does not respond to the scan clock at all.

Third, we'll connect the dangling mux *scan* inputs to each other in a serial chain; as will be seen in the lab, this will chain the flip-flops together and allow then to act as a shift register when the design is clocked in *scan* mode. All mux outputs will have been fully connected already; so, they'll need not be modified in this final step.

Now we begin the lab.

4.9 Simple Scan Lab 5

Topic and Location: An internal-scan chain formed from muxed flip-flops.

Log in and change to the `Lab05` directory.

Preview: We modify the original `Lab01` demo design, `Intro_Top`, so that it includes some design-for-test (DFT) logic as follows: We add a JTAG port and then insert sequential components (flip-flops) on every I/O. Thus, we convert our purely combinational `Intro_Top` design into a synchronous one. After that, because our design now includes sequential elements, we can replace these elements with scannable ones. We do this first by adding muxes to "short-circuit" the data flow away from the flip-flops, under control of the mux select state. We then chain the flip-flops together, two muxes for each flip-flop, so that one select state connects all flip-flops into a shift register, and the other select state just leaves the flip-flops dangling on the I/O's in the design. We add an assertion to report misuse of the chain when the design is in the scan state; and, finally, we modify our original `Intro_Top` design to insert a scan chain.

Deliverables: Step 1: A *JTAG* test port added to `Intro_Top`. Step 2: A single verilog `module`, correctly simulating, with seven independent D flip-flops connected separately to the `module` I/O ports. Step 3: Refined timing, verified by simulation and optionally by a quick synthesis. Step 4: Installed muxes, followed by simulation, and optionally, a quick synthesis. Step 5: Flip-flops chained and tied to the `ScanOut` port. Step 6: Rationally renamed components, followed optionally by synthesis and examination of the created scan chain. Step 7: Added safety assertion. Step 8: Inspection of the synthesized netlist to identify and count the resulting muxes. Optional ungrouping and flattening, followed by an area-synthesized netlist. Step 9 (optional): Synthesized original `Intro_Top`, with output flip-flops and *JTAG* ports, and verification of an inserted scan chain.

Lab Procedure:

You'll find in your `Lab05` directory a copy of the original `Lab01` design, `Intro_Top`, with minor changes. Notice the maintenance log that has been kept in each module file. This kind of commenting is recommended, not only to pass on information to other designers, but to remind the busy original designer, after a lapse of time, of what was done. The details to be logged will vary with your project coding style. There also is a synthesis script file there for use in the final, optional, scan-chained Step of the lab.

Step 1: Add a *JTAG* test port in `Intro_Top`. Do this just by adding five new one-bit ports connected to nothing: Call them `ScanMode`, `ScanIn`, `ScanOut`, `ScanClr`, and `ScanClk`. All should be inputs except `ScanOut`.

Your module header now should be something like this:

```
module Intro_Top ( output X, Y, Z, input A, B, C, D
                , output ScanOut
                , input  ScanMode, ScanIn, ScanClr, ScanClk
                );
```

Step 2: Insert D flip-flops (FFs) into the path of every I/O at the top of the design (in `Intro_Top.v`, not in `TestBench.v`), except the *JTAG* I/O's.

You may wish to use your FF design from `Lab04`; it should have a `posedge` clock and positive-asserted asynchronous clear.

To insert the flip-flops, work with the module ports and wires in `Intro_Top`; there is no need to change anything in the submodules which made up the structure of the original Lab 1 design.

Connect every preexisting design input to the `D` of a new FF; then, reconnect the `Q` output of that FF to whatever the preexisting input pin was driving. You should declare a different wire for each FF `Q` connection.

Digital VLSI Design with Verilog 81

You may find it confusing and easy to make mistakes if you do not name your wires and component instances mnemonically. For example, suppose your FF module was named DFFC ("D flip-flop with clear"). For the top-level A input port, which connected in Lab 1 to pin A of a block with instance name InputCombo, you might do this:

```
wire toInputCombo_A;
...
DFFC Ain_FF(.Q(toInputCombo_A), .D(A), ...);
```

After the inputs, do the same for the outputs: In Intro_Top, connect the Q output of a new FF to every preexisting design output; then, reconnect whatever was driving that preexisting design output to the D input of the new FF. Again, declare a new wire for each reconnection.

After wiring the data for all the FFs, connect the ScanClk input to the clock input pin of each FF; connect the ScanClr input to the asynchronous clear input pin of each FF. All your FFs then should look something like this:

```
DFFC myFFinstanceName (..., .Clk(ScanClk), .Clr(ScanClr));
```

Add drivers for the new ScanClk and ScanClr inputs in the testbench in Intro_Test.v; be sure to connect them to the top-level design instance there.

When this Step is complete, the original Intro_Top design, from Lab 1, should have been changed so that it's schematic would look like this:

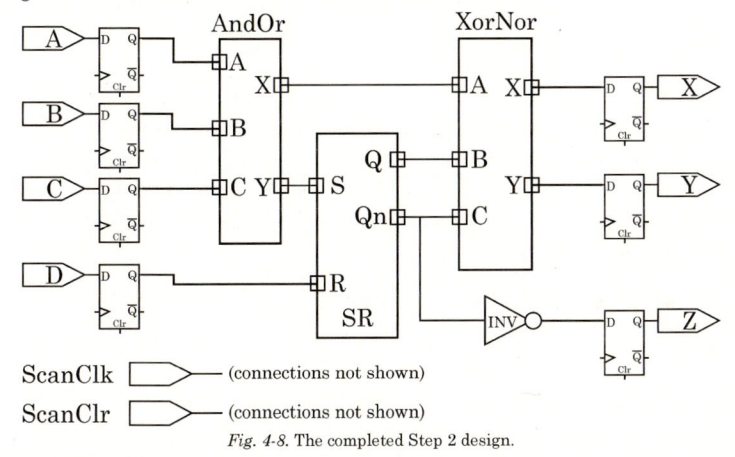

Fig. 4-8. The completed Step 2 design.

A good testbench convention throughout this course is to declare a reg for every design input to be driven; name this reg *stim, where '*' is the design input name. For example, Intro_Top input A would be driven by a testbench reg named Astim. Similarly, a good convention for design outputs is to declare a testbench wire for each one, named *watch. For example, the design output X would be connected to a testbench `wire` named `Xwatch`. You <u>stim</u>ulate the inputs and <u>watch</u> the outputs. Being consistent with testbench names makes it easy to recall which testbench variable was connected with each design variable without having to examine the testbench code itself.

After this, clocking of the FFs after every testbench stimulus change should be accomplished by applying the testbench `*stim` simulation inputs to the original design; the results of each such stimulus clock should appear as the current (not new) `*watch` outputs. After waiting enough simulation time for the combinational logic to settle, clocking the FFs a second time *without* changing inputs should clock out the new, correct `*watch` outputs determined by the current inputs and the design's combinational logic.

After completing the above, compile your changed design for simulation, from testbench on down, to check your FF and clock wiring.

Step 3: Refine the timing, if necessary. Looking at the subblocks of the `Lab05` design, there are several delays of 10 units; so, be sure to have defined the clock in your testbench with a generous period of, say, 50 units. This length of period should allow plenty of settling time.

For example,

```
...
always@(ClockStim) #25 ClockStim <= !ClockStim;
//
initial
  begin
  #0 ClockStim = 1'b0;
  ...
  end
```

The `ClockStim` in the testbench, of course, will be connected to the `ScanClk` port of the modified design. And you will want a new `ClearStim` in the testbench to drive the design `ScanClr`. Just set `ClearStim` to 0; there is no need to reset the flip-flops for this lab. Check your clocking scheme by simulating the design briefly. See figure 4-9.

Digital VLSI Design with Verilog

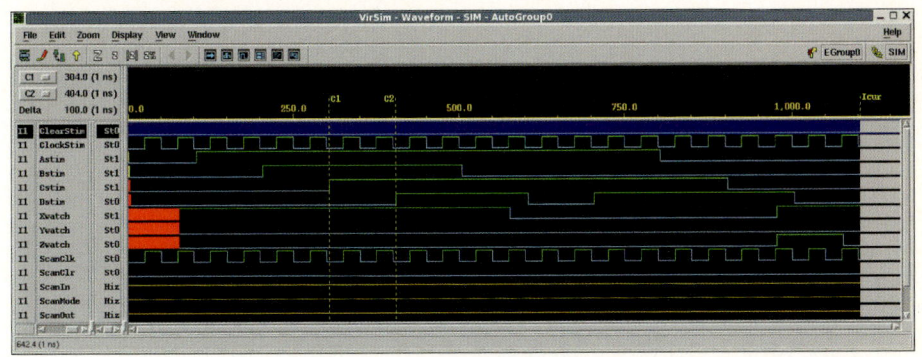

Fig. 4-9. The Step 3 synchronous implementation of `Intro_Top`.

You now have a new, synchronous design consisting of the Lab 1 combinational logic surrounded by sequential logic. It is synchronous because its operation is synchronized to `ScanClk`. Inputs are clocked in on every positive clock edge; the result is clocked out on the next positive edge.

Optionally, synthesize this design, just to see what the netlist schematic looks like.

Step 4: Install multiplexers ("muxes") to remove all synchronous behavior again. Just the muxes; don't chain anything yet. Use an always block or a continuous assignment in `Intro_Top` for a mux; do not use a (structural) component.

Each FF will require a mux driving its D input, to connect its D either to the normal D input or to its scan-chain driver. A second mux will be used to connect or disconnect the FF's Q to its normal output. The second mux is required only for this particular design; otherwise, without the second mux, the FF's Q output would contend with whatever normally is driving what is now the scan-mode's input to D. These flip-flops are present only for scan, not for our normal `Intro_Top` operation.

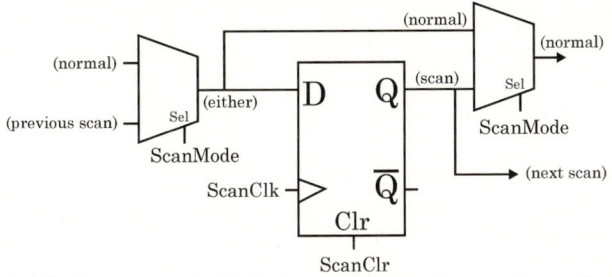

Fig. 4-10. Multiplexers around a flip-flop in `Intro_Top` to make a scan element. Usually, scan requires just one mux per flip-flop, but the one on the right is required in our design.

At this point, the leftmost "(previous scan)" input as shown in figure 4-10 should be driven with a temporarily attached, wired constant value such as 1'bx; the "(next scan)" output wire shown should <u>not</u> be added yet. It may be helpful to declare a wire with a noticeable name, such as THIS_IS_X, assign a 1'bx to it, and use it in turn to assign the 1'bx to the several mux inputs which will be set unknown at this stage in the lab.

As a result of all this, the FF shown in figure 4-10 is bypassed entirely in normal mode by its two muxes; so, in normal mode, we end up with our original, purely combinational, Lab 1 design.

While adding code for the muxes, make the select input of every mux work so that when ScanMode is '1', the FF is in the scan chain; when ScanMode is '0', the FF is in the normal operating mode. Assign a small delay to each mux input statement; say, #1 or so.

Simulate the design with a constant ScanMode = 0 (assign it in the testbench); the result should be the same as it was in Lab 1 without muxes or FFs, except for the delay added by the muxes.

We now have Lab 1 working again, but with extra, nonfunctional FFs and with muxes which merely add a delay.

Notice that we have made no change at all in any submodule. Our entire original design was combinational, so we inserted the scan elements only at the top of the design. The result looks the same as a boundary scan, but it actually is an internal scan on a design which doesn't have internal sequential gates.

Optionally, synthesize this design, just to see what the netlist schematic looks like.

Step 5: Create a scan chain. At this point in the lab, the FFs are short-circuited out of the design by muxes, and they have no function. The scan inputs to all muxes are unconnected to anything in the design. Therefore, the FFs can be connected together into a shift register, using the mux scan inputs, without affecting the design.

Refer to figure 4-10. Pick any FF and connect its Q to the design ScanOut port. After that, connect the Q of any remaining FF to the scan input pin of the mux driving the D input of the FF the Q of which just was connected. Repeat this until all FFs are chained together. The net result of this Step should be that the mux inputs which were unknown in the previous Step now are connected to FF Q drivers, except the first mux, which still is unknown. At this point, the ScanOut port should be wired to the Q output of the last FF in the scan chain.

Finally, connect the scan input pin of the mux driving the D input of the last remaining FF to the ScanIn port, making this the first FF in the scan chain. The result of all this is represented as in figure 4-11, which omits the ScanMode and ScanClr wiring for simplicity:

Digital VLSI Design with Verilog

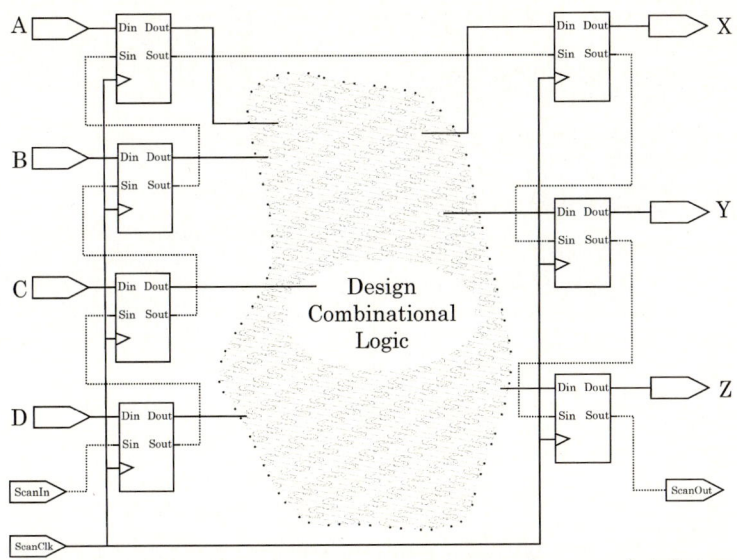

Fig. 4-11. Logical representation of Lab 5 scan insertion. In normal mode, each `Din-Dout` logically is a wire, and `Sin-Sout` is an open. In scan mode, `Sin` is a FF D input, `Sout` is a FF Q output, and `Din-Dout` is open. `Dout` and `Sout` are wired together, but this is irrelevant to the mode differences.

Recognize the scan chain on the dotted line of figure 4-11? It's just a shift register!

Step 6: Assign useful instance names. Rename the flip-flop instances so that their order in the scan chain is represented numerically: For example, `"A_FF_s01"` might be a good name for the FF on the design A input, if it were set up as the first one (`"s01"`) in the scan chain; `"FF_X_s04"` might be the FF driving the X output, if it were fourth in the scan chain.

Optionally, synthesize this design; then, try to trace the scan chain in the netlist.

Step 7: Add a simulation safety net. It takes 7 scan clock cycles to scan in or out the contents of the entire chain; generally, depending on FF ordering, it takes fewer clocks to scan in new stimuli or scan out new test results. Assume, then, that operating in scan mode for more than, say, 8 clocks in a row may represent an error. Write an assertion in `TestBench` which triggers a warning if scan mode is asserted for more than 8 scan clock cycles in a row.

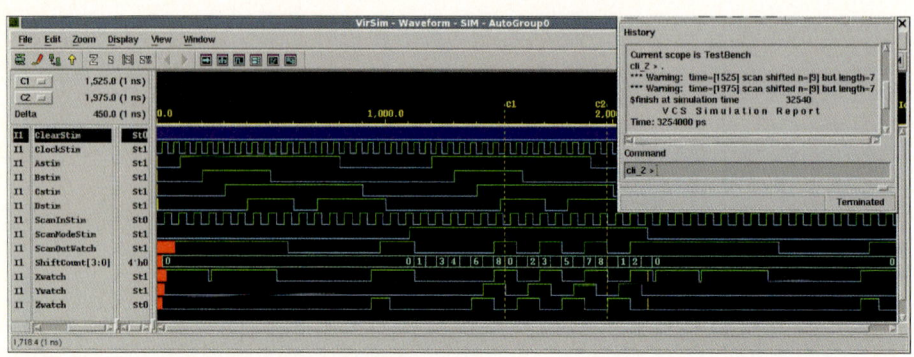

Fig. 4-12. Simulation of the completed Step 7 scanned `Intro_Top` design, with assertion safety net.

Figure 4-12 shows a simulation with a testbench which invokes the safety-net assertion. This assertion, being in the testbench, will not be seen by the synthesizer and will not have any effect on a synthesized netlist.

Incidentally, the new VCS shows the assertion messages like this:

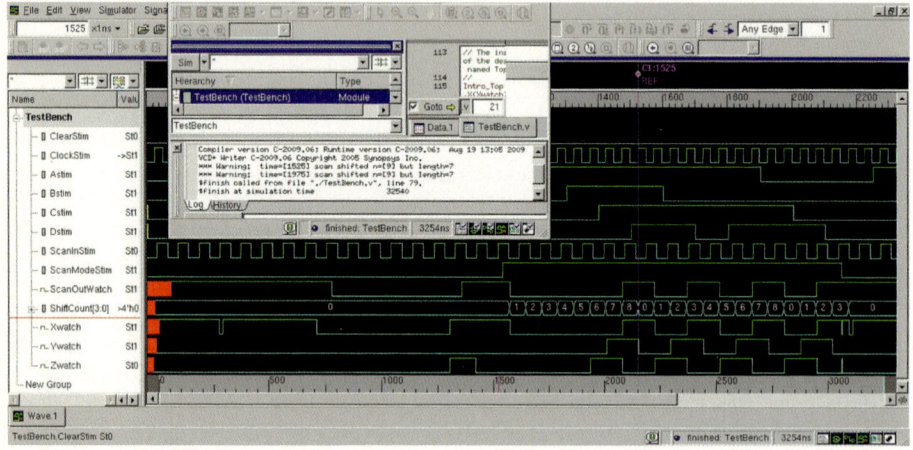

Fig. 4-13. New VCS simulation of the completed Step 7 scanned `Intro_Top` design, with assertion safety net.

Step 8: Simulate your design to be sure you understand it. Synthesize it (from the top level, not the testbench) for area. Examine the optimized netlist carefully; try to find and count the multiplexers.

Optionally, after synthesizing for area, without exitting the synthesizer, ungroup (flatten) the design and `compile` (synthesize) it again with an incremental-mapping

Digital VLSI Design with Verilog

option. Finally, still without exitting the synthesizer, do another compile on the flattened netlist for best area. The areas should improve progressively.

> **A note on "flatten" terminology**: Although "flatten" often means to remove hierarchy, this word has a special meaning to Design Compiler—for example, as used in the *DC Synthesizer Lab 1 Summary*. In DC, "flatten" refers to logical expressions, not to the design. To flatten the hierarchy in a design, the DC command is **ungroup**.
>
> In DC, one may use a `set_flatten` command to flatten combinational layers of logic to a two-layer boolean sum-of-products. The DC `set_structure` command factorizes the logic and in a sense is the inverse of `set_flatten`.

Step 9 (*optional*) : The synthesizer can insert scan automatically in a design by replacing sequential component instances with scan instances, for example by replacing plain flip-flops with muxed flip-flops. However, to use this feature, there must be preexisting sequential components. We shall start this Step by discarding the manually scanned `Intro_Top` design and going back to the original combinational one.

Keep your current `Testbench.v`, but copy in a new `Intro_Top.v` from Lab01 (you may wish to save your Step 8 `Intro_Top.v` by renaming it first).

In the new `Intro_Top.v`, interpose a D flip-flop as an output latch between each `Intro_Top` output driver and its top-level output port (X, Y, and Z); use instances of your own `DFFC` model, but modified these instances so that the Qn outputs are removed. These three new FFs will be the required sequential elements. Add a clock input port named `Clk` and a clear input port named `Clr` to operate the FFs. See figure 4-14.

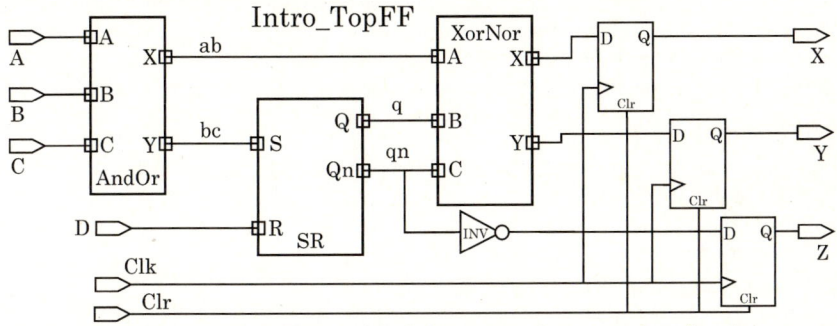

Fig. 4-14. The `Intro_Top` design with latched outputs permits automated scan insertion.

Give your flip-flops instance names beginning with "FF", for example, FFX, FFY, and FFZ; this is for consistency with the synthesis `.sct` file provided for you in your Lab05 directory. If you use other instance names, you will have to edit the `set_max_delay` command in the `.sct` file accordingly.

Then, add a set of three unconnected *JTAG* ports with these standard names, `tms` (test-mode select), `tdi` (test-data in), and `tdo` (test-data out). The other two required JTAG ports, `tck` and `trst`, will be shared with the `Clk` and `Clr` ports, respectively.

After this, rename the top module `Intro_TopFF`. Then, modify `Testbench.v` as necessary to get the design to simulate normally, with outputs latched.

You should use the `.sct` file originally prepared for you in your `Lab05` directory to insert scan automatically by means of the synthesizer. After scan synthesis, examine the synthesized netlist briefly in a text editor, VCS, or `design_vision`.

4.9.1 Lab Postmortem

In scan mode, with our design, each shift easily might cause the inputs to the combinational logic to change, leading to a lot of logic changes and possibly glitches. Is this of any concern?

How might the operational protocol be defined so as to minimize combinational-logic changes during scan chain shifting?

How might the scan chain be modified so that the design combinational logic remains in a constant state whenever it is in scan mode?

4.10 Additional Study

Read Thomas and Moorby (2002) sections 4.1 on concurrency, and 4.2 on event controls.

Read the example in 4.6 on disabling named blocks, and (optionally) 4.9 on `fork-join`. In section 4.9, note that `fork` and `join` may be used to substitute for `begin` and `end`, respectively. However, adding an explicit `begin` and `end` around every `fork-join` block makes the code more readable and more easily modified; in particular, `begin` and `end` are required to use `` `ifdef DC `` to make code with `fork-join` synthesizable.

(Optional) Do 4.10 Ex. 4.1.

(Optional) Look through Thomas and Moorby appendix G on BNF until you understand how it can be used.

Optional readings in Palnitkar (2003):

Read section 14.6.1 on coding verilog for logic synthesis.

Do Exercise 1 in section 14.9, synthesis of an RTL adder.

Do Exercises 2 and 3 of section 14.9.

Look through the on-disc synthesis example that comes with the Palnitkar CD-ROM (Chapter 14, ex. 1 directory). This example includes a presynthesized gate-level netlist and a verilog library of elementary component models.

Chapter 5
Week 2 Class 2

5 Today's Agenda:

Lecture on SerDes
 Topics: Serial vs. parallel data transfer, *PCI Express*, PLL functionality. Serializer-Deserializer and PLL design.
 Summary: We introduce our class project, a full-duplex serializer-deserializer (serdes) with a discussion of serial bus transfer advantages and the performance specifications of a *PCI Express* serdes lane. We then look at the serializer and deserializer ends of a serdes on a block level. We also see how a PLL works on a block level. We decide to start our project by designing and implementing the PLL which will be instantiated at each end of each serial line.

PLL Clock Lab 6
 After designing and testing a digital PLL, we sketch out and simulate a preliminary, generic parallel-serial converter. We finish with a parallel-to-parallel frame encoder which could be used to prepare data for serialization.

Lab Postmortem
 We discuss simulator time resolution and digital formats, comparator features, and our choice of frame encoding.

5.1 PLLs and the SerDes Project

This time, we'll design a PLL and a parallel-serial converter.

For the remainder of the course, we'll be working off and on to complete a SerDes (*Ser*ializer-*Des*erializer) design, such as is required in the *PCI Express* specification.

5.1.1 Phase-Locked Loops

A phase-locked loop (PLL) consists of an output *ClockOut* generator, a phase comparator, and a variable-frequency oscillator (VFO) such as, for example, a voltage-controlled oscillator. A PLL input signal *Clock* is provided, and the phase-difference comparator adjusts the VFO's frequency whenever a phase shift is detected between the *Clock* and the VFO *ClockOut*. The result is that the VFO is kept phase-locked to the *Clock*. In theory, a PLL could lock in to any *Clock* input; but, in practice, the VFO and the rest of the PLL is designed so that the lock is accurate and almost jitter-free only in some predictable frequency range.

Although a PLL always locks in to the *Clock* provided, if the VFO output is used to clock a counter, and a specific counter bit or value is used for the *ClockOut*, the VFO frequency will be higher than the counted-down *ClockOut* used for the lock. The direct

VFO output then may be used to provide a frequency-multiplied clock nevertheless phase-locked to the original *Clock* input.

5.1.2 A 1x Digital PLL

An important PLL application is clock latency cancellation. In this application, there is no multiplication, but the PLL is locked to a clock from a terminal branch of a balanced clock tree. This lock-in subtracts away the tree latency (clock insertion delay) and makes available a clock from the PLL at the clock-tree terminal branches which is very close to being exactly in phase with the original clock at its entry point on chip. See figure 5-1.

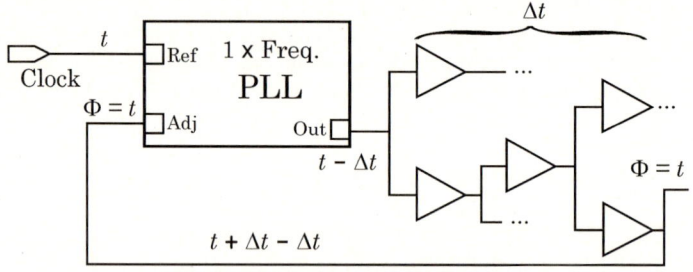

Fig. 5-1. 1x PLL used to cancel clock-tree latency.

Let us consider how we might design a verilog digital PLL for this purpose.

We don't want frequency multiplication in a 1x PLL, so we may assume two major components, a clock phase comparator and a VFO. A schematic representation is given in figure 5-2.

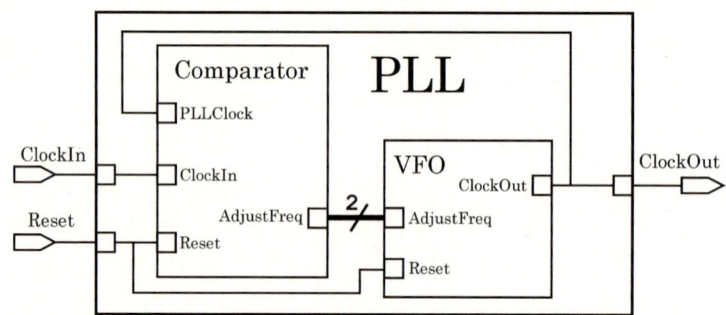

Fig. 5-2. Schematic of 1x digital PLL. The control bus from `Comparator` to `VFO` is assumed 2 bits wide.

Happily, a variable holding a verilog delay value may be changed during a simulation; this makes feasible a finely-controllable VFO period derived simply from the value of a verilog `real` variable. This PLL will not be synthesizable; but, we don't have to worry right now about synthesis.

Digital VLSI Design with Verilog

This kind of VFO frequency control may be represented by the following pseudocode:

```
real ProgrammedDelay;
...
begin
ProgrammedDelay = some delay value;
...
#ProgrammedDelay  PLLClock = ~PLLClock; // The VFO oscillator.
...
ProgrammedDelay = some new delay value;
...
#ProgrammedDelay  PLLClock = ~PLLClock; // Frequency varied.
...
end
```

Suppose now that we have decided that the Comparator adjustment code for a VFO frequency increase shall be 2'b11, and for a decrease shall be 2'b00.

Then, assume that we have chosen a delay giving us the desired VFO base operating frequency. Given this, we easily can decide upon the size of a delay increment when the Comparator signals an adjustment.

So, the actual verilog to simulate the VFO would be about like this:

```
module VFO (output ClockOut, input[1:0] AdjustFreq, input Reset);
reg   PLLClock;
real  VFO_Delay;
assign ClockOut = PLLClock;
//
always@(PLLClock, Reset)
   if (Reset==1'b1)
        begin
        VFO_Delay = `VFOBaseDelay;
        PLLClock  = 1'b0;
        end
   else begin
        case (AdjustFreq)
        2'b11: VFO_Delay = VFO_Delay - `VFO_Delta;
        2'b00: VFO_Delay = VFO_Delay + `VFO_Delta;
        // Otherwise, leave VFO_Delay alone.
        endcase
        #VFO_Delay PLLClock <= ~PLLClock;   // The oscillator.
        end
endmodule // VFO.
```

Notice the use of blocking assignments to ensure that new values are available immediately upon update. However, the VFO oscillator must use a nonblocking assignment if it is to oscillate during simulation. This unusual exception to our previously chosen rules was coded with great trepidation, and only after careful consideration of all consequences. This is a rare and unsynthesizable exception to the rule of never to mix blocking and nonblocking assignments in a single always block.

The Comparator is straightforward but a bit more complicated. We require it to issue adjustment codes as described above for the VFO, but we want a decision with a minimum of logic so that we can run it very fast in 90 nm library components. A simple verilog design would just use one clock to count the number of highs of the other clock in one clock cycle; if there was just one such high, the frequencies would be approximately matched.

We can code this as follows:

```
module JerkyComparator
        (output[1:0] AdjustFreq, input ClockIn, PLLClock, Reset);
reg[1:0] Adjr;
assign AdjustFreq = Adjr;
reg[1:0] HiCount;       // To count 1's of the PLL clock.
//
always@(ClockIn, Reset)
   if (Reset==1'b1)
      begin
      Adjr     = 2'b01; // 2'b01 or 2'b10 are no-change codes.
      HiCount = 'b0;
      end
   else if (PLLClock==1'b1)
           HiCount = HiCount + 2'b01;
        else begin
            case (HiCount)
                2'b00: Adjr = 2'b11; // Better speed it up.
                2'b01: Adjr = 2'b01; // Seems matched.
              default: Adjr = 2'b00; // Must be too fast.
            endcase
            HiCount = 'b0; // Initialize for next ClockIn edge.
            end
endmodule // JerkyComparator.
```

Blocking assignments again are used for immediate update. There is no reason to use a nonblocking assignment anywhere in this model. The various possible decisions by this comparator are shown as waveform relationships in figure 5-3.

Digital VLSI Design with Verilog

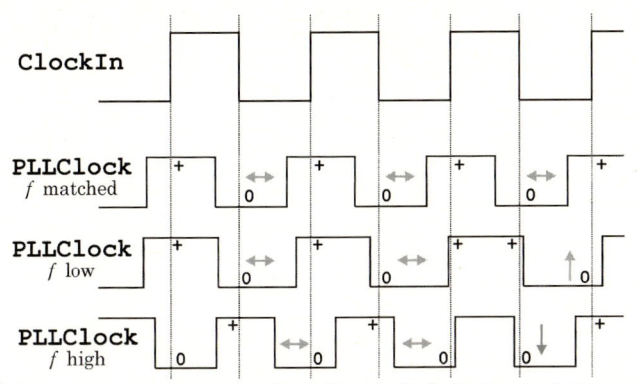

[Waveform diagram showing ClockIn, PLLClock f matched, PLLClock f low, and PLLClock f high signals with '+', '0', and arrow annotations]

Fig. 5-3. Representative `Comparator` waveforms. Up arrow for final `HiCount > 1`; down arrow for `HiCount = 0`; horizontal double-head arrow for `HiCount = 1`.

In figure 5-3, a '+' means that the `HiCount` is incremented on that `ClockIn` edge; a '0' means that it is reset to `2'b0` on the opposite edge.

This kind of model works in simulation, and we shall use one very much like it in our next lab. But, it is very slow to lock in, and it causes many spurious frequency adjustments; this last is the reason we call it "jerky". For the sake of a better lock-in, we can make several improvements:

- First, we should worry about a `HiCount` overflow and consequent wrap-around to `2'b00`, causing a spurious speed-up adjustment when the opposite is indicated. This is avoided easily by increasing the `HiCount reg` width from 2 to 3 bits.

- Second, we should improve the precision of the adjustment decisions. This can be done by averaging the high counts over several cycles; the expected value always should be 1 when the two clocks are synchronized.

  For speed, we should take a finite-difference approach to averaging in a verilog model. We should declare a fairly wide `reg` to hold the average; and, in a separate `always` block, either add 1 to it or subtract 1 from it on every clock which, respectively, has an excess high or no high at all.

As a result of these improvements, we can obtain an optimized 1x digital PLL which does a respectable lock-in, not too much poorer than could be achieved with an analogue PLL. The code follows in two parts. The first part is the same comparator as in the previous code above—but, its result is used only internally. The second part averages the comparisons of the first in order to determine the frequency adjustment to be sent to the VFO:

```verilog
module SmoothComparator
        (output[1:0] AdjustFreq, input ClockIn, PLLClock, Reset);
reg[1:0] Adjr;
assign AdjustFreq = Adjr;
//
reg[2:0] HiCount;   // Counts PLL highs per ClockIn.
reg[1:0] EdgeCode;  // Locally encodes edge decision.
reg[3:0] AvgEdge;   // Decision variable.
reg[2:0] Done;      // Decision trigger variable.
//
always@(ClockIn, Reset)
  begin : CheckEdges
  if (Reset==1'b1)
       begin
       EdgeCode = 2'b01; // The value of EdgeCode will be used to
       HiCount  = 'b0;   //        increment or decrement AvgEdge.
       end
   else if (PLLClock==1'b1) // Should be 1 of these per ClockIn cycle.
            HiCount = HiCount + 3'b1;
        else begin // Check to see how many PLL 1's we caught:
             case (HiCount)
              3'b000: EdgeCode = 2'b00; // PLL too slow.
              3'b001: EdgeCode = 2'b01; // Seems matched.
              default: EdgeCode = 2'b11; // PLL too fast.
             endcase
             HiCount = 'b0; // Initialize for next ClockIn edge.
             end
  end // CheckEdges.
// (Second part is continued below)
```

Digital VLSI Design with Verilog

The second `always` block is decision-oriented:

```
// (continued from above)
always@(ClockIn, Reset)
  begin : MakeDecision
  if (Reset==1'b1)
        begin
        Adjr    = 2'b1; // No change code.
        Done    = 'b0;
        AvgEdge = 4'h8; // 7..9 mean no adjustment of VFO freq.
        end
  else begin // Update the AvgEdge & check for decision:
        case (EdgeCode)
        2'b11: AvgEdge = AvgEdge + 1; // Add to PLL edge count.
        2'b00: AvgEdge = AvgEdge - 1; // Sub from PLL edge count.
        // default: do nothing.
        endcase
        Done = Done + 1;
        if (Done=='b0) // Wrap-around.
               begin
               if ( AvgEdge<7 )
                    Adjr = 2'b11; // Better speed it up.
               else if ( AvgEdge>9 )
                    Adjr = 2'b00; // Must be too fast.
               else Adjr = 2'b01; // No change.
               AvgEdge = 4'h8;    // Initialize for next average.
               end
        end
  end // MakeDecision.
endmodule // SmoothComparator.
```

Notice the `case` statement in the `MakeDecision` `always` block: It lacks a default and does not cover all possible input alternatives. Therefore, a strange kind of latch is implied, and synthesis of this model should not be expected to be correct. This model would be considered a simulation-only place-holder if we were to adopt it for our class project.

This 1x PLL design, with some minor changes, has been implemented for you in your `Lab06/Lab06_Ans` directory. The model was designed to be simulated with a 400 MHz clock input, which is close to the upper frequency limit possible eventually for synthesis in a 90 nm ASIC library. At this speed, the clock cycle time is 2.50 ns, with a `VFO` half-cycle delay of 1.25 ns. The testbench drifts the delay slowly upward.

Some waveforms are shown in figures 5-4 and 5-5:

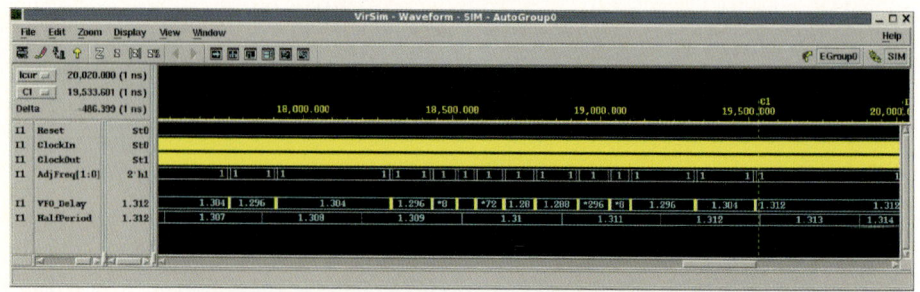

Fig. 5-4. The 1x PLL. Overview of the final 3 us of a 20 us VCS simulation.

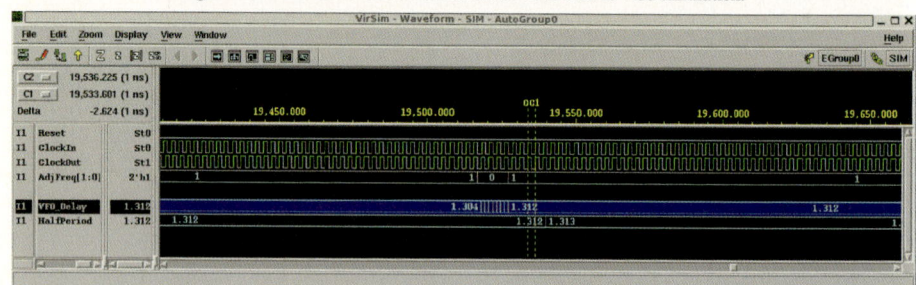

Fig. 5-5. Detailed closeup of the lock-in near the C1 cursor position of the Overview waveform.

More information on PLL's for latency cancellation, and details on static timing analysis of a design containing a PLL, may be found in the Zimmer paper in the References at the beginning of this book.

5.1.3 Introduction to SerDes and PCI Express

The *PCI Express* ("*PCIe*") bus is a serial bus meant to replace the 32-bit, parallel *PCI* (*P*ersonal *C*omputer *I*nterconnect) bus common in desktop computers between the years of about 1995 and 2005. The serial bus sidesteps the growing interbit skew problem which is the main limiting factor in wide parallel busses clocked at high speed. *PCI Express* should not be confused with the *PCIx* parallel bus standard which merely doubles the clock speed of an ordinary PCI bus.

A PCI bus operating at 66 MHz can transfer $66*10^6*32 \sim= 2*10^9$ bits/s (2 Gb/s). By contrast, each PCI Express serial link (technically, called a **lane**) can transfer 2.5 Gb/s in each direction simultaneously. This increase primarily is because of continual progress in the understanding of high-frequency digital operations in silicon. The past two decades also have seen growing automation of the use of silicon monoliths to implement what in the past were strictly discrete analogue devices.

Digital VLSI Design with Verilog

The first PCI Express specification, completed in mid-2002, allowed up to 32 PCI Express lanes, each at 2.5 Gb/s, for a total of 80 Gb/s in each direction. The current, second-generation PCI Express specification calls for 5 Gb/s per lane, in each direction, doubling the speed of the original generation. The standard document for PCIe version 3 was finished in 2012; the specified speed is 8 Gb/s per direction per lane. The packet format has been changed so that very complex deserialization algorithms will permit a format overhead of only about 2 bits per 130 bit data packet.

PCI Express, like the PCI bus, is an on-board link meant for short-range transfers of data; such operations, for example, would include transfers by the CPU to and from RAM or the video or I/O-port controller. PCI Express not only is capable of being far faster than PCI, but it is much cheaper in terms of routing area on the board. However, data management of the serialization and deserialization makes PCI Express much more complicated to design.

For example, to see the speed advantage, a typical computer terminal has a resolution of 1280 x 1024 pixels. In a full-color CMYK mode, there are 32 bits per pixel. Thus, the screen requires a buffer of 1280*1024*4 bytes, which totals about 5 MB. To refresh the screen at 70 Hz then requires a transfer rate of about 350 MB/s. A parallel video bus 128 bits wide, operated at 33 MHz, can transfer about 500 MB/s and thus can handle the screen update. However, such a bus would be running with less capacity than a single lane in second-generation PCI Express and would occupy about ten times the area on a video card. A PCI Express video bus makes a lot of sense, both in terms of performance and economy.

There are many analogue issues involved with circuits operating in the GHz range. For example, each serial line actually is a differential pair of wires, making for four wires per full-duplex lane; for our digital purposes, calling each pair a serial line is accurate enough. The mechanical implementation of a PCI Express serial line may be a simple transmission line, a twisted pair, or even a coaxial cable. We shall ignore these issues for now and work on a serdes composed of routings and gates with all analogue difficulties assumed to have been designed away. Our serdes will include a complete, self-contained serial-parallel interface the components of which do not map exactly one-to-one with those of the PCI Express standard. Unlike a PCI Express design, when we are done, ours will be entirely digital and therefore, to that extent, technology-independent. Because of lab time limitations, our serdes will not perform any packet acknowledgement, making our serdes perhaps a bit more like ethernet than PCIe.

For further insight into serdes, particularly on the analogue side of the problem, you should see the Wood (2009) article, which is one of several papers which are listed in the References near the beginning of this Textbook.

A serdes transfers data between two or more systems with busses each of arbitrary (parallel) width. These local parallel busses are so short-ranged, that interbit skew is not specifically a design consideration. The data to be transferred are clocked off a parallel bus into a buffer of some kind, often a FIFO (*F*irst-*I*n, *F*irst-*O*ut stack memory), serialized (converted to a serial format), and transferred one bit at a time by the longer-ranged serdes to their destination. At the destination, the incoming serial data are deserialized (converted to a parallel format), buffered, and clocked onto the destination parallel bus. A great simplification is that the required precisely-synchronized common clock may be generated by a low-jitter PLL from the clock image encoded in the data. Clocking in this context is the same as the defining of data-frame boundaries. If the serdes is in a stable, on-board environment, clocking can be accomplished by using a PLL synchronized to the serial stream to generate a usable clock for buffering and format conversion at both ends. A single PLL may be shared if both ends of a lane are in the same clock domain; or, two independent PLL's may be used, one at the writing end of each lane.

5.1.4 The SerDes of this course

The serdes we shall study operates, like the one in PCI Express, in a ***full-duplex*** mode; this means that transfers are possible simultaneously in both directions, with no interference or sharing required. We do this simply by having two different dedicated (verilog) data `wires`, each one capable of sending data in one direction.

Our serdes will convert 32-bit parallel data on each end to serial data in 16-bit frames. A full packet of data will be 64 bits. The 16-bit frame means that the embedded serial clock will change once per 16 bits transmitted. We shall transmit only one data byte per frame; this is less efficient than PCI Express, but it will allow our embedded clock to be extracted in an easily visible, orderly way. To serialize or deserialize the data, which must be processed one bit at a time, a PLL will be specified which multiplies the parallel-clock frequency by 32.

We shall do our design on the assumption of a relatively low parallel clock frequency of 1 MHz; our serial line then should transmit at 32 Mb/s, around 1/100 of the speed in either direction of a PCI Express lane. Because we require one pad byte for each data byte, a data packet containing one 32-bit word will occupy 64 bits in the serial stream. Thus, one framed data word will be 64 bits wide, like this:

64'bxxxxxxxx000**11**000xxxxxxxx000**10**000xxxxxxxx000**01**000xxxxxxxx000**00**000.

The 'x' bits represent values in data bytes; the pad byte values are shown as-is. We transmit serially, the MSB (on the left) first.

This means that we should expect to transmit and receive one 32-bit word on every other clock. Packets on a PCI Express bus may be as large as 128 bits, but we shall not adopt this kind of formatting in our design.

After we complete our design, we'll see how fast it can be made to run, using logic synthesis and optimization with the gate-level libraries which are available for course use.

Figure 5-6 gives a block diagram of the data flow in our serializer.

Digital VLSI Design with Verilog

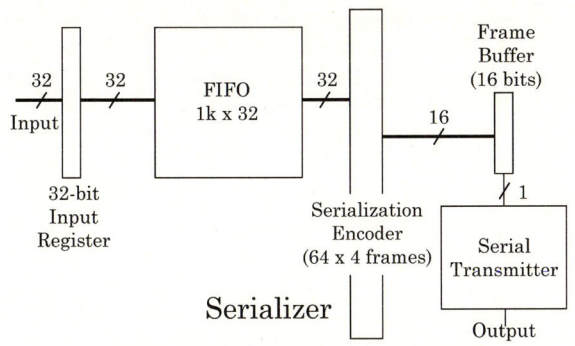

Fig. 5-6. Dataflow of the planned serializer.

Our deserializer just reverses the transformations of the serializer and is shown in figure 5-7.

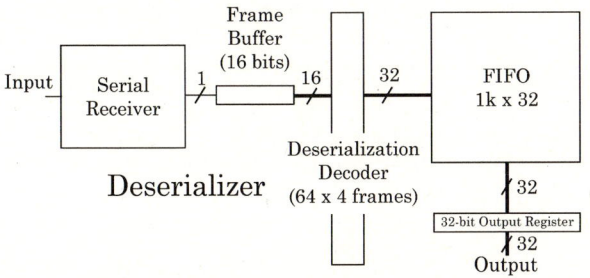

Fig. 5-7. Dataflow of the planned deserializer.

In a PCI Express or other similar serdes, the input clock on the receiving end often is extracted from the serial data stream and converted to a (analogue-approximated) square-wave for input as *Clock* to the PLL. The *ClockOut* created by the PLL is controlled to match *Clock* precisely in phase, even if the PLL *ClockOut* is at a high multiple of the frequency of the input *Clock*.

Just as a matter of side interest, figure 5-8 gives an example of a real clock waveshape for a modern memory chip.

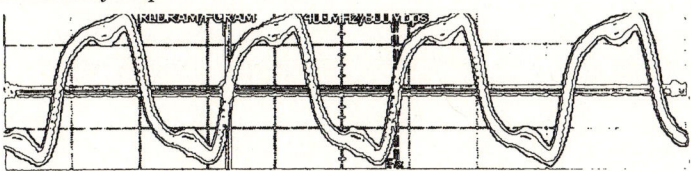

Fig. 5-8. A good 400 MHz clock. (From photo taken at the LSI Logic exhibit, *Denali MemCon 2004*).

This is a good clock which is running at 400 MHz. The horizontal grids in figure 5-8 are spaced vertically at about 300 mV divisions. The waveform is the jitter envelope of many thousands of cycles. The equivalent PCI Express serial link would be a two-wire differential pair running at a peak-to-peak voltage of about 1 volt. It's easy to see how analogue issues might arise, even at only 400 MHz.

5.1.5 A 32 x Digital PLL

In our next lab, we shall write a simple but unsynthesizable verilog model of a PLL designed to multiply frequency by 32.

In this course, we are constrained to keep to digital design; so, as for the 1x PLL above, we shall substitute a frequency lock for a phase lock in our serdes PLL. Digital synchronization will provide the equivalent phase lock. To save coding time, we shall not bother with averaging (above) for an improved lock. To cut down on comparator decisions in response to random phase misalignments, for now we shall generate an external sampling pulse; the VFO will run free and will adjust its frequency only when this pulse occurs by going to a '1'. Later, we shall eliminate the sampling pulse and use a more complex comparator.

We shall force our PLL design to be acceptable to the synthesizer just to see whether we can create a netlist, but that netlist will not be functional. Later in the course, we shall redesign this PLL so it will be correctly synthesizable to a working netlist. In the meantime, we can use this easily-written, unsynthesizable PLL to clock the serial line while we are working on other parts of our serdes.

The functionality of our PLL is represented by the block diagram of figure 5-9, with $n = 5$:

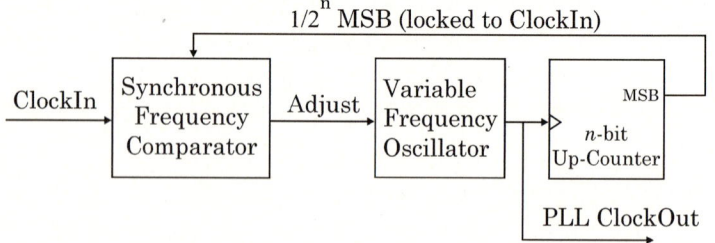

Fig. 5-9. PLL block diagram, showing `ClockIn` $\times 2^n$ frequency multiplication.

A simple, resettable verilog 5-bit up-counter can be done this way:

```
reg[4:0] Count;
always@(posedge Clock, posedge Reset)
  begin
  if (Reset==1'b1)
      Count <= 5'h0;
  else Count <= Count + 5'h1;
  end
```

Digital VLSI Design with Verilog

The lab instructions will include a schematic and further details.

5.2 PLL Clock Lab 6

Topic and Location: PLL simulation and synthesis, parallel-serial data conversion, and framing of data for serialization.

Begin this lab by logging in and changing to the `Lab06` directory.

Preview: As the beginning work of our serdes class project, we implement a verilog PLL composed of VFO, clock comparator, and counter-multiplier modules. For a rough grasp of format conversion problems, we then complete a lab-project sketch of a generic parallel-serial conversion module. Finally, we look into a way of converting parallel data into frames for later serialization.

Deliverables: Step 1: Four `.v` files with `module` and port declarations; the modules are named `PLLsim`, `VFO`, `ClockComparator`, and `MultiCounter`. We may consider a fifth testbench file which instantiates the top-level module, `PLLsim`, for simulation; however, this testbench for now may be written as a second module in the `PLLsim` file. Step 2: A verilog macro `` `include `` file named `PLLsim.inc` which contains VFO operating characteristic definitions. A working VFO simulation model provided with a temporary, fudged testbench input which emulates `ClockComparator` input. Step 3: A verilog model of the `ClockComparator`, to be simulated later. Step 4: A verilog model of the `MultiCounter`, to be simulated later. Step 5: Simulation of the completed `PLLsim` module which verifies that our PLL will track its input clock frequency, however coarsely. Step 6: A completed generic parallel-serial model based on the incomplete verilog model provided, and which simulates correctly. Step 7: A correctly simulating frame encoder; synthesized netlists of this encoder which represent optimization for area and speed.

Lab Procedure:

In this lab, we'll build a PLL and a related parallel-serial converter. Because a logic synthesizer can't synthesize delays (or delay differences), for synthesis purposes in preparation of chip layout we normally would replace this PLL with a pre-synthesized hard macro or with an analogue IP ("<u>I</u>ntellectual <u>P</u>roperty") block created by another vendor.

We shall introduce no #*delay* value anywhere in the PLL, other than testbench delays or clock-frequency durations.

To be systematic, we'll break down the PLL into three blocks, as explained above and shown in figure 5-10: VFO, comparator, and counter.

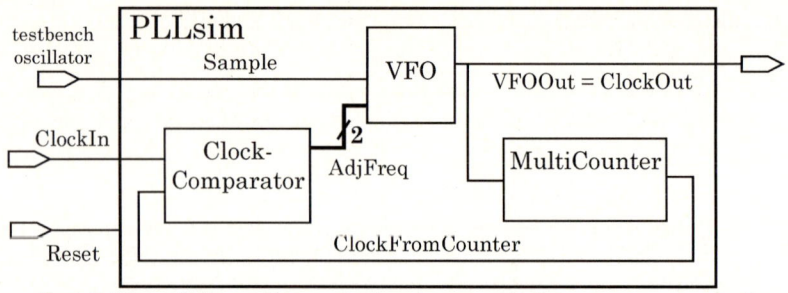

Fig. 5-10. Schematic of our verilog PLL, showing wire names. The three reset nets are omitted.

Step 1: Start by creating a top-level module named **PLLsim** (sim = simulation). Instantiate in it three submodules: a VFO, a ClockComparator, and a MultiCounter. As usual, put each module in a separate .v file named for that module. Give the top module in PLLsim just three inputs, ClockIn, Reset, and Sample, and one output, ClockOut.

Declare module header ports to match the blocks shown in figure 5-10:
- For the **ClockComparator**, declare ClockIn, Reset, and CounterClock input ports, and a two-bit AdjustFreq output port. We could use finer tuning, but two bits is enough for our purposes.
- For the **VFO**, declare a two-bit AdjustFreq input port, a SampleCmd and a Reset input port, and a ClockOut output port.
- For the **MultiCounter**, declare Clock and Reset input ports and a CarryOut output port.

Connect the ports in PLLsim with wires, consistent with the block diagram shown in figure 5-10.

You might as well add a simulation testbench module in PLLsim.v and instantiate PLLsim in it; having a testbench early in the process will allow you to try out your submodules as you complete them.

Step 2: Model the PLL VFO. We shall start with a simple verilog model of a variable-frequency output clock which adjusts itself in small increments using a two-bit correction flag on an input bus named AdjustFreq. For now, a value of 1 on this bus means no change; a value of 0 means reduce the frequency; a value of 2 or 3 means increase the frequency of the VFO.

The parallel-bus clock of 1 MHz implies a VFO base frequency of 32 MHz; so, the VFO period should be 1/32 us = 31.25 ns, and the half-period (edge delay) therefore should be 15.625 ns. These delays are fractional, but we can use only integer types (integer or maybe reg) if we wish to do synthesis later; therefore, we can't express anything smaller than 1 ns, so, we can expect our frequency control to be somewhat coarse.

We could achieve finer frequency control if we expressed everything in picoseconds rather than nanoseconds, but this would make everything we do in the course more complicated and error-prone.

In this preliminary, unsynthesizable PLL, we'll sample frequency occasionally (to be determined by the serial data in) using an external signal, Sample. We'll design to be able to adjust our VFO by 1/16 of the half-period per sample, which comes to a little less than 1 ns per sample. So, on any Sample command, if we have AdjustFreq > 2'b01, we'll decrease VFO period by about 1 ns; if AdjustFreq < 2'b01, we'll increase VFO period by about 1 ns. As above, 2'b01 means no change.

It is allowed in verilog to declare a parameter to be real and/or signed, which would be useful in a PLL testbench. But, many tools won't let us use a parameter to store a real number. Also, declaring a real port would be a serious digital implementation problem; so, we'll have to be content to control the design base-point operating frequency with defined macro constants: To accomplish this, and for flexibility in compiling individual modules separately, we shall put these lines in a separate include file, PLLsim.inc:

```
`timescale 1ns/100ps
`define HalfPeriod32BitBus 500.0     // ns half-period at 1 MHz.
`define VFOBaseDelay `HalfPeriod32BitBus/32.0 // At 32 MHz.
`define VFO_DelayDelta 1 // ns.
`define VFO_MaxDelta   2 // ns.
```

The last line is to prevent the PLL from running away or grinding to a halt: We shall use it in the VFO module to limit the size of frequency excursion from the base frequency.

Because defined macro constants can't be changed during simulation (actually, they are removed during simulator compilation and replaced by their values), these constants will have to be used to initialize verilog variables. The values above in PLLsim.inc will work whether the variables they are assigned to are synthesizable integers or unsynthesizable reals. Later in the course, we'll change the above include file to use different constant values for reals (simulation) than for integers (synthesis).

Next, in any design file using these definitions, add this line above its module header:

```
`include "PLLsim.inc"
```

The only files requiring this are the top-level one, PLLsim.v, which propagates the timescale to the rest of the design during simulation, and the VFO one, VFO.v. Notice two things about the include definitions: (*a*) Given the 32-bit to 1-bit design itself, the values depend solely on the timescale and one assigned value, `HalfPeriod32BitBus; and, (*b*) the values involved in division include decimal points. Decimal points are required in verilog to force the delays to be calculated as real numbers. Without decimal points, the divisions and rounding would be done on integers, and the result would be less accurate in simulation. However, all variables involved have to be declared as integers for synthesis anyway, so the intermediate float calculations only have a small effect.

Next, in the VFO file, use the include file to declare the following variables as integers and add this reset block:

```
always@(Reset, SampleCmd, VFO_ClockOut)
if (Reset==1'b1)
    begin
    BaseDelay       = `VFOBaseDelay;
    VFO_Delta       = `VFO_DelayDelta;
    VFO_MaxDelta    = `VFO_MaxDelta;
    VFO_Delay       = `VFOBaseDelay;
    VFO_ClockOut    = 1'b0;
    end
else (below)
```

The VFO_ClockOut is just used to initialize the output wire and thus is not included with the others.

As a minor contribution to the lab exercise, complete the delay control of this VFO design by supplying the contents of the else half of the always block which is given in part below; this else applies one of the two delay adjustment which determine changes in the VFO half-cycle delay:

```
else // as above.
    if (SampleCmd == 1'b1)
    begin
    if ( AdjustFreq>2'b01
        && (BaseDelay - VFO_MaxDelta < VFO_Delay) )
        // If floor is lower than current:
        VFO_Delay = VFO_Delay - VFO_Delta;
    else if ( AdjustFreq<2'b01
            && (fill this in)
        // else, leave VFO_Delay alone.
```

Because of the small values of the VFO_MaxDelta limits, we have replaced the earlier case statement with two levels of if.

To generate the PLL clock from the delay determined above, we complete the model by adding one more assignment statement, that of VFO_ClockOut:

```
always@(Reset, SampleCmd, VFO_ClockOut)
  if (Reset==1'b1)
      ... (as above) ...
  else begin
      ... (freq. adjustments) ...
      `ifdef DC
      // No delayed nonblocking assignments:
      #VFO_Delay VFO_ClockOut = ~VFO_ClockOut;
      `else
      #VFO_Delay VFO_ClockOut <= ~VFO_ClockOut;
      `endif
      end // main else.
```

Digital VLSI Design with Verilog   105

The oscillation inversion in the last assignment above absolutely requires a nonblocking assignment for simulation; however, the synthesizer rejects delayed nonblocking assignments; whence the `ifdef (the macro DC always is defined whenever the Design Compiler synthesizer runs). The synthesizer will use the blocking assignment to replace it with its own delay value. Some demo versions of Silos will simulate the oscillation correctly if the above was written with a blocking assignment, which actually is a verilog language error.

The reason the clock-generating always block is written to be sensitive to SampleCmd is because whenever such a command is asserted, we want the PLL clock to become synchronized to it.

After completing the VFO module as above, simulate it from PLLsim, putting a dummy clock temporarily into the comparator module to count 0 to 3 to trigger frequency adjustment events. Just do a simple simulation to check that the delay programming is working.

Step 3: Model the PLL comparator. The ClockComparator module, as in figure 5-10, compares an arbitrary input clock frequency to a variable clock frequency, that of the PLL, and issues an adjustment to make the variable frequency more closely match the input frequency.

In our design, the comparator, as implemented in this lab for the VFO model, will compare the frequencies continuously, but the VFO will not make any adjustment unless a Sample command requires it. We need not supply our comparator with a Sample command, but we must supply a Sample command periodically to the VFO.

In any case, the ClockComparator must be supplied a reset to ensure its registers are in a known state.

How will our comparator work? There are several different ways to do a purely digital frequency comparison; we shall do it in this lab by counting variable-clock edges after every input-clock edge.

There will be three possible cases:
- If more than one variable-clock edge is found, the adjustment will be set to decrease variable-clock frequency.
- If just one variable-clock edge is found, the adjustment will be set for no variable-clock change.
- If no variable-clock edge is found, the adjustment will be set to increase variable-clock frequency.

To do this, declare a 2-bit reg variable named VarClockCount to count incoming PLL clock (CounterClock) edges. The width of 2 bits allows a value of 3 to be counted, which exceeds the expected number of PLL clock edges per input clock, assuming that both clocks have very close to the same frequency and duty cycle. We wish to avoid a VarClockCount counter overflow under all reasonable conditions, but using a 32-bit verilog integer seems excessive.

It then is possible to use the following in the `ClockComparator` module to count the variable clock edges as shown next:

```
...
always@( ClockIn, Reset ) // This is the synchronizing clock.
    begin
    if (Reset==1'b1)
        begin
        AdjustFreq    = 2'b01;
        VarClockCount = 2'b01;
        end
    else // CounterClock is the clock to synchronize.  Notice that it
         // is not on the sensitivity list; the inferred latch may be
         // expected to cause synthesis problems.
        if (CounterClock==1'b1)
            VarClockCount = VarClockCount + 2'b01;
        else begin
            case (VarClockCount)
              2'b00: AdjustFreq = 2'b11; // Better speed it up.
              2'b01: AdjustFreq = 2'b01; // Seems matched.
              default: AdjustFreq = 2'b00; // Better slow it down.
            endcase
            VarClockCount = 2'b00; // Initialize for next ClockIn edge.
            end
```

Notice in this `always` block, that we are monitoring every change of logic level of the external clock. We do not trigger a reading of this block on the clock from our PLL counter, because the PLL counter-clock is what we are using this `always` block to synchronize. It would be possible to use the `VarClockCounter` count directly as a control for the VFO, but the degree of freedom added by the `case` encoding allows us to modify the `ClockComparator` somewhat without making changes in the VFO, too.

As shown in figure 5-3 above, we count every time a '1' is found for the PLL `CounterClock` when the synchronizing clock has gone to '1'. Whenever we find `CounterClock` to be '0', we check the count we accumulated since the last '0': If this count is 0, the `CounterClock` is assumed to be running too fast, so we request a speedup; if it is 1, we seem to be approximately synchronized, and we do nothing; if it is above 1, the `CounterClock` is assumed to be running so much more slowly that its positive level was sampled twice; so, we request a VFO slow-down.

With all this done, we can complete the next lab Steps to determine the adjustment: These Steps simulate both `VFO` and `ClockComparator` together from a testbench which controls `PLLsim`. We shall have to be sure to reset `ClockComparator` correctly to ensure a proper simulation.

Step 4: Model the PLL Multiplier-Counter. This will be just a simple counter which toggles its overflow bit every 32nd clock. We shall not dwell on the architecture of counters until a later chapter.

Digital VLSI Design with Verilog

Typical verilog for a simple, generic behavioral up-counter is as follows:

```
reg[HiBit:0] CountReg
...
always@(posedge ClockIn, posedge Reset)
  if (Reset==1'b1)
      CountReg <= 'b0;
  else CountReg <= CountReg + 1'b1;
```

Digression: A `parameter` is not allowed in verilog to specify the width of a literal. So, an addition in the `else` here of something like "HiBit'b1" would involve an illegal increment expression. If we wrote, `CountReg <= CountReg + 1`, that would be legal, but the increment would then be, by default, a 32-bit integer. In the above generic example, we have sized the increment, assuming that widening 1 bit to the width of `CountReg` would be easier on the compilers than narrowing 32 bits down to that width. This decision is speculative and really probably makes little difference; so, use a plain `CountReg <= CountReg + 1`, if you want.

Defining the "carry" bit as the MSB in the counter `reg` also is simple: To get a carry out every 32nd clock, our PLL's counter just has to be exactly 5 bits wide ($2^5 = 32$). So, install a 5-bit counter in the `MultiCounter` module, and wire its highest-order bit to the `CarryOut` port. The carry bit will toggle every 16 clocks, giving it a period of 32 clocks. Use a behavioral counter sensitive to `posedge` clock and `posedge` reset.

Step 5: Test the complete PLL. With the counter wired into the top-level, the PLL now should be complete and functional. Use the testbench to trigger the top-level `Sample` with a positive pulse just slightly after each positive edge of `ClockIn`. Verify the PLL by simulating from `PLLsim` (see figures 5-11 and 5-12). You should be able to change the frequency of `ClockIn` (figure 5-10) and see the `CarryClock` and the PLL `ClockOut` change (coarsely) to match it.

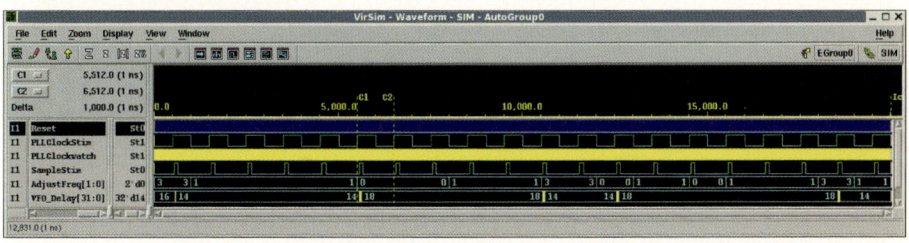

Fig. 5-11. Simulation of `PLLsim`. The "lock-in" of the VFO delay is coarse and extreme, but it does occur.

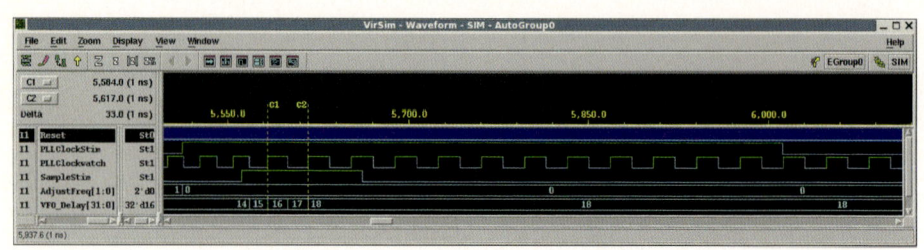

Fig. 5-12. Closeup of the `PLLsim` serial clock.

This completes our work on the PLL for now; the rest of this lab is on other, but related, topics.

5.2.1 Note on Synthesis dont_touch

In back-end design, a *hard macro* is just a block with fixed size and pin locations—for example, an IP block. This isn't a floorplanning or layout course, so we will not dwell on hard macroes. At present, instead of a hard macro to invert `VFO_ClockOut`, we have a `` `ifdef `` to permit synthesis without generating a fatal error. So, to obtain a netlist before fully completing our project, we simply can tolerate the `` `ifdef `` and its nonfunctional PLL. Alternatively, we can inhibit optimization of the VFO block by using a synthesis don't-touch directive.

Any hierarchical block instance below the top level of a design can be preserved from synthesis flattening or optimization by marking it with a ***don't touch*** comment directive (`"set dont_touch ..."`). We shall be using `dont_touch` in later labs; if you wish a glimpse of it now, you will find it used in `DC_Synth_Lab26Cmds.pdf`, in your `Lab26` directory. Instead of using it as a comment directive in the verilog, the ***don't touch*** command also may be included in a synthesis script.

When the need for a `dont_touch` arises as a synthesis convenience, locating it in the script may be preferable to inclusion in a verilog comment, because a directive in the script may be modified, or coordinated with other synthesis or optimization options, without editing the verilog in a design source file. However, when a `dont_touch` instance or net permanently is part of the design, locating it in a comment, making it part of the design, may be the better choice.

> This completes the verilog 32x `PLL`. We shall use it as-is until we redesign it for correct synthesis.

Step 6: Model a generic parallel-serial converter.

We'll now walk through the design of a converter which we shall name, `ParToSerial`. A parallel-serial converter converts paralleled data to a serial format and requires knowledge of the width of the parallel bus. Our clock input to such a converter will be named `SerClock`. We may assume that the parallel bus contents are properly clocked and synchronized externally, and that there is an input flag, `ParValid`, to indicate when data on the parallel bus are stable and valid. See figure 5-13.

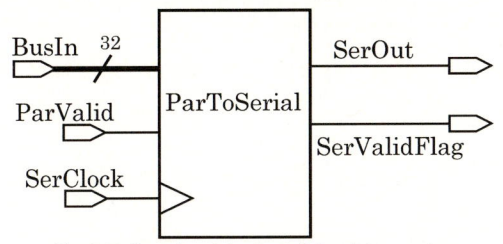

Fig. 5-13. Generic, minimal parallel-serial converter.

When `ParValid` goes to 1, its asserted value, we assume that the parallel data on `BusIn` then will remain valid until all serial data are clocked out; but, we won't clock anything out while `ParValid` is not asserted. Also, we shall not clock anything out that already has been clocked out, no matter whether the parallel data are valid or not.

The serial protocol is simple: Clock out the data high-order bit (MSB) first, one bit per serial clock, setting a `SerValidFlag` when the first bit is on the `SerOut` serial bus, and clearing it after the last valid bit is on the serial bus.

We have to adhere to the rule that no data object should be assigned from more than one `always` block; not only is this good design practice, but it is required for synthesis. Because we must clock out by the serial clock, our one `always` block therefore must be sensitive to the serial clock.

A good plan is to use a `Done` flag to terminate the valid state of the serialization. This makes the model an informal state machine: As soon as the `ParValid` is sampled asserted, move from a `Done` state to a *not*-`Done` state, and begin shifting out the serial data. If `ParValid` should be deasserted, or if the last parallel bit should have been processed, move back to a `Done` state. Remain in this `Done` state until a new `ParValid` is sampled (this means sampling a briefly deasserted `ParValid` before leaving a previous `Done`). Otherwise worded, `Done` must be cleared by the brief sampling of a deasserted `ParValid`. We shall look into state machines later in the course.

Assuming a fixed parallel width of 32 bits, one way to do the verilog for this model is below.

```verilog
module ParToSerial (output SerOut, SerValidFlag
                    , input SerClock, ParValid, input[31:0] BusIn);
   integer ix;
   reg SerValid, Done, SerBit;
   assign #1 SerValidFlag = SerValid;
   assign #2 SerOut       = SerBit;
   always@(posedge SerClock)
     begin // Reset everything unless ParValid:
     if (ParValid==1'b1)
         if (SerValid==1'b1)
             begin
                SerBit <= BusIn[ix]; // Current serial bit.
                if (ix==0)
                    begin
                       SerValid <= 1'b0;
                       Done     <= 1'b1;
                    end
                else ix <= ix - 1;
             end // SerValid was asserted.
         else begin // No start yet:
                if (Done==1'b0)
                   begin
                      SerValid <= 1'b1; // Flag start on next SerClock.
                      ix       <= 31;   // Ready to start on next SerClock.
                   end
                SerBit   <= 1'b0; // Serial bit default.
             end
      else // ParValid not 1; reset everything:
           begin
              SerValid <= 1'b0;
              Done     <= 1'b0;
              SerBit   <= 1'b0; // Serial bit default.
           end // if ParValid else
      end // always
endmodule // ParToSerial.
```

There is a copy of this model for you to modify in the Lab06 directory, named ParToSerial_unfinished.v.

As your Step 6 exercise, finish this up by introducing a parameter to set the parallel-bus width, and by writing a testbench. Write your verilog parameter (ANSI style) so that it allows the width of the parallel input bus to be varied. Be sure that nothing is hard-coded for width. Assign the default width to be 16 bits.

Also, change the module's declaration of ix to a reg declaration. Use a reasonable width, but also think about how you could make the reg width the minimum possible that will allow the model above to work correctly with a parameterized BusIn width. Simulate your model to verify that it works. Typical simulation results are shown in figures 5-14 and 5-15.

Digital VLSI Design with Verilog   111

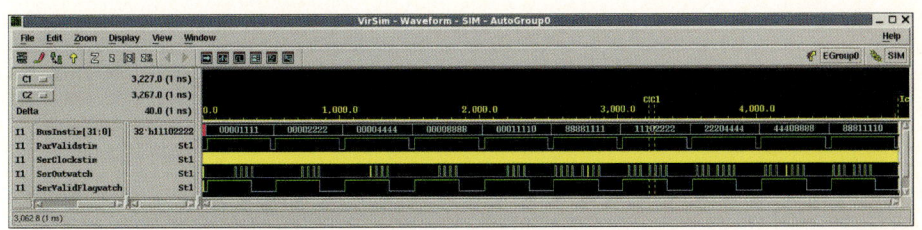

Fig. 5-14. Overview simulation of the generic parallel-to-serial converter.

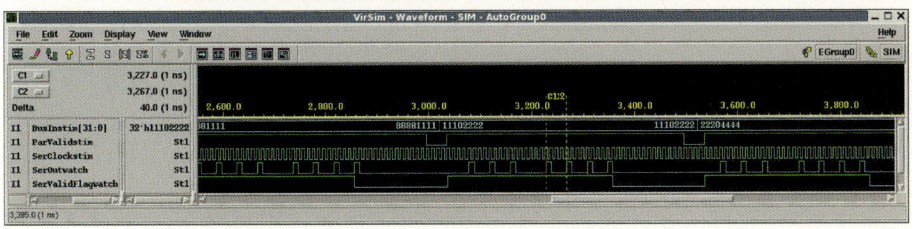

Fig. 5-15. The generic parallel-to-serial converter: Serial data closeup.

Step 7: Serialization Frame Encoder. In the final part of this lab, we shall model a parallel-to-parallel encoder which takes a generic parallel bus and converts to a wider bus which includes framing and frame boundary (pad) markers. This last format then clearly can be serialized to clock our class-project serdes PLL as well as to be transmitted serially.

Refer to the block diagram of the Serializer presented at the start of this chapter. On each positive edge of a do-the-encode input clock, the parallel bus is sampled and copied to convert it to its framed format. The input bus may be assumed 32 bits wide and the framed output bus 64 bits wide. The width increase is because we require that after each data byte on the input bus, we shall insert 8 bits of frame padding. See figure 5-16.

Data3	Data2	Data1	Data0

↓

Data3	Pad3	Data2	Pad2	Data1	Pad1	Data0	Pad0
Frame		Frame		Frame		Frame	

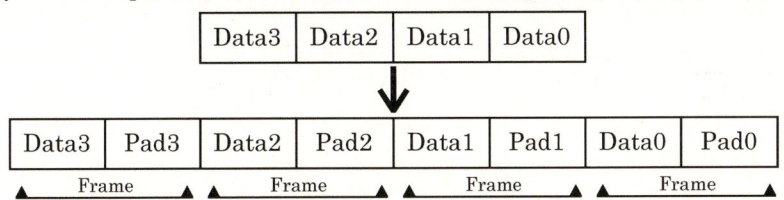

Fig. 5-16. The serdes packet format. Each data byte is framed with one identifying pad byte.

So,

  Data3 Data2 Data1 Data0

becomes

  Data3 Pad3 Data2 Pad2 Data1 Pad1 Data0 Pad0

This last can be done procedurally, if you wish; or, by continuous assignment:
```
assign Packet[63:56] = Data3;
assign Packet[55:48] = Pad3; ... etc. ...
```

Recalling the frame format for our project, to identify each byte of data, each frame boundary (8 bits) contains an ordinal number giving its place in the original 32 bits of data.

It takes 2 bits to enumerate the 4 data (bytes) in a 32-bit bus, so we shall make up each 8-bit frame boundary to include a 2-bit number padded on each side by 3 binary 0's. Our frame boundaries then will be these binary numbers: Boundary0 (below the lowest-order data byte) = 8'b000_00_000; boundary1 = 8'b000_01_000; boundary2 = 8'b000_10_000; and, boundary3 = 8'b000_11_000.

So, one sample of our framed data will look like this in binary, with x's representing the original input data:

64'bxxxxxxxx000**11**000xxxxxxxx000**10**000xxxxxxxx000**01**000xxxxxxxx000**00**000.

In this Step of the lab, design a module named SerFrameEnc which will encode an input bus this way on every positive edge of a sampling-clock input. When writing your model, use verilog parameter values to specify the input and output bus widths. After simulating it to check the result (see figure 5-17), try to synthesize your model, optimizing first for area and then for speed.

Fig. 5-17. A SerFrameEnc simulation, netlist optimized for speed.

5.2.2 Lab Postmortem

Where should `timescale be set in a multimodule design?

How does verilog distinguish integer from float (real) numerical constants?

How might the PLL comparator be modified to compare on every clock input, rather than on every edge? How would this change the constants `define'd in the top module, PLLsim?

The frame encoder seems very inefficient, using 64 bits to encode 32 bits of data. How might this encoding be made more efficient? At 2 or more Gb/s for a PCI Express serdes, does it matter? We'll look into this question again soon.

5.3 Additional Study

SerDes, PLLs, and PCIe are maturing products, so the reader can find many useful articles on them by searching the internet.

(optional) For some background on the analogue issues, try the article by S. Seat, in the References at the beginning of this Textbook.

Chapter 6
Week 3 Class 1

6 Today's Agenda:

Lecture on memories and verilog arrays
    Topics: Memory chip size descriptions, verilog arrays, parity checks and data integrity; framing for error detection,
    Summary: We look into how the storage capacity (in bits) of a memory chip can be described, and then we show how to model memory storage by verilog arrays. We study parity, checksums, and error-correcting codes (ECC) as ways of tracking data integrity in stored memory contents—or, in data storage in general.

Lab on RAM models
    We show how to model a simple single-port RAM in verilog and then to instantiate it in a wrapper to give it bidirectional I/O ports.

Lab Postmortem
    We discuss hierarchical references in verilog, the concatenation operation, and some RAM details.

6.1 Data Storage and Verilog Arrays

6.1.1 Memory: Hardware and Software Description

    Memory retains data or other information for future use. In a hardware device context, memory may be divided into two main categories, random-access and sequential-access.

    <u>Random access memory</u> (RAM) is most familiar as the storage in the memory chips used in computers. Any storage location can be addressed the same way, and by the same hardware process, as any other storage location. Every addressed datum is equally far from every other one, so far as access time is concerned. EPROM, DRAM, SRAM, and flash RAM or ROM are familiar names for implementations of this kind of memory.

    <u>Sequential-access memory</u> includes tapes, floppy discs or hard discs, and optical discs such as CD or DVD discs. Sequential access requires a process or procedure which varies with the address at which the storage is located. For example, a tape may have to be rewound to reach other data after accessing data near the end of the tape; or, the read/write head of a disc may have to be repositioned and move different distances when it changes from one address to another. So, the access time depends on the <u>sequential</u> order of addresses to be used.

    In the present course, we shall not be dealing with sequential-access memory models; they have no existence in the verilog language. However, random-access memories are very relevant to the verilog design process. Verilog has an ***array*** construct specifically intended to model RAM.

6.1.2 Definitions of Memory Size

The description of RAM chip capacity on data sheets and in the literature varies somewhat and can be confusing. A hardware databook often will describe the capacity by giving the total number of bits addressable (excluding error correcting or parity storage), and also will give the word width of the bits stored, thus: (*Storage in bits*) x (*word width*). So, a "**256k x 16**" RAM chip would store 256k = 256 x 1024 = (2^8 x 2^{10}) = 2^{18} bits. The "**16**" is a word width and does not enter this calculation; it just tells how the bits are organized in terms of address locations. To determine how many address locations are available in this chip, one divides by the word width: Clearly, in this example, there would have to be an 18-bit address bus to address by 1-bit words ("256k x 1"). This means ($2^{18}/2^3$) = 15 bits of address to address by byte ("256k x 8"); or, most relevantly, 14 address bits ("256k x 16" => $2^{18}/2^4$) to address by the hardware-defined 16-bit word width for this chip. Thus, a 256k x 16 RAM chip will require 14 bits of address to be fully utilized.

Hardware manufacturers use this method when these chips are designed for narrow 1-bit or 4-bit words and are intended to be wired in parallel to match the computer word size. For example, eight **256k x 1** DRAM chips would be required for a memory of 256k bytes. Or, nine would be required for 256k of memory with parity. No arithmetic is necessary to know the final number of addresses; 8 x 4 = 32 of these chips would be required for 1 MB (megabyte) of memory.

By contrast, in Palnitkar (2003), there is a DRAM memory model in Appendix F. The author uses a software description of this memory: It is given as (*storage in words*) x (*word width*). This memory, interpreted as a single DRAM chip, stores (256k x 16) bits = (256 x 1024) x (16) = $2^{(8+10)+4}$ = 2^{22} = 4 Mb (megabits), recalling that 1 Mb = 1024 x 1024 = 2^{20} bits.

So, to model a memory accurately, it is necessary to understand what kind of description should be used.

6.1.3 Verilog Arrays

We have worked with verilog ***vector***s up until now; these objects are declared by a range immediately following the (predefined verilog) type: For example, `reg[15:0]` is a range expression used to declare a 16-bit vector for I/O or storage.

A verilog ***array*** also is defined by a range expression; however, the array range follows the name being declared. Historically, the array range is kept separate from the vector range precisely because it was intended that an array of vector objects should be used as a memory. For example, "`reg[15:0] WideMemory[1023:0];`" declares a memory named `WideMemory` which has an address range (array) totalling 1024 locations, each such location storing a 16-bit `reg` object (word).

Notice that the above array index is not a bit position; it is the number of storage locations in the memory, minus 1 if the lower index is 0. The address bus for `WideMemory` would be $n = 10$ bits wide; because, if $n = 10$, then $2^n = 2^{10} = 1024$.

The general syntax to declare a verilog array thus is:
  reg [***vector*** *log indices*] *Memory_Name*[***array*** *location indices*];

A signed `reg` type, such as `integer`, also might be used, but this would be rare.

Some verilog assignments performing memory addressing are,

```
reg[7:0]   Memory[HiAddr:0];  // HiAddr is a parameter >= 22.
reg[7:0]   ByteRegister;
reg[15:0]  WordRegister;       // This vector is 16 bits wide.
...
ByteRegister        <= Memory[12];  // Entire memory word = 1 byte.
WordRegister        <= Memory[20];  // Low-order byte from the memory word.
WordRegister[15:8]  <= Memory[22];  // High-order byte from the memory word.
...
```

Like hardware RAM, verilog memory historically was limited in the resolution of its addressability. A 1995 CPU can only address one word at a time, and when it does, it gets the whole word, not just a single bit or a part of the stored word. It used to be so in verilog: A memory datum (= array object) could not be accessed by part or bit, unless the words it stored were just one bit wide. This is not true any more after *verilog-2001*.

For example, suppose a memory word size was 64 bits, but the system word width was 32 bits. Then, the following code would be legal after *verilog-2001* and in SystemVerilog:

```
reg[63:0] Memory[HiAddr:0];  // Assume HiAddr is a parameter > 56.
reg[7:0]  ByteRegister;
reg[31:0] WordRegister;       // This vector is 32 bits wide.
...
ByteRegister       <= Memory[57];         // Entire memory word (truncated).
ByteRegister       <= Memory[50][15:8];   // 2nd byte from a memory word.
Memory[56][63:32]  <= WordRegister;       // To the high half of memory word 56.
...
```

As shown immediately above, to declare a memory, vector and array sizes are given with ranges separated, but the resulting objects are referenced with ranges <u>all following</u> the object name. The memory address is immediately next to the declared name and references an entire array of bits of some kind; selects follow to the right of the memory location, as in selecting from a vector.

Verilog (*verilog-2001*) allows multidimensional arrays. For example,
  reg[7:0] MemByByte[3:0][1023:0];

declares an object interpretable as a memory storing 1024 32-bit objects, each such object being addressable as any of 4 bytes. Or, it might be interpreted as storing 4096 8-bit objects arranged the same way. So, using the above array declaration, "`ByteReg <= MemByByte[3][121];`" may be used as though reading the high byte stored at location

121. The "[3] [121]" part is an address, <u>not a select</u>, so the declared order of the indices is used. Also, because these are addresses, variables are allowed in the address index expressions. Variables are not allowed in part-selects anywhere in verilog; they are allowed in vector bit-selects.

In a multidimensional array, any number of dimensions is allowed, but only rarely would more than three be useful. It is possible to reduce the required dimensionality by one by using a part-select on the addressed word; of course, such a part-select would be legal only if in constant indices (literals or parameters—or constant expressions entirely of them).

For example,

```
reg[7:0]   Buf8;
reg[7:0]   MemByByte[3:0][1023:0]; // 2-D (call byte 3 the high-order byte).
reg[31:0]  MemByWord[1023:0];      // 1-D.
integer i, j;
...
i = 3;
Buf8 <= MemByByte[i][j];       // High-order byte (3,j) stored in Buf8.
Buf8 <= MemByWord[j];          // Low-order byte stored.
Buf8 <= MemByWord[j][31:24];   // Part-select; high-order byte stored.
Buf8 <= MemByWord[j][(i*8)-1:(i-1)*8]; // ILLEGAL!  i is a variable!.
...
```

Thomas and Moorby (2002) discusses multidimensional arrays in appendix E.2.

Finally, it is not legal to access more than one memory storage location in a single expression: For reg[7:0] Memory[255:0], the reference, "HugeRegister <= Memory[57:56];" is not legal, nor, for "reg[31:0] MemByByte[1023:0];", would be "MyIllegalByte <= MemByByte[121:122][31:28];", which would seem to cross address boundaries to get 4 bits from each of two different addresses. <u>Only one address per read or write is allowed</u>; but, like a bit-select of a plain vector, an address may be given by a variable.

This last implies that an array object never may be assigned directly; it has to be accessed, possibly in a loop, one address at a time.

To summarize, verilog memory access currently is associated with these limitations:

- Only one array location is addressable at a time.
- Part-select and bit-select by constant are legal after *verilog-2001*, but implementation by tools is spotty.
- Part-select or bit-select by variable never is allowed.
- Many simulators such as the *Silos* demo version can not display a memory storage waveform; new VCS, *QuestaSim*, and *Aldec* can.

Thus, currently, it is best to access memory data by addressing a memory location and copying the value to a vector; this vector value then may be displayed as a waveform and

Digital VLSI Design with Verilog

may be subjected to constant part-select or variable bit-select as desired. This approach is portable among simulators and synthesizers.

For example,

```
parameter HiBit = 31;
reg[HiBit:0] temp;               // The vector.
reg[HiBit:0] Storage[1023:0];    // The memory.
reg[3:0] BitNo;                  // Assigned elsewhere.
...
temp = Storage[Adr];   // Declared and assigned elsewhere.
HiPart = temp[HiBit:(HiBit+1)/2];   // A parameter value is a constant.
LoPart = temp[((HiBit+1)/2)-1:0];
HiBit  = temp[BitNo];   // Bit-select by variable is allowed.
...
```

6.2 A Simple RAM Model

All that is necessary for a minimal RAM is a verilog memory for storage, an address, and control over read and write. For example,

```
module RAM (output[7:0] Obus
          , input[7:0] Ibus
          , input[3:0] Adr, input Clk, Read
          );
reg[7:0] Storage[15:0];
reg[7:0] ObusReg;
//
assign #1 Obus = ObusReg;
//
always@(posedge Clk)
   if (Read==1'b0)
       Storage[Adr] <= Ibus;
   else ObusReg    <= Storage[Adr];
endmodule
```

Parity checking can be added this way:

```
module RAM (output[7:0] Obus, output ParityErr
          , input[7:0] Ibus
          , input[3:0] Adr, input Clk, Read
          );
reg[8:0] Storage[15:0]; // MSB is parity bit.
reg[7:0] ObusReg;       // Parity is not used for data.
//
assign #1 Obus = ObusReg;
assign ParityErr= (Read==1'b1)? (^Storage[Adr]): 1'b0;
//
always@(posedge Clk)
  if (Read==1'b0)
      Storage[Adr] <= {^Ibus, Ibus}; // Create & store parity bit.
  else ObusReg <= Storage[Adr];      // Discard old parity bit.
endmodule
```

6.2.1 Verilog Concatenation

At this point, it may be useful to introduce one verilog construct we have not yet discussed but have used in the above parity example; it is the *concatenation* operator. The concatenation operator, "{ ... }", may be used to append one or more bits onto an existing vector. Here, the "..." represents any value(s) being concatenated; the operator may be used in lieu of declaring explicitly a wider vector and assigning to it by part select or bit select.

All that concatenation does is save the effort of making declarations of temporary data. For permanent storage, the concatenated result always has to be assigned to a wide-enough vector somewhere.

As another example, to concatenate a parity bit to the MSB end of a 9-bit data storage location,

```
reg[7:0] DataByte;        // The 8 bit datum, without parity.
reg[8:0] StoredDataByte;  // High bit will be 9th (parity) bit.
...
StoredDataByte <= {^DataByte, DataByte};  // A 9-bit expression.
```

Likewise, two bytes stored in variables could be concatenated by an assignment such as `Word <= {HiByte, LoByte};`.

6.3 Memory Data Integrity

This topic is a huge and complex one and is an active research subject. We shall code no more than parity checking in this course, but we shall introduce the principles behind error-checking and correction (ECC).

The integrity problem is that hardware can fail at random because of intrinsic defects, or because of outside influences such as communication RF or nuclear radiation. We cannot address these failures in terms of verilog, but some background may be found in the supplementary readings referenced at the beginning of this book.

Error checking usually is done by computing parity for each storage location, by calculating various checksums, or by more elaborate schemes involving encoded parameters sensitive to the specific values of the data bits. If a bit changes from the value originally stored for it, a check is supposed to detect this and warn the user or initiate a correction process. The basic principle is to store the check value somehow so that hardware failures will change only the check or the data, but not both in a way which would mask the failure.

Parity checking is commonplace for RAM: The number of binary '1' values (or, maybe '0' values) is counted at each address, and an extra bit is allocated to each address which is set either to '1' or '0' depending on the count.

The parity bit makes the total number of '1' (or '0') values always an even number (for "even parity") or always odd ("odd parity"). It usually doesn't matter whether even or odd is used, or whether '1' or '0' is counted, so we shall from here on speak only of <u>even parity on '1'</u>. In this scheme, an even number of 1's makes the sum, the parity bit value, even—in other words, a 0; the sum of all nine bits will be 0. If a certain byte of data contains an odd number of 1's, then the parity value stored is a 1; the sum of all nine bits will be of an even number of 1's and therefore again will equal 0. So, a byte now takes 9 bits (1 parity + 8 data) of storage, but any change in a bit always is detected. Changes in two bits in the same 9-bit byte will be missed.

A sum is just an xor if we ignore the carry, so parity may be computed by the verilog *xor* (^) operator. For example,

```
reg[HiBit:0] DataVector;     // One bit narrower than DataWithParity.
reg[HiBit+1:0] DataWithParity;
...
// Compute and store parity value:
DataWithParity = {^DataVector, DataVector};
// Check parity on read:
DataVector = (^DataWithParity==1'b0) // '0' means good parity.
           ? DataWithParity // DataVector discards MSB parity bit.
           : 'bx; // Assign all 'x' on parity error.
```

Parity checking is adopted because it is fast; it incurs no speed cost when done in hardware; and, it is easy, because a simple xor reduction (verilog ^) of any set of bits yields the binary sum automatically.

Computing parity on every access to any memory location thus is modelled easily and efficiently.

Checksums usually are used with large data objects such as framed serial data or files stored on disc. The checksum often is just the sum of all '1' values (or sometimes bytes) in that object; but, unlike parity, it may be a full sum, not just a binary bit. Any change in the data which changes the sum indicates an error. If bits are summed, a change from ASCII 'a' to 'b' in a word will flag an error; if bytes are summed, a missing or additional ASCII character in a file will flag an error.

Unlike parity values, checksums generally are stored in a conceptually separate location from the data which they check. They may be used for simple error detection, as just described, or for Error Checking and Correcting (ECC) code.

A check<u>sum</u> may be calculated any of a variety of ways:
- As a sum of bytes, frames, or packets.
- As a sum of bits.
- As a sum of '1' or '0' bits (= parity, if no carry).
- As an encoded sum of some kind. For example, as a CRC (Cyclic Redundancy Check). A Linear Feedback Shift Register (LFSR) is one way to implement a CRC in hardware.

To reduce the likelihood of missing a change in, say, two bits or bytes between checks, elaborate partial encodings of stored data are used. For example, for serially transferred data objects, a linear feedback shift register (LFSR) can be used to compute a checksum-like representative number for each such object. The LFSR computes a running *xor* on the current contents of certain of its bits with a value fed back from a subsequent bit. See this idea in figure 6-1.

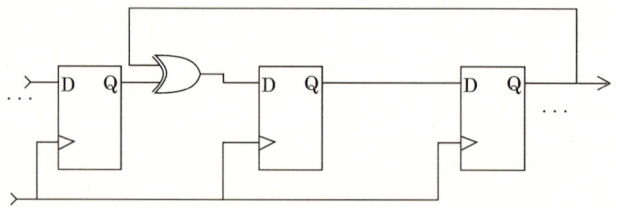

Fig. 6-1. Three stages of a generic LFSR, showing xor of fed-back data.

Every time the object is accessed, the value is shifted through this register, and the result may be compared against a saved value. The shift incurs a latency but no other delay.

Thomas and Moorby discuss CRC rationale in greater detail. An example they give is represented schematically in figure 6-2.

Digital VLSI Design with Verilog

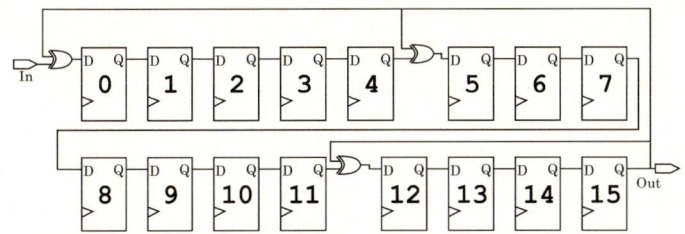

Fig. 6-2. LFSR characteristic polynomial in hardware. The value of Q[15:0] represents the CRC defined in Thomas and Moorby (2002) section 11.2.5 as $x^{16} + x^{12} + x^5 + 1$. Modulo division of valid stored data by this polynomial will return 0. The common clock to all bits, and the output taps, are omitted for clarity.

6.3.1 Error Checking and Correcting (ECC)

ECC not only checks for errors, but it corrects them, within limits. All ECC processes can find and correct a single error in a data object, such as a memory storage location. Some can detect two or more errors and can fix more than one. All corrections are made by the hardware and generally incur little or no time cost. Almost all commercial computer RAM chips include builtin ECC functions. Hard discs and optical discs always include ECC.

Basically, the ECC idea depends on a checksum: A representation of the data is stored separate from it; the data are compared with this representation whenever the data are accessed; an error is localized to the bit which was modified by the error, and the data actually seen by the computer or other device are changed to the original, correct value.

Some of the *Additional Study* reading below will explain the details, but a brief, conceptual introduction will be presented now: We shall show how to do a simple ECC using parity:

6.3.2 ECC from parity

Consider a parity bit pT representing the total, 8 bits of data, and *suppose just one bit could go bad*. Assume a specific parity criterion, such as our even-1. If a parity check failed, all that could be done would be to recognize that this specific data object (8+1 bits) was bad; no correction could be made except perhaps to avoid using those data.

But, suppose two parity bits had been computed, one ($p1$) for the low nybble (bits 0–3) and the other (pT) for all 8 bits. Then, if pT failed a check, but $p1$ didn't, the apparatus could proceed on the assumption that the error was localized in bits 4–7 or in the parity bit. If both $p1$ and pT failed, an error must have occurred in bits 0–3, or $p1$, but not pT. If $p1$ failed but not pT, one could be sure at least two errors had occurred, which we have agreed not to consider in this example.

With three parity bits, $p1$ and pT as before, and a new pE calculated from all even bits (0, 2, 4, 6), one could narrow down the error to two of the four bits in one nybble. Finally, with fourth parity bit pL on the low half of each nybble, the erroneous bit could be identified unambiguously. The correction then would be just to invert the current value of that bit and go on using the resultant, corrected datum just as though there had been no error. Cost: 12 bits to represent 8 bits of data; in this simple case, a 50% overhead in size, but no cost in speed.

The process just described may be summarized this way for one byte (**p** means parity):

> Assume only one hardware failure per word, and `reg[7:0] Word`;
>
> 1. Define **pT** = `^Word[7:0]`;
>
>   **pT** toggles if any bit in `Word` changes; system can detect this.   8'b<u>xxxxxxxx</u>
>
> 2. Define low-nybble **pN** = `^Word[3:0]`;
>
>   **pN** and **pT** toggle if any bit in low nybble changes.
>
>   **pT** toggles if any bit in `Word[7:4]` changes.
>
>   Thus, system can determine which half of `Word` is reliable.   8'bxxxx<u>xxxx</u>
>
> 3. Define even **pE** = `^{Word[6],Word[4],Word[2],Word[0]}`;
>
>   System can determine whether odd or even bits,
>   of which half, are reliable.   8'bx<u>xxx</u>x<u>xxx</u>
>
> 4. Define low half-nybble **pL** = `^{Word[5:4], Word[1:0]}`;
>
>   Using **pT**, **pN**, **pE**, and **pL**, system can determine which bit changed and flip it back during a read.   8'bxxxxxx<u>xx</u> → 8'bxxxxxxx*x*
>
>   This ECC costs 4 extra bits per 8 data bits.

Realistic ECC. Usually, data objects larger than one byte are adopted for ECC; and, for them, the size overhead can be a smaller percentage of the data to be checked. Instead of a binary search in a parity tree to recover single errors, it is statistically more efficient to use a finite element approach in which parity is replaced by an overdetermining set of pattern coefficients. This is done almost always by using LFSR hardware and applying algebraic field theory to encode regularities of the stored data in the checksums. Multiple bit errors in a large block of data can be recovered somewhat independently of where the errors occur in the block, and the checksum overhead for practical ECC of a 512-byte block of data can be less than 64 bytes. This overhead is about equal to that of simple, byte-wise parity checking without correction!

To illustrate the mechanics of a realistic ECC process, suppose we sidestep the complications of group theory and adopt a minimally complex method, again based on parity: We shall take an 8-bit byte of data and append to it a checksum made up of a vector of 8 parity bits composed as follows:

The 8 data bits will be written MSB on the left. Concatenated to the right of the data LSB will be a 8-bit checksum, making a total of 16 bits. The leftmost checksum bit will be calculated as the parity (even parity on '1') of the whole data byte; the next checksum bit will be calculated as parity of the 7 bits of data with the MSB excluded. The next checksum bit will be parity of the least significant 6 data bits, and so on down to the 8th checksum bit, which will be equal to the LSB of the data.

A simple hardware implementation of this method could be a LFSR with one storage element fed back on itself as shown in figure 6-3.

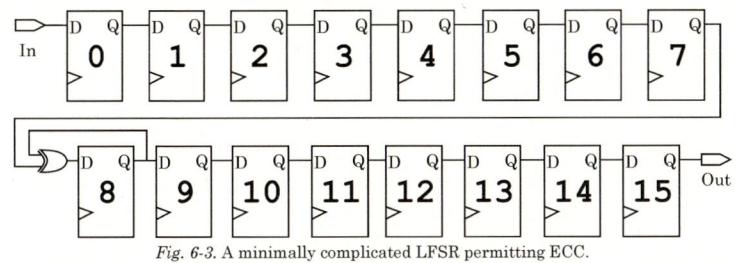

Fig. 6-3. A minimally complicated LFSR permitting ECC.

To operate this LFSR, one initializes it with `16'b0` and shifts in the data LSB first, twice (16 shifts). The result will be the desired pattern of `xor`'s in the rightmost 8 bits, and a copy of the data in the leftmost 8 bits. The result could be transmitted serially with just a latency penalty; it also could be offloaded onto a parallel bus for direct memory storage.

To see how this ECC might work, suppose the data byte was `1010_1011`; checksummed, this byte would become, `1010_1011__1001_1001`.

Now, suppose that a 1-bit error occurred during serial transmission; for example, suppose the data LSB flipped, making the received word, `1010_1010_1001_1001`.

The Rx would calculate the checksum of the received word to be `0110_0110`, clearly a gross mismatch to the received checksum. It would be unreasonable to consider the possibility that the checksum could contain an error of so many bits, making it so distant from the one calculated from the received data. Avoiding a closed-form solution in this example, the Rx could formulate 8 hypotheses to correct the data by calculating all possible checksums with 1 data bit flipped in each:

The 8 possible 1-bit corrections to a received word of 1010_101**0**_1001_1001:

Hypothesis	Corrected data	Computed Checksum
h0	0010_1010	1110_0110
h1	1110_1010	1010_0110
h2	1000_1010	1000_0110
h3	1011_1010	1001_0110
h4	1010_0010	1001_1110
h5	1010_1110	1001_1010
h6	1010_1000	1001_1000
h7	1010_1011	1001_1001

Hypothesis h7 generates the received checksum; so, our ECC should flip the data LSB to correct it.

Now let us assume two errors, say the data MSB and the data LSB. Again, we do not allow that the received checksum could contain so many errors.

The 8 possible 1-bit corrections to a received word of **0**010_101**0**_1001_1001:

Hypothesis	Corrected data	Computed Checksum
h0'	1010_1010	0110_0110
h1'	0110_1010	0010_0110
h2'	0000_1010	0000_0110
h3'	0011_1010	0001_0110
h4'	0010_0010	0001_1110
h5'	0010_1110	0001_1010
h6'	0010_1000	0001_1000
h7'	0010_1011	0001_1001

In this case, no 1-bit correction to the data yields the received checksum; however, h7' yields a checksum very close (only 1 bit away). It would be reasonable to accept h7', flip the LSB, and then try 8 more hypotheses for a second correction; this would result in a 2-bit ECC which would correct both data errors. In actual practice, the distances are quantified and minimized in closed form in the algebra of Galois fields, but this simple example shows the basic properties of a checksum permitting multibit ECC.

For more information on the algorithms and computational details of ECC checksum encoding, see the Cipra and the Wallace articles in the references at the beginning of this book. The Plank article gives an example of how to generate Reed-Solomon ECC for

Digital VLSI Design with Verilog

multiple-error recovery. While reading this last reference, keep in mind that matrix multiplication by a vector makes each term in the product vector dependent on all terms in one matrix row and on every term in the multiplier vector. Also, note that the errors being corrected in the Planck paper always have been errors localized exactly in the system (e. g., by identified crash of a specific hard disc) before the ECC is invoked.

6.3.3 Parity for SerDes Frame Boundaries

A simple parity value might be used to improve greatly the efficiency of our planned serdes serial data framing. However, we shall not use it in this course. We are interested in design in verilog, and our inefficient but obvious 64-bit packet makes it both easy and instructive to recognize verilog design errors during simulation. We do not wish to obscure a possible design error to ensure hardware we never intend to manufacture.

However, let's ignore our own project once more, for the moment. Consider the following way of determining synchronization of a local PLL clock with the embedded clock in the serial data. Instead of padding the data with 8 bits of encoded order information per byte, as we shall do in our project, suppose we added just a parity bit to each datum, extending it to 9 bits per byte. Then, a packet of 32 bits of our serialized data will look something like this:

```
36'bXxxxxxxxPXxxxxxxxPXxxxxxxxPXxxxxxxxP
```

The parity for each byte follows that byte, in the sense that we are assuming that the MSB is sent first over the serial line. Each byte's MSB is represented by an upper-case X, and the parity by P. Compare this with the Step 8 representation in our previous lab. With underscores to emphasize byte boundaries, we may write,

```
36'bXxxxxxxxP_XxxxxxxxP_XxxxxxxxP_XxxxxxxxP
```

Now, suppose we wish to synchronize a PLL clock with a stream of such frames. If we know we are on a byte boundary and are worried about a 1-bit jitter, we can calculate parity; and, if we should shift by a bit, the parity *might* change, and *maybe* we could adjust our PLL to resynchronize. If we are using even-1 parity, a '1' in the wrong frame will trigger a parity error, but a '0' won't. Detection of a framing error then would be about 50% (1/2) accurate per word. With four words, this means that for all four to be undetected, the probability would be about $2/(2^4)$ or 1/8, which seems good enough for most purposes.

Incidentally, the ratio of 10 bits per byte is the assumption usually made in actual PCI Express designs, a numerical coincidence because we have ignored many serialization complexities, such as Manchester encoding.

6.4 Memory Lab 7

Topic and Location: Single-port RAM and a simple way to make memory I/O ports bidirectional.

Preview: We exercise various kinds of verilog array assignment. Then, we use an array to design a single-port static RAM which is 32 bits wide and 32 addresses deep, with parity. We add a `for` loop in our testbench to check the addressing and data storage of this model; we use the simulator memory or `$display()` to look inside the memory. Finally, we modify the RAM from two, unidirectional input and output data busses to a single, bidirectional input-output data bus. We use the concept of a "wrapper" module to combine the original unidirectional data I/O's. Finally, we synthesize the bidirectional, single-port RAM.

Deliverables: Step 1: A verilog `module` in a file with the specified declarations and assignments, simulated. Step 2: A verilog corner-case-correct single-port RAM simulation model with parity and an assertion which reports any parity error. Step 3: Simulation of the RAM model. Step 4: A testbench including a `for` loop with exhaustive testing and display of data storage by the RAM model. Step 5: A modified, correctly-simulating RAM model with one, bidirectional data port. Step 6: Two synthesized verilog netlists of the bidirectional RAM model, one optimized for area and the other optimized for speed.

Lab Procedure:

Work in your `Lab07` directory.

Step 1: Try the following memory access statements. Initialize the RHS variables with literal constants, and then simulate to see which ones work:

```
reg[63:0] WordReg;
reg[07:0] ByteReg;
reg[15:0] DByteReg;
reg[63:0] BigMem[255:0];
reg[3:0]  LilMem[255:0];
...
BigMem[31]       <= WordReg;
WordReg          <= BigMem;
LilMem[127:126]  <= ByteReg;
LilMem           <= ByteReg[3:0];
DByteReg         <= ByteReg;
ByteReg          <= DByteReg + BigMem[31];
WordReg[12:0]    <= BigMem[12:0][0];
```

Step 2: Design a verilog 1k x 32 static RAM model (32 x 32 bits) with parity. Call the module, `"Mem1kx32"`. Check this model by occasional simulation as you do your work on it.

This RAM will require a 5-bit <u>address bus</u> input; however, use verilog `parameters` for address size and total addressable storage, so that quickly, by changing one parameter value, you could modify your design to have a working model with more or fewer words of storage. However, keep in mind that parity bits are not addressable and are not visible outside the chip.

You may model your RAM after the Simple RAM model given the preceding presentation. Use just one `always` block for read and write; but, of course, the module will have to be considerably more complicated than the Simple RAM.

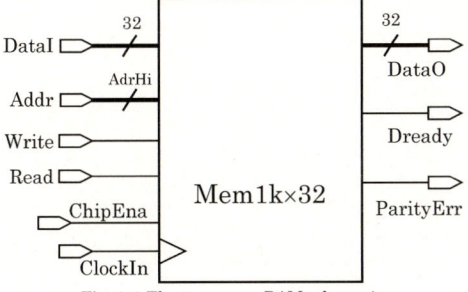

Fig. 6-4. The `Mem1kx32` RAM schematic.

Use two 32-bit data ports, one for read and the other for write., but their names should not be "read" and "write" ("DataO" and "DataI" are used in figure 6-4). Supply a clock; also an asynchronous chip enable which causes all data outputs (read port) to go to 'z' when it is not asserted (= 0), but which has no effect on stored data. The clock should have no effect while chip enable is not asserted; assertion should be when the chip enable is "1".

Supply two direction-control inputs, one for read and the other for write; these names may match their functions, as shown in figure 6-4. Changes on read or write have no effect until a positive edge of the clock occurs. If neither read nor write is asserted, the previously read values continue to drive the read port; if both are asserted, a read takes place but data may not be valid.

Assign reasonable time delays, using only delayed continuous assignments to your `module` outputs. Supply a data ready output pin to be used by external devices requiring assurance that a read is taking place, and that data which is read out is stable and valid. Don't worry about the case in which a read is asserted continuously and the address changes about the same time as the clock: Assume that a system using your RAM will supply address changes consistent with its specifications.

Also supply a parity error output which goes high when a parity error has been detected during a read and remains high until an input address is read again. Refer to the block diagram in figure 6-4.

Because this is a static RAM, of course you should omit DRAM features such as ras, cas and refresh. Design for flip-flops and not latches. Use a file named after the `module` to do all your work.

Finally, include an <u>assertion</u> to announce parity violations to the simulator screen. Of course, your simulation model can't possibly experience a hardware failure, but this message may tell you if you make a design error with the parity bit. You can force an error by putting a temporary blocking assignment in your model to confuse the *xor* producing the parity value.

Step 3: Check your RAM. Write data to an address and simulate to verify that it is read correctly (see figure 6-5).

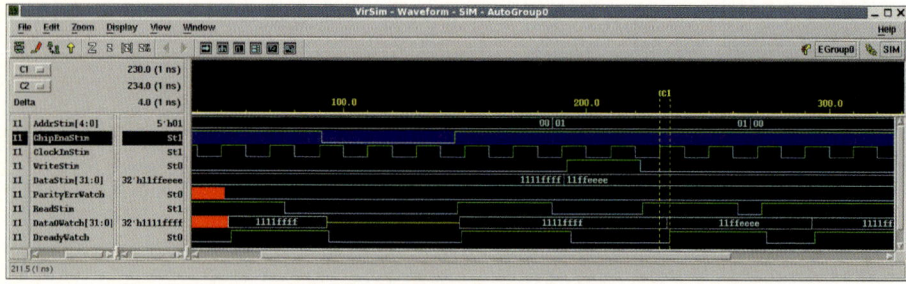

Fig. 6-5. Cursory simulation of single-port `Mem1kx32` with separate read and write ports.

Step 4: After completing the previous step and doing any simulation necessary to verify your design superficially, add a `for` loop in a `initial` block in your testbench file to write a data pattern (*e. g.*, an up-count by 3) into the memory at every address and then to display the stored value. Use `$display()` for your display. Pay special attention to the "corner cases" at address 0 and address 31. Your parity bit should be in bit 32 at each address. An example of such a loop is given just below. Notice how to address a data object ("MemStorage") in an instance, here named `Mem1kx32_inst`, in the current testbench module:

```
for (...)
  begin
   ...
   #1   DbusIn = (some data depending on loop);
   #1   Write  = 1'b1;
   #10  Write  = 1'b0;
      SomeReg = Mem1kx32_inst.MemStorage[j];
      $display("...", $time, addr, SomeReg[31:0], SomeReg[32]);
   end
```

Here, "`...`" of course represents a format specifier.

Step 5: Modify your RAM design so it has just one bidirectional data port, Do this by copying your working model (above) into a new file named "`Mem1kx32Bidir.v`". Then, declare a new, empty module in this file, named after the file. Use the exact same new

Digital VLSI Design with Verilog

module ports as in your old `Mem1kx32` model, except for only one `inout` data port. See figure 6-6.

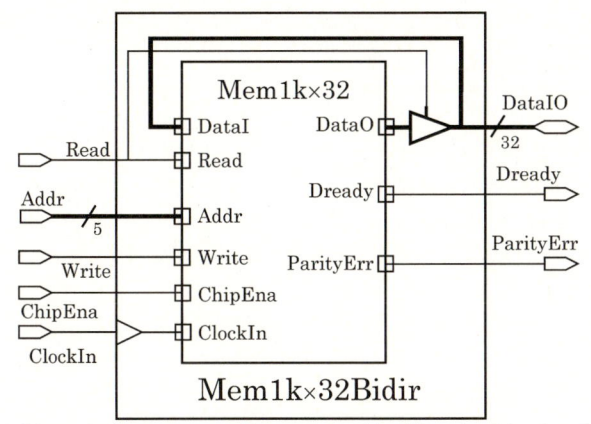

Fig. 6-6. Schematic of wrapper to provide `Mem1kx32` with a bidirectional data bus. Compare with figure 6-4 above.

Instantiate your old RAM in the new `Mem1kx32Bidir`. Connect everything 1-to-1, but leave the data unconnected.

All you have to do now to complete the connection is to add a continuous assignment in the wrapper module which turns off the `DataO` driver when `Read` is not asserted. Also, wire `DataI` to the new `DataIO` driver as shown above. Verify your new RAM by a simulation that does a write and then a read from two successive addresses, then reads again from the first address (see figure 6-7).

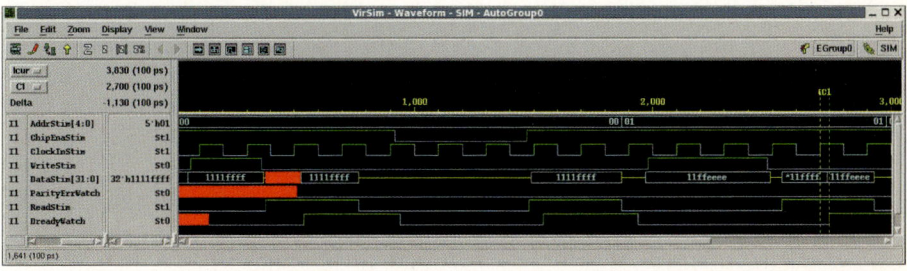

Fig. 6-7. Cursory simulation of single-port `Mem1kx32Bidir` with bidirectional read-write port.

If this were part of a much larger design project, you would separate the bidirectional and original RAM `modules` into different files. However, this exercise is simpler if you allow yourself to keep as many as three modules in one file: The original RAM model, the new bidirectional "wrapper" for that RAM model, and the testbench (or a referent to it).

Step 6: Synthesize your bidirectional-data memory design, optimize for area, and examine the resulting netlist in a text editor. Resynthesize for speed and examine the netlist again.

6.4.1 Lab Postmortem

Concatenation: When can it be useful?

How are hierarchical references made to `module` instances?

What's the benefit of a bidirectional data bus?

How might the spec for the `Dready` flag be improved? What about when a read remains asserted while the address changes? Shouldn't the RAM be responsible for all periods during which its output can't be predicted?

How would one change the RAM flip-flops to latches?

Do we really need a `ChipEna`? Why not disable outputs except when a read was asserted?

6.5 Additional Study

Read Thomas and Moorby (2002), section 5.1, on verilog rules for connection to ports.

Read Thomas and Moorby section 6.2.4, pp. 166 ff. to see how our parity approach can be adapted easily to a Hamming Code ECC format (at 12 bits per byte).

Read Thomas and Moorby appendix E.1 and E.2 on vectors, arrays, and multidimensional arrays.

Optional readings in Palnitkar (2003):

Section 4.2.3 discusses the verilog rules for connection to ports.

Look through the verilog of the behavioral DRAM memory model in Appendix F (pp. 434 ff.). It uses several verilog language features we haven't yet mentioned, so you may wish to put it aside for a few weeks. It may not work with the Silos simulator on the CD.

Chapter 7
Week 3 Class 2

7 Today's Agenda:

Lecture on counters and the verilog for them
    Topics: Kinds of counter by sequence. Counter structures discussed: ripple, synchronous, one-hot, ring, gray code.
    Summary: After defining what we mean by a "counter", we examine typical terminology used to describe HDL modelling in general. Then, we examine several kinds of counter typically used in hardware design: Ripple, synchronous (carry look-ahead), one-hot, ring, and gray code.

Counter Lab
    We write, simulate, and synthesize several different kinds of counter, showing how the rarely-used `wor` net may be applied in one of them. We finish by using our old PLL clock output to drive three of them in the same design.

7.1 Counter Types and Structures

7.1.1 Introduction to Counters

A *counter* is a data object which represents a value that is incremented or decremented in uniform steps. So, a counter may contain successive values of 0, 1, 2 ...; or, 2, 4, 6, ..., etc. A 1-bit counter alternates between '0' and '1'. A n-bit register for a binary, unsigned up-counter would count as shown in figure 7-1, in successive count values starting from 0 on the top row (first register state):

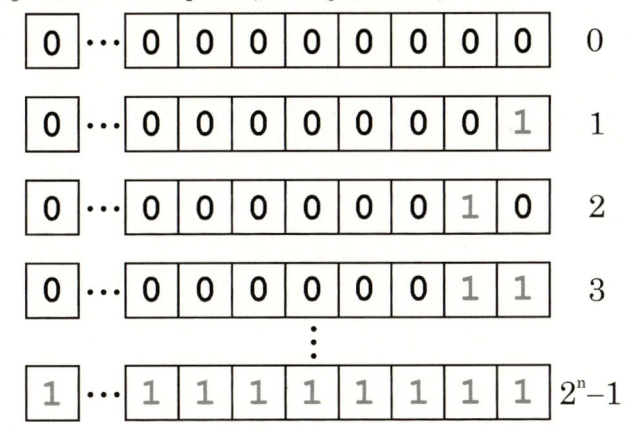

Fig. 7-1. Binary up-count register with MSB on left and successive count values. Between 1 and n bits switch on each count.

J.M. Williams, *Digital VLSI Design with Verilog: A Textbook from Silicon Valley Polytechnic Institute*,
DOI 10.1007/978-3-319-04789-8_7, © Springer International Publishing Switzerland 2014

The above storage of the count value is very compact spatially, n bits for 2^n different values, but every carry to a new bit causes simultaneous switching of several, often very many, bits at a time. This carrying and bit-switching can cause settling delays to vary from clock to clock; all the switching may generate on-chip cross-talk noise or power supply glitches, causing errors in the counter or in nearby devices.

A binary up-counter can be programmed to count just ten, instead of 16, values in at least two different ways:

1. Program the wrap-around. This is the most common way and gives an ascending count over a numerical set of values beginning with 0 and typically ending at 9 or 10.

For example, to count 0..9:

```
reg[3:0] Count;
always@(posedge Clk, posedge Rst)
if (Rst == 1'b1)
    Count <= 4'h0;
else begin
    if (Count < 9)
        Count <= Count + 1;
    else Count <= 4'h0;
    end
```

2. Preload the counter to bias the wrap-around. This gives a count over any useful number, usually ten, of contiguous, ascending values.

For example, for the ten highest values ending at 15,

```
reg[3:0] Count;
always@(posedge Clk, posedge Rst)
if (Rst == 1'b1 || Count == 4'hf)
    Count <= 4'h6;
else Count <= Count + 1;
```

Values of, say, 1, 2, 4, 8, 16, ... don't represent a count as such. However, if the value of a counter is interpreted in terms of bit position, a register content of 2^0, 2^1, 2^2, 2^3, ... 2^n, may be considered a count, in that the position of the '1' in a register holding those successive values increments in uniform steps, by one location at a time. Essentially, the $\log_2$ of the value in the register is being counted. An example of this kind of counter is the **one-hot** counter, because exactly one bit position always is "hot" with a '1'.

The hardware implementation of a one-hot counter is just a shift register which is initialized with a single '1' in it. A one-hot count would look like that in figure 7-2, in terms of successive bit-patterns in the counter register:

Digital VLSI Design with Verilog   135

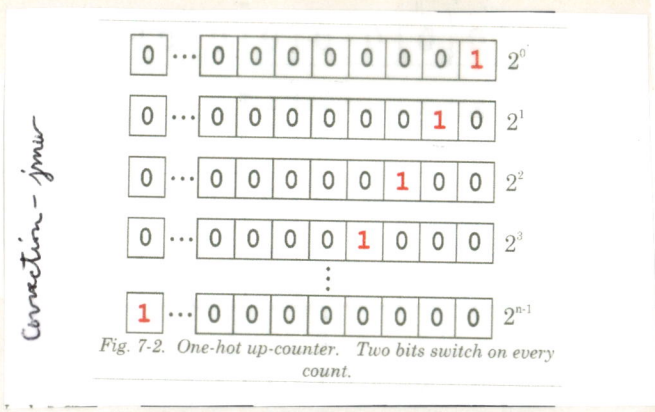

Fig. 7-2. One-hot up-count. Two bits switch on every count.

Notice that there can't be any value of 0 represented this way, if the bits in the register are to be interpreted as ordinary, verilog binary (2^n) numbers. Also, the one-hot storage capacity is very limited, requiring n bits for just n different values. This kind of count is spatially inefficient, but it is very fast and causes little (but not minimal) noise, because just two bits switch per clock. We trade storage space for time, in a sense.

7.1.2 Terminology: Behavioral, Procedural, RTL, Structural

There are at least three different counter approaches possible in verilog: behavioral, procedural, and structural. Sometimes the word, *RTL*, "Register Transfer Level", is used to describe design activity; RTL overlaps behavioral and procedural.

Behaviorally, the verilog language permits elementary operators to be used for counting, with no concern for the hardware that might result. A simple count may be written as, `Count <= Count + 1`, `Count <= Count - 1`, `Count <= Count + Incr`, etc. Behaviorally, we leave it up to the synthesizer to decide how to put these statements into our netlist. Notice that in the preceding statements, there is no way of knowing which `Count` bit might change, or which way, because of the statement alone.

Procedural assignments are distinguished by the possibility that a statement might be superceded by another one in simulation time, with no implication that multiple drivers were involved. Procedural assignments only can occur in procedural blocks. The possibility of changing the assigned value is most obvious with blocking assignments.

For example,

```
initial // Cycle the reset by time 10:
  begin
  #0   Reset = 1'b0;
  #1   Reset = 1'b1;
  #9   Reset = 1'b0;
  end
```

Behavioral statements may include procedural or continuous assignments. Procedural statements may be bit-specific and thus not behavioral.

We may assign values to counter registers in procedural statements, but the level of detail may be anywhere from that of the whole counter register, as in the behavioral example above, down to that of individual parts or bits. Composing a counter by `xor` expressions and bit-assignments could be procedural, but not behavioral. For example, in an `always` block, `Count[2] <= Count[1]^Carry` would be an example of a bit-level procedural statement. However, a continuous assignment, `assign Count[2] = Count[1]^Carry`, would make the statement nonprocedural. In this context, counters are sequential devices; and, whereas a continuous assignment statement might represent an addition, it can not represent a count.

Often the same code may be described as **RTL** (usually a special case of procedural) or **behavioral**, depending on context. Recall the simple counter for our PLL in *Week 2 Class 2*: That was an RTL counter not quite behavioral because we wrote a bit-width expression for the incrementing literal, `5'h1`.

> The present author feels justified in using the term **behavioral** when the code does not explicitly assign identifiable bit values to a port or register; thus, a structural interpretation of the statement is not specified.

This is consistent with usage in Thomas and Moorby (2002) section 1.2 and Palnitkar (2003) chapter 7. Very little in this Textbook is purely behavioral, in that it could not be called RTL.

Structurally, verilog allows us complete control over a netlist if we wish to instantiate the gates by hand from a library compatible with our intended technology. In this context, continuous assignment statements are structural connections representing simple wiring or combinational logic. Palnitkar (2003) distinguishes **dataflow** modelling as a special style including continuous assignments; however, this terminology is nonstandard when not applied to software directly defining hardware. One sure thing is that continuous assignments are not procedural (they can't be included in a procedural block in modern verilog).

Here is another example to clarify the terminology. Suppose the goal is to represent a 2-bit, binary up-count. After a few declarations, we may write verilog as shown next:

Digital VLSI Design with Verilog

behavioral or RTL:	`Count <= Count + 1;` `// Count is reg.`
RTL:	`Count[1:0] <= Count[1:0] + 2'b01;` `// Count is reg.`

RTL:	`always@(posedge Clock) Count[0] <= ~Count[0];`
	`always@(Count[0])`
	`    if (Count[0]==1'b0) Count[1] <= ~Count[1];`

structural:	`// Count is net:`
	`DFF Bit0( .Q(Count[0]), .Qn(Wire0), .D(Wire0), .Clk(Clock) );`
	`DFF Bit1( .Q(Count[1]), .Qn(Wire1), .D(Wire1), .Clk(Wire0) );`

Gate-level structural design entry should be avoided when using a synthesizer. Sometimes, when tuning a synthesized netlist, structural control may be necessary; but, for complex structures, the result may not be optimal. Furthermore, the synthesizer's logic optimizer may not be able to improve suboptimal structures which have been built by hand. Finally, gates instantiated from a specific library make the design technology-specific; porting the design to a new technology will be more complicated with hand-instantiation than when the synthesizer is allowed to choose all gates from the new technology library.

7.1.3 Adder Expression vs. Counter Statement

It is important to point out the difference here between an *adder* and a *counter*. An adder performs an expression evaluation. For example, consider the statement `X <= A + B`. Even if included in a function call, the adder is contained entirely on the right side of this statement. A counter makes a statement of the current value in a register. Thus, addition can appear in a combinational or a sequential procedural statement; counting can occur only in a sequential procedural statement. Although a count increment expression can be implemented as an addition of 1, a count can not be kept in a combinational form; it has to be stored so that the sequence of different values, *count* vs. *next-count*, can be distinguished in order of occurrence: The next *count* must depend on the previous value of *count*.

Counting requires assignment of a sum expression in a statement, which is shown schematically in figure 7-3.

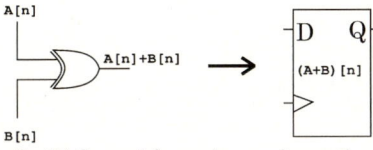

Fig. 7-3. Sequential summing permits counting.

7.1.4 Counter Structures

Let us discuss a few examples of different counter structures. In all the counters to be presented below, assume the count is the binary value taken from the Q ports of the flip-flops shown. We shall use only D flip-flops here, because they are the most common kind in current use, either in manual design entry or in synthesized logic.

The element of any binary counter based on D flip-flops is the **toggle flip-flop**, sometimes called a T flip-flop, which is just a D flip-flop with its ~Q port wired back so that it drives its D input.

For D flip-flops modelled behaviorally, toggle assignments from D to Q should be nonblocking, not blocking, to ensure that updated values are based solely on assignments from the previous clock.

```
//
// Basic toggle flip-flop:
//
wire Qn_D, Qout;
...
DFF Toggle01( .Q(Qout), .Qn(Qn_D), .D(Qn_D), .Clk(ClkIn) );
...
```

The schematic for this device hookup would be as shown in figure 7-4:

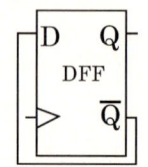

Fig. 7-4. Basic toggle (T) flip-flop.

A toggle flip-flop component may be constructed by putting a simple wrapper module around a DFF which itself is wired to toggle:

```
module TFF (output Q, input Clock, Reset);
  // DFFC is the DFF above, unconnected, with clear (reset) added.
  wire Qn_D, ToQ;
  assign Q = ToQ;
  DFFC Toggler( .Q(ToQ), .Qn(Qn_D), .D(Qn_D), .Clk(Clock), .Rst(Reset) );
endmodule
```

7.1.5 Ripple Counter

This kind of counter is minimal in size and easy to implement structurally. Its bits are flip-flops which are not clocked; instead, each one is triggered by a data edge (a carry out) of its immediately preceding stage. The flip-flops are wired to toggle on every active edge on their individual clock-pin inputs. The count must be reset to a known value; but, otherwise, no logic is required other than that of the flip-flops themselves. For example, a 3-bit ripple counter could be as simple as in figure 7-5.

Digital VLSI Design with Verilog

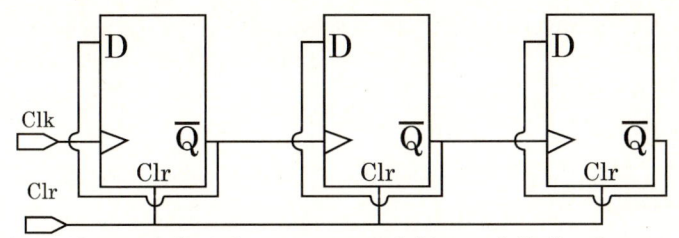

Fig. 7-5. Schematic of 3-bit ripple counter constructed of three toggle flip-flops. Outputs are not shown. MSB is farthest right. Only the LSB is clocked.

Because the higher-significance stages in a ripple counter see no clock and can not change state until all earlier bits have changed state, this is a linearly progressively slower device as the count register widens. A variety of glitch values will appear in the connecting-bit patterns during the time after each clock, while the ripples triggered by the clock are settling down. However, carry from all stages does not propagate immediately on the clock edge, so the instantaneous power consumption and noise are reduced when compared with that of a synchronous counter of the same width and clock speed. Of course, the total switching power also would be less for the ripple counter.

7.1.6 Carry Look-Ahead (Synchronous) Counter

This kind of counter completes an update of all register bits on each clock input, at the cost of additional carry look-ahead logic. Stated otherwise, all bits always are clocked and would toggle together, all at once, on every clock, except that the toggle for each bit is disabled until all lower-order bits are '1'. Thus, although every bit is clocked, no bit toggles to '1' until it becomes the carry from the previous bits.

A synchronous counter can operate at almost the same clock speed regardless of its number of bits. However, because all bit switching is synchronous with the clock, a synchronous counter occasionally briefly will consume power in big spurts, and such a spurt will create noise which might cause logic errors, as mentioned above.

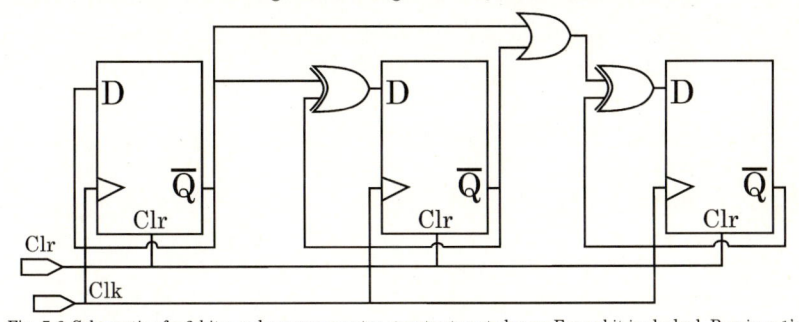

Fig. 7-6. Schematic of a 3-bit synchronous counter. Q outputs not shown. Every bit is clocked. Previous 1's (inverted) are *or*'ed, and an *xor* prevents each toggle until carry overflows into that bit.

Notice in the synchronous counter of figure 7-6 that each storage element (flip-flop) is driven by a 1-bit adder (*xor* gate). Thus, in a general sense, the flip-flops merely latch, on each clock, the sum composed by the connecting combinational logic.

7.1.7 One-Hot and Ring Counters

Unlike the ripple or synchronous counters, these counters do not use all possible bit states of a register.

The one-hot counter was described above (see figure 7-2).

A ring counter is just a one-hot counter in which the MSB is shifted back into the LSB. In a ring counter, a single '1', or other constant pattern established by reset logic or a parallel load, circulates through the register endlessly. Ring counters may be used to generate clock pulse patterns of arbitrary duty cycle. When used to create a clock, a ring counter in a sense is the inverse of a PLL; it creates a clock of lower frequency than its input, rather than a clock of equal or higher frequency.

7.1.8 Gray Code Counter

Of all common counter hardware structures, the gray code is maximally efficient in terms of switching power and noise. It is named after a designer whose name was F. Gray.

In gray-code counting, as in the simple binary up-counter, all possible register patterns are used; but, successive count patterns are such that just one register bit changes on any clock, as shown in figure 7-7. This efficiency is at the cost of some extra encoding logic which is necessary to sequence the register patterns properly. Also, determining the equivalent 2^n binary bit count value from the gray code value requires some decoding logic.

Digital VLSI Design with Verilog

0	0	⋯	0	0	0	0	0	0	0	0

0

| 0 | 0 | ⋯ | 0 | 0 | 0 | 0 | 0 | 0 | 1 |

1

| 0 | 0 | ⋯ | 0 | 0 | 0 | 0 | 0 | 1 | 1 |

2

| 0 | 0 | ⋯ | 0 | 0 | 0 | 0 | 0 | 1 | 0 |

3

| 0 | 0 | ⋯ | 0 | 0 | 0 | 0 | 1 | 1 | 0 |

4

⋮

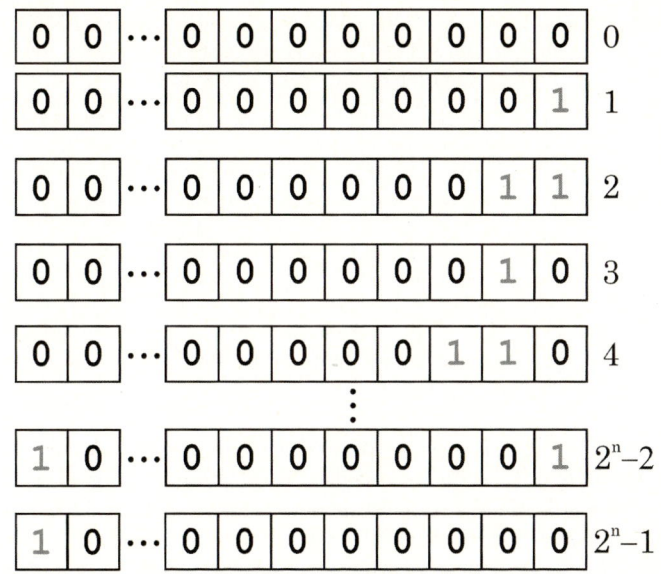

| 1 | 0 | ⋯ | 0 | 0 | 0 | 0 | 0 | 0 | 1 |

2^n-2

| 1 | 0 | ⋯ | 0 | 0 | 0 | 0 | 0 | 0 | 0 |

2^n-1

Fig. 7-7. Unsigned gray-code up-count. MSB is on the left. With the bottom pattern (2^n-1), one more count restores the 0 pattern.

We shall not be using gray code in this course, but it is a good pattern to keep in mind when minimal switching power or excessive noise is a concern.

Another application of gray code is in synchronizing the value of a count across different clock domains; for example, this would be required for generating consistent, successive addresses or state-machine state encodings. As explained in the Cummings papers listed in the References, because only one bit changes at a time, the probability of invalidly sampling an intermediate counter state (by logic in a domain other than that of the counter) is lower for gray encoding than for any other kind of count commonly used.

7.2 Counter Lab 8

Topic and Location: We write verilog counter models, simulate them, and synthesize them to compare performance of the synthesized netlists. We also create a new design in which our old PLL is used to clock three different counter structures.

Work in the Lab08 directory for this lab.

Preview: We start with verilog for a ripple counter, using flip-flop models from previous labs. We synthesize a netlist from the ripple-counter model and compare its simulation speed with a netlist synthesized from a synchronous-counter model. We also write a behavioral verilog model of a counter and simulate and synthesize it. After this, we replace or expressions in the synchronous-counter model with a wor net, to see how a wired `or` works. Finally, we create a new model by instantiating our old PLL and one each of the ripple, synchronous, and behavioral counters in a single, top-level module named `ClockedByPLL`; we end the lab by simulating this model.

Deliverables: Step 1: A correctly-simulating ripple counter and two synthesized netlists, one optimized for area and the other for speed. Also, a record of the fastest clock speed at which this verilog source model will count correctly. Step 2: The same as Step one, but for a synchronous counter. Also, a dual-netlist simulation, using speed-optimized synthesized verilog netlists for the ripple-counter of Step 1 and the synchronous counter of this Step. Step 3: A behavioral verilog counter, and netlists synthesized for area and speed. Also, a behavioral counter modified to count down and simulated correctly. Step 4: The synchronous counter model of Step 2 with `or` gates (or expressions) replaced by `wor` nets. Step 5: A correctly simulating model in which a PLL instance supplies the clock to drive instances of a ripple counter, a synchronous counter, and a behavioral counter.

Lab Procedure:

Note: For a counter clocked on the positive edge, a ***miscount*** in this lab is defined as the wrong number when the count is sampled on the opposite (negative) edge of the clock.

Step 1: Build a ripple counter. You should have a working positive-edge flip-flop (FF) from `Lab04`. If necessary, modify your FF model so it has two outputs, Q and Qn. The Qn logically will be `!Q` (or, `~Q`), obtained perhaps by clocking `Qn <= ~D` in the model. If not in your model already, assign a `#3` delay (3 ns) to changes in Q and Qn.

Then, create a verilog structural model of a 4-bit ripple counter, using your FF component and the design in figure 7-5 above. Name the counter module `Ripple4DFF`.

Simulate the model to determine how much simulation time it takes your model to count from 0 to `4'b1100` (hex `4'hc`) at its fastest possible speed. Do this by repeatedly simulating with shorter and shorter clock periods until there is a miscount. Record the time for later reference. Use a testbench in a separate file.

Note on testing counter speeds: Use `` `timescale 1ns/10ps`` and observe the VCS waveforms on each negative edge of the clock. You should be able to find a failing counter by noticing the kind of error shown in figure 7-8:

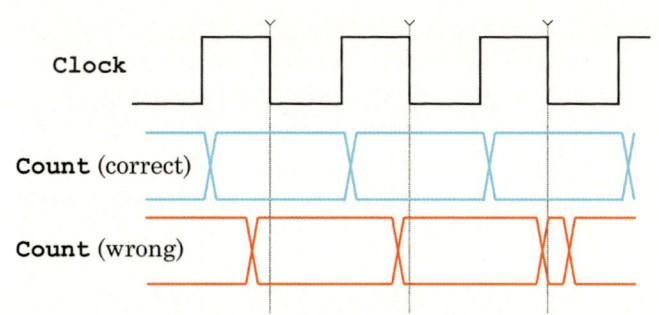

Fig. 7.8: Example failure of an overclocked counter.

> *Note*: The miscounts you find probably will depend only on the delays: Your flip-flops may toggle in simulation no matter what your clock speed. If this happens, it may be because the simulator only implements *inertial delays* for delayed statements.
>
> So, although real hardware would freeze up if clocked too fast, we may have to wait until we study path delays in `specify` blocks before your simulator will reveal this kind of problem.

Synthesize `Ripple4DFF` and try optimization for area and speed. Save the speed-optimized netlist for later in this lab.

Step 2: Build a synchronous counter. Enhance the design in figure 7-6 above to build a structural 4-bit synchronous counter model using the same FF module as in the previous Step. Name the synchronous counter module `Synch4DFF`. Use verilog continuous assignments to represent the connecting `or` and `xor` "gates", and impose a delay of `#2` for each such gate assignment.

Simulate `Synch4DFF` with the same testbench to compare it with the ripple counter in the previous Step; record the shortest possible simulation time which it takes to count from 0 to `4'b1100`, as before. Use basically the same testbench file as you did for the ripple counter.

Also, synthesize the counter for area and speed and save the speed-optimized netlist.

Then, as explained next, instantiate both `modules` for (*a*) the ripple-counter speed-optimized netlist and (*b*) the synchronous-counter netlist in a testbench module and see how fast you can run them in simulation. You will have to compile the verilog *LibraryName*_v2001.v file for this simulation, because you will require verilog models for the components in the synthesized netlists. If you use VCS, add `-v` before the library file name in your .vcs file.

You will have to use <u>different module names</u> for the DFF's in the synchronous vs. ripple counters, to get VCS to compile their synthesized netlists together in the second part of this Step. The reason is that if "DFFC" is used to synthesize both netlists, each netlist will have a declaration of a "DFFC"; and, the new VCS will reject the idea of multiple declarations of the same module, even if both are identical.

So, begin this Step by copying your standard DFFC.v to a new name, for example, DFF.v. In the new copy, change the module name to match that of the file, viz., to DFF.

Then, use DFF, not DFFC, in your synchronous counter. If you do this, after synthesis the ripple counter netlist will contain a declaration of a DFFC, and the synchronous counter netlist will contain a declaration of an otherwise identical DFF; this resolves the name conflict.

Step 3: Write a behavioral counter model. In a new module file named Counter4.v, declare a 4-bit reg and model a counter behaviorally this way:

```
reg[3:0] CountReg;
...
always@(posedge ClockIn, posedge Clear)
  begin
  if (Clear==1'b1)
      CountReg <= 0;
  else CountReg <= CountReg + 1;
  end
```

Assign a delay using a continuous assignment statement which transfers the CountReg value to an output port, CountOut. The count delay should match your #3 FF module delay (Step 1 above). Now, simulate and compare times with the two preceding structural counters.

Then, synthesize for area and speed.

Finally, do a little experiment: Make a copy of Counter4.v and change the code to CountReg <= CountReg - 1. Simulate; and, notice what happens when the count wraps below 0. A reg is an unsigned type, and the value next after 0, in a down-count, is the maximum positive value expressible in the reg.

Step 4: Use a verilog special wire type for logic. Make a copy of your Synch4DFF model in a new file named Synch4DFFWor.v, as usual renaming the module name, too. Replace the verilog or expressions in your synchronous counter with independent assignments to one or more wired-or (wor) nets. Don't assign any delay to the new or logic (we'll look into differentiating rise and fall delays later in the course). Simulate again and synthesize for area and speed, to compare with Synch4DFF.

We are doing this one exercise with a wor for instructional reasons. A wor model may not be available in a CMOS technology fabrication library, where pull up and pull down strengths usually are equal. When you code a wor net, then, the synthesizer either should replace its drivers with special library wor gate buffers or other components, or it should drive the net with a library or gate.

Digital VLSI Design with Verilog

Step 5: Use your PLL as a counter clock. Make a new directory under `Lab08` and copy into it your entire `PLLsim` model from `Lab06`. You may wish to copy the answer version from your `Lab06/Lab06_Ans` directory, if your own copy was not fully debugged. Rename the `PLLsim` module and its containing file to "`PLLTop`" to avoid confusion. Instantiate the PLL and all your counters (ripple, synchronous, behavioral) in a single module named `ClockedByPLL` and connect the PLL output clock to them.

Clock the PLL in your new module with an input named `ClockIn` at 1 MHz. See the block diagram in figure 7-9. Be sure to remove your timescale and `define code from all the old module files and put them <u>only</u> in the new top of the design (`ClockedByPLL`), and in the testbench instantiating it. Simulate briefly to be sure all three counters will count (see figures 7-10 and 7-11).

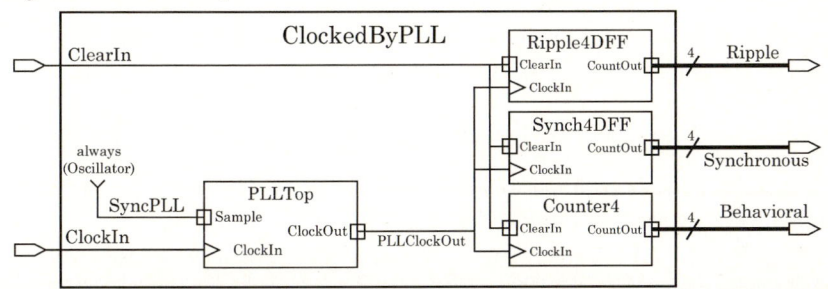

Fig. 7-9. The Lab 6 PLL adapted to clock three counters. Resets omitted.

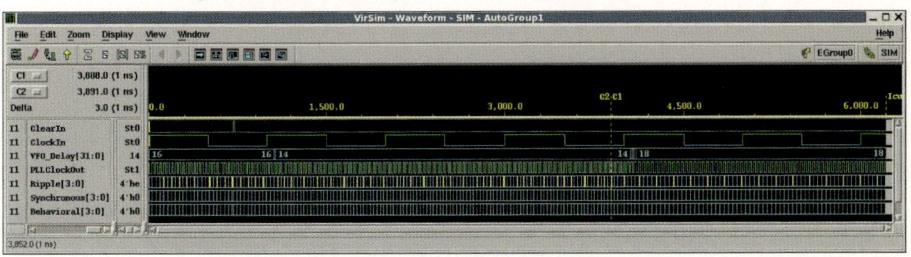

Fig. 7-10. Simulation of the three `ClockedByPLL` counters.

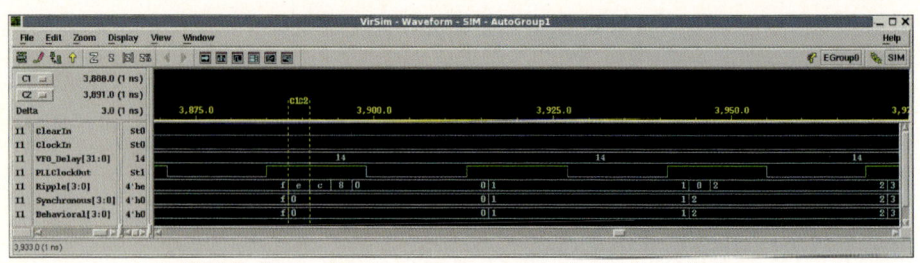

Fig. 7-11. Closeup of the `ClockedByPLL` counters, showing ripple glitches.

7.2.1 Lab Postmortem

How did your synchronous vs. ripple counter simulation times compare?

If the DFF delay is around 3 ns, why won't the ripple counter count correctly at a clock period of 10 ns?

The intermediate states which are output by a ripple counter before it settles can be confusing during simulation. They also represent brief pulses or glitches which in a real design might be current-amplified by bus drivers, or otherwise might cause unnecessary noise. How might these states be kept off an output bus?

The behavioral counter (`Counter4`) was trivial to change from an up-counter to a down-counter. How easy would it have been to change similarly the ripple counter or the synchronous counter?

7.3 Additional Study

Read Thomas and Moorby (2002) section 1.4.1 to see a counter model used in context.

Read Thomas and Moorby appendix A.14–A.17 to see a counter modelled as a state machine.

Optional readings in Palnitkar (2003):

Palnitkar models counters at several different places, to make various points about the verilog language. If you are interested in counters per se, you may wish to work the examples in his section 6.7. Solutions are given on the Palnitkar CD.

The ripple counter is described in sections 2.2 and 6.5.3.

A synchronous counter built from J-K flip-flops is described in section 12.7, exercises 5 and 6. Note: A switch-level J-K flip-flop is modelled in section 9.4 of Thomas and Moorby.

Chapter 8
Week 4 Class 1

8 Today's Agenda:

Lecture on strength, contention, and operator precedence
Topics: Verilog drive and charge strength, race conditions, and operator precedence in expressions.
Summary: Verilog charge strength and drive strength are defined and ranked in order to show how contention among different logic levels is resolved. Then, race conditions which lead to contention are discussed. Finally, the precedence of verilog operators is presented, so that the logic in complicated expressions may be handled correctly.

Lab on strength and contention
Verilog drive strengths and race conditions are simulated. Much of this lab is optional: If you have the Silos simulator provided with Thomas and Moorby or Palnitkar, you should be able to do the optional parts at home.

Lab Postmortem
Simulation of contention.

Lecture on PLL synchronization
Topics: Named blocks, and behavioral and patterned extraction of serial clocks.
Summary: Named blocks are presented to provide means of terminating execution of a procedural loop. Then, the general problem of transmitting a clock with serial data is solved, for purposes of the class serdes project, by specifying 32 bits of inert padding for every 32 bits of data transmitted. An unsynthesizable behavioral (procedural) PLL is described as an optional digression but is not used. The lecture leaves for lab the detailed development of a patterned verilog solution for PLL clock extraction.

Lab on PLL synchronization
To solve a problem in serial clock extraction, this lab begins by providing the user with some sophisticated procedural code. It then shows how a combinational-logic bit swizzle can be equally effective and much simpler.

Lab Postmortem
Extraction of a serial clock: How to deal with loss of synchronization?

8.1 Contention and Operator Precedence

Today, we shall study some new verilog: drive strengths, race conditions, operator precedence, and named blocks. We then shall move on to a discussion on how to lock our PLL in to a serial data stream.

8.1.1 Verilog Net Types and Strengths

Strength means nothing in verilog except when a net has more than one driver.

We've briefly mentioned some of the special net types (tri, wand, wor) in *Week 2 Class 1*. In the present lecture, we shall introduce the different kinds of strength but will not exercise all of them; we'll study the remainder later, when we cover switch-level modelling.

Except only the high-impedance state, which is one of the four basic logic levels in verilog and is implied by assignment of 'z', there is little value to the explicit use of strength when modelling a VLSI design. Although explicit strengths will be modelled during simulation, the synthesized will ignore them and buffer its assigned values depending on how the gates will be used.

According to the verilog standard (IEEE Std 1364 sections 4.4 and 7.9–7.13 or IEEE Std 1800 section 6.3.2), there are two kinds of strength, ***drive strength*** and ***charge strength***.

8.1.1.1 Drive Strength

The drive levels, in decreasing order of strength, are called *supply*, *strong*, *pull*, and *weak*. Recall here that 'z' is considered in the language to be a logic level, not a strength. However, a 'z' level is effectively the same as an 'x' asserted at <u>lower</u> than <u>*weak*</u> strength.

Drive strengths of *strong* ('1'; '0'; 'x') or weaker-than-*weak* ('z') are the only ones usually encountered in RTL or gate models. Drive strength proper in verilog only applies to (*a*) nets in continuous assignment statements or (*b*) gate outputs of the builtin verilog primitive logic gates. Drive strength can not be used in association with reg type declarations; however, after a reg value has been assigned to a net type, strength possibly may become relevant.

As already implied, there is a fifth drive strength named *highz* (= weaker than *weak*), as shown in the table below. A net driven with a level 'z', as in the expression 1'bz, would be resolved to a 'z' (unknown) if also driven with a *highz0* or *highz1*.

Drive strength applies only to net **contention**, when there is a connection of more than one driver to a net; otherwise, the strength of a driver is irrelevant to the logic level being driven. Although drive strength can be associated with a net when it is declared, resolution of strengths only occurs when a net is driven by at least one continuous assignment or verilog primitive gate. Instance outputs of modules always drive only with *strong* strength.

An unknown level can have any strength as a result of contention, depending on the strength of the contending drivers. However, the usual logic level 'x', as in the expression 1'bx, always has *strong* strength, as indicated below in the table.

Drive strength properties are covered in IEEE Std 1800 section 10.3.4.

8.1.1.2 Charge Strength

The charge strength levels, in decreasing order, are called *large*, *medium*, and *small*; they should be considered as equivalent to the sizes of on-chip capacitors. Switch-level models are used to represent current flow through transistors, which depends in turn on device-element capacitance.

The only data objects allowed to be declared with a charge strength are `trireg` nets. These nets are called `trireg`, because they are viewed as charge storage elements, somewhat the way `reg`'s can be viewed as logic-level storage elements.

Charge strength properties are covered in IEEE Std 1800 section 28.3.2.

8.1.2 Verilog Strength Usage

As said already, except the use of `'z'`, strength generally is of no concern in modern gate-level or behavioral design; strength is intended to be invoked only when simulating at the switch level. Switch level models include pass transistors, single nmos or pmos devices, or other gate substructure useful in simulating the elements which are combined to make up a logic gate in a technology library.

Switch-level models exist in something of a vacuum in modern verilog. Accurate validation of models at this level usually requires analogue simulation. Switch-level models can be functionally accurate but not timing-accurate. However, the major concern in verilog modelling of individual technology-library gates is timing, because other factors such as noise, crosstalk coupling strength, and leakage current, are beyond the scope of the language. Because timing usually can be expressed adequately in verilog without switch-level reference, switch-level verilog models are not commonly used in commercial library work in the industry.

Also, library models usually are not written in verilog but in ALF (IEEE Advanced Library Format) or in vendor-specific languages such as *Liberty*. This is so, because a complete library model has to include analogue-related simulation data such as slew rate under load, as well as nonsimulation data such as layout geometry, placement rules, wire-loading, and other quantities beyond the verilog language.

When it occurs, contention is resolved in verilog by strength level. Ambiguous results only occur when different logic levels of `'1'` or `'0'` contend and are of equal strength. In all other contentions, the resultant strength and logic level is that of the higher strength level. Ordinary logic gates, `module` instances, or verilog expressions act in this context as amplifiers, in the sense that they output *strong* logic regardless of the strength of their (resolved) inputs.

Here's an ordered table of verilog strengths (see also Thomas and Moorby (2002) section 10.2.2):

Strength level	Keyword(s)	Logic level(s)	Strength number
supply	supply1		7
strong	strong1	1, x	6
pull	pull1		5
large	large		4
weak	weak1		3
medium	medium		2
small	small		1
highz	highz1	z	0
highz	highz0	z	0
small	small		1
medium	medium		2
weak	weak0		3
large	large		4
pull	pull0		5
strong	strong0	0, x	6
supply	supply0		7

The verilog logic <u>levels</u> (not <u>strengths</u>) normally used in RTL or behavioral code apply only to net assignments and are defaulted as shown. This default can be changed by verilog macroes as explained below.

The strength number in the preceding table is what is used to resolve contention: Higher numbers win over lower ones, and equal numbers at different levels result in unknowns.

Strengths may be declared in parentheses preceding an instance or net name; or, it may precede the target of a continuous assignment to the net. Drive strengths must be declared in pairs (*high*, *low*), and charge strengths must be single (*size*).

Examples are,

```
wire OutWire;
...
and (strong1, weak0) and01(OutWire, in1, in2, in3);
nor (pull1, pull0)   nor31(OutWire, in1, in2, in3);
trireg (large)       LargeCapOnNet;
wire (supply1, supply0) TiedTo1 = 1'b1;
// ---
wire UpClock;
...
assign (strong1, weak0) UpClock = ClockIn;
```

In the example code above, a truth table for multiply-driven `OutWire` would be:

in1 & in2 & in3	and01 output	nor31 output	Outwire
1	strong1	(pull1 *logically impossible*)	(strong1)
1	strong1	pull0	strong1
0	weak0	pull1	pull1
0	weak0	pull0	pull0

We are interested only in the idea of *strength* at this point. We shall take up switch-level modelling in detail later.

Concerning strength, two relevant procedural assignment statements are ***force*** and ***release***. These are simulator overrides, rather than drive strengths, and they should not be used in a design—only for testbench or debug. The statement, "force *object* = *level*;", in which *object* is any `reg`, forces the design object to the given logic *level* until "release *object*;" is executed.

There also is a similar procedural ***assign*** and ***deassign***; we shall not study these and do not recommend them for design work. Along with `defparam`, these constructs are strongly discouraged in recent IEEE Std 1800 releases up through 2012 and are likely to be removed from future Std documents.

8.1.3 Race Conditions, Again

In verilog, every block in a `module` has its inputs (or right-hand side—RHS) evaluated concurrently in simulation time; this includes instantiations and `initial` blocks. Ordering of simulation events can be made meaningful (*a*) by cumulative delays from sequences of events which happen to combine to cause events to occur at unique simulation times; or, (*b*) by location of statements within procedural blocks (`always` or `initial` blocks).

There is an important ordering of evaluations within each time step, but it is not visible in simulation waveforms. We shall study this ordering, which is enforced in the form of the *verilog simulator event queue*, later in the course.

Synthesis tools currently can not use concurrency in procedural blocks very well; they also ignore design delay expressions. To create a synthesizable sequence of events, it is necessary to make the later events depend on the earlier ones (*a*) by defining the earlier ones as inputs (RHS) and the later as outputs (LHS); or, (*b*) by making the desired sequence depend on the sequential state of design variables, primarily `reg` types.

Order can be considered either unique or multiple. If an event (such as a hardware reset) occurs only once in the simulation lifetime of the device being modelled, it may be ordered uniquely with respect to other events. Obviously, something which happens on every clock, or on every number of them, can't be ordered uniquely; but, it may perhaps be ordered within the clock cycle or relative to some other repetitive event.

A <u>race condition</u> occurs when some state depends on the relative order of two or more preceding events, and the order of them can't be predicted. Clearly, this implies that both of the preceding events can occur at the same simulation time. The preceding events are said to be in a race to determine which one comes before the other(s). This is a very undesirable condition for the hardware, because unpredictable hardware usually is not functional. If the synthesizer doesn't reject such constructs as errors, it may synthesize the wrong hardware.

The simplest kind of race condition is repetitive and caused by assignments to the same data object from two or more different concurrent blocks. If such blocks were not concurrent, the race would not occur, and the hardware would be functional.

For example,

```
always@(posedge clockIn)
  begin
  #1 a = b;
  (other statements)
  end
...
always@(a, b, c)   #2 ena = (a & b) | c;
...
always@(a, b)      #2 ena = a ^ b;
```

In each of the three assignments shown in the example, the delay should be interpreted this way: If a variable in an event control expression changes to '1' or '0' (to '1' for a *posedge* and to '0' for a *negedge*), that always block is read and thus the delay is initiated; after the delay has lapsed, the RHS is evaluated and the result is assigned.

The race in the code above is to assign ena. We assume other statements or blocks not shown are assigning to a, b, and c. This race is fairly obvious; but, it's not hard to imagine a bigger module which might cause confusion and conceal a race because of complicated concurrency.

One might imagine that a designer just looking at the details would think of the simulation of the above race example this way:

> Start time = 0 at the clock edge. Then, at time 1, a gets the value of b. Suppose a thus has been changed to '1' or '0'; then, this triggers reading of the two other blocks. At time 3, ena goes to some value because of the second always statement—then again, depending on what was the value of c, perhaps ena has its value replaced by 0 at time 3, because of the third always statement? Hmmm . . . may be a problem?
>
> But, worse: What if c changes after time 1? Then this triggers the second always but not the third! Thus, ena may have its value changed arbitrarily either to the value of (a&b)|c or to a^b. Not only that, but the clock might intrude at any time to sample any arbitrary value of whatever is controlled by ena.

Also, what if a has been set equal to b before the clock edge? The clocked block won't change the value of a; then, neither of the other blocks shown will be read right after the clock; rather, they may be read at any arbitrary time because of changes in a or b from other blocks not shown. Yuk! This won't work.

The way to avoid this kind of race condition within a module is never to assign any data object from more than one `always` block. When orderly execution is required, a simple solution is to have the `always` blocks read on opposite edges of the same clock, and to have all delays total less than half of a clock period in every block involved. Another way would be to provide two or more well-defined clocks derived at different phases from the same original clock. A third way would be to assign one of the RHS's to a gate or second `always` which would delay the ena on that side so that the other predictably (but glitchily) could be assigned first.

Returning to the example, the first step in correcting the preceding race condition without using clocking schemes becomes obvious: Put all assignments to ena in a single `always` block.

This would result is something like the following:

```
always@(posedge clockIn)
   begin
   #1 a = b;
   (other statements)
   end
...
always@(a, b, c)
   begin
   #2 ena = (c==1)? 1 : a & b;    // An if could be used here.
   #2 ena = a ^ b;
   end
```

The sequencing of the two assignments to ena now is obvious, whereas it might have gone ignored or unnoticed in a big collection of independent `always` blocks. One here might question why the exclusive-or assignment should occur after the one above it, causing at most a glitch ignored by anything reading the outputs of this module. But, at least now there is no race condition: The upper assignment to ena will occur first, because of the blocking assignments used. Perhaps the (inconsistent) intent of the original code would be realized even better by not using multiple assignments at all.

We then would end up with a nice, consistent

```
always@(posedge clockIn)
   begin
   #1 a = b;
   (other statements)
   end
...
always@(a, b, c)   #2 ena = (c==1)? 1 : a ^ b;
```

The `initial` kind of block shouldn't be overlooked in a discussion of race conditions. Verilog allows any number of `initial` blocks in a module, and the same races can occur as with `always` blocks, except that usually such races will be nonrepetitive. The synthesizer ignores `initial` blocks, so they can cause no synthesis problem—except a simulation/synthesis mismatch!

In general, `initial` blocks should be used only in testbenches, or, in a design, for special purposes such as denotation of an SDF back-annotation file. However, we should nevertheless point out the following:

In a simulation, both `always` blocks and `initial` blocks are equally concurrent, and, at time 0, either one might be the first to cause evaluation of what should be assigned to something. Because it is unavoidable to assign some objects both in one `always` block and in an `initial` block, brief race problems may be overlooked or tolerated at the startup of a simulation as a minor nuisance.

In the following example, if the `always` block is read before the `initial` block, the Reset value of 1'b0 will be missed at time 0, and Abus may not be set to 0 until Reset is toggled later:

```
initial
  begin
  Reset    = 1'b0;
  #0 Reset = 1'b1;
  ...
  end
always@(ClockIn, Reset)
  if (Reset==1'b0)
       Abus <= 'b0;
  else Abus <= Inbus;
```

Again, in synthesis, only the `always` block will be executed.

Another problem can be caused by mixing blocking and nonblocking assignments in a single procedural block: Don't do it. Determine whether any of the logic will be clocked; if so, use only nonblocking assignments. For combinational logic, we want setup time for clocking, and, any change has to be propagated throughout the whole block. Because nonblocking assignments will update with the earliest, not the immediately new, values, if all the logic will be combinational, use blocking assignments. When the logic is sequential but unclocked (latched), be very careful of correct synthesis, and use either blocking or nonblocking assignments, depending on setup requirements. If necessary, repartition your design (within a `module`) so that the sequential and combinational logic are more cleanly separated.

8.1.3.1 Inertial Delay

In closing this discussion of race conditions, it should be mentioned that verilog simulators use the same pulse-propagating process as most of the other available digital simulators; it is called <u>inertial delay</u>. An input pulse has to have a certain amount of

Digital VLSI Design with Verilog

"inertia" to get through a gate and influence the output. By default, if a statement in a module has been assigned a propagation (or path) delay, and a new level in an input lasts for less time than the propagation delay, the change will be ignored. This is a form of glitch filtering: Only glitches which are wide enough will make it through the gate. Physically, it is a coarse model of energy response: A brief pulse is assumed not to have enough energy to switch the gate; and, the path delay value being available, it is used to decide this energy. Unlike other HDL's, in simulation, verilog allows the threshold inertial pulse width to be controlled by the designer; we shall study this later in the course.

Simulators intended for VLSI design often do not simulate inertial delay properly, unless the delay is given in a `specify` block. So, unless you have tested it, do not depend on your simulator to swallow glitches based on hand-entered delays in continuous assignments or in procedural blocks. We shall study `specify` blocks later.

8.1.4 Unknowns in Relational Expressions

We'll return to this later in more detail, but, for now, it is enough just to say that relational expressions always evaluate to 'x' when a bit neither is '1' nor '0'. Such values logically are read as *not true* and thus are used as equivalent to *false*. Thus, any 'x' or 'z' causes failure of *true*.

For example,

```
reg[3:0] A, B;
...
A = 4'b01x1;
B = 4'b0001;
if ( A > B )
    X = 1'b1;
else X = 1'b0;
```

The result is that X goes to 1'b0. Perhaps surprisingly, with values assigned as above, the same result occurs for this:

```
if ( A == A )
    X = 1'b1;
else X = 1'b0;
```

To handle the four verilog logic levels literally and individually, one may use a `case` statement instead of an `if`. With A equal to 4'b01x1 as above,

```
case (A)
4'b0000:   X = 1'b0;
4'b0011:   X = 1'b0;
4'b01x1:   X = 1'b1;
default:   X = 1'bx;
endcase
```

we get X set to 1'b1.

The `case` statement only does exact matches; it does not permit wildcards. There exist two special wildcard variants of the `case` statement named `casex` and `casez`; we shall ignore them for now.

8.1.5 Verilog Operators and Precedence

We've already used the verilog bitwise operators most often seen in a design. Here is a list of all the verilog operators, as given in the Thomas and Moorby (2002) Appendix C, or in IEEE Std 1364 section 5.1. The equivalent list may be found in IEEE Std 1800 section 11.3; however, bear in mind that SystemVerilog includes operators not available in many verilog implementations.

Symbol	Type	Symbol	Type	Symbol	Type
~	bitwise	*	arithmetic	>	relational
&	bitwise	/	arithmetic	<	relational
~&	reduction	+	arithmetic	>=	relational
|	bitwise	-	arithmetic	<=	relational
~|	reduction	%	arithmetic	==	equality
^	bitwise	**	arithmetic	!=	equality
~^,^~	reduction	!	logical	===	equality (case)
>>	shift	&&	logical	!==	equality (case)
<<	shift	||	logical	? :	conditional
>>>	shift (arith)	{ }	concatenation		
<<<	shift (arith)	{ n{ } }	replication		

There are two alternative ways of writing the *xnor* reduction operator; the present author prefers ~^.

Corresponding logical operators and bitwise operators yield the same results on one-bit operands. '1' for true is the same as 1'b1, for all purposes. When manipulating bits, busses, or registers, or when expecting to synthesize gates to perform operations, it is best to provide an explicit width, just to keep in mind what one is doing.

The arithmetical shift right operator (>>>) shifts in the value of the preoperation sign bit instead of a leading '0'. The other shift operators shift in a '0'.

Synthesizable verilog generally imposes limits on the exponentiation operator (**). Examples of such limits are that both operands must be constant; or, that the expression must evaluate to a power of 2. As a rule, wherever practical, avoid exponentiation entirely by writing "1<<n" to evaluate 2**n.

The case equality operator (=== or !==) is not used in a `case` statement; it may be used in an `if` or a conditional expression. It works like the comparison done in a `case` statement, which distinguishes 'x' from 'z'. Also, it can not return a value of 'x', as can the other equality, the bitwise, or the relational operators. The case equality operators do not allow wildcarding and do not interpret 'x' or 'z' as wildcards. Case equality is **case** equality, not **casex** or **casez** equality.

The replication operator may be used to assign a value or pattern to a part select or a whole vector. For example,

```
Xbus[15:0] <= {Abus[4:1], 6{1'bz, 1'b0}};
```

will set the 12 low-order bits of Xbus to an alternating pattern of 'z' and '0'.

All operators except the conditional operator associate left to right when of equal precedence. All unary operators are of the highest precedence.

Here is a table of operator precedence, after IEEE Std 1364 section 5.1 or IEEE Std 1800 section 11.3.2:

Precedence	Operator
lowest = 0	? : (conditional) { } or {{ }} (concatenate)
1	||
2	&&
3	|
4	^ ~^ ^~
5	&
6	== != === !===
7	< <= > >=
8	<< >> <<< >>>
9	+ - (binary)
10	* / %
11	**
highest = 12	+ - ! ~ & | ^ ~& ~| ~^ ^~ (unary)

This ends our discussion of verilog operators. Now you know them all . . . .

8.2 Digital Basics: Decoder and Three-State Buffer

Before starting the next lab, let's look at a couple of design basics as they are realized in verilog.

First, let's introduce the verilog three-state buffer component shown in figure 8-1:

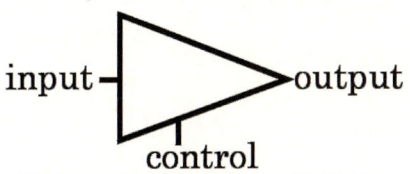

Fig. 8-1. bufif1, with pin functionality labelled.

The **bufif1** is a verilog primitive gate which, when on, simply amplifies the power of, or buffers, its input logic level. When off, it outputs a 'z'. Its port list includes one <u>output</u> bit, one <u>input</u> bit, and one <u>control</u> bit, in that order from the left:

    bufif1 *optional_InstName*(*out*, *in*, *control*);

For example,

    bufif1 CtlBuf01(OutBit_1, InBit_1, Ctl_On);

When the control bit is a logic '1', the gate is on; so, it is a ***buffer if*** control is '**1**', and this explains its name. We shall use this gate to exercise what we have learned about contention among different driving strengths.

Next, how do we write a verilog decoder? For a 2-to-4 decoder, the simplest way is to use a `case` statement. We can use a `case` as though it were a lookup table to assign the one-hot '1' depending on which 2-bit count is in the `case` expression.

With a binary count in Sel, to run through all possible bit patterns, and the decoded result in xReg, this is our `case` decoder for the *X* buffer selection in the next lab:

```
reg[3:0] Sel, xReg;
...
always@(Sel)
    begin
    case (Sel[1:0])
      2'b00: xReg = 4'b0001;
      2'b01: xReg = 4'b0010;
      2'b10: xReg = 4'b0100;
      2'b11: xReg = 4'b1000;
      default: xReg = 'bx; // e. g., to handle an 'x' in Sel.
    endcase
    end
```

This will be our first explicit use of the `case` statement in a lab. It is very important not to leave open an unassigned alternative; this means always to add a `default` statement to cover leftover alternatives. Setting the default to assign 'x' values tells the logic synthesizer you don't care about leftover alternatives, and this permits better area optimization than otherwise. Also, overlooking an alternative may create a latch of some kind, and this may not be what you want. We'll look later at other implications of this very useful verilog construct, the `case` statement.

8.3 Strength and Contention Lab 9

If you have either the Thomas and Moorby or the Palnitkar hardcover books, they come with a demo version of the Silos simulator, which may be used for the optional Steps of this lab, a `wire` strength exercise. The demo version of Silos is not intended for large designs.

Digital VLSI Design with Verilog

The optional steps of this lab probably will not work as described in a simulator which is optimized for use in CMOS VLSI design; such a simulator never would be used to resolve strengths; instead, speed and capacity for synthesizable verilog would be its goal. Strength is not synthesizable as a netlist gate output; for gate assignments, the synthesizer estimates the required strength and just chooses driving gates of adequate strength. Contention, except for 'z', typically is assumed in a big design to imply only unknowns.

Lab Procedure:

Topic and Location: Verilog three-state primitives, contention between different drive strengths, race conditions, and bitwise vs. logic operators.

Work in your `Lab09` directory.

The Silos simulator may be used at home for the optional Steps of this lab, a `wire` strength exercise. If you do these optional Steps, you may want to bring in copies of your results and keep them in `Lab09` for synthesis in optional Step 5 below and for possible later reference.

Preview: The optional Steps use two decoders in one module to apply all possible pairs of verilog strengths, at opposite levels, to a single `module` output bit. A testbench counter guarantees exhaustive exercise of all possible contentions. The lab returns to nonoptional Steps with coding examples of race conditions. A final exercise shows the important difference between bitwise and logical operators in verilog.

Nonoptional Deliverables: Step 1: A 4-to-1 decoder modelled by a `case` statement, which handles unknown input bits and simulates correctly; then, two of these implemented in one `always` block in a module named `Netter`. Steps 2 through 5 are optional. Step 6: A race condition simulation model. Step 7: A simulation model showing the difference between '`&`' and '`&&`' when applied to vector operands.

Optional Deliverables: These require Silos or a similar old-style simulator. Step 4: The `Netter` model from Steps 2–3, with all `bufif1` outputs tied together to a single net, driving a one-bit module output port. Step 5: A correctly simulating `Netter` model demonstrating correct contention output; synthesized netlists for area and speed.

Lab Procedure:

In this lab, we shall create a module named `Netter`. This module will contain logic as shown in the block diagram of figure 8-2.

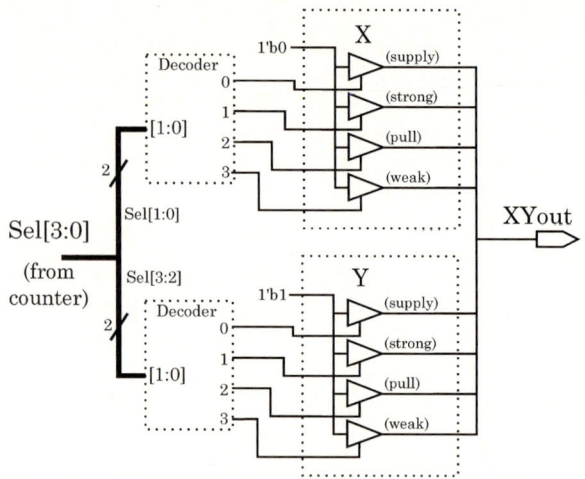

Fig. 8-2. Netter functionality. Blocks with dotted borders indicate logic clouds, not submodules.

A four-bit counter, located in the testbench, is not shown. It should be implemented this way, with a short delay to prevent clock/counter edge races:

```
...
reg[3:0] CountStim;    // An output from this will drive Netter.
...
always@(ClockStim) #10 ClockStim <= ~ClockStim;    // The clock.
always@(ClockStim)  #1 CountStim <= CountStim + 1; // The counter.
...
```

In figure 8-2 above, notice that the Netter output drivers are all three-state buffers. It is illegal in verilog to drive an output with a simple wired logic net (wor, wand, etc.), for example, one connected directly between a module's inputs and outputs. This probably is because no localized delay can be associated with the logic of such a net (as opposed to a path on a net originating at a gate output).

We wish to run a simulation which will demonstrate the result of contention of each of the four drive strengths against the others, including itself. To do this, we shall connect together some buffers and turn them on and off selectively so that only two are on at a time. One of the two always will be at logic level '1', and the other at '0'. Thus, we can see during simulation which strength wins a contention of two opposite levels.

This means 4 x 4 = 16 different pairwise connections. So, to match the schematic of figure 8-2, we define four bufif1's as the *X* buffers and four others as *Y*. All the *X* buffers receive a '0' input, and all the *Y*'s a '1'. Each buffer in *X* or *Y* is assigned a different one of the 4 drive strengths. Then, we may use a 4-bit binary counter and assign the lower two bits (2 bits = 4 choices) to control *X* and the upper 2 bits to control *Y*. If we decode the

Digital VLSI Design with Verilog

count, we then can control each of X and Y to have just one buffer in its "on" state at a time. By *decode*, it means here that the binary count is represented "one-hot"—in other words, a 2-bit count is represented by a logic '1' in one of 4 positions on a 4-bit bus.

Because a 4-bit binary counter goes through all possible combinations as it counts, if we tie all the X and Y buffer outputs together, we shall get all possible combinations of one buffer on among the X buffers, and, correspondingly, one on among the Y buffers.

Step 1: Enter two decoders. Begin the Netter module by writing the header; it should have a 4-bit Sel input (from the testbench counter) and a 1-bit output named XYout. Use the schematic in figure 8-2. Enter two different decoders as shown in the text immediately preceding this lab, one selecting from Sel[1:0] and outputting a 4-bit buffer xReg, and the other selecting from Sel[3:2] and outputting a 4-bit buffer yReg. The outputs should be declared, of course, as reg types. Locate the two decoders in the same Sel-sensitive always block. Declare two 4-bit wires named xChoice and yChoice; use continuous assignments to drive these wires individually with xReg and yReg.

Don't bother to connect the two Choice wires to anything yet.

The incomplete Netter, at this point, should be as represented in figure 8-3.

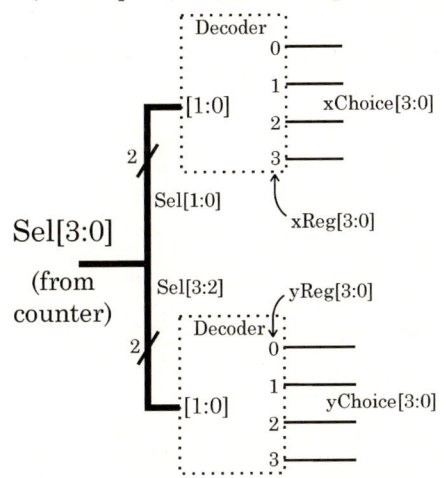

Fig. 8-3. Wiring of the two Netter decoders.

Because the decoded outputs will be assigned from constants, there is no reason to make the always block sensitive to anything but the Sel input bus.

To check your model, connect a testbench counter and simulate it. Use a testbench which sometimes provides an 'x' bit in the input.

VCS will not simulate either the verilog source or the synthesized netlists correctly for these Steps: VCS is not designed to distinguish strengths in contention and simply will produce unknowns ('x'). QuestaSim probably will do the same. As for the netlists, DC does not synthesize gates capable of reliable resolution of contention, and the synthesis libraries don't include such gates, anyway. Contention almost always is an error in VLSI design, except only contention vs. 'z'—which VCS, QuestaSim, and DC handle correctly. The only reason for running VCS on the verilog source in these Steps is for a good syntax check.

Optional Step 2: Using your module from Step 1, you should add bufif1's as shown in figure 8.2 and in the code below, tying their inputs to logic '1' or logic '0' and their controls to the decoder outputs.

Define the strengths by assigning them in the bufif1 instantiations; you should instantiate two each of the following bufif1's in Netter. See Step 3 for a hint on how to fill in the wire names and the instance names:

```
bufif1 (supply1, supply0)  inst_name( , , );
bufif1 (strong1, strong0)  inst_name( , , );
bufif1 (pull1,   pull0)    inst_name( , , );
bufif1 (weak1,   weak0)    inst_name( , , );
```

Optional Step 3: After declaring the 8 bufif1's, enable one buffer at a time by wiring them this way to the decoder outputs:

```
wire[3:0] xChoice; // Just to rename xReg.
    ...
assign Xin = 1'b0;
assign Yin = 1'b1;
    ...
assign xChoice = xReg;
//
bufif1 (supply1, supply0) SupplyBufX(SupplyOutX, Xin, xChoice[0]);
bufif1 (strong1, strong0) StrongBufX(StrongOutX, Xin, xChoice[1]);
    ... (2 more X and 4 more Y)
```

Optional Step 4: After this, tie all the buffer outputs together; this may be done by a set of continuous assignments all to the same wire. Such assignments of course are concurrent, so there is no need to worry about their order in the verilog source code:

```
    ...
assign XYwire = SupplyOutX;
assign XYwire = StrongOutX;
    ... (6 more) ...
```

Then, the contention may be examined by looking at XYwire in a simulator such as Silos, which can display strength differences correctly; the result also may be assigned to a Netter output port, but if it is assigned to a reg type, the strength will be lost and only the logic level at strength = *strong* will remain.

Digital VLSI Design with Verilog

Don't clock Netter itself, because all you want is combinational (muxed and decoded) nets. You had to declare reg's to use the case statement, but these reg's will be assigned to nets continuously and never will be allowed to save state when one of their drivers changes; so, they will represent combinational logic—if you have all your case alternatives covered!

Optional Step 5: Clock a counter in your testbench (see the beginning instructions for this lab) to assign the value of Sel and step through all contention alternatives.

After viewing the simulation at home, you may bring in the model to the lab for the rest of this Step.

Synthesize Netter with no area optimization and examine the netlist. What did the synthesizer do to implement the decoders? Was strength preserved? Optimize for area and examine the netlist again.

This completes our exercise in wire strength and contention, and it ends our dependence on the use of the Silos simulator for this lab.

Step 6: Race condition exercise. Create a new module named Racer in a new file and put in it two always blocks as follows:

```
always@(DoPulse)
  begin
  #1 RaceReg = 1'b0;
  #1 RaceReg = 1'b1;
  #1 RaceReg = 1'b0;
  end
//
always@(DoPulse) #1 RaceReg = ~RaceReg;
```

Instantiate Racer in a testbench which toggles the value of DoPulse a few times. The toggles should be 10 ns or more in duration.

Simulate: (Note: This will not synthesize). The first block should cause a 1 ns positive pulse lagging every DoPulse edge by 2 ns. But, what effect has the second always block? If it inverts after the first '0' assignment, then it should advance and widen the pulse; if it inverts before the first '0', it should be superceded by the '0' and should not be noticed. Reverse the locations of the two always blocks in Racer.v—put the one-line DoPulse block on top. Does this change the result? It should, because concurrent blocks may be read in any order, according to the simulator developer; and, generally, the order of appearance in the file will be used consistently, one way or the other.

Figures 8-4 and 8-5 show the waveforms in VCS or QuestaSim. What about Silos or Aldec?

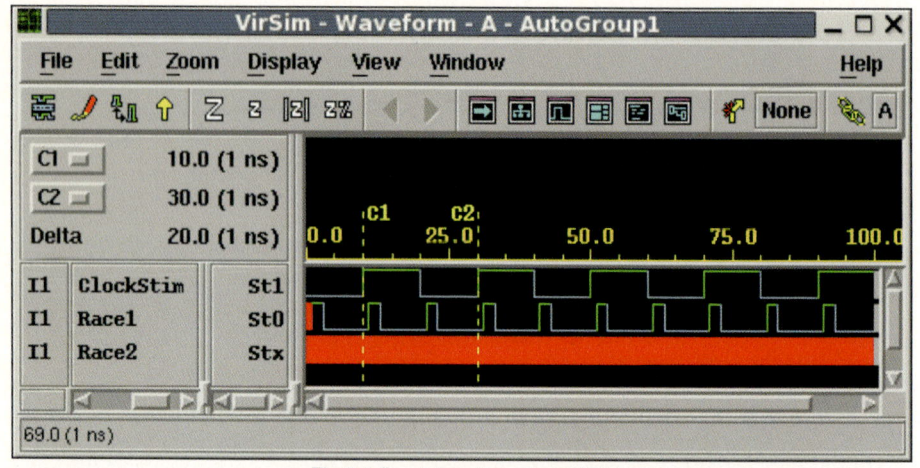

Fig. 8-4. Case 1: Inverting `always` first.

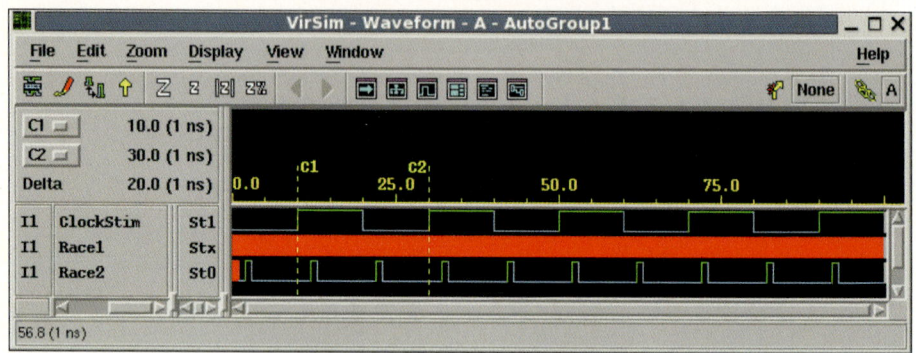

Fig. 8-5. Case 2: Inverting `always` second.

Step 7: Operators and precedence. It is easy to get confused about the *and* and *or* logical vs. bitwise operators. For one-bit variables, they produce the same results. But for multiple-bit variables, they differ importantly.

For example, consider the three different bit-masks in these expressions (the left value "masks" the right one):

```
3'b110 & 4'b1101,
3'b010 & 5'b00111,
3'b101 & 3'b111.
```

Digital VLSI Design with Verilog

The '&' is bitwise; so, the three results, in order, numerically are 100, 10, and 101, widened on the left by 0-extension to the width of the destination vector (which is not shown).

However, "&&" is a logical operator, and it returns a '1' (= 1'b1) if both operands are nonzero; otherwise, it returns a '0'. So, the expression

   3'b010 && 6'b111000

evaluates to '1', again widened by 0-extension to the width of the destination vector.

By contrast, the bitwise expression

   3'b010 & 6'b111000

evaluates to '0', widened to the destination width.

Look at the expressions above and imagine them assigned to an 8-bit bus. What happens if one bit is 'x' or 'z'?

Does the logical operator return an 'x'?

Are logical and equality operators different in the way they handle an 'x'?

For this Step, test your insight by coding a small module named Operators in its own file and simulating the three bit-masked expressions at the start of this Step. Use destination operands of 3 and 8 bits width. Then, replace the bitwise operators '&' with logical operators '&&' and simulate again.

Optional: If you have the time, use the simulator to evaluate the following two results, for a and b both '1', and c and d both '0':

```
NoParens  = a&!d^b|a&~d^a||~a^b^c^a&&d;
Parens    = ((a&(!d)^b) | (a&(~d)^a)) || ((~a)^b^c^a) && d;
```

The point of this is that to understand someone elses mess, break the expression at the lowest-precedence operator(s); insert parentheses, and break again at the next lowest, etc. To prevent your own mess, use parentheses and spacing to clarify the meaning of complicated expressions.

8.3.1 Strength Lab Postmortem

How do VCS and QuestaSim handle resolution of contention?

Is contention important when two concurrent statements assign the same logic level to a net?

8.4 Back to the PLL and the SerDes

8.4.1 Named Blocks

Several verilog constructs, which we shall study more closely in the next chapter, depend on being able to <u>name a block</u> of code. A name is assigned to any verilog block simply by following the `begin` with a colon and a valid verilog identifier. Of course, only procedural code can be in blocks, so only it can be named. A block without `begin` can't be named; so, if necessary, one simply inserts `begin` and `end` around anything to be named. No semicolon follows the identifier. Any reasonable alphanumeric string is a valid identifier if it does not begin with a decimal numeral.

For example,

```
always@(negedge clk)
  begin : MyNegedgeAlways
  ...
  end
```

Or,

```
begin : Loop_1_to_9
for (j=1; j<=9; j = j + 1)
  begin
  ...
  end // for j.
end // Named loop block.
```

Naming a block has no functional effect. However, a block name, like a `module` name, can be used by a tool and thus may be found in a synthesized netlist or a simulator trace list. Thus, naming blocks can help in debugging or optimizing a netlist.

Although the name itself changes nothing, it can be used to create new functionality. For example, naming the block in a looping statement such as `for`, `while`, or `forever` allows the loop to be exitted by the **disable** statement. Such a `disable` is similar to a C language `break` statement.

8.4.2 The PLL in a SerDes

Recall the SerDes design schematically described as blocks in *Week 2 Class 2*. Our planned full-duplex system will require a serial transmitter (`Tx`) and receiver (`Rx`) in each direction. Each end of the serdes will be defined in a different system clock domain, these clocks being the clocks used to manipulate the parallel bus data at the outer boarders of the two systems.

To make our design reflect the most general case, there will not be any synchronization between the two system clock domains. A serial clock will have to be generated for each `Tx`, to transmit the data over the serial line; and, of course, each `Rx` then will have to have a deserializing clock running at the same speed as the serializing `Tx` clock. We have decided that the system clock speed in each domain will be 1 MHz, and that all serial clocks will run at 32 MHz, these clocks being generated by four independent PLLs, one for each `Tx` and one for each `Rx`. A clock ratio of 32:1 easily is achievable by an analogue PLL currently in use in the industry.

Digital VLSI Design with Verilog 167

To avoid unnecessary complexity, we shall not address error detection or correction, encryption, compression, edge detection, or packet retries, of the serial data transferred. The protocols involved would be too time-consuming for the present course.

The design on the Tx side of our serdes is straightforward: The parallel data are clocked in on a 1 MHz system clock, and the Tx PLL is used to serialize and clock them out at 32 Mb/s. The Tx PLL can track its system clock directly, the PLL phase being constrained only by that of the well-defined system clock. Because we have decided to transmit packets 64 bits wide per 32 bit data word, the maximum speed of transmission on each serial line will be one word on every other system clock; thus, each receiver will run at a maximum deserialization rate of one word per two system clocks.

The design on the Rx side will have to be more complicated. Because the two systems are clocked independently (see figure 8-6), the Rx PLL will have to extract the Tx clock from the serial data. This is equivalent to saying that the receiver will have to identify incoming packet boundaries and use their arrival rate to provide a 1 MHz clock for the Rx PLL. Given the extracted 1 MHz clock, the Rx PLL can synchronize to it and generate the 32 MHz Rx serial clock required to clock in the data and deserialize it, one word per two extracted 1 MHz clocks. After deserialization, the incoming data words will have to be clocked out using the receiving system clock.

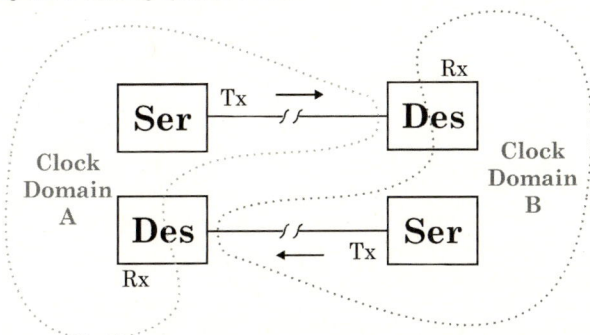

Fig. 8-6. Full duplex SerDes and the two system clock domains.

To allow slack for Tx synchronization of the PLL serial clock and the transmitting system clock, each serializer will include a FIFO to buffer incoming data. This FIFO will reduce loss of data words by averaging out PLL-caused variations in serialization speed to match the average rate of arrival of valid parallel data from the transmitting system.

Each deserializer also will include a FIFO to take up the same kind of slack. In addition, the Rx FIFO will reduce data loss because of possible delays in synchronization of the extracted Tx clock with the Rx clock. In many systems, only the receiving end would be designed with a FIFO; however, we shall add one also to the transmitting end, perhaps to allow for irregularity in the sending rate.

We shall discuss more general clock-synchronizing technology later in the course.

8.4.3 The SerDes Packet Format Revisited

To construct a serial data source allowing synchronization, we shall use the data packet format previously given as,

64'bxxxxxxxx00011000xxxxxxxx00010000xxxxxxxx00001000xxxxxxxx00000000,

in which each 'x' represents one data bit in a serialized 8-bit byte value.

The format shown is sent and received MSB first, on the left. To provide a synchronizing, incoming clock for our Rx PLL, we shall look for a pad pattern in the input data stream containing a 4 valued down-count from 2'b11 (= 3) to 2'b00 (= 0).

To make things easy for a start, we'll set up a pattern of data values of the same 4 bytes and send it repeatedly: For this, we shall pick the ASCII codes for 'a', 'b', 'y', and 'z', in that order. These codes are, respectively, 8'h61, 8'h62, 8'h79, and 8'h7a.

At this design stage, then, our serial data source therefore repeatedly will send this binary data stream, left to right:

```
//       'a'       pad 3      'b'       pad 2      'y'       pad 1      'z'       pad 0
64'b01100001_00011000_01100010_00010000_01111001_00001000_01111010_00000000
```

The alphabetical interpretation of the 8 bytes x 8 = 64 bits is shown above the stream representation.

To test our deserializer as we write the verilog, we have to generate the above data stream repeatedly at about 32 Mb/s. This is trivial: We just leave the data pattern above as a constant and pretend it is being received repeatedly over a serial line.

With the date stream above as a start, we can decide how to synchronize the PLL to it. At 32 MHz, our delay of 31 ns x 64 bits = 1.984 µs period gives us about half of the data we want for our approximately 1 MHz embedded clock. Actually, this half is just about right to handle one parallel word every <u>other</u> clock.

The entire two-clock, 64-bit packed is shown here:

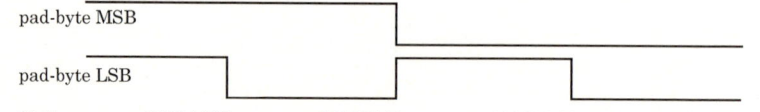

pad-byte MSB

pad-byte LSB

64'bxxxxxxxx00011000xxxxxxxx00010000xxxxxxxx00001000xxxxxxxx00000000

Fig. 8-7. The embedded clock in the serial stream can be at twice the packet frequency. MSB and LSB refer to the 2-bit pad numbers above.

8.4.4 Behavioral PLL Synchronization (Language Digression)

At this point, we digress a little to show how a software-oriented, behavioral (or, procedural) verilog model can be developed to synchronize a PLL to an embedded serial clock. This model will not be synthesizable; so, we shall return later to our previous frame-encoder approach to devise a synthesizable model. The next discussion is a language excursion meant to demonstrate software-oriented coding in verilog.

A ***behavioral*** synchronization, or perhaps more accurately, *procedural* synchronization, is done easily in verilog, using (*a*) a `for` statement to sample the stream, and (*b*) a `while` statement containing the `for` which has no exit criterion. The construct `for (j=j; j==j; j=j)` would be preferred to a `while` for our learning purposes, but the synthesizer requires the `for` iteration variable to be initialized with a constant, and this would not work in our application. So, we shall use `while(1)`.

In our procedural approach, you will notice, as we proceed, that the model is not purely behavioral; it includes bit-level RTL procedural assignments. We shall call our new program `FindPatternBeh`, for "*find pattern behaviorally*".

An outline of the model's loop would look something like this:

```
integer i, j;
reg CurrentSerialBit; // A design data object.
reg[63:0] Stream;     // For now, this is a fixed data object.
...
        // Concatenation used for documentation purposes, only:
        //      'a'                  'b'
Stream = {32'b01100001_00011000_01100010_00010000,
                //    'y'                  'z'
                32'b01111001_00001000_01111010_00000000};
begin : While_i
while(1) // Exit control of this loop will be inside of it.
  begin
  for (i = 63; i >= 0; i = i - 1) // Repeats every 64 bits, if no disable.
    begin
    #31 CurrentSerialBit = Stream[i];  // The delay value will be explained.
    ( do stuff with CurrentSerialBit ... then disable While_i )
    end // for i.
  end
end // End named block While_i.
... // Pick up execution here after disable While_i.
```

The real control of the outer `while(1)` is by the name on the block containing it, `While_i`. When "`disable While_i;`" is executed anywhere in the module containing the code above, everything scheduled between the named `begin` and `end` is terminated immediately, and execution continues below the end of the `While_i` block. As mentioned before, this works like a *C* language `break`.

Use of the disable statement: In verilog, any procedural block may be placed between `begin` and `end`, the `begin` labelled as above with a verilog identifier, and `disable` called on the block by name. Like `goto` in *C*, this should be used only when unavoidable, because it leads to complicated execution which may become difficult to debug when anything goes wrong or when it is desired to modify or improve the code.

To continue with the synchronization, we know from a previous lab, and figure 8-7 above, that, by design, our PLL clock is free-running at a nominal frequency of 32 MHz and continually monitors the frequency of its incoming 1 MHz `ClockIn` clock. In the present design, whenever the PLL receives a positive-edge `Sample` command, it uses the `ClockIn` edges it has been monitoring to make a small frequency adjustment toward that of `ClockIn`. So, we must supply a `ClockIn` nominally at 1 MHz (= 32 Mb/s ÷32) to the PLL, and we must issue regular `Sample` commands.

We therefore choose to issue a `Sample` command on every data packet received, and to supply the PLL with an embedded clock (`EClock`) defined by the toggle of the LSB in the 2-bit frame padding down-count. So, we shall sample on every other `EClock`; `EClock` will be connected permanently to the PLL `ClockIn` input. See figure 8-7.

To extract the embedded clock, `EClock`, we can use the pad pattern in the data stream; we should look first for 3 successive '0' bits, followed by any two more bits nn, which are read as a count value, followed by another 3 successive '0' bits. Each of these successive, padded nn values is separated from the previous one by 8 ignored (data) bits. The nn values will count down by 1 after every data byte, as the data stream is traversed from MSB toward LSB.

To set a somewhat arbitrary synchronization criterion, we shall do nothing until the down-count pattern has been established at least for 4 nn (pad) values which end in `2'b00`. We then set the logic level of our `EClock` equal to that of the nn LSB; this initially will be `1'b0`. This should allow the PLL to determine the direction of a frequency correction, if any, and so we shall begin issuing `Sample` commands to the PLL.

If we lose synchronization as defined by any nn down-count miscount, we stop issuing `Sample` commands and leave `EClock` at its most recent level until synchronization is again established by the criterion above. Our PLL, of course, continues to provide an output clock at a frequency based on the most recent period of synchronization; it is up to other logic to decide what to do with this clock.

Given these synchronization and desynchronization criteria, the question remains of how we should identify the pattern, `8'b000nn000`, where nn represents a 2-bit count.

To answer this question, the preceding outline of the model's loop may be expanded in detail and expressed as a flowchart in figure 8-8.

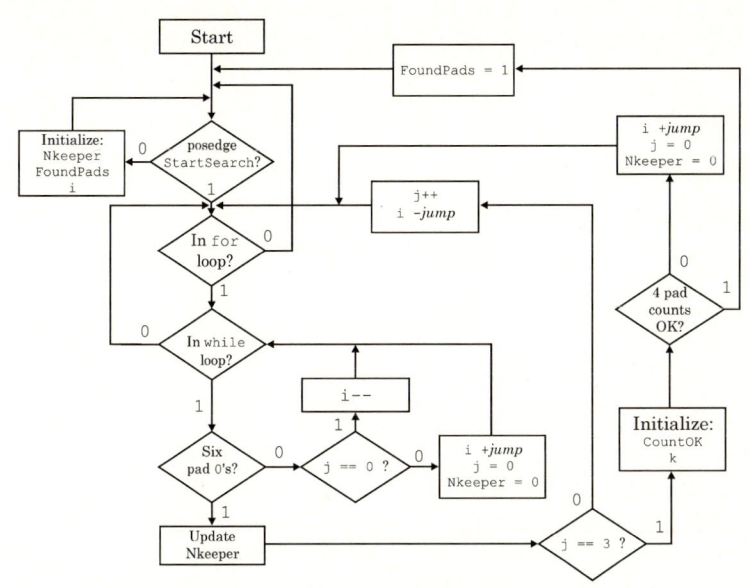

Fig. 8-8. `FindPatternBeh` flowchart used to develop the code fragments below.

From the flowchart of this procedure, the data are arriving serially, so we can start the identification of a packet with a `for` statement that is triggered every `i`-th serial bit received, with `i` stored in a 6 bit (64-value) `reg`. A verilog `reg` is unsigned by default, so counting up or down in this `reg` connects 0 with 63 in either direction.

Let's look for the pad pattern assuming the *current* value of `i` has put us on the MSB + 1 byte = `SerVect[63-8]` = `SerVect[55]`, which is the first '0' in a correctly identified padded count-byte.

An implementation of this in verilog might be by the following code fragment:

```
// Assume SerVect is a saved, serial 64-bit vector:
reg[5:0] i;        // i traverses the serial stream (64-bit vector).
reg n1Bit, n2Bit;  // Two 1-bit regs to hold the expected nn pad count.
...
FoundPads = 1'b0;
begin : While_i  // Name the block to exit it.
while(1)
  begin
  if (   SerVect[i]==1'b0    && SerVect[i-1]==1'b0 && SerVect[i-2]==1'b0
      && SerVect[i-5]==1'b0 && SerVect[i-6]==1'b0 && SerVect[i-7]==1'b0
     )
    begin
    #1 FoundPads = 1'b1;      // 1 means true here, for later use.
    #1 n1Bit = SerVect[i-4];  // Save the padded nn = {n2, n1} value.
    #1 n2Bit = SerVect[i-3];
    disable While_i;          // Exit the while block, if found.
    end // if.
  ...
  i = i - 1;
  end // While_i statement.
end // While_i named block.
```

Study the code above; notice that the j of a previous example is not present. Be sure you understand what this code does in relation to our framing pattern if the *current* value of i puts it on the first '0' of the MSB pad byte (bit 55 in the vector above). Then, move i: Assume that the *current* value of i now is at bit 23, for example, in that pattern. If so, with the big && expression in the if, we are looking at all the 0's in the byte in the part select, SerVect[23:16], to see whether we can capture its *nn* count value.

The i = i - 1 in the while(1) loop control means that the && pattern match test will be run on the next less significant (i-1) position in the SerVect vector if a pattern match does not succeed at the current (i) position. This is because we assume the data stream is arriving MSB first; thus, over time, we assume that less and less significant bits will be at the center of our attention. Of course, in the special case of bit 23 of our SerVect above, the match will succeed, the While_i block will be disable'd, and no new if will be run at i-1 = bit 22, given the code fragment shown.

The assignments shown in the verilog above all are blocking; we are not modelling hardware but are creating behavior. The delays are just to space out edges in the simulation waveforms, but we want every assignment to take effect before the next sequential line is read, just as in a software program. Later, perhaps some of this model will be seen as clearly sequential and thus perhaps will be implemented using undelayed nonblocking assignments.

Also, it might seem that the above pattern search would be speeded up by nesting six if's:

```
if (SerVect[i]==1'b0)
   if (SerVect[i-1]==1'b0)
      ...
```

Digital VLSI Design with Verilog

In terms of the verilog simulation language, this is a false impression, because the multiple `&&` expression is evaluated left-to-right; and, on the first equality failure, the `&&` will go false just as quickly as would a nested `if` in that position. However, the logic synthesizer might produce a different netlist for the nested `if`'s, so perhaps a rewrite in the form of nested `if`'s should be kept in mind.

Anyway, the code above will not set `FoundPads = 1` unless what *probably* are the six pad '0' bits we are seeking have been found. Such a pattern match might, however, be a coincidence, and we might have matched by mistake on an `i` in the middle of a data byte. So, let's elaborate on our search further.

Let's declare a vector `Nkeeper` which is 4 x 2 = 8 bits wide, and use the code above to collect 4 successive, 2-bit nn values (which we think are nn values, anyway).

This adds some more functionality to the code above; it can be written as follows:

```
reg[7:0] Nkeeper;  // Stores 4 2-bit nn values.
reg[5:0] i;        // Indexes into a saved 64-bit SerVect vector.
reg[2:0] j;        // Counts which of 4 assumed nn's we are on.
...
FoundPads = 1'b0;
i = starting value;
for ( j=0; j<=3 ; j=j )  // The for j increment is nonfunctional,
  begin : While_i      // so, the for does not terminate this loop.
    while(1)           // Control of j will be added below.
      begin
        if ( six pad 0's; same as above )
          begin
            #1 Nkeeper[2*j+1] = SerVect[i-3];   // MSB of nn.
            #1 Nkeeper[2*j]   = SerVect[i-4];   // LSB.
            #1 j = j + 1;
            #1 i = i - 16;  // Jump ahead to the assumed next pad byte.
            #1 disable While_i;
          end // if.
        i = i - 1;
      end // while.
  end // While_i block.
```

One minor point here: The delays are located here as placeholders; they make the sequence of simulation events easier to see in a waveform display. When the model is working, these delays should be removed; ultimately, some assignments in the containing module might be changed to nonblocking ones, too. The only caution in adding such delays is that, if blocking, they add up, and the device clock must be slow enough to let them all time out under all conditions. If not, some events might be cancelled during simulation, and the model might fail to work.

The code above is the same as the preceding verilog, except that, assuming four `SerVects` are correct, it repeats until `j` exceeds its limit. The termination of `j` is shown below and requires that j be tested as greater than 3. Therefore, the width of j must be 3 bits or more; if it were just 2 bits, j would count past 3 to 0, and there never would be a count greater than 3; so, the outer for never would exit.

One problem with the above code fragment is that it might get stuck in the data and repeatedly jump past the pad bytes. Also, Nkeeper might get filled up with random values whether or not they came from pad bytes. For example, what if the stream contained no packet at all and was mostly '0' and only an occasional '1'?

We want to check for four <u>successive</u> *nn* pad values, separated by exactly 8 (data) bits each; so, we should modify the above. We shall decrement i by 1 only while searching for a pattern match and not finding it; when we find a match, we'll jump ahead by 16 bits (decrement i by 16) to look for another one. If we ever fail to find a subsequent match, we shall restore j to 0, increment i by 15 to get the first bit which was ignored by jumping ahead, and start the search again after advancing by 1 bit in the stream.

The result comes out something like the following:

```
reg[7:0] Nkeeper;  // Stores 4 nn values.
reg[5:0] i; // Indexes into a saved 64-bit SerVect vector.
reg[2:0] j  // Counts which of 4 assumed nn's we are on.
...
FoundPads = 1'b0;
i = starting value;
for ( j=0; j<=3; j=j ) // No increment.
  begin : While_i
    while(1)
      begin
        if ( six pad 0's; same as above )
            begin
            #1 Nkeeper[2*j+1] = SerVect[i-3]; // MSB.
            #1 Nkeeper[2*j]   = SerVect[i-4]; // LSB.
            if (j==3) // We're done:
              begin
              #1 FoundPads = 1'b1;
              disable While_i;
              end
            // If j wasn't 3, jump ahead for another look:
            #1 j = j + 1;
            #1 i = i - 16;
            end
        else // No six-0 match this time.
            if (j==0)
                 #1 i = i - 1;  // First nn not found yet.
            else begin       // We found at least one nn, but now failed:
                 i = i + 15; // Drop back after jumping by mistake.
                 j = 0;      // Reset nn counter; we're not in a pad byte.
                 end
      end // while(1).
  end // While_i.
```

So, if the preceding block of code exits, we shall have stored four 2-bit values in Nkeeper on the assumption that they will be a sequential, binary down-count in our packet format. Really, we should have dropped back by j*16 - 1, not just by 15; we'll fix that when we revisit our code below. But, is it true that in the above we have a way to find our pad pattern?

No, we haven't checked the counts. To be very sure of having found the pad bytes, and thus the packet boundary, we should add a check for the down-count. If we don't confirm a down-count, we have not actually found our packet framing; so, we should continue searching. We only allow the code to exit if we have found what seems to be a properly padded packet of data.

The easiest place to check for a down-count is in the block above which sets FoundPads to 1. We should add a branch there on whether j is 0 or not; if it is 0, there is only one *nn*, so, there will be no use checking it. But, if j>0, we can see whether the current 2 nn bits is exactly 1 less than the previous 2 *nn* bits.

First, let's break out the initialization of the above into a new always block which is read on the opposite edge of StartSearch. This way, even if we assign to the same variables from the two always blocks, there can't be a race condition unless the delay during one block extends to the other edge. This won't happen in our design, because StartSearch is not a clock but a search-enable signal.

Our new always block:

```
always@(negedge StartSearch)
  begin
  #1 Nkeeper   = 'b0;    // Init count keeper every search start.
  #1 FoundPads = 1'b0;   // Move this init here.
  #1 i         = StartI;
  end
```

Then, the final searching block is as shown here:

```verilog
always@(posedge StartSearch) // Clocked, maybe sequential logic?
begin : AlwaysSearch
for ( j=0; j<=3; j=j ) // No increment. '<=' is relational operator!
  begin : While_i
  while(1)
  begin
    if ( six pad 0's; same as above )
        begin
        #1 Nkeeper[2*j+1] = SerVect[i-3]; // MSB of pad count.
        #1 Nkeeper[2*j]   = SerVect[i-4]; // LSB of pad count.
        // Check whether done:
        if (j==3) // We have 4 apparent nn values; do they count down?
            begin
            #1 CountOK = 2'b00;
            for (k=1; k<=3; k=k+1)
              begin
              // Use concatenation to get 2-bit nn values:
              #1 nPrev = { Nkeeper[2*k-1], Nkeeper[2*k-2] };
              #1 nNow  = { Nkeeper[2*k+1], Nkeeper[2*k] };
              if ((nNow+1)==nPrev) #1 CountOK = CountOK + 1;
              end //for k.
            if (CountOK==3) // Total of 4 were OK; so,
                begin     // issue a pulse and stop everything:
                #1 FoundPads = 1'b1;
                #1 disable AlwaysSearch;
                end
                else begin // If not a down-count, start all over:
                #1 i = i + 16*j - 1; // See * below:
                #1 j = 0;
                #1 Nkeeper = 'b0;
                end // if CountOK.
            end // if j==3
        else begin // j not 3:
            #1 j = j + 1;
            #1 i = i - 16; // Jump ahead, for another padded nn.
            #1 disable While_i;
            end // else not j==3.
        end // if 6 zero matches.
    else if (j==0) // First nn not found yet:
        #1 i = i - 1;
        else begin // * This was not first apparent nn found.
        #1 i = i + 16*j - 1; // Drop back after jump by mistake.
        #1 j = 0;            // Reset nn counter.
        #1 Nkeeper = 'b0;    // Reinit count keeper.
        end
    end // while(1) block.
  end // While_i labelled block.
end // always AlwaysSearch.
```

The preceding is a complete RTL design which correctly locates the packet pad bytes in our fake piece of serially streamed data. It would work as well, slightly adapted, to a serially varying input which was formatted to conform with our required framing protocol.

A copy of the RTL search code above, with a testbench, is provided in the `Lab10` directory and is in a file named `FindPatternBeh.v`.

However correct it might be, we now replace this digressive model with a simpler, synthesizable one.

Digital VLSI Design with Verilog

8.4.5 Unsynthesizability of the Behavioral PLL Code

The problem with the model above is that it almost is a C model; it is not efficiently synthesizable. In fact, the code in `FindPatternBeh.v` probably will not synthesize at all. If it did synthesize, the result would be quite large and complex.

One hint of synthesis problems is the presence all over of blocking assignments with delays. A test: Remove the delays: If the code still simulates correctly, it may be synthesizable.

For example,

```
if (CountOK==3)
    ...
else begin // If not a down-count, start all over:
    #1 i = i + IJump*j - 1;
    #1 j = 0;
    #1 Nkeeper = 'b0;
    end
...
```

If the above #1 delays were necessary to proper functioning of some other `always` block or `module`, this verilog would not synthesize to logic which was functionally correct.

Also, if one replaced all the blocking assignments in the code fragment with nonblocking ones, there would be a race condition, with the delays shown. If the assignments were changed to nonblocking, and selectively longer delays were provided, say #2 j <= 0, it might simulate correctly; but, again, synthesis probably would produce defective logic. For this fragment, the best solution would be to remove all delays entirely.

Delays aside, it can be difficult to rewrite a behavioral model in synthesizable form if the model is of any complexity and requires specific delays. Complex models are best divided up into smaller, simpler ones to prepare them for synthesis. The reader may consider spending a little time thinking about how to convert the behavioral model above, but we shall take a different tack in writing a synthesizable version.

8.4.6 Synthesizable, Pattern-Based PLL Synchronization

Let's go back and look again at the problem: We have a stream of serial data and we wish to synchronize a clock to the framing pattern. In our project, the pattern is 64 bits wide, so it makes no sense to worry about anything less than 64 bits in width.

As an alternative to the behavioral, C-like coding presented previously, we can approach the problem this way: Assume that we shall have a dynamic sampling "window" of the serial stream which is 64 bits wide and is stored in a register. We use the window to check for a framing (pad) pattern; if we find it, we synchronize our clock to a boundary which is well-defined in the frame; if we don't find the pattern, we shift the window by one new bit (= serially shift in a new bit) and check again. If our window only changes one bit at a time, we cannot fail to find the frame boundaries, if they are there.

So, all we need do is check a completely static window of data. This can be done concurrently by checking every pad bit, all 32 of them, in the 64-bit sample we have. There is no need to shift or change anything, so there is no sequencing or delaying of anything at all in simulation time. All we have to do is agree not to change anything in the register while we examine the sample for our frame boundaries.

To solve our problem, first, we define our pad boundary patterns; then, we decide where to look for them in the 64-bit window. And, that's all there is to it.

Actually, there's less to it than that: We can agree only to look for the entire 64-bit pattern centered in the window (serial packet MSB at the window MSB position). Why not? If the actual serial stream is not found to be centered on one sampling, we just shift a new LSB serial window bit in, shift the old window MSB out, and check; shift and check; and just keep doing this until we have the framed pattern centered; then, we recognize it, and off we go! Actually, we could shift this way either up or down.

Using this approach, all we have to do in our model is choose bits to which we should attach comparator logic (*xor*'s in the netlist, maybe).

The whole process might go something like this:

```
reg[63:0] SerVect = // The current 64-bit window to scan.
/*        60         50         40         30         20         10         0
 *  32109876 54321098 76543210 98765432 10987654 32109876 54321098 76543210 */
64'b01100001_00011000_01100010_00010000_01111001_00001000_01111010_00000000;
...
localparam[PadHi:0] p0 = 8'b000_00_000; // The pad patterns.
localparam[PadHi:0] p1 = 8'b000_01_000; // localparams can't be overridden.
localparam[PadHi:0] p2 = 8'b000_10_000;
localparam[PadHi:0] p3 = 8'b000_11_000;
...
   if (    SerVect[55]==p3[7] && SerVect[54]==p3[6] && SerVect[53]==p3[5]
        && SerVect[52]==p3[4] && SerVect[51]==p3[3]
        && SerVect[50]==p3[2] && SerVect[49]==p3[1] && SerVect[48]==p3[0]
        && SerVect[39]==p2[7] && ... (total of 32 compares)
      ) Found = 1'b1;
   else Found = 1'b0;
   ... (etc.)
```

This process will be much easier to do after we have studied functions, which will be next time. So, for now, we leave this problem as-is and shall return to it later.

8.5 PLL Behavioral Lock-In Lab 10

Topic and Location: Reorganization of PLL modules; behavioral extraction of a serial clock from a serial stream.

Work in the `Lab10` directory.

Preview: We'll develop some understanding which will help us later in our serdes class project. We first copy in our PLL model from `Lab08`, renaming some things and isolating the PLL from its `Lab08` counters. Then, we'll edit a copy of the `FindPattern` model (developed above) so that it works to detect serial packet boundaries in a faked, constant serial stream test pattern. We shall not yet hook the PLL to the packet-boundary detector; extraction of the packet boundaries and thus of the serial embedded clock will be completed later.

Deliverables: Step 1: A cleanly isolated and renamed `PLLTop` module representing our fully functional PLL model of `Lab08`. Step 2: A correctly simulating counter clocked by `PLLTop`. Step 3: A behavioral model which simulates correctly the extraction of our packet boundary patterns for the fixed test-pattern serial stream supplied.

Lab Procedure:

In this lab, we shall exercise use of the `for` statement as a way of sampling a data stream. Thomas and Moorby (2002) recommends varying among `for`, `while`, and `forever`, depending on the context; however, in practice, the `for` statement can do anything these others can; we shall use it extensively here. The others usually are simpler, so we shall exercise `for` also to understand its complexity (and its benefits) better.

Step 1: Make a subdirectory in `Lab10` named `PLLsync`. Copy the entire `PLL` design from the `Lab08/Lab08_Ans/Step05_ClockedByPLL` directory into the new `PLLsync` directory. This design will include a PLL clock which is applied to three different counter structures. The top of the design should be a module named `ClockedByPLL`.

You probably have a complete design of your own for this, from your Lab 8 work, but the lab instructions will work best if you use the answer files provided.

Reorganize the files. Assume there are too many files in one place in this design; so, make a new subdirectory named `PLL`, and move the five PLL module files, which are named next, into the new `PLLsync/PLL` directory.

The files to be moved are named after the `ClockComparator`, `MultiCounter`, and `VFO` submodules making up the PLL, and `PLLTop.inc`, a defined-constant include file for the VFO module. Finally, the `PLLTop.v` file also should be moved into your `PLL` directory and should contain the top level module which connects the PLL submodules. The top level `module` should be named `PLLTop`; rename it, if it is not already so.

Create a simulator file-list file, `PLLsync.vcs`, one level up (in the `PLLsync` directory), and invoke the simulator briefly so you are sure it can be run with the PLL design moved into the `PLLsync/PLL` directory. This is just to check the file locations.

Step 2: Rename and revise the design top. Change the module and file names of the three-counter design from `ClockedByPLL` to `PLLsync`. Remove all counters except the behavioral one, so that the only count output is the behavioral one. Delete the `DFF.v` model and any other unused files from `Lab08`; you always can make new copies, if you want them.

The block diagram of your new `PLLsync` design, which in `Lab08` once was called `ClockedByPLL`, now should look like figure 8-9.

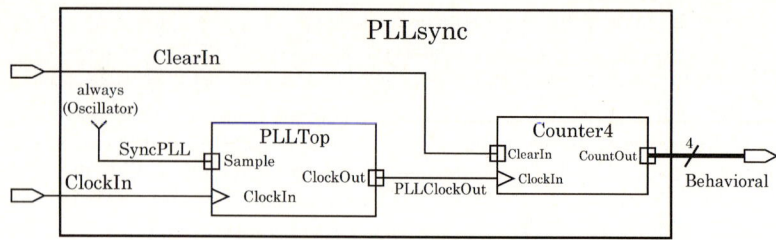

Fig. 8-9. The PLLsync design block diagram.

Simulate PLLsync to verify its correct functionality; you may alter the testbench already in that file to save some time. Use a 1 MHz ClockIn. This completes our use of PLLsync in this lab.

Step 3: Modify a pattern-finder to extract a clock and a PLL sample command. In your Lab10 directory, you have a file named FindPatternBeh.v; make a copy of it named EClockSample.v ("E" for *extract*). Rename the module in the copy to EClockSample.

See the left side of figure 8-10:

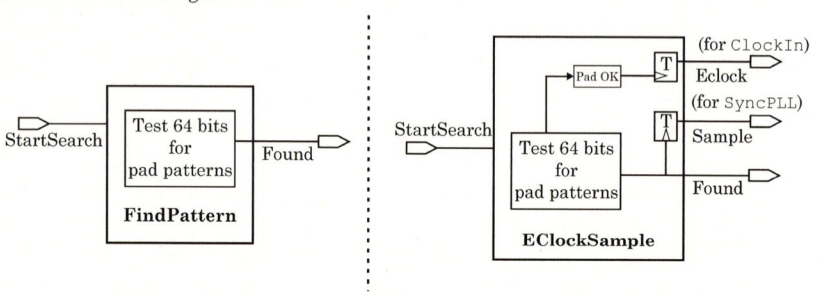

Fig. 8-10. FindPattern can be modified trivially to EClockSample, (right) to provide inputs for PLLsync. T = toggle; StartSearch would be triggered on every new received serial bit.

Note: As already explained, FindPattern, now named EClockSample, is a simulation model, only. It can not be synthesized.

As shown on the right side of figure 8-10, modify the module contents in the new EClockSample so that it outputs our required extracted clock and sample command, as we specified above. Recall that a D flip-flop with an output of its ~Q tied back to its D often is called a ***toggle*** flip-flop, or T flip-flop, as shown in the right side of figure 8-10.

Do ***not*** try the pattern-based approach instead of EClockSample; the pattern-based approach, as you may recall, was a loopless, synthesizable design introduced at the end of today's serdes presentation. It won't work for this exercise. Do not worry about actual sample speed in your test bench; just extract the EClock and the Sample command as you search through the serial bit pattern with the unsynthesizable FindPattern looping approach.

Digital VLSI Design with Verilog

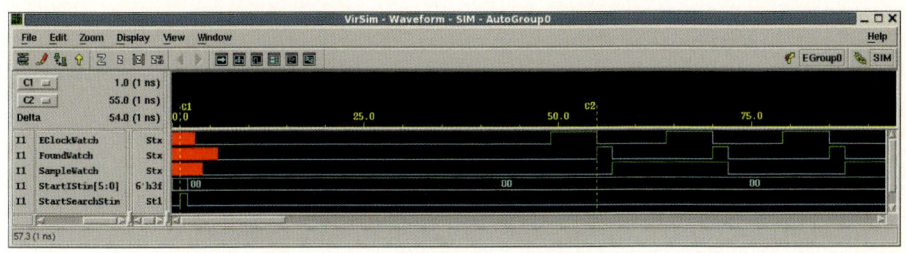

Fig. 8-11. Simulation of `EClockSample`, showing the extracted clock and the sample command.

Thus, `EClockSample` simply should output the value of the pad-pattern counter LSB continuously, renamed to `EClock`, which was wired to the `EClockWatch` in the testbench used for figure 8-11, where *Stim are module inputs and *Watch are outputs. While Found (= FoundWatch) is asserted, the `EClock` is good (synchronized); while Found is not asserted, the `EClock` can not be depended upon.

Do not try yet to attach the `ClockIn` from `PLLsync` to the `EClock` from `EclockSample`. However, keep in mind that, later, we may use a source of `EClock` to adjust the PLL frequency when `EClock` is known good and let the PLL oscillator run free when `EClock` is not known good.

8.5.1 Lock-in Lab Postmortem

How should one deal with establishing and losing synchronization?

8.6 *Additional Study*

Read Thomas and Moorby (2002) section 4.6 on **disable**.

Thomas and Moorby section 6.5.1 gives a truth-table for the `bufif1` primitive.

Read Thomas and Moorby section 10.2 on strength and contention.

Read Thomas and Moorby Appendices C.1 and C.2 on verilog operators and precedence. The operator functionality is very important to know thoroughly; precedence is less important, because it can be defined or overridden by parentheses.

Optional readings in Palnitkar (2003):

Section 3.2.1 and Appendix A on verilog builtin net types and strengths.

Section 6.1.2 on *implicit* net declarations.

Look through sections 6.3 and 6.4 on verilog operators.

Chapter 9
Week 4 Class 2

9 Today's Agenda:

Lecture on FIFO state-machine design
Topics: Tasks, functions, `fork-join` blocks, state machines, and FIFOs.
Summary: After introducing tasks and functions, we describe the procedural concurrency of the `fork-join` block. We then discuss the best use of verilog in state machine design. After that, we describe operation of a FIFO in detail, focussing on the read and write address controls. We end by describing how to code a FIFO controller state machine in verilog.

Lab on FIFO design
After a warmup on tasks, a simple FIFO state machine is coded.

Lab Postmortem
FIFO design across clock domains; gray code counters.

9.1 State Machine and FIFO design

Our biggest topic today will be FIFO design and the state machines required for it. First, though, we require some tools for making the verilog more effective.

9.1.1 Verilog Tasks and Functions

The structure of a verilog design is defined by its hierarchy of `module` instances. Any common functionality can be put into a module and then instantiated as many times as desired. However, sometimes, repetitive functionality is required which is procedural and should not be visible as part of the design hierarchy. As we have seen, associating new values to numbers or logic states by using operators such as '+', '&', or "==" is far more convenient than wiring up adders, *and* gates, or comparators. Likewise, defining a shift-register in an `always` block with one statement is much quicker and less error-prone than wiring one from individual gates—or even wiring one in as a module instance. In verilog, convenient handling of this kind of commonplace, repetitive, procedural functionality is made available to the user in the form of ***tasks*** and ***functions***.

Both of these constructs are declared in the module in which they are intended to be used. Although they may be called remotely by a hierarchical reference, such references are bad design practice and are discouraged. If reuse in different modules is required, the declarations may be put in a file and the file then may be introduced into a module by means of a `` `include ``. Because neither tasks nor functions represent design structure, their declarations can not contain local `wire`s to wire their contents to anything; however, they may contain local `reg` variables for temporary or persistent data manipulation.

The difference between `tasks` and `functions` has to do with complexity and, more importantly, with timing. We refer here only to <u>user-defined</u> `tasks` and `functions`, not to the verilog language's builtin system tasks or system functions.

<u>Complexity</u>. A `function` is simpler. A `task` can call a `function` or another `task`; a `function` can not call a `task` but may call another `function`. So, `tasks` potentially are more complicated than `functions`. In addition, a `function` just changes one thing when it is executed: its return value; in effect, a `function` merely defines an expression which is evaluated each time it is called. A `task` can modify external `reg` objects as it executes, possibly leaving a diversity of changes after it is done.

A `task` may include delay expressions or nonblocking assignments, although delayed nonblocking assignments probably will not be synthesizable. A `function` may contain only undelayed blocking assignments.

<u>Timing</u>. A `task` can include scheduling delays and event controls; a `function` can not. Because `tasks` can be executed over some arbitrary simulation time period, they can be used to represent the concurrency typical of hardware components.

All `functions` are just complicated substitutes for expressions; `functions` execute in zero simulation time and return a value, very much like a simple or concatenation-grouped expression.

9.1.1.1 Task and Function Declarations

A ***task*** is declared with a port list much like a module header; its declaration begins with the keyword `task`, and it ends with the keyword `endtask`.

Here is an example of a `task` declaration and call:

```
task SwizzleIt (output[3:0] SwizOut, input[3:0] SwizIn, input Ena);
  begin
  if (Ena==1'b1)
      #7 SwizOut = { SwizIn[2], SwizIn[3], SwizIn[0], SwizIn[1] };
  else #5 SwizOut = SwizIn;
  end
endtask
...
always@(posedge GetData)
   SwizzleIt( Bus2, Bus1, SwizzleCmd );
```

A ***function*** is effectively the name of a temporary `reg` type that expresses the value returned by the `function`. For this reason, a `function` is declared as an object with a specific width, and with inputs only. Even though a `function` is called with inputs only, and can not have `output` or `inout` identifiers declared for it, its input(s) must be declared with the keyword `input`. A `function` may be called in a continuous assignment statement, although this is done rarely in practice. A `function` may call timing-independent system tasks or functions, such as `$display`.

When called, a `task` or `function` must be passed real params (arguments) by position, only; port-mapping by name is not allowed.

A `function` is declared between the keywords `function` and `endfunction`. For example, here is a `function` declaration and call:

```
reg[7:0] CheckSum;
function[7:0] doCheckSum ( input[63:0] DataArray );
   reg[15:0] temp1, temp2;    // Just to illustrate local declarations.
   begin
   temp1 = DataArray[15:0] ^ DataArray[31:16];
   temp2 = DataArray[63:48] ^ DataArray[47:32];
   doCheckSum = temp1[7:0] + temp2[7:0] ^ temp1[15:8] + temp2[15:8];
   end
endfunction
...
#2 CheckSum = doCheckSum(Dbus);
```

Statements to be run by a `task` or `function` must be placed between its `begin` and `end` keywords. Local variable declarations (`reg`, `integer`, `real`) must appear before the `begin`.

A `task` exits, returning nothing, after its last statement has been executed.

A `function` exits when one of its statements assigns to its name, the value assigned being returned. If the name of the `function` is not assigned during its run, then the value of the last expression assigned is returned.

9.1.1.2 Task Data Sharing

Because `tasks` affect declared data objects as they run, a possible problem arises when two calls are made to the same `task` during the same simulation-time running period: Both `task` calls operate on the same data, because, the `task` being declared only once, both executions must operate upon the same, unique objects declared in the *one* `task` declaration. This means that concurrent execution may lead to unpredictable behavior which may be fatal to the hardware being simulated.

To prevent sharing of declared internal data, and to allow recursion, `tasks` or `functions` may be declared as **automatic**, which means that their data are copied and pushed on the (simulator CPU) stack as the sequence of recursive executions proceeds. The `automatic` declaration prevents sharing of local data among different `task` or `function` calls; the rationale becomes the same as that of the handling of local variables when making recursive function calls in *C*.

You may wish to look at the on-disc example in the `Lab11` directory, `Find3Mod.v`, to see how the `automatic` keyword is used. Implementation of `automatic` is not consistent across all EDA tools, so its use should be avoided if portability is important.

9.1.1.3 Tasks and Functions Are Named Blocks

Finally, `tasks` and `functions` are considered named blocks, and either one also may contain other named blocks. A `task` can be disabled by name from within itself or from anywhere that it can be called. A `function` can't be disabled except from within itself, because it executes in zero time; however, named blocks in a `function` can be disabled by statements within the function.

9.1.2 A Function for Synthesizable PLL Synchronization

We outlined a synthesizable rewrite of the `FindPattern` module in the previous chapter; the idea was to do one giant, 32-expression `if` to check format in a 64-bit data stream. However, an equally valid way would be to run four `function` calls, each checking just 8 bits at a time.

For example, below is a `function` named `checkPad()`, which uses a `for` loop to do such a check. A `function` executes procedurally in zero simulation time, so it can be written in software style, with no worry about simulator-synthesis scheduling inconsistencies. No delay is allowed, of course; but, we already have assumed that our synthesizable version of `FindPattern` would require no simulation-object update.

The `CheckPad function`:

```
function       // 64-bit vector      8-bit pad pattern    offset in the stream
checkPad ( input[VecHi:0] Stream, input[PadHi:0] pad, input[AdrHi:0] iX );
   reg OK;      // Flags pattern match.
   integer i    // i is the data stream offset.
         , j;   // j is the pad-byte pattern offset.
   begin
   i   = iX;   // Init to stream MSB to be searched.
   OK  = 0;    // Init to failed state.
   begin : For // Capitalized, this is not a verilog keyword.
    for (j=PadHi; j>=0; j = j-1)
      begin
      if (Stream[i]==pad[j])
          OK = 1;
      else begin
          OK = 0;
          disable For; // Break the for loop.
          end
      i = i - 1;
      end // for loop.
   end // For.
   checkPad = OK;
   end
endfunction
```

Notice the "`disable For;`" statement, which stops all activity inside the named block, requiring that the final line, "`checkPad = OK;`" be executed next. This last ends execution, because a verilog function returns as soon as it is assigned a value.

Digital VLSI Design with Verilog

Two comments on the `function` declaration:

(a) the index variables are declared `integer`, because the `for` loop exit variable must be used in a signed comparison: To exit this `for` loop, a value less than 0 must be expressed. The synthesizer's optimization routines will remove unused bits from any 32-bit `integer`, leaving only a register big enough to do the job.

(b) The `for` loop is put inside a block named, perhaps humorously, "For"; this is legal, because verilog keywords all are lower-case, and verilog is a case-sensitive language. The block is named to allow `disable` to trigger an early exit (during simulation) on a mismatch.

If we assume that our pad-byte patterns are to be local (permanent) parameters, they will be declared about like this:

```
...
localparam[PadHi:0]  pad_00  =  8'b000_00_000;
localparam[PadHi:0]  pad_01  =  8'b000_01_000;
localparam[PadHi:0]  pad_10  =  8'b000_10_000;
localparam[PadHi:0]  pad_11  =  8'b000_11_000;
...
```

With such pattern declarations, the function `checkPad()` may be called this way:

```
...
if (     checkPad(Stream, pad_11, OffSet-(1*PadWid))
     && checkPad(Stream, pad_10, OffSet-(3*PadWid))
     && checkPad(Stream, pad_01, OffSet-(5*PadWid))
     && checkPad(Stream, pad_00, OffSet-(7*PadWid))
   )
     FoundPads = 1;
else FoundPads = 0;
```

The `checkPad` four-statement test above is much more readable (and reusable) than 32 equality comparisons all globbed into one `if` expression.

See the new rewrite of `FindPattern.v` in the `Lab11` directory for a full implementation of the `function` calls above.

9.1.3 Concurrency by `fork-join`

In preceding chapters, we have looked at isolated, concurrent, independent execution by different `always` or `initial` blocks. We also have looked at procedural execution, line by line, within sequential blocks. We have seen how procedural evaluation of delayed blocking assignments differs from concurrent evaluation of delayed nonblocking assignments. Verilog also provides a construct which allows controlled concurrency between statements. This is called the parallel block, or `fork-join` construct. It is modelled after the *fork* system call in the unix operating system.

A `fork-join` currently (2011) is not synthesizable, but it is used in simulation. Because IP blocks are not normally synthesized, the `fork-join` may be used to represent hardware in a simulation model of an IP block to be included in an otherwise synthesizable netlist. Thus, understanding of fork-join is more important to a designer than one might initially guess.

A fork-join block is allowed wherever a procedural block is allowed, and it may contain any number of statements which are executed concurrently by the simulator, with the usual unknown, effectively random, order of initialization of each such statement relative to the others. What is special about the fork-join block is that it provides for a waiting point during which all its concurrent statements are allowed to catch up with one another and resynchronize.

The statements following the fork are effectively independent threads of execution; they are gathered together to a single time point, indicated by join, when the last of them finishes its execution. This gathering is not simultaneity of the forked statements but rather applies to the next sequential statement following the join in the procedural block containing the fork-join.

For example, in the following code, DataBus will see several glitches during simulation. If the delay times shown can't be modified to fix this, the glitches will have to be tolerated.

In this particular example, OutBusReg is updated 4 ns after clock, and the change in DataBus[3] is not seen until the next clock:

```
always@(posedge Clk)
  begin
  #1 DataBus[0] <= 1'b0;
  #2 DataBus[1] <= 1'b1;
  #3 DataBus[2]  = 1'b0;
  #4 DataBus[3] <= 1'b1;   // OutBusReg misses this one.
  #1 OutBusReg = DataBus;  // Updated after DataBus[2].
  end
```

In verilog, the first two delayed nonblocking assignments above should be concurrent, but they may not simulate correctly in some simulators. Regardless, by collecting the result inside a fork-join block, the glitches can be prevented and OutBusReg will be updated with all changes 5 ns after clock.

The collected result:

```
always@(posedge Clk)
  begin
  fork
  #1 DataBus[0] <= 1'b0;
  #2 DataBus[1] <= 1'b1;
  #3 DataBus[2]  = 1'b0;
  #4 DataBus[3] <= 1'b1;
  join
  #1 OutBusReg = DataBus;  // Updated after DataBus[3].
  end
```

Even though the above assignment to `DataBus[2]` is a blocking assignment, because of the `fork` it is scheduled beginning at the same time as all the others. When the longest `fork`ed delay lapses, the `join` allows the assignment to `OutBusReg` to be scheduled.

When you want several assignments to occur simultaneously during simulation, and if you have to use procedural delays, consider using a `fork-join` block to accomplish this. Otherwise, avoid this construct.

9.1.4 Verilog State Machines

There are two main kinds of state machine, Mealy and Moore. In a Moore machine, the outputs depend solely on the current state; in a Mealy machine, the current state has some control over outputs, but additional control comes directly from inputs and may be independent of the state. For example, a state machine might cycle through various conditions, but a separate device might not always be ready to accept the state machine's outputs; if so, the other device may put some or all of the state machine's outputs into a high impedance state or may switch the outputs using a multiplexer; this machine then would be a Mealy machine. We shall not consider this distinction further here, because for our purposes, the essence of the state machine is in how it changes state, not in how its output logic is controlled.

A state machine consists minimally of a state register which changes value over time in a deterministic way. A simple state machine is a toggle flip-flop, such as the one we used in our ripple counter lab exercise: If the current state is *set* ('1'), then on the next clock the state changes to *clear* ('0'); if the current state is *clear*, then on the next clock the state changes to *set*. Any digital counter is a simple state machine; but, usually, state machines which are designed as such are much more complicated. To describe the complicated functionality of such machines, a bubble diagram or a flow chart is used. Such diagrams may be found in abundance in the data book of a modern microprocessor.

Good verilog design of a simple state machine is to separate the state register control logic from most of, or all, the combinational logic in the machine. This separation is not required by the language, but it simplifies the understanding and maintenance of the state machine. Such separation also (usually) reduces the amount of coding required. See figure 9-1.

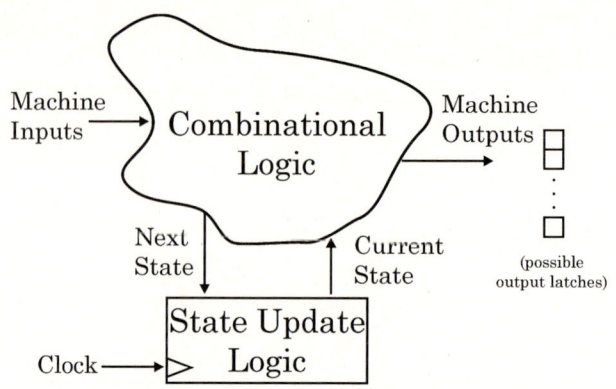

Fig. 9-1. Abstraction of a verilog state machine. Separation of the functionality allows blocking assignments to be used throughout the combinational code.

The state update logic of such a machine includes the state register and is not accessible to external devices. The combinational logic uses inputs and the current state to (*a*) assign outputs and (*b*) determine the next value of the state register.

9.1.5 FIFO Functionality

FIFO is an acronym for First-In, First Out: The first value written into a FIFO is the first one to be read out, somewhat like a pipeline in which the fluid sent in first is the first to be delivered.

A FIFO is a kind of register or memory stack. A stack structure is typified by storage which is addressed by numbers calculated from recent numbers, as opposed to being calculated independently as some arbitrary pointer or address value. It is, then, very reasonable that the complement of a FIFO should be a LIFO, Last-In, First Out. A LIFO is like a standard microprocessor stack: The last value pushed onto the stack is the first one popped off of it.

A LIFO can be controlled by a single register, a stack pointer, because the location used to push data into a LIFO is the same as the location from which data would be popped. However, a FIFO is more complicated, because, like a pipeline, it has two ends to be controlled. Whereas a LIFO can have just one stack pointer, a FIFO has to have two different pointers, one to write data in, and the other to read data out.

Digital VLSI Design with Verilog

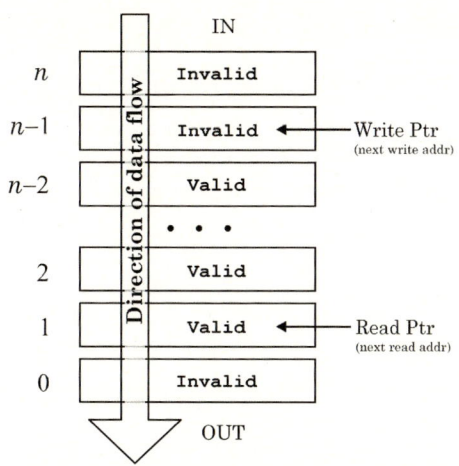

```
                    IN
                    ↓
   n      ┌──────────────┐
          │   Invalid    │
          ├──────────────┤
  n−1     │   Invalid    │ ←── Write Ptr
          ├──────────────┤       (next write addr)
  n−2     │    Valid     │
          ├──────────────┤
                • • •
          ├──────────────┤
   2      │    Valid     │
          ├──────────────┤
   1      │    Valid     │ ←── Read Ptr
          ├──────────────┤       (next read addr)
   0      │   Invalid    │
          └──────────────┘
                    OUT
      (Direction of data flow)
```

Fig. 9-2. First-in, first-out functionality gives a FIFO register file a specific directionality.

As shown in figure 9-2, the FIFO storage consists of some number n of registers, indicated by horizontal rectangles, and a predefined direction of data flow. Data are written in at some rate per unit time; they are read out at some perhaps different rate. One application of a FIFO is as a buffer between two different clock domains on a chip; bursts of reading or writing are cushioned by the FIFO so that devices in both domains can move data at some useful rate without waiting on every clock tick for the other domain to become ready. Other FIFO applications are in RS-422 or ethernet serial links, both of which usually communicate between independent clock domains.

There are two main conventions in representing data flow in FIFO read and write pointers: (*a*) These pointers can be viewed as pointing to the next valid address (register) in the FIFO; or, (*b*) they can be viewed as pointing to the most recently used one. In this course, we shall adopt the former convention: We always represent a pointer so that it points to the **next** address.

Our read pointer, as shown in figure 9-2, then must point to valid data in the FIFO, the next datum to be read; the write pointer must point to the unused register into which the next new datum will be written. Thus, the write pointer points to currently invalid data, generally data which already have been copied (read) out, so that that register's contents are of no further value.

If the data flow is as shown in figure 9-2, from top to bottom, it must be that the two pointers move upward after each use. If we call "writing" a process that changes invalid data addresses to valid ones, what happens when the write pointer shown reaches register n, at the "top" of the FIFO? Simple: It wraps around and positions itself at the bottom register shown, if that register's contents are invalid. We have seen the same thing when one of our unsigned `reg` counters wraps around from its maximum value to 0 again; and, in fact, unsigned counters can be used to control FIFO read and write pointers.

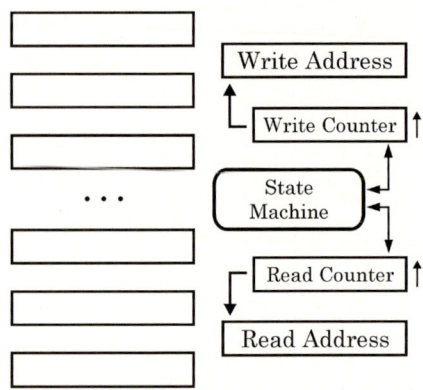

Fig. 9-3. FIFO component parts.

The component parts of a FIFO are as shown broken down in figure 9-3. There is a set of data storage registers as shown on the left; two pointers storing the read-address and write-address; and an unsigned counter to control each pointer value. The state machine enables or disables counting and monitors the current count values.

Address translation for the FIFO data storage register file could be used to map the counts to and from Gray code addresses. Thus, the register read and write addresses can be encoded values rather than simple, sequential binary counts. The small vertical arrow on the right of each counter indicates that it is allowed to count just one way, here depicted as "up". If the counters could count both the way shown and the opposite way, which is to say both "up" and "down", there would not be any well-defined direction of data flow in the FIFO, and it would not be a FIFO. By encapsulating count-to-address translation in verilog *task*s, the translation rationale, Gray code or whatever, can be changed arbitrarily without need to alter anything else in the FIFO implementation.

9.1.6 FIFO Operational Details

Before attempting to write the verilog describing a FIFO, let us look more closely at how it must operate. First, suppose the FIFO was "empty", which is to say, that all data had been read out of it and no new data had been written since. Our way to describe this depends on where the write pointer was when the read pointer was used to read out the last valid datum.

Suppose, for example, that the write pointer was at the second register from the top in figure 9-3. Then, the FIFO would look something like that in figure 9-4, labelled $Empty_1$.

Digital VLSI Design with Verilog

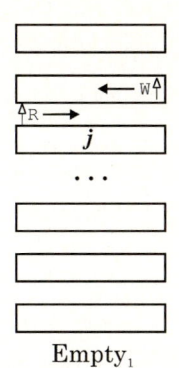

$Empty_1$

Fig. 9-4. Register addressing for an empty FIFO.

The horizontal arrow sits on the register or other location at which the pointer is pointing; the small vertical arrow indicates the only allowed direction of change. The write pointer, labelled with a "W", is free to move upward; the read pointer ("R") has just read from register j and is not allowed to move upward to $j+1$, because those data are invalid, as explicitly indicated by the position of W. Thus, we must depict R as itself being invalid, pointing nowhere useful. The R stands for "Read"; so, this FIFO can not read anything and therefore is empty.

In the state just described, suppose, now, that one new data value was written. The FIFO no longer would be empty. This datum would be written to register $j+1$, and W would move to point to the next invalid register, $j+2$. We are now allowed to read from $j+1$, so R should be pointing there.

This is shown in figure 9-5, labelled $Empty_1+W$.

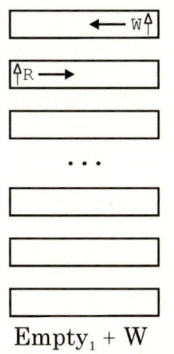

$Empty_1 + W$

Fig. 9-5. Register addressing for an almost-empty FIFO.

Let's look at a different, but equally "empty", condition. Suppose the last valid datum had been read while the write pointer was pointing to the bottom register ($n=0$, in figure

9-2). Then, the read pointer must be invalid and again must point nowhere. We may indicate this by putting R right below W, because when W does a write, R will be ready to read from that exact, same register. Therefore, this second empty FIFO may be depicted as shown in figure 9-6, labelled *Empty₂*.

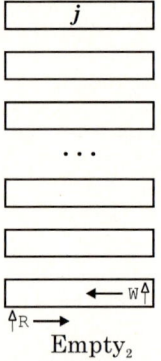

Fig. 9-6. Register addressing for an empty FIFO.

In this case, when the next write occurs, making the FIFO no longer empty, W will move from 0 to point to register 1, and R will become valid, pointing to register 0. This is shown in figure 9-7, labelled *Empty₂+W*.

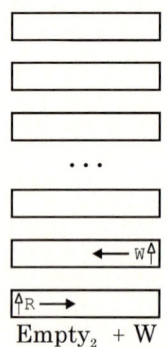

Fig. 9-7. Register addressing for an almost-empty FIFO.

Very good. We might now guess from this that R must be less than W at all times, because R never can move upward to point to the same register as W; therefore, R never can cross over W and get above it.

However, now let us look at the opposite condition of the FIFO, when it is "full". This means that not enough has been read recently, and that all possible registers have been written, so that no more writes are allowed, and therefore the W pointer must be invalid.

Digital VLSI Design with Verilog

Suppose this condition occurred just after W had written its last datum into some register j, as shown in figure 9-8, labelled $Full_1$.

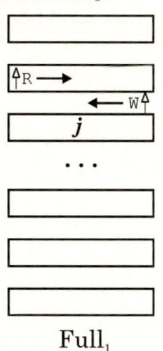

Full$_1$

Fig. 9-8. Register addressing for a full FIFO.

Oops! We guessed wrong. As easily seen, it is R which is now just ahead of the next possible position of W. W is not allowed to move up past R, because writing to the already valid register $j+1$, pointed to by R, is forbidden.

If a read now takes place, the FIFO no longer will be full, register $j+1$ will become invalid, and the condition will change to the one shown in figure 9-9, labelled $Full_1 - R$.

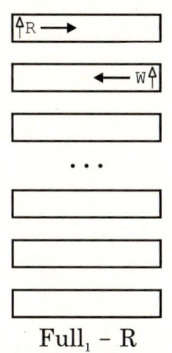

Full$_1$ − R

Fig. 9-9. Register addressing for an almost-full FIFO.

After the read, R points to $j+2$, and W is allowed to point to $j+1$, where a write now can occur.

Finally, let's look at another full condition, where R is at the bottom of the FIFO, at register 0. If so, it must be that W is invalid but will point to register 0 as soon as a read has allowed the FIFO to contain an invalid register. The full condition is shown in figure 9-10, labelled $Full_2$.

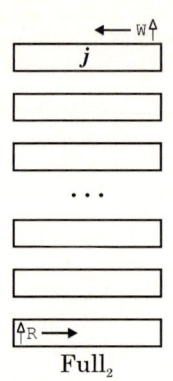

[Figure: boxes stacked vertically, top box labeled "j" with "← W" arrow above it, bottom box with "R →" arrow; ellipsis in middle]

Full$_2$

Fig. 9-10. Register addressing for a full FIFO.

The position of W in *Full$_2$* actually is nowhere, because W is invalid; but, because we know its first possible valid location, it is shown ready to wrap around from the other end of the register set to register 0; all that this wrap requires is one read.

After that one read, the FIFO no longer is full and the situation is shown in figure 9-11, labelled *Full$_2$-R*.

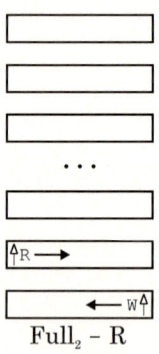

[Figure: boxes stacked vertically with ellipsis; second-to-last box has "R →" arrow, last box has "← W" arrow]

Full$_2$ - R

Fig. 9-11. Register addressing for an almost-full FIFO.

We see that R now points to register 1, and that W has become valid and points to register 0, at the bottom of the FIFO as shown.

The conditions of "full" and "empty" are of great concern to the other devices depending on the FIFO for data transfer. A full FIFO means that the supplier of data must stop trying to send data; an empty FIFO means that the consumer of data must stop trying to receive data. Thus, a FIFO in general must provide output flags that indicate when it is full or empty.

Digital VLSI Design with Verilog

9.1.7 A Verilog FIFO

Our FIFO will be composed of a memory (register file) and a control module. The control module will be designed to depend upon a state machine, but most of it will be combinational logic. In different terminology not used in this Textbook, our register file would be described as a ***datapath***; similarly, our state machine would be called a ***control unit***.

Proceeding in our own terms, we shall separate the state machine sequential transition logic from the combinational logic of the rest of the design. A schematic of the particular design breakdown we shall use is given in figure 9-12. The FIFO full and empty flags are omitted for simplicity.

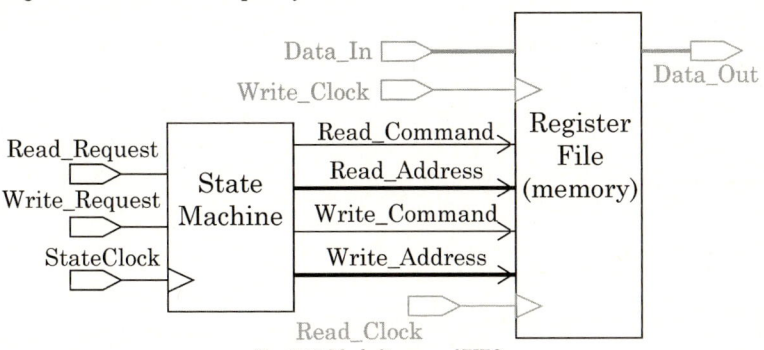

Fig. 9-12. Block diagram of FIFO.

We assume that everything will be in a single `module`, except the memory storage register file (RAM) itself. In the following, we shall refer to "read" and "write" <u>commands</u> by the state machine. These terms refer to commands issued by the state machine to the RAM, at the read or write address being output by the state machine.

By contrast, as indicated above, there will be an external device which is using the FIFO and which will be sending read and write <u>requests</u> to the FIFO; the state machine will honor these requests by issuing commands to the RAM. However, when the FIFO is full, a write request will have to be ignored; when the FIFO is empty, a read request will have to be ignored.

The problem in the detailed description of the FIFO above is in the read and write pointers when they are not usable; one of them is unusable in the "empty" condition, the other in the "full" condition. At face value, we can't write verilog to handle a pointer which is pointing nowhere. On the other hand, study of the preceding details reveals that the R and W pointers, as well as their next transitions, always are well defined numerically if we stay just one step away from the FIFO states of empty or full.

So, we shall design around "almost-empty" and "almost-full" states. We shall treat "empty" and "full" as special cases directly derived from those two and each of which includes one invalid pointer. We don't allow of the possibility that anything could cause two invalid pointers in a single FIFO.

For brevity, let R and W indicate the numerical positions n of the FIFO registers to which they point. Then, when R == W-1, the FIFO is one read away from empty; and, when W == R-1, the FIFO is one write away from full. These are easily described arithmetic relations and may be considered as numerical descriptions of what we showed above in figures 9-7 and 9-11, respectively. The condition R==W is, of course, operationally forbidden for *two* valid pointers; one must be invalid. All other values of R and W allow simple arithmetic describing correct operation with no special concern: The pointer simply is incremented by 1 each time, after it is used, as was shown above in figure 9-2.

Given this, we may describe our FIFO as a state machine with a possible state transition only on a read or write and with the following five simple states:

normal	Normal. Read may transit to almost_empty; write may transit to almost_full. No other transition is allowed. A 4-state FIFO will move directly between a_empty and a_full, with no normal state.
a_empty	Almost empty. Read transits to empty; write transits to normal.
a_full	Almost full. Write transits to full; read transits to normal.
empty	Empty. Read is forbidden; write transits to a_empty.
full	Full. Write is forbidden; read transits to a_full.

A state transition diagram ("bubble diagram") for this machine would be as shown in figure 9-13.

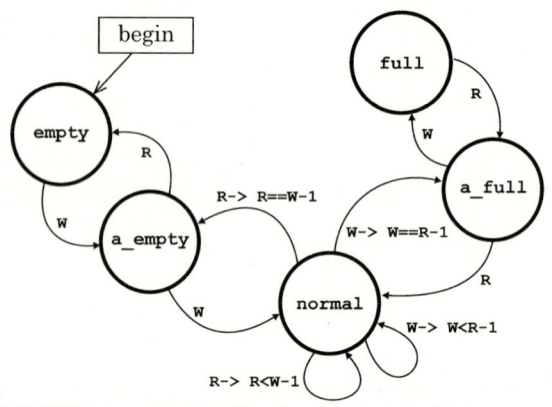

Fig. 9-13. FIFO state machine transition diagram. "begin" is the power-up or reset transition. Assumes a FIFO with five or more storage states and R and W the addresses of possible next action.

Digital VLSI Design with Verilog

We'll start by defining our state register and its states. For 5 states, the major alternatives would be a 3-bit binary register or a 5-bit one-hot register. Let's use a binary encoding.

The machine must retain state, so we require a block of clocked sequential logic for state transitions. Applying our previous rules for coding, we want nonblocking assignments separated from blocking ones, so we shall isolate state transitions in a small, clocked block with nonblocking assignments only. Combinational logic will determine the next state from the current one, so we need only pass the sequential block a variable with the next state encoded in it. The transition logic can be blocking assignments. As shown in figure 9-12 above, we shall code just the state assignment logic in the state machine module; the FIFO storage registers themselves will be in a separate module.

We need two other sequential elements, the read and write counters. They have to be clocked so that state changes take place only while the FIFO pointers are in a known, consistent state. We'll therefore only allow the combinational logic to be read on the opposite clock edge from the one which updates the address counters and the state register. If we update the state register on the positive edge of the clock, this means that we should update the address counters on the same edge but read the combinational block, which programs all updates, only while the clock is low.

There are several ways of implementing the counter updates, but a `task` called in the combinational block perhaps is the simplest:

```
task incrRead; // Called while clock is low.
  begin
  @(posedge Clk)
    ReadCount = ReadCount + 1;
  end
endtask
```

When this `task` is called, it stops at the event control and waits on a positive clock edge; when that edge occurs, it increments the read counter address and then exits. A similar `task` may be declared for the write counter.

Using `tasks` to increment the counters, and perhaps also to compare values, would make modifying the counter easier and less error-prone during early development than having to go around and edit isolated stretches of combinational code, for example if we decided to change to a one-hot or gray-code address counter instead of a binary counter. Another reason to encapsulate the address increments in `tasks` would be to better handle address wrap-arounds for a register file with a number of words not equal to a power of 2.

Turning to the state encoding and ignoring state transition logic, so far we have this:

```
//
reg[2:0] CurState;
//
// Don't allow any other module to affect the state encoding
// in our 3-bit CurState counter; so, use localparams:
//
localparam  empty   = 3'b000, // all 0 = empty.
            a_empty = 3'b010, // LSB 0 = close to empty.
            normal  = 3'b011, // a_empty < normal < a_full.
            a_full  = 3'b101, // MSB 1 = close to full.
            full    = 3'b111; // all 1 = full.
//
// The clocked sequential block controlling state transitions:
//
always@(posedge Clk, posedge Reset)
   if (Reset==1'b1)
       CurState   <= empty;
   else CurState  <= NextState;
// End sequential state transition block.
```

`CurState` is fine, but we cannot reset `NextState` in the clocked block, because the synthesizer would object to the contention of the clocked assignment to `NextState` in the sequential logic combined with the inevitable other unclocked assignments to `NextState` in the combinational logic; so, we shall have to plan for a reset of `NextState` within the combinational logic. This synthesizer behavior is a good thing in general; it helps prevent race conditions.

We have to process requests to read and write, so our combinational block for controlling state transitions should be sensitive both to state changes and to these requests.

A first cut at the combinational block might be as follows, with variables named "*xx*Reg" being the reg used for procedural assignment to output wire *xxx*:

```
always@(ReadReq, WriteReq, CurState, Clk)   // NOTE: Request, not Register.
if (Clk==1'b0) // Only read after a negedge of clock.
  begin
  case (CurState)
    empty: // Combines reset unique conditions
           // with a simple empty state during operation:
           begin
           if (Reset==1'b1)
             begin
             // Reset conditions:
             FullFIFOReg = 1'b0; // Clear full flag.
             WriteCount  = 'b0;
             WriteCmdReg = 1'b0;
             ReadCmdReg  = 1'b0;
             NextState   = empty;
             end
           //
           // Generic empty conditions:
           EmptyFIFOReg = 1'b1; // Set empty flag.
           ReadCmdReg   = 1'b0; // Disable RAM read.
           // One transition rule:
           if (WriteReq==1'b1 && ReadReq==1'b0)
             begin
             ReadCount   = WriteCount; // Could also init to Adr 0.
             incrWrite; // Call task, which blocks on posedge Clk.
             WriteCmdReg  = 1'b1;  // Issue a RAM write.
             EmptyFIFOReg = 1'b0;  // Clear empty flag.
             NextState    = a_empty;
             end
           else ReadCount = 'bz; // Nowhere.
           end   // empty state.
    a_empty: begin
             ...
             end // a_empty state.
    normal: begin
             ...
             end // normal state.
    a_full: begin
             ...
             end // a_full state.
    full: begin
             ...
             end // full state.
    default: NextState = empty; // Always handle the unexpected!
  endcase
  end // always.
```

A problem which might be overlooked here is that the value of WriteCount is being read in this block, but that the variable itself is not on the sensitivity list, risking creation of an unwanted latch during logic synthesis. To avoid this, in our next approximation we shall change `always@(ReadReq, WriteReq, CurState, Clk)` to `always@(*)` to avoid any possible future unpleasant surprise.

Let's consider one more state and set up transition rules for it. The `normal` state would seem to be a good candidate; so, we proceed as follows:

Transitions in our state machine only occur on a read or write. When a clock occurs, but the FIFO performs neither read nor write, it is possible to consider the machine in an "idle" state. For some state machine problems, implementation of an explicit idle state is useful; but, for ours, it is unnecessary, because we don't consider occurrence of a clock to be an event relevant to the machine state. We don't care about anything but read or write. The two clock edges just provide orderly execution as well as adequate time for logic to settle.

In the `normal` state, on any read or write, we only have to check to see whether the next state should be `a_empty` or `a_full`. There is no way in our design that we could go directly from `normal` to `empty` or `full`, barring a machine reset. We also know from above that

$$(write\ counter\ value) < (read\ counter\ value) - 1 \text{ and}$$
$$(read\ counter\ value) < (write\ counter\ value) - 1;$$

both are required to be true to remain in the `normal` state.

So, the simplest way to look for a transition out of `normal` is to add another 1 to every new counter value and compare with the current value in the other counter, checking for an equality instead of a greater-than. If an equality has occurred, then the machine must exit the `normal` state.

Digital VLSI Design with Verilog

Taking all this into consideration, the combinational block case alternative for normal should be something like the following:

```
normal: begin
        // On a write:
        if ( {WriteReq,ReadReq}==2'b10 ) // Concatenation.
          begin
          ReadCmdReg  = 1'b0;   // Disable RAM read.
          WriteCmdReg = 1'b1;   // Issue a RAM write command.
          incrWrite; // Call task, which blocks on posedge Clk.
          //
          // Transition rule:  Check for a_full:
          if ( ReadCount == WriteCount+1 )
              NextState = a_full;
          else NextState = normal;
          end
        // On a read:
        if ( {WriteReq,ReadReq}==2'b01 ) // Another concatenation.
          begin
          WriteCmdReg = 1'b0;   // Disable RAM write.
          ReadCmdReg  = 1'b1;   // Issue a RAM read.
          incrRead; // Call task, which blocks on posedge Clk.
          //
          // Transition rule:  Check for a_empty:
          if ( ReadCount+1 == WriteCount )
              NextState = a_empty;
          else NextState = normal;
          end
        end // normal state.
```

To reduce the complexity of comparing 2 bits, we concatenate them into a single, 2-bit expression. The read if statement might have been put into an else, making it the alternative to write; for now, the else has been omitted because it seemed of limited value and would have added lines of code for the designer to read, understand, and possibly modify.

We are planning for conditions in which the request bits might be reversed by an external device while the write operation was being processed. In a different context, the independent if's (write and then read) in the normal state code above might be seen as a possible race-condition hazard. If so, an if-else or a nested case definitely would be a good idea.

Notice that the state transition is assigned last; this is to ensure that all operations to be completed in the current state are scheduled before the state can be changed.

Finally, notice that in the code above we are comparing *some*CounterValue+1 with *someother*CounterValue. The "+1" might be a problem, because the width of the expression becomes ambiguous: Is it the expression of an integer ('1') or of a rather small reg ("*some*Counter")? When a verilog variable has been assigned, the result takes on the width and type of the destination variable; but, this expression will not be assigned to anything. Integers are <u>signed</u> types, and the last thing we would want is for a negative count to be expressed. To avoid potential problems caused by different simulator implementations, in our next approximation of this model, we shall assign the sum to a reg of the same width as *some*Counter before making the equality comparison.

Delay times are omitted in the code above because blocking assignments guarantee the evaluation sequence, and preceding each statement with, say, "#1", would make no obvious difference. Furthermore, synthesis will introduce a variety of unpredictable delay differences in each branch of the logic, and backannotated delays from floorplanning and layout would supercede programmed delays in this model, anyway.

One reason to add delays in the code might be to make the simulation protocol behaviorally accurate with regard to other devices to be connected to the state machine controller; however, such delays far better would be added in the final continuous assignments, as shown in FIFOStateM_header.v, available in your Lab11 setup directory.

Notice the continuous assignments:

```
module FIFOStateM
        #(parameter AdrHi=4)
         ( output[AdrHi:0] ReadAddr, WriteAddr
         , ...
         , Clk, Reset
         );
    reg[AdrHi:0] ReadAddrReg, WriteAddrReg;
    ...
    reg           EmptyFIFOReg, FullFIFOReg
                , ReadCmdReg, WriteCmdReg;
    //
    assign #1 ReadAddr   = ReadAddrReg;
    assign #1 WriteAddr  = WriteAddrReg;
    assign #1 EmptyFIFO  = EmptyFIFOReg;
    assign #1 FullFIFO   = FullFIFOReg;
    assign #1 ReadCmd    = ReadCmdReg;
    assign #1 WriteCmd   = WriteCmdReg;
...
```

A possible reason to add delays in the individual statements would be to spread the simulator waveform display to displace edges and perhaps make the sequence of events easier to see; however, such modifications then would have to be marked somehow in the simulator so that they could not be forgotten and become part of the design. Such delays might be justified occasionally; but, generally, the only delays in a design should be those as shown, near the top of the design file, in continuous assignment statements to the module outputs.

9.2 FIFO Lab 11

Topic and Location: Task coding, and the design of a FIFO in verilog.

Do all work for this lab in the `Lab11` directory.

Preview: We start with a minor practice run to add a `task` to a previously completed module. Then, we code a useful, generic error-handling assertion, using a `task` to make it easily available in any verilog design. After that, we code a simplified verilog FIFO state machine controller based on today's lecture. Finally, we connect our previously coded verilog RAM model to our state machine to make a functional (but imperfect) FIFO.

Deliverables: Step 1: A modified `FindPatternBeh` module, using a `task` to collect repeated lines of code. Step 2: Simulation of a `task` containing the error handling assertion specified. Step 3: A correctly simulating FIFO state machine controller (minus register file), and. if DC permits, two synthesized netlists, one optimized for area and the other for speed. Step 4: The same as Step 3, but with a register file attached. Step 5 (optional): Check the state machine combinational block for race conditions

Lab Procedure:

Step 1: Use of a `task` in a minor modification of the Lab 8 design. Copy the file, `FindPatternBeh.v`, from the `Lab10` directory to `Lab11`, renaming it `FindPatternTask.v`. As usual, rename the `module` so that it corresponds to the file name.

In this verilog, there is a repeated stretch of code:

```
#1 i = i + IJump*j - 1;
#1 j = 0;
#1 Nkeeper = 'b0;
```

Clearly, the two identical occurrences are supposed to do exactly the same thing; and, if it became necessary to change one, the other should be changed the same way. This is an ideal situation in which a `task` or `function` should be used. Time delays are involved, so it has to be a `task`.

Enclose the code above in a `task` and call the `task` at the two places where the code above appears. Comment out the original code which is replaced by the `task`. Briefly run a simulation to verify correctness.

Step 2: An assertion task.

Use the `ErrHandle` code below as an assertion somewhere in each of the Steps in this lab, including the previous one, to warn the user of some error or dubious condition. Test it for each of the possible actions (see figure 9-14 below). There is an example file, `ErrHandleTask.v`, containing this task in your `Lab11` directory:

Looking at details of the example file, the `Sts` type is 4 bits, which allows for 16 different status conditions. Because a `reg` is unsigned, this also means that there is plenty of room for improvision: We shall be cautious and avoid trying to declare one of our regs as a `signed` type; so, let us treat Sts values of `4'b1000` or above as though they had been declared two's complement numbers to represent negative status values passed to the task. There are four possible Action values:

- 0: The Sts is 0 or the Msg is null. This condition is normal, and the task silently returns, with no message.
- 1: The Sts is -1 (`4'hf`). The condition is a fatal error and the simulation is finished (`$finish`).
- 2: The Sts is more negative than -1 (`4'hf>Sts>=4'h8`). The condition is an error and the simulation is stopped (`$stop`) but may be continued.
- 3: The Sts is positive (`4'h1` - `4'h7`). The condition is a warning or an informative one, the Msg is printed to the simulator console, but no other action is taken.

Here is the `task` supplied, in its entirety:

```
task ErrHandle(input[3:0] Sts, input[255:0] Msg);
reg[1:0] Action;
  begin
  if (Sts==4'h0 || Msg=='b0)
                         Action = 2'b00; // == 0.
    else if (Sts==4'hf) Action = 2'b01; // Sts == -1, 2's complement.
    else if (Sts>=4'h8) Action = 2'b10; // Sts < -1, 2's complement.
    else                Action = 2'b11; // Sts >  0.
  //
  case (Action)
    2'b00: Sts = 0; // Do nothing.
    2'b01: begin
           $display("time=%4d: FATAL ERROR. %s"
                    , $time,                  Msg);
           $finish;
           end
    2'b10: begin
           $display("time=%4d: ERROR Sts=%02d. %s"
                    , $time,          Sts,    Msg);
           $display("\nYou may continue the simulation now.");
           $stop;
           end
    default: $display("time=%4d: Sts=%02d. NOTE: %s"
                      , $time,    Sts,           Msg);
  endcase
  end
endtask
```

And, here is how the `task` responds in the simulation model appended to the design in your `Lab11` directory:

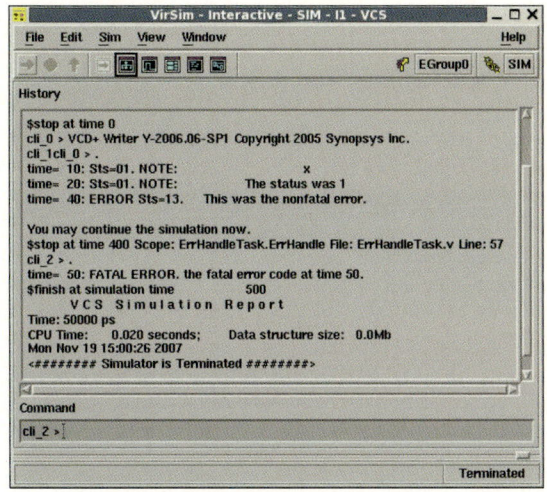

Fig. 9-14. Test simulation of `ErrHandleTask`.

Of course, there is no effect of these messages on simulation waveforms; however, the messages asserted will appear in the simulator console window, as shown in figure 9-14.

Step 3: FIFO state machine design. A module header and partial design has been provided in the file, `FIFOStateM_header.v`; a sketch of the code for a testbench also is included. Copy this file to a new one named `FIFOStateM.v`, and complete the design as follows:

A. Verify that the combinational `always` block's sensitivity list has been coded so that it will be sensitive during simulation to any change in a variable which is read within that block.

B. Complete the assignments and transitions for the remaining states; use today's lecture for guidance. Simulate the machine to check correct address generation. Use `for` loops in your testbench to write the FIFO full, then read it empty, to verify correct address, state register, and flag operation.

C. Consider synthesizing the design, optimizing first for area and then for speed. However, it may be a good idea here to skip the synthesis: Versions of DC later than about 2009 probably will not synthesize a `task` which contains an event control. Older versions reported a warning but proceeded with synthesis. It also should be noted that this FIFO design is a prototype and includes several design weaknesses, of course including synthesis failure, which will be overcome later.

Step 4: Attach a register file to the FIFO state machine, making it a complete, although flawed, functional FIFO. We shall improve the memory and controller functionality in a later lab. For now, proceed as follows:

Copy your `Mem1kx32.v` static RAM design from the `Lab07` directory into the `Lab11` directory. Create a new verilog module named `FIFO_Top` in its own new file in the `Lab11` directory. In `FIFO_Top.v`, instantiate your `FIFOStateM` and `Mem1kx32` and connect them together.

Supply top-level input and output data busses to read and write to the FIFO. You will have to combine the state machine's separate read and write address busses to provide a single address for the memory; use a top-level continuous assignment statement and a conditional operator to do this.

After simulating enough to satisfy yourself that your FIFO is working (see figures 9-15 and 9-16 below), you may consider synthesizing the FIFO in random logic and optimize first for area and then for speed; however, first recall the Step 3C instructions above.

Step 5 (*optional*): If time permits, and if you have not already done so, double check your state machine combinational block for possible race conditions as discussed in the lecture notes above. Race conditions can be avoided by using a single statement to check at one point in simulation time, in each state, to decide what the requested operation (or state transition) shall be.

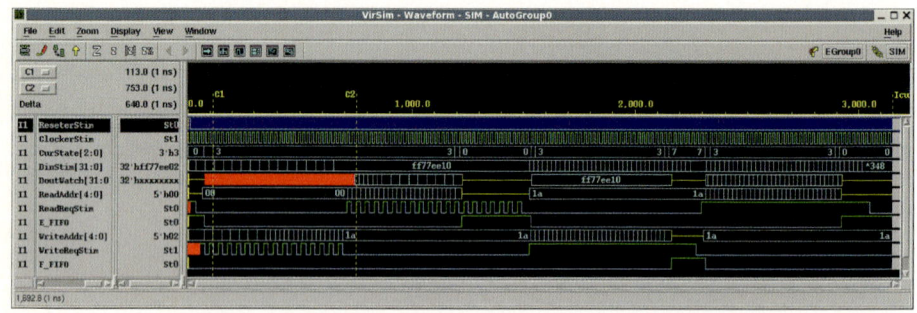

Fig. 9-15. First-cut FIFO simulation—limping, but with obvious race conditions avoided.

Digital VLSI Design with Verilog

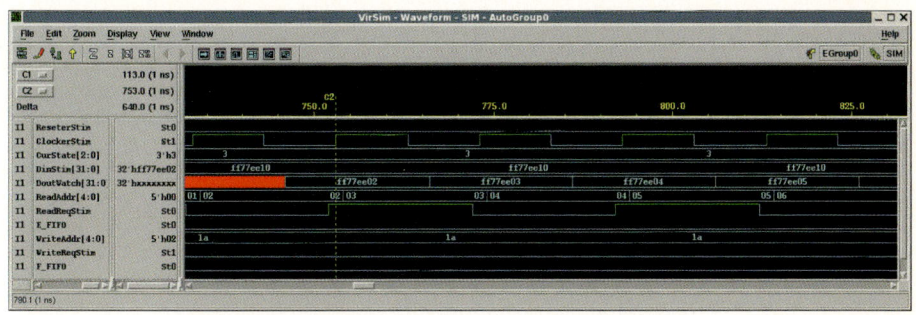

Fig. 9-16. Close-up of the FIFO simulation, showing the first read-out of the register file.

9.2.1 Lab Postmortem

What is the relationship of FIFO design across clock domains to the use of Gray code counters?

9.3 Additional Study

Read Thomas and Moorby (2002) section 3.5 on `functions` and `tasks`.

Read Thomas and Moorby section 4.9 on `fork-join`.

Read and understand the simple memory model in the Thomas and Moorby section 6.3.2, Example 6.9.

(Optional) Thomas and Moorby contains several different perspectives on state-machine modelling. If you would like to know more about this, try (re)reading sections 1.3.1, 2.6–2.7, chapter 7, and appendix A.14–A.17. See the optional Palnitkar readings below, too.

One note: We do not view *datapath* as useful terminology in this book; however it is introduced in Thomas and Moorby's sections 2.6.2 and 2.7.2. The contrast between datapath and control logic can be very useful in chip design and fabrication.

Optional readings in Palnitkar (2003):

Read chapter 8 on tasks and functions.

The code in Appendix F.1 represents a synthesizable FIFO. The use of counters to track the FIFO state is important; some of our References discuss gray-code counters and other FIFO subtleties. Notice that Palnitkar's Appendix F FIFO has only four storage registers.

There are state machine models in sections 7.9.3 and 14.7. Look through these to see how the control is implemented.

Chapter 10
Week 5 Class 1

10 Today's Agenda:

Lecture on rise-fall delays and the verilog event queue
   Topics: Regular and intraassignment delays, rise vs. fall, event controls, and the verilog simulation event-scheduling queue.
   Summary: We describe <u>regular</u> vs. <u>intraassignment</u> delay statements inline in the verilog code, which represent gate-to-gate or module-to-module delays, only. We then see how to specify different rise and fall delays, and different delays for high-impedance assignments, in these statements. After that, we look into how the verilog language requires a simulator to queue up, evaluate, and schedule assignments to variables. We finish with the two different verilog event controls, '@' and 'wait'.

Lab
   Delay statements and event scheduling.

Lab Postmortem
   Q & A on the event queue, lab results, and some fine points on delays.

10.1 Rise-Fall Delays and Event Scheduling

This time we shall look into delays, timing, and the way they are used in verilog statements.

10.1.1 Types of Delay Expression

We'll deal with path delays, component internal delays, and switch-level delays later in the course. For now, we shall concentrate only on gate-level (gate-to-gate) and procedural delays.

Regular vs. Intraassignment Delay. Two kinds of delay expression are <u>regular</u> and *intraassignment*. We use the word "regular" here to mean "standard" or "usual", not as a defined technical term. Although the Thomas and Moorby (2002) authors seem to find *intraassignment* delays useful, for example in section 8.3.1, in practice they rarely are so. A regular delay appears to the left of the target of a statement; an *intraassignment* delay appears to the right of the assignment operator ('=' or '<=').

For procedural statements, the possibilities are,

```
#(delay) variable1 =  value;            // 1. regular blocking.
#(delay) variable2 <= value;            // 2. regular nonblocking.
variable3 =  #(delay) value;            // 3. intraassignment blocking.
variable4 <= #(delay) value;            // 4. intraassignment nonblocking.
#(delay) variable5 =  #(delay) value;   // 5. both, blocking.
#(delay) variable6 <= #(delay) value;   // 6. both, nonblocking.
```

The <u>regular delay</u> statement delays the statement until the specified delay value of time has lapsed; then, and only then, the RHS expression is evaluated, and the statement is executed.

All nonblocking delay statements are flagged as errors or with serious warnings by the synthesizer. Blocking delays are synthesized with the delay values ignored. As we have mentioned before, the synthesizer can't synthesize delay values, especially those scheduling concurrent, future events; and, from the synthesizer's perspective, delayed nonblocking assignments might be intended to be executed in any arbitrary order.

The <u>intraassignment delay</u> statement causes immediate evaluation of the RHS of the statement, and it delays assignment of the result until the specified value of time has lapsed; then, the statement is executed. This is equivalent to a VHDL simulator's *transport delay* scheduling mode. Intraassignment delays are not allowed in continuous assignments.

The "both" delay statement above combines regular and intraassignment delays. We shall examine it a little more in lab.

> ***Terminology Differences:***
>
> Thomas and Moorby (2002) introduces the term *regular event* in section 8.4.3; this is meant to refer to something different from our <u>regular delay</u>. The Thomas and Moorby *regular event* is what this book and the IEEE Std 1800 sections 4.4.2.2 and 24.4 would call an *active event*, as explained below.
>
> Our own terminology here also is nonstandard: Thomas and Moorby, IEEE Std 1364 section 9.7.7, and we use *intraassignment delay* to refer to what IEEE Std 1800 section 14.16 calls it a *cycle delay*.

Why Not Intraassignment Delays. It is strongly discouraged to use the intraassignment delay for anything. In effect, this is an analogue, active-element RC delay-line delay which almost always will be unrealistic in the simulation of on-chip digital logic. Furthermore, this construct in effect creates a new concurrent thread of execution within a procedural block, negating the value of sequence to control functionality. If you wish to simulate concurrency, instead use independent `always` blocks or continuous assignment statements; if you are not planning to synthesize the logic, you may use a `fork-join`. However, a `fork-join` should be a last resort. Keep in mind that not only is it not synthesizable, but a `fork-join` can not allow delays to time out independently; rather, the join restores sequence only after all forked delay statements have timed out.

It might seem reasonable to use intraassignment delays for appearances: To separate simultaneous edges in a simulator waveform display for better visibility. However, no simulator known to the author will display intraassignment-delay displacements differently from regular-delay displacements; thus, the verification tool may become risky for verification, by confusing modelled delays (setup or hold timing slacks, for example) with intraassignment delay displacements which are not part of the design.

For example, suppose a D flip-flop was clocking in data, but the setup delay was not evident (see figure 10-1, part *A*). A intraassignment delay would make this clearly visible, but only if everything went well, and no design problem developed:

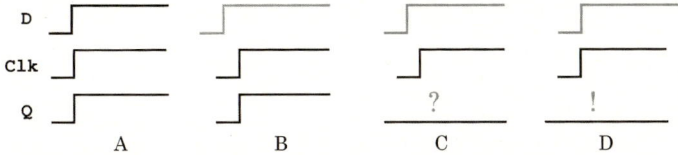

Fig. 10-1. Intraassignment delays can cause confusion. *A*, original display; *B*, appearance fixed by intraassignment delay on Clk and Q; *C*, external events cause loss of setup, but an ***intraassignment*** delay conceals the reason; *D*, same external events as *C*, and omission of intraassignment delays exposes the setup loss.

As pointed out in Thomas and Moorby (2002) section 8.4.1, an intraassignment delay also may be viewed as though it created a temporary reg variable which is invisible and thus protected against other assignment statements: For a delay value ***d***, "x <= #***d*** y;" is the same as, "temp = y; #***d*** x = temp;" in addition to which this particular temp instance is protected against any other assignment during the intervening #***d*** delay. This can be wasteful of simulator memory, and it may introduce design bugs where otherwise none would be. The hidden value can't be cancelled by input changes, making realistic inertial scheduling impossible.

Multivalue Wire-Delay Expressions. Even in CMOS technology, in which gates are relatively symmetrical, rise vs. fall, there is some difference in performance between the delay to a '0' and the delay to a '1'. For example, capacitive interaction with ground or power planes can change rise and fall timing differently. Also, NMOS transistors can be very slow to pull up to a '1', and PMOS to a '0'.

To model this kind of difference, verilog allows primitive-gate or wire-connecting timing expressions to include two, and sometimes three, values, thus:

```
assign #(3, 5) OutWire_001 = newvalue;
bufif1 #(3,5,7) GateInst_001(OutWire_001, InWire_001, Control_001);
```

The two values are in (*rise, fall*) order, and the three values are in (*rise, fall, toz*) order and represent the rise delay from '0' to '1' and the fall delay from '1' to '0'. For assignments, or for components, capable of it, the third value represents high-impedance delay. We shall study these differences in detail later in the course. They are ***not*** allowed in procedural assignments; a procedural statement only is allowed one delay value.

In summary, a connecting delay expression may have one, two, or three delay values:

# (*every_delay*)	Simulator uses this for all scheduling.
# (*rise_delay*, *fall_delay*)	Simulator uses *rise_delay* to schedule changes to '1'.
	Simulator uses *fall_delay* to schedule changes to '0'.
# (*rise_delay*, *fall_delay*, *z_delay*)	Simulator uses *rise_delay* and *fall_delay* as for 2 values.
	Simulator uses *z_delay* to schedule changes to 'z'.

In terms of target and source unknown values, delays *to* 'x' use the shortest delay in the expression; also, when the transition is *from* 'x' *to* 'z' and just two LHS values are given, the shorter delay is used. These are facets of the principle of "delay pessimism". In contrast, when the transition simply is *from* 'x', the specified delay to the target level is used.

Multivalue delay expressions may be used in:
- Continuous assignment statements.
- Primitive component instantiations.

Scheduling Surprises. Here are a few perhaps surprising points to ponder, before we look into the verilog event queue. In the following, keep in mind that many simulators designed for VLSI netlists will not simulate inertial delay properly, when the delays are hand-entered in continuous assignments or procedural blocks:

First, let's see how blocking assignments may be affected by #0 delays. The following is poorly written and unsynthesizable, but it is legal verilog:

```
...
reg   x, y, z;
//
always@(x)  z = x;
always@(y)  z = y;
//
initial
  begin
      z = 1'bz;
   #2 x = 1'b0;
      y = 1'b1;
   #5 $finish;
  end
```

The way that the preceding assignments are written, y will be updated to 1'b1 at time 2 after x is updated to 1'b0, which will leave z at 1'b1. However, changing the first always block to

```
always@(x)  #0 z = x;
```

will cause z to be left at 1'b0 at time 2. Why should a zero delay do this?

Digital VLSI Design with Verilog

Second, when a list of <u>un</u>delayed statements is encountered, they are guaranteed to be run in the order listed, in an interval of zero time duration. However, nonblocking statements assign the right-hand value read, whereas blocking statements block evaluations, update the left-hand sides in order, and use only the updated values.

This is shown here:

```
x <= 1'b1;
y <= 1'b0;    // z gets the old value of x (not shown).
z <= x;       // x gets the 1'b1 originally scheduled.
```

Of course, if the designer only writes synthesizable code, this problem never arises.

Third, only one evaluation can be held scheduled for a future time. Thus, when conflicting assignment statements are read, the last one read determines the delay and the value to be scheduled.

We have,

```
#2 x <= 1'b1;
#2 x <= 1'b0;    // x will be scheduled for 1'bz
#2 x <= 1'bz;    //    at 2 time units from current time.
```

This is pathological code, and it will not be simulated correctly by any simulator currently available to the class. This is because modern designers rarely if ever write delayed nonblocking assignments; so, modern simulators usually ignore the possibility and treat delayed nonblocking assignments as delayed blocking assignments. Again, writing only synthesizable code avoids this whole issue; but, it is good to be aware of it— for example, when writing a testbench.

Now, let us proceed to an understanding of why we should not have been surprised by anything above.

10.1.2 Verilog Simulation Event Queue

Specifications for simulation are built into the verilog language, including specification of how events shall be queued for execution and scheduled for future simulation times. For the rest of this presentation, the word *time* will refer only to <u>simulation time</u>, unless explicitly stated otherwise.

As described in IEEE Std 1364, section 11, the verilog language depends on a ***stratified event queue***. A conforming simulator initially starts with a list of undelayed events, executes them, sets time to 0, and then proceeds forward in time in a completely deterministic way. However, when two or more events are scheduled for the same time, the language may not specify an order of execution; in other words, all may be found in the same ***stratum***. Thus, when events are simultaneous, different simulator implementations sometimes may produce logically different, but equally correct, results. It is up to the designer to write verilog avoiding such differences, when they matter to the validity of the design. Thus, it is important to understand the stratified event queue.

The verilog stratified event queue is illustrated in figure 10-2. Before 2005, the verilog IEEE Std 1364 defined additional, PLI-related strata which now are obsolete. SystemVerilog, in IEEE Std 1800 chapter 4, retains the events described below and adds new ones related to testbenches and extended transitional controls.

Keep in mind that an **event** is a change in value, caused by an output driver (structurally) or by execution of an assignment statement. Mere evaluation of an input, or of the expression on the right-hand-side of a statement, is not an event.

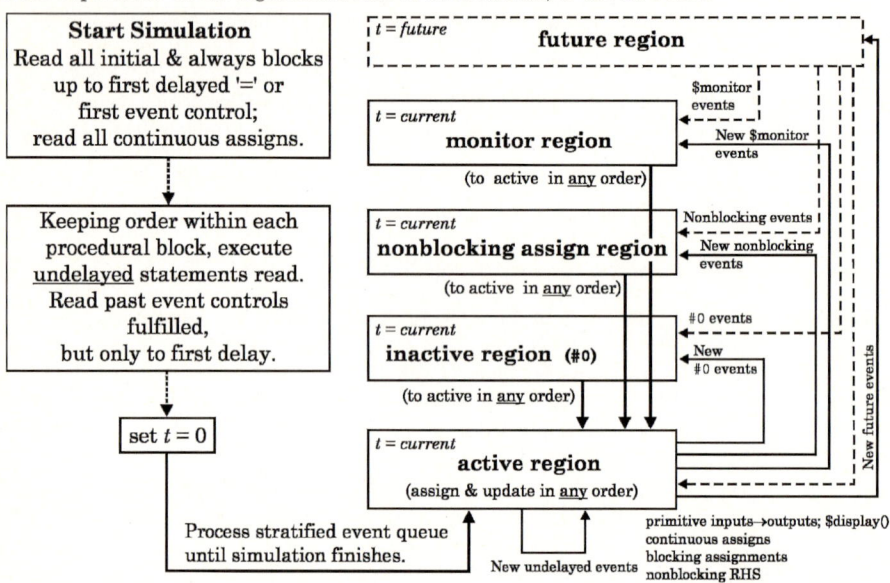

Fig. 10-2. Verilog Event Queue. The five-region queue proper is on the right, separated from simulator initialization actions. Solid lines represent current-time flow or processing order; dotted lines represent input obtained upon advancing the current time to the next (future) time.

Statements not preceded by a procedural delay and controlled by edge-sensitive event controls will be executed in the simulation initialization block. Use a #0 delay in an `initial` block to ensure initializations which must be present at simulation time 0 (this is the only good use of #0 if one is coding in the recommended style).

The solid lines in figure 10-2 are meant to represent current-time accesses to data, or the flow of control in the current time. For example, one of the solid lines ("To active in any order") represents *active* access to the *inactive* region of the stratified queue after all *active* events have been executed (after all new values are assigned to their variables). That solid line means that all *inactive* events found are moved to the *active* region. Any ordering of statements in the verilog source code, or any ordering which occurred while moving events into the *inactive* region, is lost during the transfer from the *inactive* to the *active* region.

It is interesting that, even though #0 means zero time delay, events scheduled at the current time with #0 are placed initially in the *inactive* region, until all *active* events have been processed. The *active* events can include undelayed blocking assignments, changes in verilog primitive inputs resulting in undelayed output changes, or other events which had been scheduled for the current time because of being delayed from some previous time.

Other solid lines in figure 10-2 show that the *active* region may trigger new *inactive* events ("New #0 events"), new nonblocking assignments, and so forth. When all *active* events have been exhausted, and no new event can be created in the current time (by transfer from the *inactive, nonblocking assignment,* or *monitor* region, in that order), the time is advanced to that of the next event, in the future, which was read and scheduled. The dotted lines in figure 10-2 then represent the flow of new data from the new current time into the stratified queue.

The often-useful $strobe system task is executed from the *monitor* stratum, which is above (after) the *nonblocking assign* stratum; this is why $strobe can report the result of a nonblocking assignment.

10.1.2.1 Reading of always Blocks

The concurrent always ***always*** causes its block to be read. The verilog designer controls this behavior by using the '@' event control immediately after the always to halt reading of procedural code until the event expression is fulfilled by a change (or edge) in the signal(s) named.

In addition, to prevent the simulator from hanging up on every procedural event, rereading of each always block is suppressed (disabled) in two main ways:

1. Rereading of all always blocks is disabled each time the active stratum is being processed. Rereading is reenabled each time the simulator completes execution of all events in the active stratum, before moving new events from other strata into the active region. This allows reread always events to join other new events in the active region.

2. Rereading of each always block is disabled while the statements in that particular block are being read.

Concerning the second way, recall these examples:

```
always@(Clk) #10 Clk = !Clk;
```
Clk is updated while always is disabled; always only runs once.

```
always@(Clk) #10 Clk <= !Clk;
```
Clk is updated while always is enabled; always runs on every update.

10.1.3 Simple Stratified Queue Example

There is an example named InactiveStratum.v in your Lab12 directory. You may wish to simulate it to understand the reasoning here.

The verilog in that example, with some of the comments and extra spacing omitted, is as follows:

```
`timescale 1ns/100ps
module InactiveStratum;
reg Clk, A, Z, Zin;
always@(posedge Clk)
   begin
      A = 1'b1;
   #0 A = 1'b0;
   end
`ifdef Case1
// Case 1:   Z inactive:
always@(A)  #0 Z = Zin;   always@(A)  Zin = A;
`else
// Case 2:   Zin inactive:
always@(A)  Z = Zin;   always@(A)  #0 Zin = A;
`endif
//
initial
   begin
   #50 Clk = 1'bz;
   #50 Clk = 1'b0;
   #50 Clk = 1'b1;
   #50 $finish;
   end
endmodule
```

For either `ifdef condition of compilation, there are two always blocks sensitive to change on A.; they are written on the same line, but their positions will determine which one is read first by the simulator. There is only one active Clk edge in the testbench, and that triggers two successive changes in A: The first change is from 'x' to '1', and the second, without advancing simulation time, is a change of A from '1' to '0'.

Here is how the stratified event queue determines this simple simulation:

1. Nothing happens (except to Clk) until time 150, when Clk goes to '1', which is a positive edge in the *active* region during time 150.

2. At time 150, the top always block is desensitized, and the simulator puts the one new *active* event, the top assignment statement, into the t=150 *active* region.

3. Evaluating and finishing the one *active* assignment statement event, the simulator assigns the value '1' to A.

4. This assignment allows the blocked, second statement in the top always block to be read and placed in the time=150 *inactive* region.

5. The always blocks sensitive to the change in A to '1' now must be read.

The remaining events depend on whether we are in Case 1 or Case 2.

Digital VLSI Design with Verilog   219

Case 1:

5. Both other `always` blocks are desensitized, and the undelayed assignment of A to Zin becomes a new event in the time=150 *active* region. The #0 delayed assignment to Z in the other `always` block becomes a new event in the time=150 *inactive* region.

6. There is only one *active* event, so A is evaluated to '1', and Zin is assigned '1'.

7. There are no more *active* events, so the two events in the time=150 *inactive* region are moved in random order into the *active* region: These events are:

    A to 1'b0; Z to Zin;

8. We know Zin definitely is 1'b1, so Z goes to 1'b1. Whether or not executed first, A goes to 1'b0.

9. The change in A again triggers both of the Case 1 `always` blocks, and this puts the assignment to Zin into the *active* region and the assignment to Z in the *inactive* region.

10. The value of Zin is changed to 1'b0, and this empties the active region. The one event in the inactive region then is moved into the active region and executed, changing Z also to 1'b0.

11. All statements in all `always` blocks have been read; so, there are no further *active* events. All `always` blocks are resensitized. The simulator moves all *nonblocking* assignments, without reevaluation, to the *active* region of time=150 (there are none). After that, all events from the time=150 *monitor* region are moved into the *active* region (there are none).

12. The simulator locates the first event in future time, the $finish, puts it in the time=200 *active* region and executes it, terminating the session.

The final values in Case 1 then should be: A=0, Z=0, Zin=0. Notice that delaying the assignment to Z has guaranteed that it will be determined by the value of Zin, whenever A changes.

Case 2:

5. Both other `always` blocks are desensitized, and the undelayed assignment of Zin to Z becomes a new event in the time=150 *active* region. The #0 delayed assignment to Zin in the other `always` block becomes a new event in the time=150 *inactive* region.

6. There is only one *active* event, so Zin, never yet assigned, is evaluated to 'x', and this value also is transferred to Z.

7. There are no more *active* events, so the two events in the time=150 *inactive* region are moved in random order into the *active* region: These events are:

    A to 1'b0; Zin to A;

8. We know A definitely will go to 1'b0, but we can not tell how the race to assign Zin will be resolved. However, we know that either the old or the new value of A will be used, so Zin definitely will become either 1'b1 or 1'b0, not 1'bx.

9. The change in A again triggers both of the Case 2 always blocks, and this puts the assignment to Z into the *active* region and the assignment to Zin into the *inactive* region.

10. The value of Z is changed either to 1'b0 or 1'b1, and this empties the active region. The one event in the inactive region then is moved into the active region and executed, changing Zin to 1'b0.

11. All statements in all always blocks have been read; so, there are no further *active* events. All always blocks are resensitized. The simulator moves all *nonblocking* assignments to the *active* region of time=150 (there are none). After that, all events from the time=150 *monitor* region are moved into the *active* region (there are none).

12. The simulator locates the first event in future time, the $finish, puts it in the time=200 *active* region and executes it, terminating the session.

The final values in Case 2 then should be: A=0, Z=(0 or 1), Zin=0. Notice that delaying the assignment to Zin by #0 has created a race condition on Z.

The verilog of this example was unusually complex to decipher and understand manually; it would be an error to write such code in a real design, because maintenance or modification would be difficult, time-consuming, and error-prone, especially if there was more to the module than three simple always blocks. Also, as commented in the on-disc Lab12 file, two very widely used and well-written simulators produce different simulations!

Application of even one of our rules of thumb would mitigate the complications of this instructional example. The example purposely ignores three of our coding rules of thumb:

- Use nonblocking assignments in a clocked block.
- Never put a #0 delay in a design.
- Don't allow strange latches (the two always@(A) blocks).

10.1.4 Event Controls

There are just two kinds of event control, other than (re)location of a statement in a specific procedural block: The @ statement and the wait statement. All other statements merely are executed or not. There also is a declared event statement.

The @ statement is allowed only in (or at the start of) a procedural block, and it causes a wait in further reading of the block until the event expression changes. This change makes @ effectively edge-sensitive. We already have used this statement extensively, so examples should not be necessary.

The @ event expression may be the name of a data object or the choice of an object edge (`posedge` or `negedge`). When the expression changes from some other logic <u>level</u> either to a logic '1' or '0', the expression comes true, the @ statement is executed, and subsequent procedural statements may be read and possibly executed. When the <u>edge</u>-chosen expression changes to the specified level, up to '1' (`posedge`) or down to '0' (`negedge`), the @ statement likewise is executed. The @ statement itself changes nothing in the simulation data.

Introducing @ with the concurrent `always` ("`always @`") allows an @ statement concurrently to initiate reading of a procedural block of other statements.

The declared `event`, although rarely used in practice, is mentioned here for completeness. A declared event is introduced by the verilog keyword **`event`**, and it allows for the assignment of names to nondesign objects called *events*, so that event controls (@ statements) might be triggered remotely from anywhere in the same `module`. The syntax is, "`event MyEventName;`" followed somewhere else by an event statement using the name. For example, suppose a `task` included the event-control line,

@(MyEventName) *do_something*;

in which *do_something* was some statement. The event control would be triggered by a statement consisting of an arrow (*hyphen* plus *greater-than*) and the declared name.

For example, "`if (some expr) -> MyEventName;`" could trigger `@(MyEventName)`.

Declared events effectively are *goto* constructs, and designers have avoided using them; they depend on an unusual syntax and thus provide complexity with no redeeming special functionality. Better to declare a function or task and simply call it under event control.

The `wait` statement is a level-sensitive construct, and it depends on a logical expression, not a design object name. The syntax is

wait (*expr*) *statement;*

When *expr* comes true, `wait` executes its statement. For example, `wait (x>5) x = 0;` Although a construct of the same name is used frequently in VHDL, `wait` in verilog rarely is seen; this probably is because of availability of the more versatile @ statement event control.

10.1.5 Event Queue Summary

All the preceding considered, it's a good idea not to use #*n* delays in procedural code or anywhere else in a design module. If necessary, put #*n* delays only on `module` output drivers.

If this advice is taken, then the event queue complexities can be ignored in synthesizable coding; and, the following simple considerations are the only ones which will remain:

1. Nonblocking assignments in a `begin-end` block always use old values and are executed <u>after</u> all other statements at a given simulation time.
2. Blocking assignments in a `begin-end` block always use updated values, as in *C*, and are completed <u>before</u> the first nonblocking statement at a given simulation time.
3. Therefore, under ordinary circumstances, use only nonblocking assignments in clocked (edge-sensitive) blocks, and use only blocking assignments in other blocks. This simulates normal setup and sequential activity properly and tells the synthesizer where setup is required.

> Note: Because understanding of the stratified event queue is necessary to an understanding of verilog, we shall be using procedural delays frequently for instructional reasons. But, in general, one should not use procedural delays in ones design.

10.2 Scheduling Lab 12

Topic and Location: Event scheduling by the simulator and rise-fall delay statements.

Do this work in the `Lab12` directory.

Preview: We start with a few simple puzzles, asking when different events will be scheduled to occur. We then look into problems caused by mixing blocking and nonblocking assignments. We revisit the *Week 2 Class 1* latch-synthesis problems caused by omission of variables from a sensitivity list. Finally, we look at statements to schedule different delays on rising vs. falling edges.

Deliverables: Steps 1–3 are observational and require no deliverable in fixed form: Step 1: Eight answers to the questions posed. Step 2: Three answers. Step 3: Simulation of the file provided; your comments on the results. Step 4: A synthesized verilog netlist containing instances of the two modules provided. Step 5: A completed module which simulates the statements given with the three delay values provided.

Digital VLSI Design with Verilog

Lab Procedure:

Note: Answers to Steps 1 and 2 are in the Lab Postmortem below.

Step 1: Two scheduling and delay examples.

```
module SchedDelayA;
reg a, b;
initial
  begin
  #1 a <= b;
  #1 a  = 1'b1;
  #2 a  = 1'b0;
  #2 a  = 1'b1;
  #1 a  = 1'b0;
  #0 b  = 1'b1;
  #0 b  = 1'b0;
  #0 b <= 1'b1;
  #5 $finish;
  end
//
always@(b) a = b;
//
always@(a) b <= a;
//
endmodule // SchedDelayA.
```

```
module SchedDelayB;
reg a, b;
initial
  begin
  #0 b  = 1'b1;
  #0 b  = 1'b0;
  #0 b <= 1'b1;
  #1 a <= b;
  #1 a  = 1'b1;
  #2 a  = 1'b0;
  #2 a  = 1'b1;
  #1 a  = 1'b0;
  #5 $finish;
  end
//
always@(a) b <= a;
//
always@(b) a = b;
//
endmodule // SchedDelayB.
```

Here are a few questions to ponder, based on the two examples above. You may simulate to answer them, but try first to guess without simulating.

In `SchedDelayA`, at what time does a first rise? b?

In `SchedDelayA`, at what time does a last change? b?

Answer the preceding questions for `SchedDelayB`.

Step 2: Scheduling and delay with rise-fall timing:

```
module SchedDelayC;
wire a;
reg b, c, d;
initial
  begin
    #0   b = 1'b0;
         c = 1'b1;
         d = 1'b1;
    #5   d = 1'b0;
    #10  c = 1'b1
    #20  $finish;
  end
//
assign #(1,3,5) a = c;
assign #(2,3,4) a = d;
//
endmodule // SchedDelayC.
```

A. When does a first get a well-defined logic level ('1' or '0')?

B. Is the order of first assignment of b, c, and d predictable? If not, why?

C. What is the final value of a?

Step 3: Scheduling with mixed-up assignments. First, using the file, Scheduler.v in your Lab12 directory, simulate to see the difference in stratified order of evaluation between events from blocking assignments, zero-delay (#0) assignments, nonblocking assignments, and monitor events. See figure 10-3. The comments in the verilog file explain the simulation result.

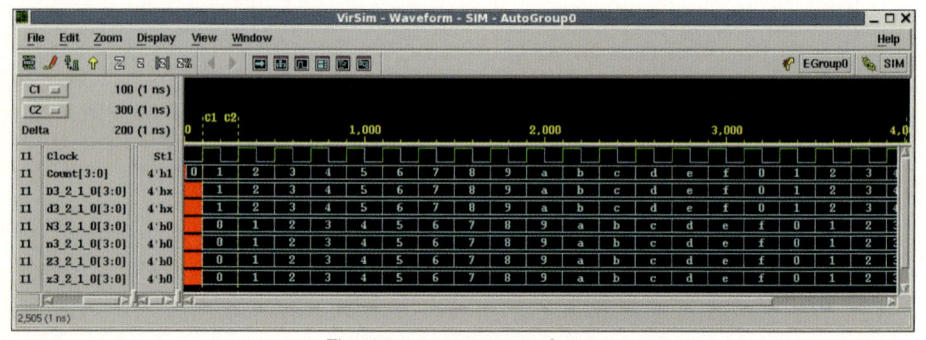

Fig. 10-3. Scheduler.v simulation.

Optional: For the rest of this Step, use Silos, if you wish to see exactly correct verilog, because simulators designed for serious coding do not handle mixed blocking and nonblocking assignments, or nonblocking concurrency, properly. If you don't wish to simulate, just read through the verilog files provided. This exercise makes a point about using regular delays, only. In the Lab12 directory, open the file BothDelay.v and read through it. It is based on the code example immediately following the topic above in the section entitled, "Regular vs. Intraassignment Delay".

Try to predict when each change will take place. If you simulate this module in Silos, you will find that the initial block ends with what may be a surprise: Can you explain it?

Finally, suppose you encountered these assignments in a new module:

```
initial
begin
    Y <= 1'b0;
#0  Y <= 1'b1;
#0  Z <= 1'b1;
    Z <= 1'b0;
  $display("display: %04d: Y=%1b Z=%1b", $time, Y, Z);
  $strobe( "strobe:  %04d: Y=%1b Z=%1b", $time, Y, Z);
  $finish;
end
```

What would be the final values of Y and Z whenever this block is run? Which system function, $display or $strobe, will report them correctly? Should the simulator compiler place the delayed assignments initially in the inactive region or in the nonblocking delay region? What on Earth would be the reason for writing such code!? By now, it should be clear why the synthesizer rejects, or warns about, delayed procedural statements, especially nonblocking ones.

Step 4: Event control sensitivity. Below are two different modules. Instantiate both of them in a third, containing module named EventCtl, of course in a file named EventCtl.v. Simulate to check functionality (see figure 10-4); then, synthesize.

```
module EventCtlPart(output xPart, yPart, input a, b, c);
reg xReg, yReg;
assign xPart = xReg;
assign yPart = yReg;
always@(a,b)
  begin:   PartList
   xReg <= a & b & c;
   yReg <= (b | c)^a;
  end
endmodule // EventCtlPart.
```

```verilog
module EventCtlLatch(output xLatch, yLatch, input a, b, c);
reg xReg, yReg;
assign xLatch = xReg;
assign yLatch = yReg;
always@(a)
  begin: aLatcher
  if (a==1'b1)
    xReg <= b & c;
  end
always@(b)
  begin: bLatcher
  if (b==1'b1)
    yReg <= (b | c)^a;
  end
endmodule // EventCtlLatch.
```

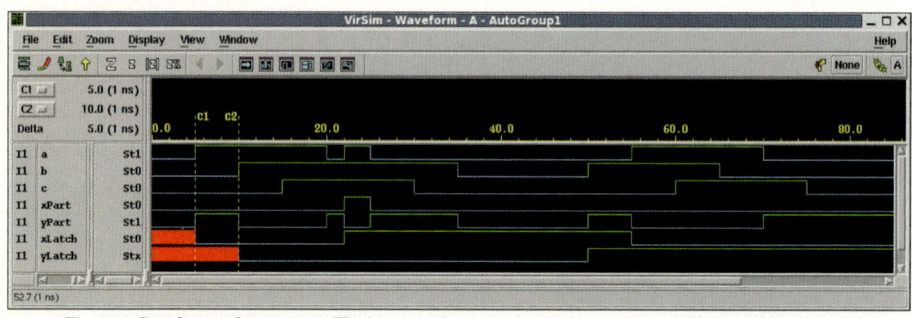

Fig. 10-4. Simulation of `EventCtl`. The *`Part` and *`Latch` wires respectively are from its two instances.

The omission of one variable from the sensitivity list should cause synthesis of a latch. The module names are to help keep track of what is what in the synthesized netlist.

The synthesizer is biased not to infer latches. If you want latches in addition to an incomplete sensitivity list, then for guaranteed success, you should code for instantiated images of latch `modules` with 1-bit output ports. If you do this, examine the synthesized verilog netlist to see whether the expected latches have been created. In the synthesis library being used, latch components all are named beginning with an '*L*'. For example, an "`LHQxx`" in the netlist would be a transparent latch.

Step 5: Rise-fall delays. Rise-fall delays mainly are used with structures in a netlist, such as gates or IP ("Intellectual Property": large, predesigned blocks); however, they work with RTL code, too, with some adaptation.

Use parameters tR, tF, and tZ, with ***regular*** delays, in a module named RiseFall to schedule the following assignments on each of the *Reg variables shown in the code block below:

1. a rise (tR) after 4 time units;

2. a fall (tF) after 3; and

3. a high-impedance (tZ) after 5 time units.

All rise-fall delays specify delay characteristics of a wire or port, not a statement, so it won't work to insert them in procedural code, for example in an always block. They work well in an assign (continuous assignment).

Therefore, the values of the regs declared below will have to be assigned to wires in continuous assignment statements to see the effect of specifying different delays for rise and fall.

Declare these wires (use the names given, without "Reg") and assign them:

```
reg[3:0] OutBusReg;
reg[7:0] DataBusReg;
reg Out2valReg, Out3valReg;
...
always@(negedge Clk)
  begin
  OutBusReg   <= 4'bzz01;
  DataBusReg  <= 8'b1111_0zzz;
  Out2valReg  <= 1'b1;
  Out3valReg  <= 1'bz;
  end
always@(posedge Clk)
  begin
  OutBusReg   <= 4'b0101;
  DataBusReg  <= 8'b1zzz_0000;
  Out2valReg  <= 1'b0;
  Out3valReg  <= 1'b0;
  end
```

Simulate a module which includes the code above, and the specified delays. You should obtain waveforms as in figure 10-5.

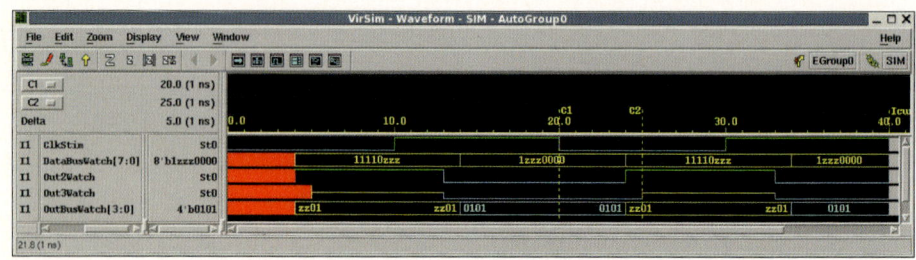

Fig. 10-5. Simulation of `RiseFall`, showing the delay differences.

Concerning these delays, according to the IEEE Std 1364-2005 section 6.1.3 and IEEE Std 1800 section 10.3.3, assignments to vector nets must be simulated according to this three-part rule for multivalue delays:

(*a*) If the RHS makes a transition to 0, the falling delay shall be used.

(*b*) If the RHS makes a transition to z, then the turnoff delay shall be used.

(*c*) In all other cases, the rising delay shall be used.

However, this rule is ambiguous, because "makes" can be interpreted bitwise or to mean the numerical value in the vector. Both Aldec and VCS seem to treat vector delays consistent with the numerical value interpretation: Thus, they require a numerical value of 0 for case (*a*), all bits 'z' for case (*b*), and any other pattern for case (*c*). Other simulators may use the transition on the LSB, only.

10.2.1 Lab Postmortem

Lab 12 answers:

Keep in mind that "rise" and "fall" refer to ***changes***, not to initial assignments.

Step 1 `SchedDelayA`: First rise a = t 5; b = same.   Last change a = t 6; b = same.
      `SchedDelayB`: First rise a = t 0; b = same.   Last change a = t 6; b = same.

Step 2 `SchedDelayC`: **A.** t = 1;   **B.** Predictable (continuous assigns);   **C.** a = 1'b1.

The answers for the other Steps are in your Lab12 subdirector, somewhat expanded from those explicitly given above.

Some questions:

Should the verilog event queue have fewer strata? More?

Mull over and explain the `BothDelay.v` result.

What is the value of named blocks in synthesis?

What is the use of `@(*)`?

Why not use nonblocking assignments in combinational blocks?

What if a '1' goes to 'z'? Is that a rise or a fall? How about '0' to 'z'?

10.3 Additional Study

Read Thomas and Moorby (2002) section 6.5 on rise, fall, and three-state net-connection delay specification.

Read Thomas and Moorby chapter 8 on scheduling of procedural and behavioral events.

Optional: Thomas and Moorby explains intraassignment delay usage in section 4.7. This construct usually should be avoided, but it is worth understanding what it is supposed to do. The authors point out that even where apparently useful, a intraassignment delay can not do any better than a regular nonblocking delay in fixing an otherwise malfunctioning design.

Optional readings in Palnitkar (2003):

Read chapter 7, with special attention in 7.2.2 and 7.3 to the use of a # delay expression on the RHS of an assignment statement ("intra-assignment delay"). We shall avoid this usage, but it is important to understand what it is intended to do.

Read section 5.2.1 on rise-fall delays, which primarily are used when describing the timing in netlists.

Chapter 11
Week 5 Class 2

11 Today's Agenda:

Lecture on verilog built-in gates and net types
  Topics: Verilog built-in elementary gates, net types, implied wires, and a port and parameter review.
  Summary: A netlist consists of gate instances and wiring. We enumerate the elementary logic gates which are built into the verilog language, pointing out their input and output port organization. We then enumerate all verilog built-in net types useful in VLSI design, saving switch-level constructs for later. We mention the `default_nettype` control, and we then discuss the rationale for verilog rules for mapping of net and reg types to a port. We point out that parameter and delay values both may be preceded by '#' but that they should not be confused with each other.

Lab
  We construct several verilog netlists by hand, and we simulate and synthesize them.

Lab Postmortem
  After general Q&A, we discuss synthesis results for a netlist counter *vs.* one built with behavioral flip-flops.

11.1 Built-in Gates and Net Types

In this chapter, we shall study verilog netlists in some detail. Netlists are purely structural implementations made up of elementary gates and hierarchical (sometimes large) other components.

11.1.1 Verilog Built-in Gates

Thomas and Moorby (2002) section 6.2.1 and Appendix D discusses all the verilog built-in primitive gates; they are specified in IEEE Std 1364 section 7 and IEEE Std 1800 section 28. We list them below for convenience, omitting for now the switch-level primitives. Notice that none of them is sequential logic, although sometimes a three-state output can be considered as saving briefly its most recent non-z logic state.

Multiple Inputs	Multiple Outputs	One Input & Output	One Output
and	buf	bufif1	pullup
nand	not	bufif0	pulldown
or		notif1	
nor		notif0	
xor			
xnor			

We already have used `bufif1` in a lab; `notif`x is just an inverting `bufif`x.

In their port connection lists, these built-in gates always have their output(s) as the first, leftmost pin(s); they are not allowed named port lists; and, they are allowed instantiation without an instance name. Verilog builtin gates are the only gates which can be instantiated without an instance name. As shown already in lab, any of the gates above may be instantiated with a strength specification.

The `pullup` and `pulldown` components should be avoided, because they are not synthesizable (as of 2011). Representing a tap into one of the chip power rails, the strength of a `pullup` or `pulldown` gate is *source* strength.

11.1.2 Implied Wire Names

It isn't always necessary explicitly to declare a net if it is of `wire` type. When one end of a connection is a module port connection, merely providing the name of the port connected is enough to declare implicitly the name of a net.

For example, this is a complete `module` declaration:

```
module NoNets (output Xout, input Ain, Bin);
   and And01(Xout, Ain, Bin);
endmodule
```

The following verilog unidirectional buffer or pass-through module also is complete and legal, although it probably should be removed during logic optimization:

```
module NoThing (output Xout, input Ain);
   assign Xout = Ain;
endmodule
```

Implied wires (one bit wide, only) also may be used to interconnect pins of instances in a netlist, thus permitting port-mapping without explicit wire declarations.

It's probably not a good idea to use any net without declaring it specifically. Such declarations take little time, and they provide an explicit definition which then can be considered for later change, should design problems arise depending on what otherwise would be an undeclared, and therefore invisible, type.

11.1.3 Net Types and Their Default

There is a refinement of the idea of implied net names: It is legal to change the default type of implied connection nets, as explained in IEEE Std 1364 section 19 or IEEE Std 1800 section 22. This is done by the compiler directive, `default_nettype, which affects all implied nets following it in compilation order. The type named may be any net type; the default type is `wire` under the usual circumstance of no directive. A `default_nettype directive changes this default.

The types which may be defaulted are as follows:

Type	Default Net Functionality
`wire`	Connection only.
`tri`	Connection; identical to `wire` except in name.
`tri0`	Pull down to logic '0' level, with resistive (`pull`) strength.
`tri1`	Pull up to logic '1' level, with resistive (`pull`) strength.
`wand`	Logical *and* of driver logic levels.
`triand`	Logical *and*; identical to `wand` except in name.
`wor`	Logical *or* of driver logic levels.
`trior`	Logical *or*; identical to `wor` except in name.
`trireg`	Unique. Storage of a (capacitive) charge at a given strength level when its driver(s) all are in high-impedance ('z') logic state.

We have used the `wor` type in a lab, and those others which have logic function work the same way.

There are three other net types not in the list above: The default net type **none** is discussed below. The two remaining verilog net types, `supply0` and `supply1`, are power-supply elements and may not be used as default implied net types. Note that all these net types are gate, RTL, and behavioral modelling constructs and are independent of verilog strength expressions and of the verilog switch-level constructs we shall study later in the course.

Use of the `default_nettype directive may incur a big risk for very little advantage: What if a `module` simulates correctly because of implied nets in a default state of, say `wor`, and then is reused or moved elsewhere, so that it is compiled <u>before</u> `default_nettype wor appears? The simulation probably then will fail, possibly with mysterious symptoms.

Most importantly, the default type may be set to <u>none</u>. After `default_nettype none is encountered, no implied net connection of any type is allowed; all nets must be declared explicitly.

The safest way of using `default_nettype is to avoid it. If this directive has to be used, it is recommended to use it only to set the default type to `none`.

11.1.4 Structural Use of Wire *vs.* Reg

Thomas and Moorby (2002) section 5.1 explains the verilog port connection rules which are detailed in IEEE Std 1384 section 12.3 and in IEEE Std 1800 section 23. The simulator will enforce them when they are overlooked. It is easy to overlook these rules, especially when writing a testbench, in which design considerations often are not a priority.

Here's another perspective on the connection rules:

- First, drivers of a `module` may be a `reg` or any net type. This is shown in figure 11-1, in which declarations in the containing `module` are assumed to include "`reg InReg;`" and "`wire InWire;`".

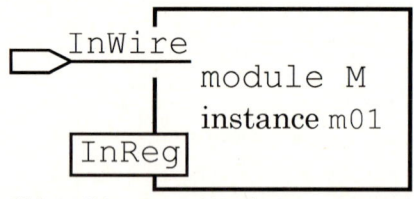

Fig. 11-1. M m01(..., .In1(InWire), .In2(InReg));

- Second, anything driven by a `module` output and external to that `module` must be a net. The reasoning here is very simple: A `reg` may hold a value or be a driver (first rule); if a `module` output could be connected directly to a `reg` type, there would be contention between the `reg`, as a driver holding a value, and the `module` output port.

Therefore, `module` outputs must be connected to net types only. This is shown in figure 11-2; the `reg` connection to an output port is illegal—and, the dual-meaning '**x**' emphasizes the reason why. However, see the final note below.

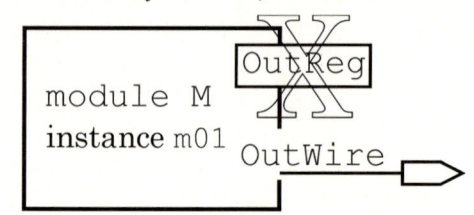

Fig. 11-2. M m01(.Out1(OutWire), .Out2(OutReg), ...);

- Third, anything driving internally from a `module` input must drive a net; in other words, internal connections to input ports must be nets. There is no way to avoid this, because driving anything with a `module` input, including a (internal) `reg`, uses the implied net associated with every input port.
- Fourth, these considerations also mean that a bidirectional port (an `inout`) must be connected on both sides to a net. This is because an `inout` port is subject both to

Digital VLSI Design with Verilog

the input and the output restrictions just described; these leave only a net type as a way to connect on either side of an `inout` port. Thomas and Moorby (2002) says that an `inout` port must be connected to a three-state gate; however, a plain `wire` may be used, assuming that internal-external contention can be resolved in a meaningful way because of strength differences.

One final note on the structural use of `wires`: It is legal in verilog to drive an `input` port from inside the `module`, using a continuous assignment statement to create an inside-outside contention on a `wire`. This, of course, is not very advisable! In other languages, such as VHDL or SystemVerilog, driving a port declared as `input` is not legal except by connection from outside the `module`.

11.1.5 Port and Parameter Syntax Note

We have done many port and `parameter` connections already in lab. We point out that both ports and `parameters` are declared similarly in ANSI format; and, in instantiation, the connection format also is very similar. All `parameter` declarations, like delay values, are introduced by the '#' token. They are easily distinguished using ANSI declarations, because the keyword, `parameter`, always appears in a `parameter` declaration.

For example, there is no delay involved in this declaration (the `parameter` named `Delay` might be used for anything):

```
module DeviceM
    #(parameter BusWidth=8, Delay=1)  // No resemblance to a delay time.
     (output[BusWidth-1:0] OutBus,
    , input[BusWidth:1] InA, InB, input Clk, Rst);
...
```

A `parameter` declaration or value-list always immediately follows the identifying `module` name. In another `module` instantiating the one above, `parameter` overrides by name also obviously are not delays:

```
DeviceM #( .BusWidth(16), .Delay(5) )   // Default overrides, not delays.
    DevM_01 ( .OutBus(Dbus), .InA(ArgA), .InB(ArgB)
            , .Clk(ClockIn), .Rst(Reset)
            );
```

A confusion can occur when assigning a `parameter` value by position, rather than by name. Suppose in the preceding example that we overrode `parameter` values by position and not by (ANSI) name. The following might look like a delay assignment:

```
DeviceM #(16, 5)   // Default overrides may resemble delays.
    DevM_01 ( .OutBus(Dbus), .InA(ArgA), .InB(ArgB)
            , .Clk(ClockIn), .Rst(Reset)
            );
```

For comparison, here is a delay associated with the output of an instantiation of a verilog built-in gate:

```
    xnor #(16, 5) Adder_U12 ( Z, A, B, C );
```

All `parameter` overrides, like the majority of delays, immediately precede the target instance name. To avoid misunderstanding, not only between delays and `parameters` but also among different `parameters`, it is strongly recommended to override `parameters` only by name.

11.1.6 A D Flip-Flop from SR Latches

In our next lab, a gate-level, simple D flip-flop will be constructed from *nand* gates assembled as three SR latches.

To understand the logic, start with an output and choose output states which imply fully defined inputs. A *nand* gate outputs '1' if any input is '0'; however, when the output is '0', all inputs are fully determined to be '1'. So, look first at the case of a *nand* output of '0'.

Why not begin by considering an SR latch with a small modification in which both SR inputs are tied together?

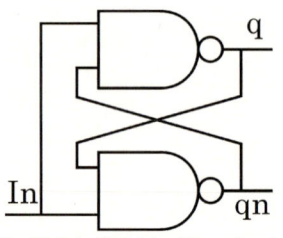

Fig. 11-3. An SR latch with one input.

In figure 11-3, arbitrarily labelling the outputs q and qn, suppose q = 0. Then, qn must be 1 and `In` must be 1. We could have had qn = 0, also, if `In` was 1. Therefore, the latched state is with `In` = 1. If `In` = 0, then both q and qn must be forced to 1.

This design doesn't permit any specified input value to be latched, but it does exhibit sequential behavior.

Now let us add an input X which we hope might supply the value to be latched:

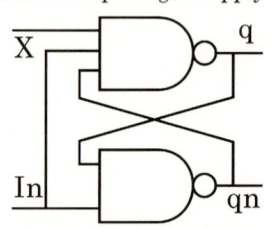

Fig. 11-4. A one-input SR latch with a second input.

To latch data, we now must have both `In` = 1 and X = 1. With X = 1, if `In` goes to 0, q and qn go to 0; if `In` then returns to 1, there is a latched value, but it is indeterminate,

Digital VLSI Design with Verilog

because if either q or qn shifts to 0 sooner than the other, it can clamp or reverse the other to its original 1 value.

From the latched state, if X goes to 0, q is forced to 1 and qn to 0. If X returns to 1, the latched state will remain q = 1 and qn = 0.

In this second design, X does act somewhat as a data input, with In a clock or maybe an asynchronous clear, if we look only at qn as the stored bit. Suppose we hold X at 0 and toggle In from 1 to 0 and back to 1: We get qn = 0. If we could invert X and store its qn of 0 in another SR latch which always would be interpreted as inverted data, maybe we could get closer to a real D flip-flop?

For starters, we need something more loosely tied to the In. Let's go back and look at a real SR latch again, as in figure 11-5:

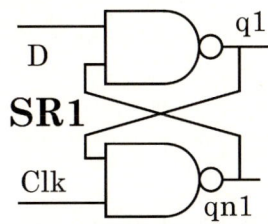

Fig. 11-5. An SR latch which acts like a flip-flop, but only when D is '0'.

Truth-table for the device in figure 11-5
time	D	Clk	q1	qn1			
0	0	0	1	1			
1	0	1	1	0	qn1	follows	D
2	1	1	1	0	qn1	latches	D
3	1	0	0	1	q1	follows	Clk
4	1	1	0	0	q1	latches	Clk
5	0	1	1	0	qn1	follows	D

Just as above, we see that it acts like a positive-edge flip-flop, but only when the data input is at '0'.

To make this work for data input '1', we can invert the data as in figure 11-6:

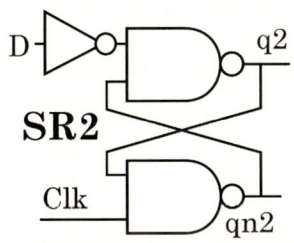

Fig. 11-6. An SR latch which acts like a flip-flop when D is '1'.

Truth-table for the device in figure 11-6						
time	D	!D	Clk	q2	qn2	
0	1	0	0	1	1	
1	1	0	1	1	0	qn2 follows !D
2	0	1	1	1	0	qn2 latches !D
3	0	1	0	0	1	q2 follows Clk
4	0	1	1	0	0	q2 latches Clk
5	1	0	1	1	0	qn2 follows !D

So, from the two preceding tables, all we need do now is to build a flip-flop which assigns qn1 to its Q when D was latched as a '0' and assigns qn2 to its Q when D was latched as a '1'. We can't use a mux for this, because a mux is combinational and can not retain the past state of D while selecting the current assignment to Q.

However, we can use a third SR latch in a latched state whenever clock is low; we do this simply by driving the inputs of the third SR latch with the outputs of the *nands* which receive the clock.

The result is our D flip-flop, shown in figure 11-7:

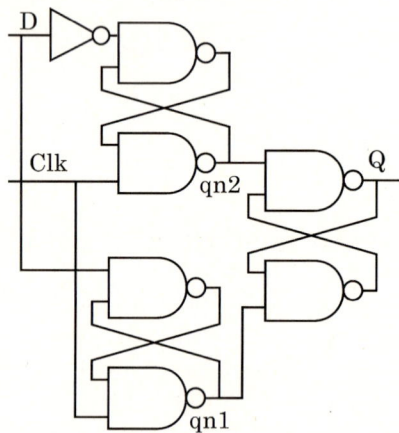

Fig. 11-7. Two SR latches on D, one for '1' and the other for '0', with a third SR to latch the result.

Digital VLSI Design with Verilog

Truth-table for the device in figure 11-7						
time	D	!D	Clk	qn1	qn2	Q
0	1	0	0	1	1	?L
1	1	0	1	1L	0	1
2	0	1	1	0	0L	1
3	0	1	0	1	1	1L
4	0	1	1	0	1L	0
5	0	1	1	0	1L	0
6	0	1	0	1	1	0L
7	1	0	0	1	1	0L

L = latched

11.2 Netlist Lab 13

Topic and Location: We construct by hand several verilog netlists and simulate and synthesize them.

Work in the `Lab13` directory.

Preview: Using elementary combinational gates, we construct a netlist representation of a D flip-flop, connect several of our flip-flops in a synchronous counter, and simulate and synthesize the verilog to library gates. We also construct a similar counter from behavioral D flip-flops, to compare the synthesis result.

Deliverables: Step 1: A correctly simulating model of a D flip-flop consisting of a hand-entered netlist of verilog built-in combinational gates. Step 2: A correctly-simulating synchronous counter created by replacing the behavioral D flip-flops of the `Synch4DFF Lab08` exercise with your flip-flops of this Step 1. Step 3: Four synthesized netlists: One each, optimized for area and speed, from the old `Synch4DFF` counter and from your new counter made from structural D flip-flops. Step 4: Optionally, simulate a speed-optimized netlist.

Lab Procedure:

Step 1: A gate-level D flip-flop. Figure 11-8 below gives a gate-level schematic of a D flip-flop, with a reset.

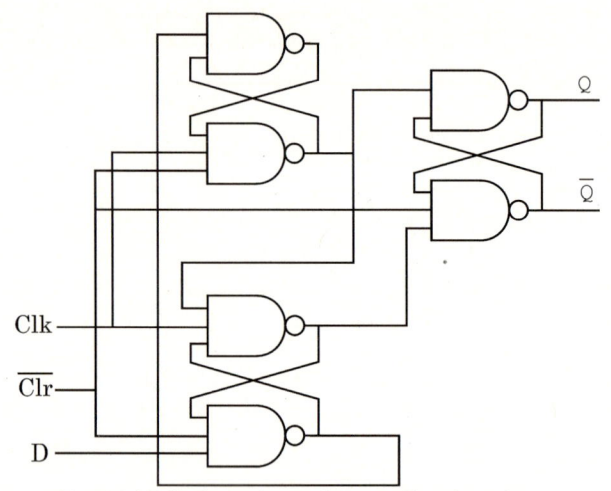

Fig. 11-8. A D flip-flop implemented structurally with *nand* gates.

Use this schematic to create your own D flip-flop `module` named `DFFGates`; use verilog built-in `nand` gates for the gates shown. In your design, make the D flip-flop positive-edge triggered and with asserted-high clear. As usual, put `DFFGates` into a file named the same as the `module`.

Suggestion: Start by assigning names to the *nand* gates in figure 11-8, and use those names as instance names in your netlist. Also, consider the possibility of breaking down the D flip-flop into S-R latch elements and connecting latches instead of *nand* gates. The on-disc answer for this version of the exercise includes a PDF schematic of the S-R latch approach.

Simulate your structural D flip-flop to verify its functionality.

Step 2: A gate-level synchronous counter. Go back to `Lab08` and find your Step 5 `Synch4DFF` design, a synchronous counter assembled from behavioral D flip-flops. If you did not complete this `Lab08` step, use the answer files provided.

Copy both your behavioral `DFFC.v` and `Synch4DFF.v` into the `Lab13` directory. Duplicate `Synch4DFF.v` as `Synch4DFFGates.v`, renaming the `module` inside correspondingly.

Replace the `DFFC` instances in `Synch4DFFGates.v` with instances of your Step 1 `DFFGates`. You can do this by replacing the instance structure with a `module` containing the entire `DFFGates` structural D flip-flop. Simulate to verify your completely gate-level design (see figure 11-9).

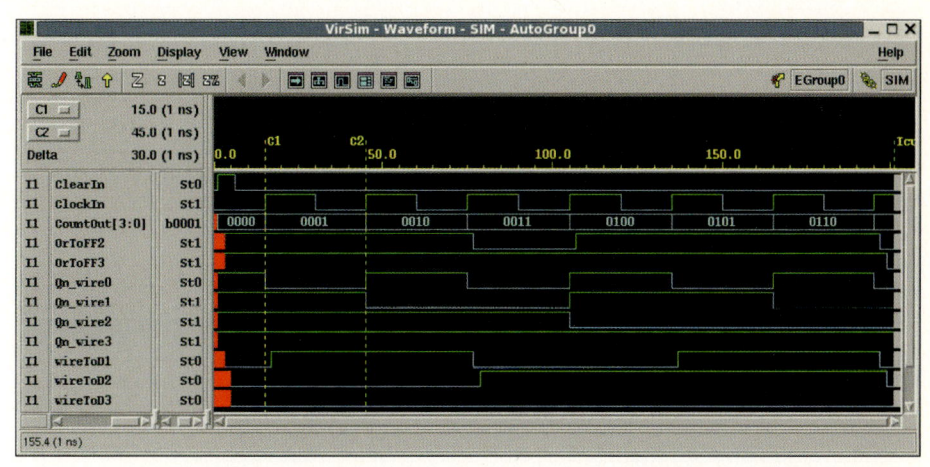

Fig. 11-9. Simulation of the synchronous counter of DFF's defined structurally by *nand* gates.

Step 3: Synthesis at gate level. Synthesize both the `Synch4DFF` and `Synch4DFFGates` designs. Keep your synthesis design rules the same for all conditions, but vary the constraints as follows:

 Optimize both for area and then for speed. When doing the area optimization, use no constraint except one for area; set that to 0.

 When doing the speed optimization, impose no constraint except one for maximum output delay and one for clock period. Adjust the result so that both of these constraints are <u>unfulfilled</u> but are no more than 1 ns too small for the actual netlist result reported by the synthesizer.

Compare the speed and area netlist sizes: Did the design make any difference?

Step 4 (*optional*): Simulate the speed-optimized netlist (see figures 11-10 and 11-11).

This should be tried with the speed-optimized `Synch4DFFGates` netlist, only. You may wish to try simulation with and without SDF back-annotated delays. Recent versions of DC should produce a working netlist with simulation timing from our training library (`tcbn90ghp_v2001.v`); however, SDF timing may be required.

If you run the netlist simulation with correct timing, you will be able to zoom in VCS to see the SR latch's internal process of reset: For example,

  `4'b1111 --> 4'b0111 --> 4'b0101 --> 4'b0100 --> 4'b0000.`

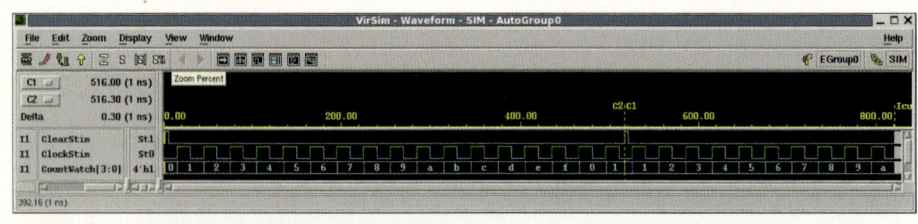

Fig. 11-10. Simulation of the speed-optimized `Synch4DFFGates` synthesized netlist.

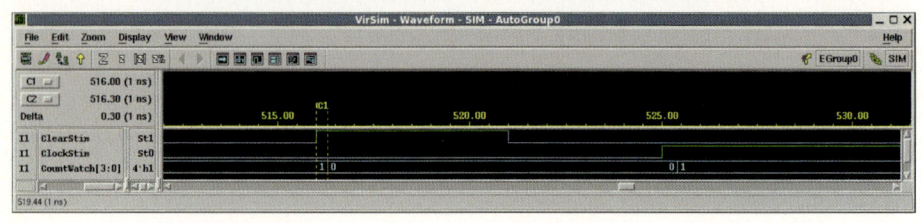

Fig. 11-11. Closeup of the `Synch4DFFGates` synthesized netlist simulation near a counter reset.

11.2.1 Lab Postmortem

Do you understand the relationship of built-in gates and strengths?

What is the relationship of structural design to logic optimization and technology mapping?

In synthesizing from a behavioral *vs.* gate-level design, as in the lab Step 3, should the technology mapper do better when the D flip-flops are synthesized from behavior or from gates? Why?

11.3 Additional Study

Read Thomas and Moorby (2002) sections 6.2.1–6.2.3 on primitives and net types.

Read Thomas and Moorby chapter 10 on strength and switch-level modelling.

Read Thomas and Moorby section 5.1 on port connection rules.

Optional readings in Palnitkar (2003):

Section 4.2.3 shows the verilog port connection rules.

Read chapter 10.1 on assignment of delays to gates in a netlist. Look through the rest of chapter 10 if you are interested; we'll study component internal delays later in the course.

Do the exercises in section 5.4.

Digital VLSI Design with Verilog 243

Work through the verilog of the examples in section 5.1.4; the code is available on the Palnitkar CD, so these examples can be simulated without any keyboard entry. The simulator will display waveforms more realistically if you add a `timescale.

There are procedural models of the devices of section 5.1.4 in section 6.5, so it may be interesting to contrast the differences. Palnitkar gives a gate-level schematic for a D flip-flop in his Figure 6-4 (section 6.5.3).

Chapter 12
Week 6 Class 1

12 Today's Agenda:

Lecture on procedural control and concurrency
Topics: Procedural control constructs, procedural concurrency, and verilog name space.
Summary: We deepen our understanding of the procedural control constructs `forever`, `repeat`, `while`, `case`, and the use of the ***case equality*** operator. We also study `casex` and `casez`, but we warn against them. Then, we change topic somewhat to revisit procedural control of concurrency by `fork-join` blocks. We end with a few new observations on the scope of verilog names.

Lab on concurrency
After exercises in common procedural control constructs, we design a watchdog device which runs concurrently with the CPU it guards.

Lab Postmortem
We review the problem of `casex` and devise a bubble-diagram for our watchdog-CPU design.

12.1 Verilog Procedural Control Statements

The `for` is the most versatile looping control statement in verilog; and, synthesizers typically can make better use of a `for` statement than the others, with the sole exception of `if`. For completeness, now, we shall describe the others. Except in this lab, you are encouraged to prefer `for` as much as possible in your work.

We already have used `if` extensively, and we shall discuss the `case` variants below. The remaining control statements are `forever`, `repeat`, and `while`.

forever. The syntax simply is `forever`, followed by a statement or possibly a block of them.

The `forever` is a looping statement which is not intended to be terminated once it is executed; however, other verilog constructs may be used to terminate it. It is used almost always in an `initial` block, for obvious reasons. Usually, a `forever` loop will continue until `$stop` or `$finish` is executed by the simulator.

```
initial
  begin : Clock_gen
  Clock1 = 1'b0;
  forever #10 Clock1 = ~Clock1; // Works, but don't do this!
  end
...
initial
  begin : Test_vectors
  Abus = 32'h0101_1010;
  Dbus = 'b0;
  #5 Reset = 1'b0;
  ...
  #220 $finish;  // Terminate the simulation after about 10 clocks.
  end
```

A concurrent clock generator such as, "`always@(Clock1) #10 Clock1 <= ~Clock1;`", is not equivalent to the `forever` above. How do they differ?

The main difference is that the above `forever` clock has been implemented with a blocking assignment, preventing reliable setup of clocked variables unless special delays are added to related combinational logic.

Another difference is that the `forever` schedules an assignment to `Clock1` every 10 units regardless of anything. By contrast, the `always` concurrent clock can not schedule a new assignment until a change in the current value has been simulated. Thus, the `always` clock can not oscillate if the assignment is a blocking one, while the `forever` clock can.

Also, `always` is itself a concurrent construct, not a procedural one. A looping `always` can not be included in a procedural block.

The `forever` example is not synthesizable; `initial` blocks in general have no corresponding hardware. Of course, the `always` concurrent clock generator described above is not synthesizable, either.

Two different termination techniques are available for `forever`:

```
// Technique 1:
  forever
    begin
    #1 Count = Count + 1; // Must include a delay or @!
    if (Count>=1000) $finish;
    end
```

```
// Technique 2:
begin: Mod16_Counter
  forever
    begin
    Count = Count + 1;       // Count is an integer or wide reg.
    #5 Dbus[4:0] = Count%16; // This delay avoids hanging the simulator.
    end
end // Mod16_Counter.
            (somewhere else, or in the loop, put disable Mod16_Counter)
```

A restartable clock generator is easy to do:

```
always@(posedge Run)
  begin : RunClock1
  Clock1 = 1'b0;
  forever #10 Clock1 <= !Clock1;
  end
always@(negedge Run) disable RunClock1;
```

Again, one should consider the setup and hold implications of the use of a blocking vs. a nonblocking assignment in any clock generator.

repeat. The syntax simply is repeat (*number_of_times*), followed by a statement or a block of statements. The repeat is strictly a loop-counting construct. When the repeat is read, the value of *number_of_times* is stored, and the statement is executed that number of times. Unlike the for iteration value, the value of *number* can not be reread or changed during loop execution. The repeat statement(s) may be terminated early the same way as shown above for forever.

Example of a repeat:

```
...
i = 0;
repeat(32)
  begin
  #2 Abus[i] = LocalAbus[i+32] + AdrOffset;
  i = i + 1;
  end
```

Notice that the coding overhead for useful invocation of repeat is similar to that of for in a loop-counting construct; but, the code in a disable'd repeat would be spread out and thus may be more prone to error. In a block of several hundred lines, the control, "i = i + 1" in the example above, might be difficult to find. The author has worked with state machine designs of over a thousand lines in a module. Usually, for will be preferred to repeat, unless no loop control is to be exercised.

while. The syntax is while (*expression*), followed by a statement or block of statements. The *expression* is evaluated on first reading, and it is reevaluated each time after that, as soon as the controlled statement ends. On any evaluation, if the *expression* is not equal to *false* (a logic-level or numerical '0'), the while executes its statement(s).

Otherwise, the `while` terminates, and execution picks up after the statement(s) controlled by the `while`.

The `while` arguably is a reasonably efficient alternative to a `for` in many contexts. The `while`'s *expression* may be a relational one involving a count, so `while` is more versatile than `repeat`. Also, a `while` may be almost interchangeable with `for` in a loop-counting context involving a delay.

A couple of `while` examples:

```
// Example 1:
...
SleepState = 1'b1;
while(SleepState==1'b1)
  begin
  slowRefresh;   // A task.
  #SleepDelay Status = CheckUserInput;     // A function call.
  if (Status!=Asleep) SleepState = 1'b0;  // Exit the while.
  end
// Example 2:
Count = 10;
while(Count > 0)
  begin
  ... (do stuff)
  #3 BusA[Count] = BusB[Count];
  Count = Count - 1;
  end
```

The following two `for` statements are functionally equivalent to the `while` examples above:

```
// Example 1:
...
for (SleepState = 1'b1; SleepState==1'b1;
     SleepState = (Status!=Asleep)? 1'b0: 1'b1
    )
  begin
  slowRefresh;
  #SleepDelay Status = CheckUserInput;
  end
// Example 2:
...
for(Count=10; Count>0; Count=Count-1)
  begin
  ... (do stuff)
  #3 BusA[Count] = BusB[Count];
  end
```

Assignment of an updated control variable has to be located away from the `while` header, in the `while` statement block. This brings out an advantage of the `while`: It works the same way whether or not the control variable is changed with a delayed assignment. A `for` control header is not allowed to include a delay; so, iterator updates sometimes have to be moved out of the `for` control header.

In the `for`, usually all control is up-front, in the header; this makes it easier to understand the control or to debug errors. For this one reason, it is recommended to use `for` in preference to any other loop control, whenever it is reasonably possible.

12.1.1 Verilog `case` Variants

The main advantage of an `if` statement is that relational expressions can be used, so whole ranges of values may be covered in one expression. By comparison, in a verilog `case` statement, only specific values are allowed in the alternatives. These values may be variables and may be combined with comma separators, but they still have to be enumerated explicitly. However, the table-like arrangement of `case` alternatives often makes a `case` more readable than a chain of `if` ... `else`'s. In addition, `if` and `case` differ importantly in the way they handle expressions containing 'x' or 'z' logic levels.

An `if` attempts equality matches on all logic states at once, including 'x' and 'z'. If a bit pattern includes an 'x', then `if` interprets the 'x' as a combined '1' and '0' and returns 'x'; thus, it can not express a match, even if that 'x' is included explicitly in the expression. If a variable X contains any bit at 'x' or 'z' level, then even "`if (X==X)`" will fail to match, and the `else` (if any) will execute!

A `case` attempts a match on the specific pattern of bits, whether or not any of them is 'x' or 'z'. If a vector in the `case` expression contains some 'x' or 'z' levels, and one of the alternatives contains the same pattern, `case` evaluates to a match and will selectively execute the statement for the matching alternative.

For example,

```
X = 4'b101x;
Y = 4'b101z;
//
if (X==4'b101x) ...; // Evaluates as 'x'; won't match the value above.
if (X==Y) ...;       // Won't match; and if (X!=Y) won't, either.
if (X==X) ...;       // This won't match, either!
//
case (X)
  4'b1010: ...;
  4'b1011: ...;
  4'b101z: ...;
  4'bxxxx,
  4'bzzzz: ...; // Comma separation is legal.
  default: ...; // This will execute because nothing else did.
endcase
case (X)
  4'b1010: ...;
  4'b101x: ...; // This matches and will execute.
  4'b101z: ...;
  4'bxxxx,
  4'bzzzz: ...;
  default: ...;
endcase
```

One other thing to keep in mind: Although the ***conditional operator*** (`?:`) can be used as a substitute for `if` under some circumstances, it is an operator and not a statement; it returns an expression. The logic is the same as that of `if` for 1-bit alternatives. For multibit vector alternatives, when an 'x' or 'z' appears in the `?:` condition, this causes the return value of the expression to go bitwise to 'x', unless corresponding bits in the '?' and ':' vector values agree. The case equality operator may be used with this operator; an example is below.

A special operator is provided to allow `if` to match unknowns, the ***case equality operator***, "`===`" (negation is "`!==`"), which we briefly have mentioned before. This operator can not cause an evaluation to be 'x'. Using this operator, 'x' and 'z' match themselves, just as do '1', or '0', wherever they appear, just as in a `case` statement.

Consider these examples:

```
X <= 4'b101x;
Y <= 4'b101z;
//
if (X===4'b101x) ... else ...; // expr = 1; the if executes.
if (X!==Y) ... else ...;       // expr = 1; the if executes.
if (X!=Y) ... else ...;        // expr = x; the else executes.
//
// The conditional operator is not an if but accepts case equality:
Z = (X==Y )? 1'b1: 1'b0; // Assigns 1'bx.
Z = (X===Y)? 1'b1: 1'b0; // Assigns 1'b0.
Z = (X!==Y)? 1'b1: 1'b0; // Assigns 1'b1.
```

Digital VLSI Design with Verilog

To take advantage of the table-like `case` syntax, two variants of that statement have been defined in verilog, ***casex*** and ***casez***.

casex. A `casex` expression matches an alternative as though 'x' and 'z' were wildcards: Any bit matches an 'x' or a 'z', including an 'x' or a 'z'. The statement ends with `endcase`, the same as for the `case` statement we have been using. It can be very confused to allow a specific logic level such as 'z' become an "anything" character; so, to clarify intent, a 'z' wildcard in an alternative may be written as '?', rather than arbitrarily writing either 'x' or 'z', or even mixing them.

Examples of `casex`:

```
X <= 4'b101x; Y <= 4'b101z;  // X and Y are reg[3:0].
// Example 1:
casex (X)
  4'b100z: ...;  // No match.
  4'b10xx: ...;  // Executes, because it matches and comes first.
  4'b11xz: ...;  // Can't execute on this value of X.
  4'bxxxx: ...;  // Would execute if 4'b10xx didn't.
  default: ...;  // Would execute if nothing else did.
endcase
// Example 2:  Some sort of bit-mask or decoder:
casex (Y)
  4'b???1: ...;    // Executes, because of casex LSB wildcard 'z' in Y above!
  4'b??1?: ...;    // Would execute if the first one didn't.
  4'b?1??: ...;    // Can't execute on this value of Y.
  4'b1???: ...;    // Would execute if nothing above did.
  default: ...;    // Would execute if no '1' or wildcard in Y.
endcase
```

There might seem to be an advantage in using `casex` when decoding or searching for patterns in data, especially in sparse matrices of data. In Example 2 above, either a chained `if` would have had to be used, or a `case` would have had to be written which individually rejected 12 of the 16 patterns of '1' and '0' possible in a 4-bit object.

Unfortunately, `casex` is extremely dangerous and error-prone. In the code above, the designer's intent presumably was a prioritized search for 1's in the four bits of Y. However, any failure in an assignment to Y during simulation, or a high-impedance value as actually shown, might cause the `casex` to switch and execute on a Y without a '1'.

Thus, if the intent was to look for a '1' in well-defined but arbitrary data, chained `if`'s would be a better way:

```
Y <= 4'b101z; // Y is reg[3:0].
//
if      (Y[0]==1'b1) ...;  // 'z' means no match (but === 1'bz would match).
else if (Y[1]==1'b1) ...;  // This one executes.
else if (Y[2]==1'b1) ...;  // No, not '1' and below the first match.
else if (Y[3]==1'b1) ...;  // No, below the one that first matches.
else                 ...;  // Executes if no '1' in Y.
```

Notice that indenting the chain this way makes it fairly readable, although the expression is changing on every line, risking a typo. Also, the if's are a little risky, because the evaluations would be wrong if Y had been declared reg[0:3] instead of reg[3:0], and this is harder to see when using the if's instead of a case.

The consequences of using casex are even worse when one considers synthesis. The "don't care" entries in the casex alternatives might seem advantageous in synthesis, because usually a synthesizer uses 'x' or other don't-cares to its advantage: The designer doesn't care, so the synthesizer is free to do its best. What the synthesizer does, then, is ignore wildcarded alternatives. In the first casex example above, then, the second and third alternatives are folded into one, the earlier statements are synthesized but not the later (depending on the synthesizer implementation), and the resulting netlist probably will not simulate the way the original verilog does, assuming that Y can vary during design operation. In addition, in that example, the default alternative probably will be folded into the 4'bxxxx one above it during synthesis, which may or may not have been foreseen by the designer.

Because of unexpected consequences of too much wildcarding, <u>it is not recommended ever to use casex for anything</u>. When data are sparse, or when wildcarding is unavoidable for other reasons, use casez in place of case or if. At least with casez, an unexpected simulator 'x' will not be wildcarded, and the synthesizer will jump on the don't-cares more lightly.

casez. A casez expression matches an alternative with any 'z' in the expression or in an alternative treated as a wildcard. No 'x' is a wildcard. Also, as for casex, a '?' may be used as a wildcard character in an alternative instead of 'z'.

For example,

```
X <= 4'b1x00;    // X is reg[3:0].
// Example 1: No match for any of the alternatives:
casez (X)
   4'b100z: ...;
   4'b10xx: ...;
   4'b10xz: ...;
   4'bxxxx: ...;
   4'b0zzz: ...;
   default: ...; // Executes.
endcase
// Example 2:   '?' is same as 'z'
casez (X)
   4'b???1: ...; // No match.
   4'b??1?: ...; // No match.
   4'b?1??: ...; // No match.
   4'b1???: ...; // Executes; X[3] matches 1==1, and others are wild.
   default: ...; // Can execute on X == 4'b0000, or on anything
endcase          // with no '1' or 'z' anywhere, etc.
```

The `casez`, like `casex`, should be avoided; when one is tempted to use it, first see whether a simple `case` or chain of `if` can be used instead. If there is no efficient other way, `casez` may be used. The `casex` ***never*** should be used, because it has no wildcarding advantage over `casez`, and it is far more prone to create results in which simulation of the synthesized netlist can not be made to match simulation of the original design.

12.1.2 Procedural Concurrency

We shall use the word *thread* somewhat generically and intuitively here; there is no intentional reference to the similar concept in software parallelism, in which a "thread" is differentiated from a "process".

We already have studied parallel blocks (`fork-join` blocks) and have worked with them a little in lab. It is possible to take advantage of parallelism not only in individual statements, as we have done, but in entire threads of execution of the simulator. This most easily is done by parallelizing `tasks`.

Assigning two or more parallel threads each to its own `task` has many advantages; we give four of them here:

(*a*) the thread is completely well-defined, because it consists just of the statements in each `task`.

(*b*) Multiple `tasks` parallelized in a given procedural block easily can be identified for design debugging purposes; one can know where they start (`fork`) and where they finish (`join`). This can't usually be said of a collection of concurrently executing `always` blocks.

(*c*) The block names (`task` names) generally will be preserved in the synthesized netlist; otherwise, one must dedicate special attention to naming `always` blocks or whatever else one has chosen instead of `tasks`.

(*d*) Modification of the statements in a `task` can be relatively free of unintended side effects, because statements in a `task` are localized and thus can be changed independent of any of the design constructs which cause the `task` to execute. By contrast, parallelizing a collection of unnamed statements requires that any change be made in a collection of code containing all the statements parallelized, all at once.

12.1.2.1 Task Concurrency and **fork-join**

Let's next look at a typical example of concurrency implemented with the concurrent construct we have been using most throughout this course, the `always` block. Here is the essential code of a device which detects toggling bits on a 4-bit bus and reports the bit number on a 2-bit output bus. The reason for the concurrency is to simulate nonprioritized evaluation of the bus bits; checking them in a procedural block would imply a priority of some bits, which would be checked before others.

The code:

```
...
task CheckToggle(input[1:0] BitNo);
  begin
  @(InBus[BitNo]) #1 ToggledReg = BitNo;
  end
endtask
//
always@(posedge CheckInBus) CheckToggle(0);
always@(posedge CheckInBus) CheckToggle(1);
always@(posedge CheckInBus) CheckToggle(2);
always@(posedge CheckInBus) CheckToggle(3);
...
```

Concerning this `CheckToggle` example: It is to illustrate `task` instance concurrency, only. It is not synthesizable (as of 2011) because of the '@' statements which it contains. Also, our examples here do not exemplify good modelling.

A simpler and better way to solve the above bit-checking problem concurrently and synthesizably would be,

```
// Use bit-selects:
always@(InBus[0]) TempToggledReg = 2'h0;
always@(InBus[1]) TempToggledReg = 2'h1;
always@(InBus[2]) TempToggledReg = 2'h2;
always@(InBus[3]) TempToggledReg = 2'h3;
always@(posedge CheckInBus) #1 ToggledReg = TempToggledReg;
```

A procedural solution, using a `for` loop over `InBus[i]`, i = 0 to 3, also would be efficient and synthesizable.

Very similar functionality can be achieved by replacing the multiple `always` blocks with a single `always` block containing a `fork-join`:

```
always@(posedge CheckInBus)
  begin    // begin/end not required.
  fork
  CheckToggle(0);
  CheckToggle(1);
  CheckToggle(2);
  CheckToggle(3);
  join
  end
```

However, the `fork-join` is not entirely equivalent to the multiple `always` blocks, which is the reason we prefer to call it a *fork-join* rather than a *parallel* block: The `fork-join` block does not exit until all four input bits have toggled, meaning that all four concurrent tasks have been run to completion. If one input bit never toggles, then each change of `CheckInBus` will spawn a new `fork-join` block event in memory, which will

Digital VLSI Design with Verilog

join the others in the simulator event queue, waiting for all four bits to toggle so it can exit. In effect, this creates a simulator software memory leak, and it might crash the simulator or cause it to become very slow at evaluating new events. The first `always` block example above also suffers from this problem, but to a lesser extent.

The point here is that, when using a `fork-join` block, one must be certain that the `join` eventually will be reached.

The effect of `fork-join` may be seen easily by its effect on mixed blocking and nonblocking assignments. The following example is for illustration only; it is a bad idea to mix blocking and nonblocking assignments in a real design:

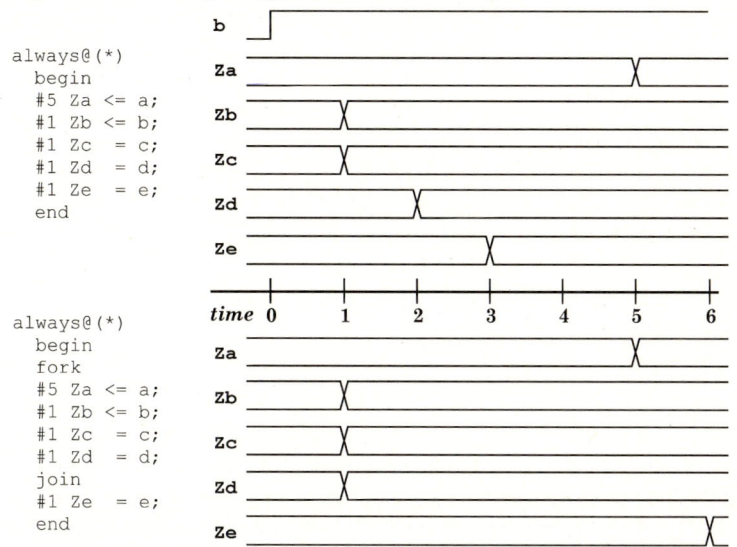

```
always@(*)
  begin
    #5 Za <= a;
    #1 Zb <= b;
    #1 Zc  = c;
    #1 Zd  = d;
    #1 Ze  = e;
  end
```

```
always@(*)
  begin
    fork
      #5 Za <= a;
      #1 Zb <= b;
      #1 Zc  = c;
      #1 Zd  = d;
    join
    #1 Ze  = e;
  end
```

Fig. 12-1. A `fork-join` radically can change scheduling of delayed events.

There is another very obvious way of achieving concurrency in verilog: Put the concurrent functionalities into separate `module`s or different `module` instances. Events in different `module` instances are scheduled concurrently, just as are the `always` and `initial` blocks within them, and just as are the different hardware chips on a board.

12.1.2.2 Task Concurrency and **automatic** Keyword

As in the `CheckToggle` discussion above, we are concerned here with `task` instance concurrency; our examples are not synthesizable (in tools up to 2011) because of the '@' statements which they contain.

Variables declared in a `task`, like those in any other named block, are static and hold their most recent values no matter how many times the `task` is called. Also, concurrently running instances of the same task share its declared variables. This holds except for tasks declared **automatic**: The declaration, task automatic Mytask gets private copies of its local variables for each instance. Because of this, an automatic task may be called recursively.

In all the CheckToggle code above, the examples actually will not work as expected, because the input BitNo would have been declared implicitly as a reg just once, and this one variable will be shared among all executing instances of CheckToggle.

Thus, all executions of CheckToggle above will be checking the same bit of InBus! The effect is exactly the same as though reg[1:0] BitNo had been declared in the module but external to the task declaration, and the task declaration included no I/O. Task local variables are *static* reg-type variables shared among all calls of a given task; so, of course, they share the same declaration. Sometimes this sharing may be desired, because it allows different running instances of the task to communicate with each other.

In the CheckToggle examples above, and in general when a task may be called more than once in the same simulation time interval, one would want independent variables, not shared ones. Thus, the example task declaration above should have included an automatic keyword.

The improved declaration should look like this:

```
...
task automatic CheckToggle(input[1:0] BitNo);
  begin
   @(InBus[BitNo]) #1 ToggledReg = BitNo;
   end
endtask
...
```

A verilog function also may be declared automatic for purposes of recursion. A simple example of useful function recursion is calculation of a factorial:

```
...
function automatic[31:0] Factorial(input[3:0] N);
  begin
  if ( N>1 )
       Factorial = N * Factorial(N-1);
  else Factorial = 1;
  end
endfunction
...
```

Notice the location of the function width index expression, and the use of the declared input as an implied reg. The externally-called Factorial function can not return until Factorial internally has been called with N==1, resulting in evaluation of N*(N-1)*(N-2) ... 2*1, which is equal to N!. Recursive functions or tasks may not be synthesizable; so, they should be avoided when not essential to simulation of the design.

12.1.3 Verilog Name Space

One final thing: Up until now, all the examples have built upon assumed experience with *C* language, or other programming languages, in which the name declarations generally precede the programming. Verilog does not allow "global" names, the way *C* does, declared outside a module; so, we have assumed the below sort of layout in the verilog source file:

```
module module_name ( I/O's );
reg   name declarations; ...
wire name declarations; ...
(programming stuff, using previously declared (or implied) names in statements)
endmodule
```

In many other languages, it is bad style, or illegal, to add declarations anywhere but before the programming: Declare everything first (to avoid name conflicts); then do the programming. However, verilog only requires that a variable name be declared before it is used. Furthermore, `functions` or `tasks` may be declared <u>anywhere</u> in a `module`, and any variable declared there is accessible to them.

We have seen that verilog variable names may be declared in `modules`, or locally in `functions`, `tasks`, or `always` blocks. These local names do not conflict with one another because the name is visible to the compiler only in the local block of code in which it was declared.

Verilog allows declarations in a <u>named region</u>. So, it is possible to declare new `reg` or net names anywhere in a `module`, which is a named region, even between `always` or `initial` blocks, and have those names usable anywhere in the file following the declaration.

This feature should be used with caution; but, in our next lab, it may be useful to declare a few variables close to the `always` blocks in which they are used, rather than far away, at the beginning of the source file. This is a feature which makes verilog more object-oriented than *C*: Objects which do distinct things can be more self-contained than otherwise.

For example:

```
module ...
...  (500 lines of verilog) ...
reg[7:0] ClockCount;
always@(negedge ClockIn)
  begin : Ticker
  if (StartCount==1'b1)
      ClockCount = 'b0;
  else ClockCount = ClockCount + 8'h1;
  if (ClockCount >= 8'h3a) (do something);
  end
```

It should be mentioned that the synthesizer may not synthesize this last, because the `always` block contains more than just one top-level `if` statement.

Notice in the next example that the count will be preserved:

```
always@(negedge ClockIn)
  begin : Ticker
  reg[7:0] ClockCount;  // A new declaration, no matter whether some
  if (StartCount==1'b1) //    other "ClockCount" is declared elsewhere.
      ClockCount = 'b0;
  else ClockCount = ClockCount + 8'h1;
  ...
end
```

This shows that a variable declared this way in an `always` block will retain its value between block executions. Incidentally, VCS will simulate the example above correctly as an up-counter. It should be mentioned that `ClockCount` is local to the named `always` block and can't be accessed directly by an ordinary, simple name. However, `ClockCount` can be accessed from outside its `always` block by a hierarchical name, as will be studied in the next day's lecture.

12.2 Concurrency Lab 14

Topic and Location: Procedural control constructs, concurrency in procedural code, and a CPU watchdog design.

Do this lab in the `Lab14` directory.

Preview: We do a few elementary exercises with `forever`, `repeat`, `while`, and `case`. Then, we take on a fairly complex design which requires concurrent execution of a watchdog device and another device, in this case a CPU.

Deliverables: Step 1: A simple module containing a correctly simulating `forever` loop. Step 2: The same design with a `repeat` loop added. Step 3: The same design with a `while` loop added. Step 4: A new, small design of an encoder implemented with a `case` statement. Step 5: A correctly simulating design containing a minimal emulation of a CPU and a concurrently-running CPU watchdog.

Lab Procedure:

Step 1: A `forever` block. Write a small simulation model with a testbench clock implemented by means of a `forever` block. Consider a clock generator an exception to the general rule that one should not have more than one `initial` block in a `module`. Simulate the clock to verify functionality.

Step 2: A `repeat` block. To the Step 1 design module, add a clocked `always` which contains a `repeat` block to initialize a 32-bit bus with a fixed pattern of alternating '1' and '0' ("...010101..."), one bit at a time. Simulate it.

Digital VLSI Design with Verilog

Step 3: A `while` block. To the Step 2 module, add a `while` in its own `always` block to check a 32-bit input bus, one bit at a time, for a specific pattern of '1' and '0' (anything you want). If the pattern should be found, cause the `while` to set an output flag bit. Do this <u>without</u> using a `disable` statement. Simulate the combined module.

Step 4: A `case` application. Write an encoder which receives as input a one-hot bit pattern in an 8-bit register and which outputs the 3-bit binary value giving the numerical position of the '1' in the register (starting at 0). The encoder also should have a second output equal to the ASCII code for the bit position ('0' = 8'h30, '1' = 8'h31, ...). Use a `case` statement to do the encoding. Simulate the model to verify it.

Step 5: Concurrency exercise. This is what may be a somewhat difficult exercise in understanding simulation; don't worry about making the design synthesizable.

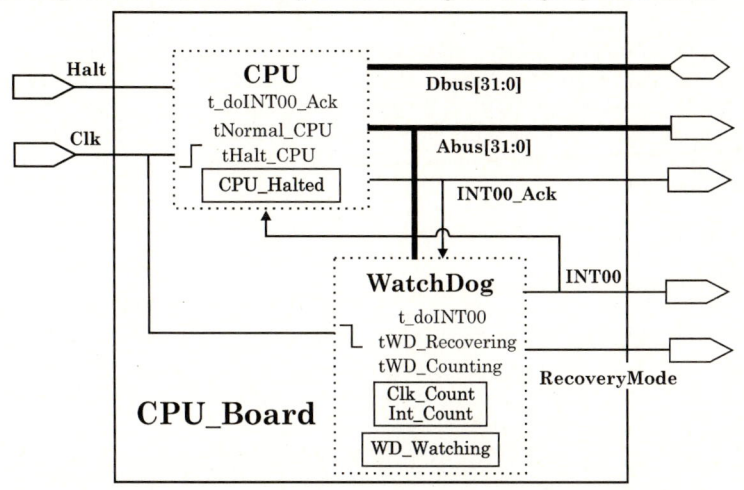

Fig. 12-2. The `CPU` and `WatchDog` design. Dotted blocks represent logic, not hierarchy. The "t" in the various "t_(*command*)" actions stands for <u>task</u>.

Set up a top-level `module` named `CPU_Board` with two named `always` blocks, `CPU` and `WatchDog`. For this exercise, implement the two `always` blocks as described below; normally, a good design would put the `CPU` and `WatchDog` in separate `module`s.

The containing `module` should have a 32-bit `Abus` output, a 32-bit `Dbus` bidirectional bus, and `Halt` and `Clk` inputs. It also should have three other 1-bit outputs, `RecoveryMode`, `INT00`, and `INT00_Ack`. Supply the same clock to both `always` blocks, but different edges (as shown in figure 12-2) may be used.

We don't care much about `CPU` functionality in this exercise, so we'll use the verilog `$random` system function to supply bit patterns on the `CPU` busses, something like this:

```
...
always@(posedge Clk)
  begin : CPU
  ...
  #1 Dbus = $random($time);   // $time repeats only between simulations;
  #1 Abus = $random($time);   // so should the (32-bit) random patterns.
  ...
  end
```

A watch-dog device is a simple and essentially nonfunctional component of a system. The watch-dog device monitors activity of a more complex device and helps the system recover if that activity should indicate a malfunction.

Typically, the watch-dog starts a timer whenever activity of some kind ceases; when the timer lapses, the watch-dog device interrupts or reboots the system. In this way, a high-reliability system can be protected from hardware or software defects which cause a communications deadlock. One of the shortcomings of distributed or parallelized systems is that they can deadlock if two or more components persist each in waiting for the other(s) to release shared resources.

A. The `WatchDog` always block. It should do these four things: (*a*) It should count clocks after each change on the `CPU` address bus, resetting the count to 0 after each such change; (*b*) when the count exceeds 10 cycles, it should interrupt the `CPU` with an `INT00` pulse to attempt to restore it to activity; and, it should repeat the `INT00` pulse periodically until the `CPU` acknowledges; (*c*) it also should assert a `RecoveryMode` output on the `CPU_Board` module, to signal external devices of its action; (*d*) on assertion of `INT00_Ack` by the `CPU`, the `WatchDog` should deassert the `RecoveryMode` output, stop issuing `INT00`, and resume counting clocks.

B. The `CPU` always block. It should do these four things: (*a*) It normally should output a variety of `Dbus` and `Abus` patterns, as described above; (*b*) it should halt on command, the command coming from an external (testbench) `Halt` input pulse; (*c*) it should service an `INT00` interrupt by resuming activity (assuming `Halt` not asserted); and (*d*) it should assert `INT00_Ack` briefly upon resuming activity, or while continuing activity if the `INT00` interrupt should occur while the `CPU` was not halted.

Implement the `CPU_Board` module to fulfill these requirements, using two named `always` blocks and at least one verilog `task`. It is suggested that each "t" action in figure 12-2 be implemented as a separate `task` instance; however, the assignment only requires one `task` to be used.

It is suggested to implement one `always` block completely before worrying about the other one. The `CPU` should be easier than `WatchDog`, so maybe try it first. Simulate the result to verify it (see figure 12-3).

Digital VLSI Design with Verilog

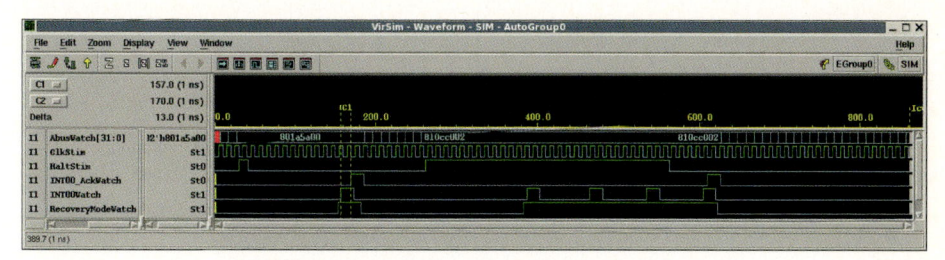

Fig. 12-3. Simulation of an implementation of the `CPU_Board` design.

The clock counts in this design are for lab exercise; a real embedded CPU probably would require many more clocks than 5 or 10 to service an interrupt.

12.2.1 Lab Postmortem

What's wrong with `casex`?

How should the `CPU_Board` design be described by state-machine bubbles? One or two machines?

12.3 Additional Study

Read Thomas and Moorby (2002) sections 3.1–3.4.4 on procedural control constructs. Note that the `repeat` statement *is* synthesizable with current software. Also, keep in mind that `casex` never should be used in design, and that `casez` should be avoided unless absolutely necessary.

(Re)read Thomas and Moorby section 4.9 on `fork-join`.

Optional. Read the paper by Mike Turpin on the dangers of `casex` and `casez` in synthesis: "The Dangers of Living with an X (bugs hidden in your Verilog)", in the References cited for this Textbook.

Optional readings in Palnitkar (2003):

Read sections 7.5–7.7 on the topics presented this time.

Chapter 13
Week 6 Class 2

13 Today's Agenda:

Our purpose today will be to introduce several different constructs useful to create regular patterns of hardware structure.

Lecture on hierarchical names and `generate` blocks
Topics: Hierarchical names, arrayed instances, compiler macroes vs. conditional `generate`s, looping `generate`s.

Summary: We lay groundwork by presenting in detail the rationale for hierarchical names in verilog. We then show how to create arrays of instances wherever a single instance could be used, assuming the connectivity is fairly simple. After that, we introduce the `generate` statement. First, we discuss the conditional `generate`, which may be compared with conditional compilation using verilog macro definitions. Next, we discuss the unique looping `generate`, which allows creation of arrays of instances and of wiring of any complexity among them. We expand on the looping `generate` construct by stepping through the design of a large decoder.

Lab
After brief exercises in arrayed instances, we use `generate` to design a RAM, which we then substitute into our previous FIFO design.

Lab Postmortem
We discuss indices in arrayed instances and `generate`; then, we compare the effect of writing a certain `generate` block three different ways.

13.1 Hierarchical Name Access

We introduced this briefly in an earlier lab. It is possible in verilog to access any element in a design tree from any other location in that tree. The rationale is very much like that of access to files in a modern filing system by means of path names: The file is referenced by a path either beginning at the root directory (top of the design) or beginning in the current, working directory (current `module` instance). In a unix or linux filing system, the name separator is a slash '/'; in a verilog design, the separator is a dot '.'.

```
┌─────────────────────────────────────┐
│  module A                           │
│   ┌──────────────────┐              │
│   │ Dtype D ...      │   ┌────────┐ │
│   │  ┌────────────┐  │   │ DFFC C │ │
│   │  │ Etype E ...│  │   └────────┘ │
│   │  └────────────┘  │              │
│   └──────────────────┘              │
│                                     │
│          ┌────────┐                 │
│          │ DFFC B │                 │
│          └────────┘                 │
└─────────────────────────────────────┘
```

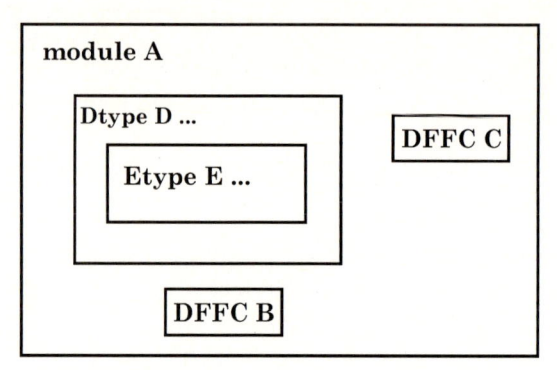

Fig. 13-1. Module A and its design hierarchy as a "floorplan".

For example, consider the arrangement of module instances shown in figure 13-1. In this instance, `module A` instantiates `modules B, C, and D`; and `module D` in turn instantiates `module E`. Here, A is a module name; the others are instance names, the instances being of various different types of declared `modules`.

This kind of system is called a hierarchy, because each module instance is contained in only one other ("higher") one. A pictorial representation of this sort of containment structure resembles the floorplan of a building, as can be seen in figure 13-1.

Another way of representing a hierarchy is as a sort of tree—a family tree, or any other one which grows with its root spreading outward and downward. In a design tree, as in a tree's root system, each element has only one precedent (parent) element. This representation is shown in figure 13-2.

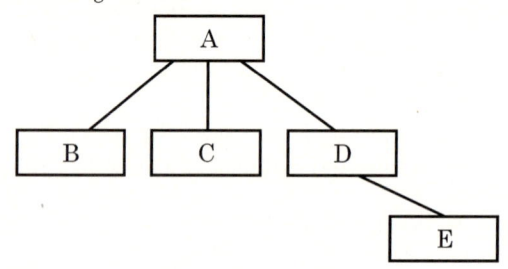

```
              ┌───┐
              │ A │
              └─┬─┘
          ┌─────┼─────┐
        ┌─┴─┐ ┌─┴─┐ ┌─┴─┐
        │ B │ │ C │ │ D │
        └───┘ └───┘ └─┬─┘
                    ┌─┴─┐
                    │ E │
                    └───┘
```

Fig. 13-2. Module A and its design hierarchy "tree".

Verilog hierarchical names are implied when any object is named; such names do not apply only to `module` instances. So, naming a `task` or a `fork-join` block, or declaring a variable, establishes a name which may be used hierarchically. Within a module, any procedural `begin` also may be named.

In verilog, hierarchical name access is allowed anywhere in the design by use of the full path in the hierarchy, using module names or instance names or both. This isn't to

Digital VLSI Design with Verilog

say that such access is advisable as a general design practice. In particular, we strongly discourage any use of the defparam construct, a faulty approach which we shall discuss in detail later.

To see how hierarchical names are used, consider this example: Suppose that in figure 13-1 above, instances named DFFC were flip-flops. This would mean that module A contains two DFFC flip-flop instances named B and C, as shown in figure 13-1 and explicitly in figure 13-2. Suppose that C is driving B, a relationship which is not shown in the figures. Given this, we can describe the Q output port of instance C in figure 13-2 by the hierarchical name, C.Q; the D input port of B similarly can be described as B.D. We then can use hierarchical addressing to describe the C to B connection by the verilog statement,

    assign B.D = C.Q;

Notice that the preceding statement is written with instance names B and C; so, it follows that the implied location of the statement being made must be wherever B and C are located, namely, from figure 13-2, in module A. Because A is the design module name and is not present in the continuous assignment, we also see that instance access is allowed without specifying a full path in the design.

When a lower-level statement is made, hierarchical references to containing (parent) objects only may be made through the containing module names; instance names are not allowed. For example, if the C.Q-to-B.D relationship above had been expressed in module D and not in A, the relationship to A would have to have been explicit, and the preceding assignment statement would have to be written as,

    assign A.B.D = A.C.Q;

In instance D of module D_Type, writing assign B.D = C.Q would not be legal, because B and C are instance names of objects not instantiated in D.

As another example, suppose module E_Type had an output port named OutBus, and suppose module DFFC had a net OutWire. Given this, to connect the net OutWire to that output port, using a statement in module DFFC, one might write,

    assign A.D.E.OutBus = OutWire;

The same connection made by a statement in module A might be written,

    assign D.E.OutBus = C.OutWire;

Obviously, hierarchical names are a complex and dangerous feature, and they are outlawed in most design projects except only under very restricted circumstances—usually, they are permitted in the design only for downward reference to objects within the current module. <u>Upward</u> hierarchical name references are similar to the infamous *goto* in software engineering; a rarely useful feature which generally creates bugs.

Indiscriminate use of the hierarchy is not recommended as a general practice; proper design would dictate that wiring be routed through the ports of modules, and not across the verilog name space. However, it is allowed by the language; and, it can be useful in a testbench or assertion, in which reference across the design does not imply a breach of the design structure.

<u>Downward</u> hierarchical name reference can be useful and consistent with good design when it accesses only *programatically created* substructure within a module. We shall see soon how this kind of downward hierarchical access can help us create and use `generate`'d structures.

13.1.1 Verilog Arrayed Instances

An interesting feature of the verilog language is that it permits instances to be arrayed, in a way very much like the memory arrays we have studied before. The range specification, like that of an array, again is an enumeration range, not a bit width.

For example,
```
and #(3,5) InputGater[2:10](InBus, Dbus1, Dbus2, Dbus3);
```
inserts an and gate into nine (= 10-2 + 1) `InputGater`'s so that it drives the corresponding nine `InBus` bits with the *and* of three data bus bits, each corresponding in bit number.

The preceding statement is equivalent to nine instantiations:
```
and #(3,5) InputGater[2]  (InBus[2],  Dbus1[2],  Dbus2[2],  Dbus3[2]);
  ...
and #(3,5) InputGater[10](InBus[10], Dbus1[10], Dbus2[10], Dbus3[10]);
```
Notice that the array number remains as an index value in brackets []; also, the port connections remain correspondingly indexed, but incrementally.

Instance arrays may be applied to user-defined `primitives` or `modules`. The range may be parameterized. Also, the indexed port connections may be replaced by a concatenation to change the connection bit order.

For example,
```
MyBuffer Xbuff[1:3](.OutPin(OutBus), .InPin({InBus[5], InBus[3], InBus[7]}));
```
is equivalent to,
```
MyBuffer Xbuff[1] (.OutPin(OutBus[1]), .InPin(InBus[5]));
MyBuffer Xbuff[2] (.OutPin(OutBus[2]), .InPin(InBus[3]));
MyBuffer Xbuff[3] (.OutPin(OutBus[3]), .InPin(InBus[7]));
```

Note: The Silos demo version simulator may not be able to compile arrayed instances of user-defined `modules`; apparently, it can compile arrays of verilog primitive gates with ascending-order array indices, only.

Next, we shall study the more flexible `generate` statement to create bussed design objects. However, simple bussing can be accomplished easily and transparently by the arrayed instance construct.

13.1.2 `generate` Statements

A `generate` statement is a concurrent statement invoking one of the verilog procedural control statements. A `generate` can be conditional and be based on an `if` or `case`; but, more frequently, `generate` is used with `for` to create arrays of design objects. A `generate` statement can contain design structure such as gate or `module` instances, `regs`, nets, continuous assignment statements, or `always` or `initial` blocks. A `generate` statement is not allowed to contain another `generate`.

These different functions imply that there are two somewhat different kinds of `generate` statement; they are called *conditional* and *looping*.

To characterize the two kinds of generate:

Conditional generate: (*a*) Modifies design structure and thus possibly functionality; and, (*b*) involves no unrolling of instances or of structure.

Looping generate: (*a*) Creates new design structure; (*b*) may contain multiple levels of loops; (*c*) is unrolled by synthesizer—like arrayed instances; and, (*d*) is more flexible and complex than arrayed instances—may include almost any concurrent statement other than another `generate` block.

We shall discuss the conditional kind of `generate` first, by introducing it in the context of conditional compilation:

13.1.3 Conditional Macroes and Conditional `generates`

In *C* or *C++*, there is a collection of directives, also called macroes, for the compiler preprocessor; these are used for conditional compilation. For example, in the following, each line beginning with '#' introduces a preprocessor directive:

```
#define MacroName1
...
#ifdef MacroName2
...  (compile something)
#else
...  (compile something different)
#endif
```

Verilog has similar directives, but they are introduced with an accent grave '`', also called a "backquote", and not a '#'; in verilog, the latter are reserved for quantification by delay or parameter. We have seen `` `timescale `` in almost every verilog source file, and we used `` `include `` in the first lab exercise of the course. We discussed `` `default_nettype `` at some length.

Verilog's compilation directives include `` `define ``, `` `ifdef ``, and so forth; see IEEE Std 1364 section 19, IEEE Std 1800 section 22, or our *Week 12 Class 1* chapter for a complete list. Compiler directive applications are not especially subtle or informative, and we shall introduce them in this course only as they become useful to us. However, it is important to distinguish their meaning from that of a conditional `generate`: Compiler directives can create alternative simulations by acting outside the simulation language to rearrange it; conditional `generate`s create language-based alternative structures.

A `generate` can be parameterized just as can be any other verilog statement; so, the `generate` statement can facilitate parameterized modelling of large IP blocks.

For example, here is a conditional `generate` fragment which instantiates a multiplexer named `Mux01` either with latched or unlatched outputs:

```
parameter Latch = 1;
. . .
generate
if (Latch==1)
    Mux32BitL Mux01(OutBus, ArgA, ArgB, Sel, Ena);
else Mux32Bit  Mux01(OutBus, ArgA, ArgB, Sel, Ena/*unused internally*/);
endgenerate
```

Note that our `generate` and the conditional compilation examples here use port mapping by position in order to fit on the printed page; this is not a recommended design practice. We should be writing, for example,

```
Mux32BitL Mux01(.O(OutBus), .I1(.ArgA), .I2(ArgB), .Sel(Sel), .Ena(Ena));
```

Anyway, our example above assumes that the `Mux32Bit` module has been declared with an unused input port for `Ena`; this would make it possible for unused wiring in the `module` to be connected and not dangling.

Equally valid would be the following example, which assumes that other conditional `generate`s in the `module` have eliminated or rerouted the `Ena` wire:

```
localparam Latch = 1;
. . .
generate
if (Latch==1)
    Mux32Bit Mux01 (OutBus, ArgA, ArgB, Sel, Ena); // As in Textbook.
else Mux32Bit Mux01 (OutBus, ArgA, ArgB, Sel);
endgenerate
```

A result similar to that of the first example above might be achieved by conditional compilation directives:

```
`define Latch 1 // Macro name, but no macro; used as flag for `ifdef.
...
`ifdef Latch     // NOT `ifdef `Latch; that substitutes a '1'!
  Mux32BitL Mux01(OutBus, ArgA, ArgB, Sel, Ena);
`else            // If no `define Latch
  Mux32Bit  Mux01(OutBus, ArgA, ArgB, Sel, Ena/*unused internally*/);
`endif
```

However, one of the preceding conditional `generate`s should be preferred. The reason for such a preference is that `` `define ``'d token states persist during compilation from the point at which they occur, over all modules subsequently accessed by the verilog compiler.

If the compilation order of modules using the `` `define``'d `Latch` above should change, (*a*) it might not be defined yet (or even might have been `` `undef``'ed) when the `` `ifdef`` above was encountered; or, (*b*) `Latch` might be `` `define``'d for unrelated reasons before the `` `ifdef`` above was encountered. Either alternative could cause unexpected results <u>not</u> accompanied by any warning.

A similar problem was described with `` `default_nettype`` previously. Avoid compiler directives when possible, except `` `timescale``. If they are necessary, try to use them only within a single module file, and then `` `undef`` every new one at the end of that file.

13.1.4 Looping generate Statements

A looping `generate` may be used for conditional hardware alternatives, but its primary benefit is its applicability in building parameterized, repetitive hardware structures which would be time-consuming and error-prone if done manually (even by cut-and-paste). A looping `generate` statement itself can be generated conditionally, but we shall not concern ourselves here with that level of complexity.

The generate **genvar**

There is a special, nonnegative integer type required for use in a looping generate and which is guaranteed not to be visible in synthesis or simulation; this type is **genvar**. A `genvar` is used for indexing and for instance name generation by the compiler as it converts the `generate` statement to the structure it represents. The conversion may be viewed as an unrolling process: The looping statement is unrolled, like a carpet, to a netlist, which then may be simulated.

One or more `genvar`s may be declared inside the `generate` statement, provided each is declared before the looping construct using it. Different `genvar`s must be used at different levels in nested loops. The looping construct almost always is a `for` statement.

13.1.5 **generate** Blocks and Instance Names

The block containing the looped `generate` statements must be named; this name is propagated, indexed by one or more `genvar` values, and becomes the hierarchical root path to the instances generated.

In effect, the generation block is a structural element, a submodule, in which are located the instances. This makes it possible to use object names within each block without indexing or modifying them to make them unique.

A `generate` block, looping or not, may not contain another `generate` block; the verilog language forbids it. Such a block also may not contain module I/O declarations, a `specify` block, or `parameter` declarations. However, multiple levels of looping (`for` loops) are allowed.

As our first example, suppose we want to generate an 8-bit bus of wires, each driving the D input of a flip-flop the output of which should be gated by an inverting three-state buffer.

Also, suppose these other requirements: The library type of the flip-flop is DFFa, and the instance names all should be FF-*something*.

In addition, the data input bus should be named DBus; the buffered output should be DBusBuf. There should be common Clk, and Rst; and there should be individual QEna signals; all with the obvious functionality.

This is shown in figure 13-3.

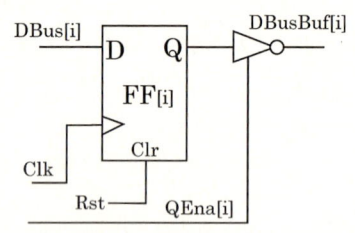

Fig. 13-3. Generic schematic of one member of the required array of flip-flops.

The code for the `generated` array is as follows:

```
// module I/O's include 8-bit DBusBuf output, and DBus & QEna inputs.
...
generate
  genvar i;
  for(i=0; i<=7; i=i+1)
    begin : BuffedBus       // Here is the block name.
    wire QWire, QWireNot;
    DFFa    FF (.Q(QWire), .D(DBus[i]), .Clr(Rst), .Clk(Clk) );
    not     Inv(QWireNot, QWire);
    bufif1 Buf(DBusBuf[i], QWireNot, QEna[i]);
    end
endgenerate
```

We used a verilog primitive inverter (not) in this example to illustrate local wiring; a notif1 instead of the not plus the bufif1 would have been fine.

The result of the above generate loop, unrolled, is an array of 8 named blocks, BuffedBus[0], BuffedBus[1], ..., BuffedBus[7]. Each unrolled block contains its own nets named Qwire and QWireNot. It also contains its own three gate instances, one DFFa, one bufif1, and one not, each with the instance name exactly as given in the loop statement. Thus, logic levels are propagated properly among the unrolled, i-indexed array of named blocks.

Digital VLSI Design with Verilog

To clarify the process, consider the code example immediately above. The first generate, with index number 0, would unroll as a set of instances with the following identifiers:

```
BuffedBus[0].QWire
BuffedBus[0].QWireNot
BuffedBus[0].FF(.Q(BuffedBus[0].QWire), .D(DBus[0]), .Clr(Rst), .Clk(Clk) )
BuffedBus[0].Inv(BuffedBus[0].QWireNot, BuffedBus[0].QWire)
BuffedBus[0].Buf(DBusBuf[0], BuffedBus[0].QWireNot, QEna[0])
```

The index numbers 0 for DBus[0], QEna[0], and DBusBuf[0] are from externally indexed vectors, not from the genvar i.

We don't get the desired flip-flops named "FF[i]", but we do get something just as good, "BuffedBus[i].FF". If we wanted, we could rename "begin : BuffedBus" to "begin : FF", but that would create FF[i].FF instances, which are not worth the resulting possible confusion.

There are eight instances each of a DFFa, a not, and a bufif1 generated. The unrolled loop is shown in figure 13-4.

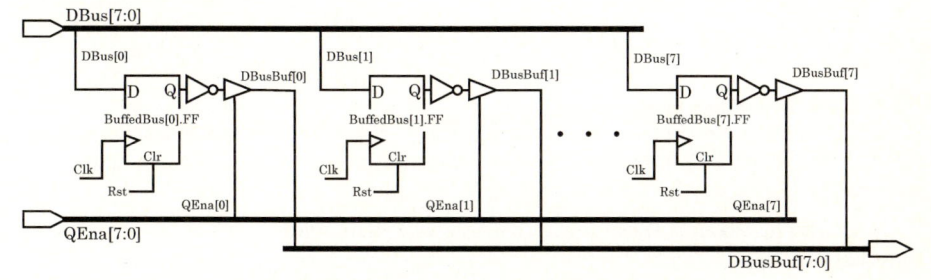

Fig. 13-4. Generated array of flip-flops with individual output enables.

Notice the use of the genvar in the verilog coding above: It generates the instance names of the unrolled blocks, and it is used to connect generate'd instances to outside indexed objects. There is no special reason to index components or nets within a generate loop statement, because the differently named, unrolled blocks create their own, unique new local instances and nets. For example, the not instance in the 3rd block unrolled would be addressed in the simulator or in a synthesized netlist as, BuffedBus[2].Inv. If the block name was changed in the loop statement to DB, then that not would be named, DB[2].Inv.

So, if an external object is vectored or arrayed, that object must be referenced in the generate loop statement using a genvar index value. However, something in the loop must be indexed in order to create a mapping to more than one indexed block name. External objects (nets) which are not vectored or arrayed, are not replicated within the unrolled loop and instead are fanned out within it. We can see this last being done with the external nets, Clk and Rst, in the example above.

Declarations allowed inside a generate loop include net types, reg, and integer. A task or function declaration, being allowed in a module, also may be located inside a generate block, but these declarations are <u>not</u> allowed inside a generate loop statement.

It is easy to understand how the unrolled naming would work, should the above generate loop statement be used with externally generated vectors of local wiring. If so, then for connectivity the genvar index has to be referenced within the loop block.

For example,

```
...
wire[7:0] QWire;
//
generate
  genvar i;
  for(i=0; i<=7; i=i+1)
    begin : IxedBus   // Here is the block name.
    DFFa    FF (.Q(QWire[i]), .D(DBus[i]), .Clr(Rst), .Clk(Clk) );
    notif1 Nuf(DBusBuf[i], QWire[i], QEna[i]);
    end
endgenerate
```

In this example, the first, i=0 inverting buffer would be named, IxedBus[0].Nuf. It would be driven by IxedBus[0].FF.Q through external Qwire[0] and would drive external bit DBusBuf[0]. It would be enabled externally by QEna[0].

The following example shows an equivalent generate fragment which does not use an external wire to drive Nuf:

```
for(i=0; i<=7; i=i+1)
  begin : IxedBus   // Here is the block name.
  wire Qwire;
  DFFa    FF (.Q(QWire), .D(DBus[i]), .Clr(Rst), .Clk(Clk) );
  notif1 Nuf(DBusBuf[i], QWire, QEna[i]);
  end
```

An alternative approach to this whole buffered FF example would be to put the DFF and its buffer in a new module, perhaps named FF, and then to generate (or instance-array) on the module.

Finally, here is an example of a combinational decoder perhaps unnecessarily implemented by `generate`:

```
parameter NumAddr = 1024;
...
generate
  genvar i;
  for (i=0; i<NumAddr; i=i+1)
    begin : Decode
    assign #1 AdrEna[i] = (i==Address)? 1'b1: 1'b0;
    end
endgenerate
```

This would create a structure with an input bus of the same width as the variable `Address`, which may be assumed to be $log_2(1024) = 10$ bits wide, minimum, and an `AdrEna` bus with 1024 one-bit outputs. Whenever the value of `Address` equalled that of some number n in the range 0 to 1023, the n-th bit in `AdrEna` would go high after 1 unit of delay, and all other bits would go (or stay) low. This could be used to enable output in reading a word from a RAM.

An alternative procedural (RTL) way of implementing this decoder, which is very simple and includes no `generate` or component instance, would be something like this:

```
parameter NumAddr = 1024;
integer i;
...
reg[NumAddr-1:0] AdrEna;
always@(Address)
  begin : Decoder
  for (i=0; i<NumAddr; i=i+1)
    #1 AdrEna[i] = (i==Address)? 1'b1: 1'b0; // PROBLEM!
  end
```

The block name, `Decoder`, is not required in this RTL model. More seriously, however, the apparently innocent delay value causes a serious simulation problem: The blocking assignment causes the delays to accumulate, so that `AdrEna[1023]` is not assigned until 1024 ns after an `Address` change. Another good reason not to put delays in procedural code. In regard to this, keep in mind that the delays in the preceding `generate` example were structural, not procedural.

A different and efficient way to implement a large decoder procedurally would be,

```
parameter NumInputs = 10;
parameter NumOutputs = 1<<NumInputs;
input[NumInputs-1:0] Address;
reg[NumOutputs-1:0] AdrEna;
//
always@(Address)
  begin : Decode
  AdrEna = 'b0;
  #1 AdrEna[Address] = 1'b1;
  end
```

This last approach does not cumulate delays, although neither does it match the `generate` loop structure originally given.

13.1.6 A Decoding Tree with `generate`

Let's look further at the design of a big decoder. If the designer had a preferred library component such as a 4-to-16 decoder available, the desired fanout tree might be shown abstractly as in figure 13-5.

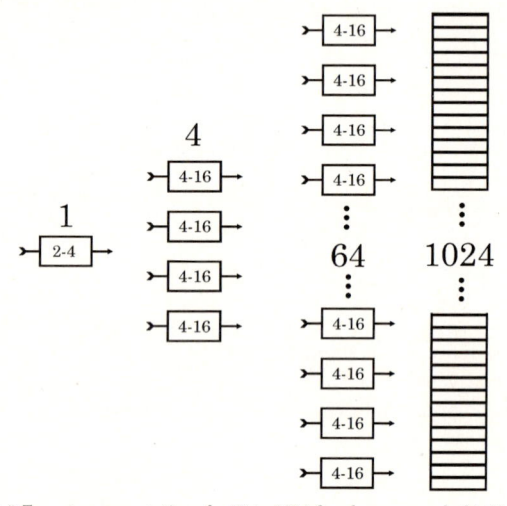

Fig. 13-5. Fanout representation of a 10-to-1024 decoder composed of 4-16 decoders.

The leftmost, level 1 decoder, selects one of the four level 2 decoders. Each of them selects one of 16 others, totalling 64. The 64 level 3 decoders each selects one of 16 addresses, for a total decode of 1024 addresses, which then would employ 10 bits on the address bus.

For this arrangement to work, decoders with an *enable* input must be available; a disabled decoder should put '0' on every one of its output pins, thus decoding nothing. A simplified schematic showing this idea is given in figure 13-6.

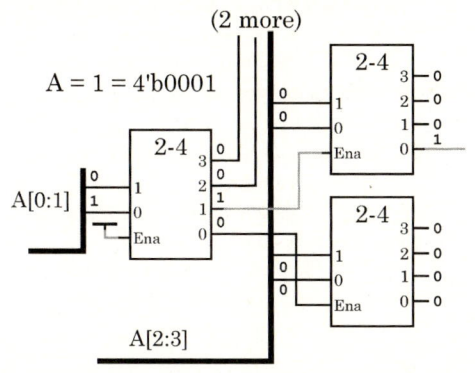

Fig. 13-6. Detail of the select logic for A = 1, showing only 2-4 decoders for simplicity.

This kind of arrangement easily is mapped to verilog operations: In the full-sized tree, the level 1 selection can be represented by the state of the two LSB's on the required 10-bit bus.

Each subsequent level then adds 4 bits of information, because the number of outputs is just 4 times the number of (binary) inputs. Therefore, the level 2 states map to the next 4 bits, bits 2–5; and, the level 3 to the remaining bits, 6–9, which include the MSB.

To fix the idea, figure 13-7 shows the path of enabled decoders in the tree when address 0 is selected:

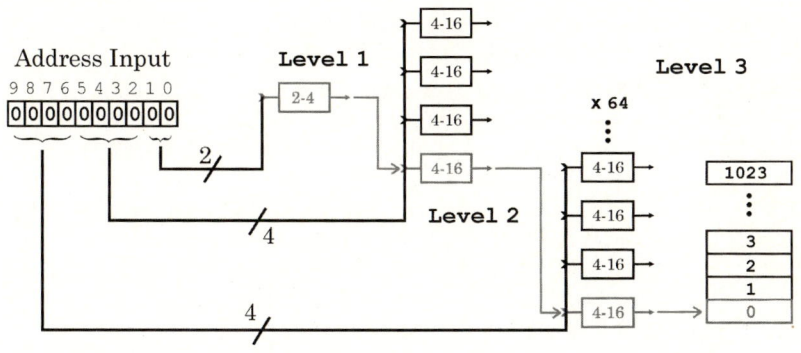

Fig. 13-7. Address 10'h0 decoded. Large arrows (->) indicate the enables, not addresses.

The trick in doing this kind of thing easily is to ignore whether the numerical value of the address which is decoded happens to equal the bit position on the decoded address bus; in almost all applications, it won't matter, because generating an address to write data always will result in the same decoded value as when a <u>read</u> is intended at the same address. What is important is that each decoded bit position map uniquely (1-to-1) to the encoded input value. A given bit's encode and decode must match. For example, looking at figure 13-7, which shows the decode of address 10'h0, we can see that address 10'h1 would enable Level 2 decoder 1, which is shown second from the bottom of the Level 2 list. Assuming all other address bits 0, this would select Level 3 output location 256. But, the fact that address 1 is *schematically* decoded to location 256 would be completely irrelevant in the majority of applications.

Given that the level 1 decode above requires only one quarter of a 4-to-16 decoder, there isn't any need for a `generate` to implement it. Let's assume that an optimized 4-to-16 decoder component is available in the synthesis library and that it is named `Dec4_16`. We'll keep track of the progress of our decoder construction process with the help of mnemonic identifiers as follows: Each input wire will be named in*something*, and each output wire will be named Decoded*something*.

Our first decode then will be,

```
// Level 1 decode is trivial:
wire[3:0] DecodedL1;
wire[1:0] inL1;
wire[15:4] DecodedUnused; // Stub to suppress warnings.
//
// Get Address LSB's:
assign inL1 = Address[1:0];
// The level 1 decode, using verilog concatenation:
//              output[15:0]           , input[3:0]
Dec4_16 U1({DecodedUnused, DecodedL1}, {2'b0, inL1}); // Done!
...
```

Next, we'll see that `generate` comes in handy for the level 2 decode, but that it still could be cut-and-pasted conveniently.

What is truly valuable about using `generate` at level 2, is that the level 2 `generate` provides a template for implementation of the level 3 decode, which, with 64 decoder instances, would be a difficult thing to do correctly any way but by generation.

Digital VLSI Design with Verilog

This is how the level 2 decode might be done:

```verilog
// The Level 2 decode, which requires 4 decoders, fully utilized:
wire[16*4-1:0] DecodedL2;
wire[3:0]   inL2;
assign inL2 = Address[5:2];
//
generate
  genvar i;
  // Generate the 4 4-16's:
  for(i=0; i<=3; i=i+1)
    begin : DL2
    wire[15:0] tempL2;
    Dec4_16 U2(tempL2, inL2); // Each gets bits 2 - 5.
    //
    // Compose the decoded address from L1 and L2, and assign the bit:
    assign DecodedL2[(16*(i+1)-1):16*i] =
                (DecodedL1[i]==1'b1)? DL2[i].tempL2: 'b0;
    end
endgenerate
```

The level 3 decode, requiring 64 decoders, is hardly any more complicated than level 2, when done with `generate`; it assigns everything to the 1024-bit address-enable bus, which might represent logic in a memory chip used to select a word for read or write.

Here is level 3:

```verilog
// The Level 3 decode requires 64 x 4-16's:
wire[(16*4)*16-1:0] AdrEna;
wire[3:0]   inL3;
assign inL3 = Address[9:6];
//
generate
  genvar j;
  // Generate the 64 4-16's:
  for(j=0; j<=4*16-1; j=j+1)
    begin : DL3
    wire[15:0] tempL3;
    Dec4_16 U3(tempL3, inL3); // Each gets bits 6 - 9.
    //
    // Compose the decoded address from L2 and L3, and assign the bit:
    assign AdrEna[(16*(j+1)-1):16*j] =
                        (DecodedL2[j]==1'b1)? DL3[j].tempL3: 'b0;
    end
endgenerate
```

The most important rule to keep in mind when writing a generated structure, is that the result will be structural. The loop index vanishes, so that nothing in the generated structure will be allowed to change with the loop index during simulation time. Every generated object is concurrent—usually a data object or a connection.

13.2 Generate Lab 15

Topic and Location: Synthesis comparison of arrayed instances and generated instances; coding of a RAM using `generate`; and, use of a `generate`'d RAM in a FIFO.

Work in the `Lab15` directory, and be sure to save the results of Step 1 and Step 2 separately, so they can be compared in Step 3.

Preview: First, we ensure that the alternative construct, arrayed instances, is understood by doing a small design (above) with them; we save synthesis results for later comparison. Next, we do a very similar design using `generate`, again saving synthesis results. Then, we compare area and speed optimizations for one of the designs previously synthesized. After these warm-ups, we provide requirements for a new RAM design which is to be coded using `generate`. Once this RAM has been simulated superficially, we include it in a FIFO from a previous lab and simulate and synthesize the result.

Deliverables: Step 1 and 2: A synthesized netlist, each. Step 3: Two synthesized netlists, one optimized for area and the other for speed. Step 4: A correctly simulating RAM design, according to requirements given in the lab instructions. Step 5: A correctly simulating FIFO which uses the generated RAM of Step 3; also, two, and, optionally, three, synthesized netlists of the FIFO design.

Lab Procedure:

First, we'll do some `generate`. Then, we'll reimplement our `Mem1kx32` memory using generated structure; we then can use this memory in our `FIFO`.

Step 1: Arrayed instances. Implement the second `generate` example in the discussion above which follows figure 13.4 and has `IxedBus` and `notif1s`; use arrayed instances and no `generate`. Simulate your result to verify it. Synthesize a netlist and examine the resulting schematic. Keep the netlist for the next Step.

Step 2: Generated instances. Use the earlier `generate` example in the discussion above, the one just after figure 13.3, with `BuffedBus`, `bufifs`, and `nots`; put it into a modified `module` with `parameters` defining all indexed widths. Use a `parameter` value in the `generate` loop to ensure consistency of bus widths with the `genvar` variable. Simulate the result to verify it. Synthesize a netlist with your same constraints as in Step 1, and compare the netlist schematic with the one from Step 1. Do the results differ?

Step 3: Area *vs.* speed in a small design netlist. Pick a design from Step 1 or 2, synthesize it to compare areas when optimized for area (no speed constraint; max area 0) instead of speed. For the speed optimization, set some short time limits, leave the max area at 0, and run the `compile` command with a `-map_effort high` option.

Step 4: RAM `generate` exercise. In our Memory Lab 7 (*Week 3, Class 1*), we designed a verilog RAM behaviorally; it was called `Mem1kx32`. After the instructions in Step 4 **F** below, you will find a slightly modified copy of the RAM requirements.

Digital VLSI Design with Verilog

After reading the suggested procedure below in paragraphs **A - F**, implement this RAM again, this time in a `module` named `Mem1kx32gen`. Use a generated vector of 33 D flip-flops (33rd for parity) for storage at each address. Simulate to verify your work.

Consider using the following procedure for this exercise:

A. Delete the old memory array declaration. Don't copy anything but `module` I/O's and their associated variables from `Mem1kx32`; the old design is not very compatible with a structural `generate`, so attempting extensive reuse may waste considerable time. However, your testbench should work with this lab's design, so keep a copy of the Lab 7 testbench handy.

B. Implement a simple behavioral D flip-flop with `D` and `Clk` inputs, and a `Q` output. No clear or inverted `Q`. Put it in a `module` named `Bit`.

C. Use flip-flops to `generate` a memory word storing 33 bits: In the `module` from Step **B**, use a loop `generate` statement with `genvar j` to generate a single vector of 33 FFs.

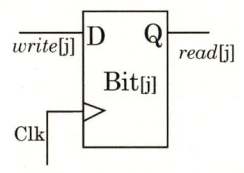

Fig. 13-8. The j-th element of the vector `generate` loop.

Every `Bit` instance should be connected to a common clock, as shown in figure 13-8. One bit (j) of a 32-bit data-in `wire` vector should be connected to each D; likewise connect a data-out `wire` vector to each Q. The unrolled word structure should be as in figure 13-9.

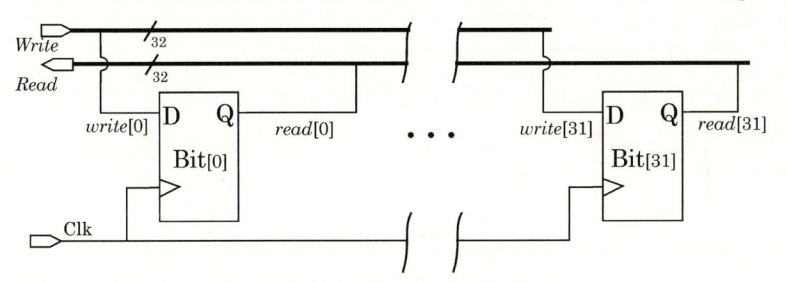

Fig. 13-9. The j-`generate`'d vector structure.

After getting the 32 bit vector model to simulate, add a 33rd bit for parity; however, leave the 33rd bit unused for now. At this point, you should have a `generate` which creates a single 33-bit vector.

D. Use the preceding Step 4 **C** memory words to `generate` a 4-word memory. For this, use another loop in your `generate` statement with `genvar i`, with i = 0 to 3, to generate just four RAM storage locations (words), each one consisting of the unrolled vector loop `generate` statement from Step 4 **C**. To select one word rather than another, the simplest way is to add a buffer to each read bit, so that words not addressed will have their read output (flip-flop Q pin) disabled. The disable can be by a decode of the word address.

The result will be a prototype of the final 32-word memory. Your testbench can exercise all corner cases conveniently on just four words. A schematic representation of this prototype is shown in figure 13-10.

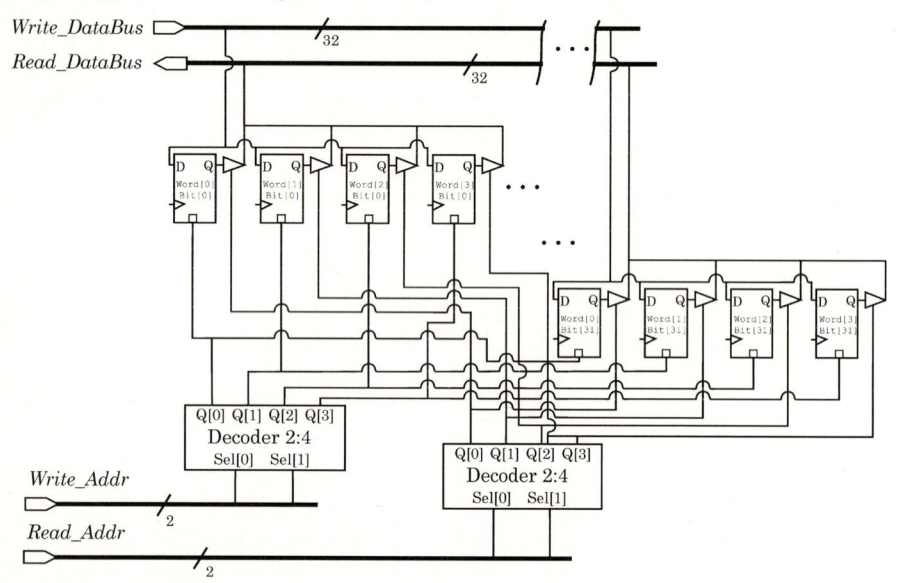

Fig. 13-10. Sketch of the 4-address `generate`'d memory prototype. Parity is not shown; naming is descriptive, only.

The clock inputs to each of the flip-flops are quite ordinary and are omitted here. Notice that the flip-flops include enable inputs (see Lab 4; *Week 1 Class 2*) which are supplied by the decoders.

E. Check that both `generates` (4 **C** and 4 **D**) are under control of `parameters`, not literal constants. If you haven't done this already, define a module-level `parameter` named `DHiBit` to control indirectly the value of `genvar j`. You may do this somewhat as follows:

If the data bus is declared directly by use of the `parameter DHiBit` as a `reg[DHiBit:0]`, the number of data wires (bits) then may be calculated by the compiler as

```
localparam DWidth = DHiBit+1;
```

Then,

```
genvar j; ... for (j=0; j<DWidth; ...
```

Likewise, you may define `AHiBit` to set the number of address lines. The compiler then can calculate,

```
localparam AWidth = AHiBit+1;
```

Given this, the number of storage locations (addresses) is,

```
localparam NumAddrs = 1<<AWidth;   // Same as 2**AWidth.
```

You then may use the simulator compiler to set the upper value of `genvar i`,

```
genvar i; ... for (i=0; i<NumAddrs; ...
```

For example, with `AHiBit` at 1, the address bus width is 2, and the number of addresses is 1<<2 = 4. It's also fine to calculate `AWidth-1`, etc., for the upper limit of the memory array index range.

F. Add functionality according to the requirements below, until your RAM model seems complete. Note that these requirements are to a great extent redundant upon the suggestions in the preceding Steps **A** - **E**.

Simulate with just a few addresses for verification. Then, set `AHiBit` up to 4 to simulate a full 1kx32 RAM (see figure 13-12 below). Synthesis and optimization of the full RAM may take quite a while; see the answer synthesis script files for why. You may wish to defer synthesis until after including the RAM in the FIFO of the next Step.

RAM `Mem1k×32gen` Requirements:

This will be a verilog 1k x 32 static RAM model (32 x 32 bits) with parity. Call it, "Mem1kx32gen"; its port names are below.

Create a structural memory core of D flip-flops using `generate` statements, one for generating the vector of flip-flops, and another, operating on the first, to generate the addressable array of vectors. You may use behavioral or RTL verilog to manipulate this core, which may be put in its own submodule, even in a separate file, if you like.

Use verilog `parameters` for the size and total addressable 5-bit address bus (`Addr`) storage. Parity bits are not addressable and are not visible outside the chip. Of course, parity errors are so visible.

Use two 32-bit data ports, one for read (`DataO`) from the memory and the other for write (`DataI`) to the memory. Supply a clock (`ClockIn`) and also an asynchronous chip enable (`ChipEna`). The latter causes all data outputs on the read port to go to 'z' when it is not asserted but has no effect on stored data. The system clock should have no effect unless the chip enable is asserted.

Supply two direction-control inputs, one for read (`Read`) and the other for write (`Write`). Changes on read or write should have no effect until a positive edge of the clock occurs. If neither is asserted, the most recent values continue to drive the read port; if both are asserted, a read takes place but data need not be valid.

Assign some time delays for data and address bus changes, and supply a data ready (`Dready`) output pin to be used by external devices requiring assurance that a read is taking place, and that data which is read out is stable and valid. Don't worry about the case in which a read is asserted continuously and the address changes about the same time as the clock: Assume that a system using your RAM will supply address changes consistent with its specifications.

Also supply a parity error (`ParityErr`) output which goes high when a parity error has been detected during a read and remains high until an input address is read again.

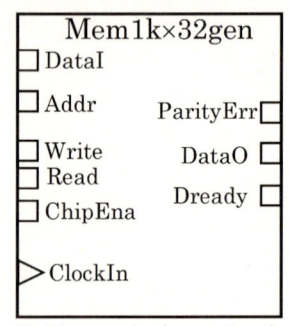

Fig. 13-11: Port names for `Mem1kx32gen`, a reimplementation of a memory from *Week 3 Class 1*.

Digital VLSI Design with Verilog

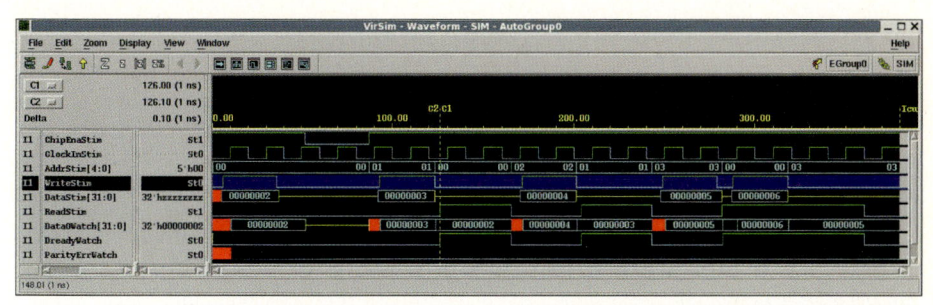

Fig. 13-12. Cursory simulation of the `generate`'d and synthesized 32-word `Mem1kx32gen` RAM.

Step 5: Implement a FIFO using a `generate`'d RAM. Create a new subdirectory in the `Lab15` directory, and copy your FIFO model from `Lab11` into it. This model should consist at least of three files, `FIFO_Top.v`, `FIFOStateM.v`, and `Mem1kx32.v`. You should use the answer version from that lab if you are not completely confident of your own work.

Use your new verilog `generate`'d RAM to provide the storage for a FIFO by substituting `Mem1kx32gen` for `Mem1kx32`. This model might be too big for the demo version of Silos.

Simulate. After verifying the functionality, it would be feasible to synthesize the memory alone; but, why not synthesize the whole FIFO as an exercise in synthesis constraints?

First, synthesize with area optimization only, and no design rule constraint (fanout, load, or drive limits) or speed constraint of any kind. This is a fairly large module for synthesis (about 30,000 transistor equivalents); a register file this large usually would be implemented as a hard macro rather than being synthesized in random logic.

Notice that the synthesizer will report that none of the parity bits are connected to anything. We have omitted parity detection in this model, so we should expect these warnings.

Second, synthesize with no design rule, a zero-area constraint, a 500 ns clock period, and just a modest 50 ns max delay speed constraint on the outputs. A substantially stiffer speed constraint or any design rule constraint will prolong optimization time noticeably in this design; the cause is the `generate`, not the design itself. Do not apply an input or output delay constraint on clocked data, because this also will lengthen the optimization time greatly.

Third, *optionally*, if time permits, try synthesis using the following procedure: Define a 500-ns clock and apply a zero-area constraint; use the synthesizer to compile this with no other constraint. Do not exit the synthesizer when this compilation has completed. Then, in the same synthesis script, set a don't-touch on the memory module, apply the design rules and other constraints just below, and compile a second time with incremental mapping, only.

Optional rules and additional constraints for incremental compile:

```
set_drive          10.0  [all_inputs]
set_load           30.0  [all_outputs]
set_max_fanout 30        [all_designs]
#
set_max_delay      50 [all_outputs]
set_output_delay   1  [all_outputs]  -clock Clocker
set_input_delay    1  [all_inputs]   -clock Clocker
```

The generated FIFO netlist will not simulate correctly—not because of the `generate` but because we have not yet designed a synthesizable FIFO state machine. Among other things, the present FIFO state machine includes `task`s with event controls; this causes DC to skip entirely any synthesis of the `module` for `FIFOStateM`. We shall postpone netlist simulation of the FIFO until later in the course.

Final comment: A single, full synthesis of this FIFO, with or without `generate` assistance, very likely would take a considerable time to finish, perhaps more than an hour on a fast machine. This means that a production-oriented synthesis of this FIFO alone, which generally would be expected to require multiple reruns to tune the size and gate speeds, might take a week or more and probably would not be as good as a commercially available IP model.

13.2.1 Lab Postmortem

Contrast the use of index values in arrayed instances *vs.* `generate`'d instances.

Can a `genvar` name conflict with that of an integer in the same module?

Does a `generate` block create its own name space?

What difference in results do the following three different `generate`s make?

```
generate
for (i=0; i<Max; i = i+1)
  begin : Stuff
  reg temp;
  and A(Abus[i], InBus[i], InBus[i+1]);
  always@(Abus) #1 temp <= &ABus;
  assign #1 OutBit1 = temp;
  end
endgenerate
```

or,

```
reg temp;
always@(Abus) #1 temp <= &ABus;
assign #1 OutBit1 = temp;
generate
for (i=0; i<Max; i = i+1)
  begin : InAnd
    and A(Abus[i], InBus[i], InBus[i+1]);
  end
endgenerate
```

or,

```
generate
reg temp;
for (i=0; i<Max; i = i+1)
  begin : InAnd
    and A(Abus[i], InBus[i], InBus[i+1]);
  end
always@(Abus) #1 temp = &ABus;
assign #1 OutBit1 = temp;
endgenerate
```

13.3 Additional Study

Read Thomas and Moorby (2002) section 3.6 on scope and hierarchical names.

Read Thomas and Moorby sections 5.3–5.4 on arrayed instances and `generate` blocks.

Optional readings in Palnitkar (2003):

Look over Appendix C for the various verilog compiler directives. You may be able to guess what half of them do just from their names.

Read section 7.8, on `generate`.

Chapter 14
Week 7 Class 1

14 Today's Agenda:

Lecture on serial-parallel conversion (deserialization)
   Topics: Generic requirements for serial-parallel conversion; uses of functions and tasks.
   Summary: We briefly discuss serial-parallel conversion as a generic problem and then move on to specific the usefulness of verilog functions and tasks in simplifying this kind of conversion. Continuation of our serdes class project is reserved for the lab work.

Lab on serial-parallel conversion
   We first do two related, generic deserializer designs; then, we move on to recall our serdes class project. We implement a first cut on the required `Deserializer` module of our project.

Lab Postmortem
   We discuss reusability issues of our class serdes deserializer.

14.1 Serial-Parallel Conversion

A while ago, we put aside our serdes project in order to strengthen our understanding of verilog; now we return to the serdes to implement another part of the deserializer.

14.1.1 Simple Serial-Parallel Converter

We have implemented the serialization frame encoder (Lab 6, Step 8 of *Week 2 Class 2*), and we studied the processing, which is to say, the decoding, of the incoming serial stream when we developed a way of synchronizing a PLL with our framed packet protocol; this was toward the end of Lab 10 of *Week 4 Class 1*.

So, in terms of our project, we have only a little more to do beyond putting together some things we have done already, and modifying a couple of modules for synthesis. However, at this point, let's look some more into the parallelizing of a serial stream.

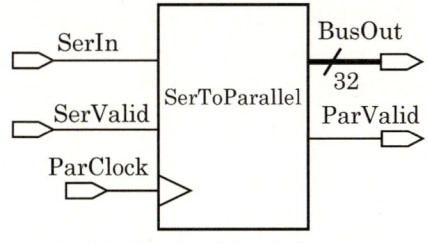

Fig. 14-1. Generic serial-parallel converter.

The general functionality is shown in figure 14-1. We should have a parallel bus of output latches or flip-flops, a clock or other synchronizer, and at least one serial input. A purely combinational deserializer is possible, but it would be difficult to use for anything without output synchronization.

Such a converter has to have a serial clock, but this may be the same as the parallel clock in use for other purposes. The serial clock, if not the same as the parallel clock, may be generated either by copying the sending clock; by synchronizing with an independently running clock on the serial side; or, as in our serdes project, it may be derived from the stream of serial data.

Concerning deserialization, there should be an internal register on the receiving side to hold partially deserialized data, but this shouldn't be the same as the one latching the parallel output, unless the design is intended to allow the fully parallelized data to be available on the output for a duration lasting less than one clock.

Optional functionality might include a flag announcing when parallel data on the output bus are valid; however, this functionality in principle could be achieved by clock-counting. There should be a `SerValid` input from the serial side to flag when serial data are available for conversion. This `SerValid` flag might toggle with each serial bit or byte received; or, instead of such a flag, we might provide a serial clock based on the assumption that the serial stream can be synchronized with it. If we want to be able to start up the device with a zeroed shift register, we may provide an optional parallel-side reset to do this. For our purposes, we shall assume that the converter can have no control over the serial line's transmitter, so it would not be meaningful to provide a serial reset.

Summary of serial-parallel conversion functionality:
- Latched parallel data out.
- Parallel-side clock in.
- Serial data in.
- Serial-side clock in (optional).
- Internal deserialization register.
- Parallel-valid output flag (optional).
- Serial-valid flag (optional).
- Parallel-side reset (optional).
- No serial-side reset.

In our serdes project, the serial clock will be embedded in the serial stream and must be decoded by the receiver without any `SerValid` flag. This has the obvious disadvantage of requiring some startup overhead before the clock phase and frequency can be established; it has the more-than-redeeming advantage of not requiring its own separate clock, with attendant differential phase-lag, separate cross-talk, or related noise issues. So long as the data are usable, the embedded clock will be usable, too. This permits a data stream independent of any other signal in the design; in practice, this is an

Digital VLSI Design with Verilog

important factor permitting GHz frequency-range data transmission almost with a zero error rate.

14.1.2 Deserialization by Function and Task

If we don't worry about frame boundaries, parallelization is almost trivial in verilog: We shift the serial data in until we have enough for a parallel word; we unload the shift register onto the parallel bus; and, we continue shifting. We may assume that the parallel clocking is fast enough that no serial data will be lost.

Here's a verilog representation of a simple, generic deserialization:

First, we do the shift. This can be implemented by a function, which can return the shifted value at some reasonable delay time, but which can be allowed to perform the shift itself in zero simulation time.

Verilog has a built-in shift operator, so all the function has to do is be sure to append the new bit being shifted in:

```
parameter ParHi = 31;
...
function[ParHi:0] Shift1(input[ParHi:0] OldSR, input NewBit);
  reg[ParHi:0] temp;
  begin
  temp     = OldSR;
  temp     = temp<<1;   // MSB goes lost.
  temp[0]  = NewBit;
  Shift1   = temp;
  end
endfunction
```

When calling the Shift1 function, we pass it the current shift register contents and the new serial bit to be shifted in. The function above, it should be mentioned, can be simplified; it is written as above for instructional reasons (did you notice that it could be simplified? How?); a simplification will be adopted below. Keep in mind that verilog function declarations are not allowed to contain delays or nonblocking assignments.

Second, we do the conversion. Unloading the shift register onto the parallel bus involves a `ParValid` flag and probably a few delays, so why not implement it as a `task`?

```
parameter ParHi = 31;
...
reg ParValidFlagr;
reg[ParHi:0] ParSR, ParBusReg;
//          rise, fall
assign #(  1,     0  ) ParValidFlag = ParValidFlagr;
...
task Unload32; // Copies the parallel SR to the output bus.
  begin        // Also clears the SR for the next word.
     ParValidFlagr = 1'b0;  // Lower the parallel-valid flag.
     ParBusReg     = ParSR; // Transfer the data.
  #5 ParSR         = 'b0;   // Clear the SR.
     ParValidFlagr = 1'b1;  // Raise the flag.
  end
endtask
```

The delays are included just for illustration. As we have said before, delays in procedural code should be avoided; to account for delays which for some reason must be included in synthesizable procedural code, one should use lumped delays in module output continuous `assign` statements.

Third, we regulate the shift. We have to put things together in a way that guarantees that the shift-register will shift in the bit on the serial line on each clock, provided only that the serial bit is valid and the device is not being reset.

Actually, there was no design reason in the second step above to clear the shift register after each conversion; this is because a shift-counter will be used to determine unloading of new data, and not left-over data, to the parallel bus. However, starting each conversion with a clear register does make it easier to see new data arriving during simulation.

The `temp` register in the `Shift1 function` above was overdone a little for instructional purposes; `Shift1` could be written more minimalistically this way:

```
function[ParHi:0] Shift1(input[ParHi:0] OldSR, input NewBit);
  begin
  OldSR  = OldSR<<1;
  Shift1 = {OldSR[ParHi:1],NewBit};
  end
endfunction
```

An `always` block to assemble our preceding code fragments is shown next:

```
always@(posedge ParClk, posedge ParRst)
  begin : Shifter
  if (ParRst==1'b1)
        begin
        N             <=    0;    // N counts the bits shifted.
        ParSR         <=   'b0;   // The shift register.
        ParBusReg     <=   'bz;   // The parallel out bus.
        ParValidReg   <= 1'b0;    // ParValid.
        end
  else if (SerValid==1'b1)        // Ignore the serial line if 0.
          begin
          ParSR <= Shift1(ParSR, SerIn); // function called.
          N     <= N + 1;
          if (N>ParHi)   // If 32 bits shifted.
            begin
            Unload32;    // task called.
            N <= 0;
            end
          end // SerValid.
  end // Shifter.
```

In this code block, the "`ParValidReg`" is the same as the "`ParValidFlagr`" of the third code block above. The `Unload32 task` briefly lowers the `ParValid` flag each time it updates the parallel bus.

This same `always` block which calls the `Unload32` task also lowers the `ParValid` flag on reset (`ParRst`); this is consistent with synthesizer requirements, although any delay on an assignment within the `always` block or the `task` will be ignored by the synthesizer.

14.2 Lab Preface: The Deserialization Decoder

The next lab will begin with a couple of very important exercises on deserialization. After that, there will be a first cut at implementing the deserialization decoder of our serdes class project.

It is essential to understand what the deserialization decoder (`DesDecoder`) is supposed to do: ***It identifies data frames which were encoded in the serial input data stream.*** It also generates a 1 MHz clock meant to be synchronized with the sender's clock.

Also, it is crucial to understand the two clocks involved in the deserialization. The PLL comparator's ***two clocks*** are:

1. The PLL's own 32 MHz free-running serial clock, which is divided down to 1 MHz for input to the `ClockComparator`; and,

2. The `DesDecoder`'s 1 MHz clock, created by toggling on extracted packet boundaries from the incoming stream of serial data.

If the two above 1 MHz clocks were perfectly in phase, deserialization could be perfect in the sense that no incoming data would be lost because of resynchronization.

In our design, there are four frames to a data packet, and each frame is 16 bits wide, 8 bits of data ending with an 8-bit pad pattern. The `DesDecoder` decodes the packets; it does not do a simple logical decode, such as did the 4-16 decoder of our previous lab. In decoding the packets, our `DesDecoder` also extracts the sender's clock, which is implied by the format of the incoming serial data stream.

Our deserializer spans two independent clock domains. The serial stream comes into the deserializer at an approximately known frequency which depends solely on the clock in the sender's clock domain. However, the deserializer also clocks data out of its FIFO using a clock in the receiver's clock domain. The FIFO input is in the sender's domain; the FIFO output is in the receiver's domain. To see this again, recall figure 8-6 of *Week 4 Class 1*.

Looking at our `Deserializer`'s PLL, it is clocked by a free-running 1 MHz clock which the `DesDecoder` attempts to synchronize to the sender's clock. The free-running clock emulates, in verilog, an on-chip clock controlled by a variable-capacitance oscillator. This clock is concerned only with the deserialization and with the FIFO input. The PLL's comparator receives two different 1 MHz clocks: One is the free-running one which is multiplied by 32 to clock in the serial data; the other is a clock extracted by the `DesDecoder` from the serial data stream. As already mentioned, if the free-running clock was perfectly synchronized with the serial data stream coming in, it would be perfectly in the sender's clock domain, and every packet could be identified correctly.

A `DesDecoder` failure to decode a packet means that there is no guarantee any more that the free-running clock is indeed in the sender's domain. The PLL comparator keeps track of any frequency difference between the incoming stream and the free-running clock; however, the PLL does not adjust its frequency unless the `DesDecoder` can confirm that it has identified a packet. Whenever the `DesDecoder` identifies a packet, it allows the PLL to adjust its frequency slightly, and it synchronizes an edge of the free-running clock with the identified packet boundary.

Our design uses integer arithmetic to equate the two sender-domain clocks, the one actually in use by the sender, and the free-running clock of the `Deserializer` PLL. Because 1/32 of the period of a 1 MHz clock is not an integer value in ns, and because in our design we must round to the nearest integer, our PLL never will be truly locked in to its input the way an analogue PLL could be. However, our PLL can be locked in approximately, and this approximation can be close enough that several packets can be extracted correctly, even with no intermediate frequency adjustment, before the PLL serial stream drifts out of synchronization.

Before beginning the lab, you may wish to review our original plan for the serdes project as described for *Week 2, Class 2*. In the present lab, we shall not extract the sender's clock (that will be later); however, we shall decode the packets being sent.

Here again, in figure 14-2, is the block diagram of the conceptual serial-parallel converter (deserializer) as described in *Week 2 Class 2*. We shall implement a simplified, first-approximation Deserialization Decoder in this lab.

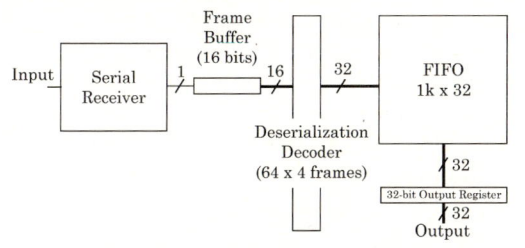

Fig. 14-2. Week 2 serdes `Deserializer` data flow.

Recall what was done in Step 4 of Lab 6 in *Week 2 Class 2*: In that lab, we decided on a packet format in which a frame of 16 bits would be used to represent each 8-bit byte of serial data.

We decided that the data bits would be contiguous and bounded below (= later in the stream, which arrives MSB first) by a pad byte consisting of three '0' bits, then a 2-bit count locating the data byte in its 32-bit word, then another three '0' bits.

A packet of 32 bits of data then would look like this, each 'x' representing a data '1' or '0':

   64'bxxxxxxxx00011000xxxxxxxx00010000xxxxxxxx00001000xxxxxxxx00000000,

with serial arrival being from left (earliest) to right (latest).

14.2.1 Some Deserializer Redesign—An Early ECO

The abbreviation "ECO" stands for *Engineering Change Order* and is used widely in the industry. On a big project, design changes have to be approved by the authority issuing the ECO. An ECO is a formal modification (update or correction) to the design specifications.

According the the block diagram of figure 14-2, which was meant to be conceptual, we can count on the incoming data's having been deserialized and collected into 16-bit words before it is received by the Decoder. However, given our serial protocol, no framing can be done in the Frame Buffer until the serial clock has been decoded; so, as shown, it is not obvious that the 16 bit buffer would not contain data which crossed the boundaries of incoming encoded bytes and thus was not valid.

The Frame Buffer, then, interpreted as a serial-parallel converter, first has to be aligned on data+pad boundaries. This might be done in the diagram of figure 14-2 by including a PLL feedback from the Decoder to the Serial Receiver.

But, it is unclear how this kind of feedback could be separated from deserialization; so, probably, for data flow purposes, the best plan would be to fold the Frame Buffer into the Decoder logic and no longer view it as an independent block.

Therefore, the Decoder should manipulate the digitized serial stream using its own registers and buffers. Our `Deserializer` design data flow block diagram then should be amended as shown in figure 14-3.

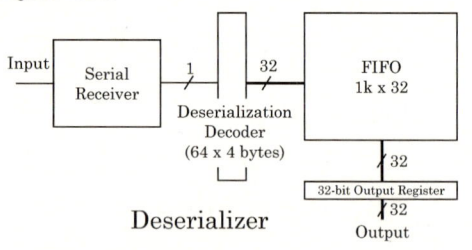

Fig. 14-3. Amended serdes `Deserializer` data flow.

We are not interested in PLL clock synchronization in this lab exercise. If we assume that the PLL is located in the Serial Receiver block, we can delete the Frame Buffer block; we then have no reason to worry about distribution of the PLL clock at the block level, and we need not show any feedback from the Decoder.

This block diagram, after all, is just data flow and need not include any representation of the clocking scheme.

Because it takes 64 bits to encode 32 bits of data, it will take two `ParClk` cycles, at 1 MHz, to process each 32-bit decoded word. These considerations were discussed in detail in Step 4 of Lab 10. We only assume here that we can shift in the serial data from port `SerIn` at 32 Mb/s, using an externally supplied 32 MHz serial clock named `SerClk`.

14.2.2 A Partitioning Question

Going beyond data flow, another question here is one of second thoughts: Should we indeed establish frame synchronization here, in the Deserialization Decoder, or should we leave it to PLL-related clock-extraction code in the Serial Receiver block? The verilog already mostly has been written (see the final code example in Lab 10 of *Week 4 Class 1*). Frame synchronization is essentially equivalent to serial clock extraction.

If we leave serial clock extraction <u>in the Serial Receiver</u>, then the PLL will be localized entirely there, and there will have to be considerable digital structure, maybe a shift register or small register file, to handle the pad-pattern extraction. But, serial data arriving at the Deserialization Decoder (`DesDecoder`) will be accompanied by frame-

Digital VLSI Design with Verilog

boundary flags from the Serial Receiver, and parallelization in the `DesDecoder` block could be very simple and generic.

On the other hand, if we put serial clock extraction in the `DesDecoder`, the PLL may be located either in the Serial Receiver block or in the `DesDecoder`. If the former, synchronization feedback will have to be provided by the `DesDecoder` to the Serial Receiver; if the latter, we have mixed digital and analogue in the `DesDecoder` block, requiring special layout procedures and other `DesDecoder` design overhead.

The answer we shall adopt here, is to put the PLL in the Serial Receiver block, and to do the frame synchronization (serial clock extraction) in the `DesDecoder`. This will isolate analogue functionality to the Serial Receiver block and will reduce duplication of digital effort in deserializing the data. The `DesDecoder` will have to extract a (nominally) 1 MHz clock from the serial stream being parallelized and provide that clock to the PLL. This is shown in figure 14-4.

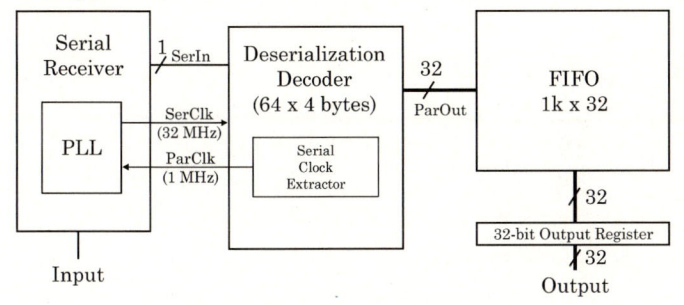

Fig. 14-4. Final `Deserializer` detailed block diagram.

A 1-MHz clock is a low-frequency signal in our design, so no special precaution will have to be taken, except possibly to account for the transport distance delay, a small phase lag, from the `DesDecoder` to the PLL.

With this ECO completed, we now move on to the lab exercise.

14.3 Serial-Parallel Lab 16

Topic and Location: A generic serial-parallel converter, a modification sensitive to bit-patterns in the serial stream, and a first-cut at the `Deserializer` module of our class serdes project.

Do this work in the `Lab16` directory.

Preview: We first warm up by doing two related, generic deserializer designs: (*a*) A generic serial-parallel converter which ignores content of the incoming serial stream; and then, (*b*) a modification of it which requires use of bit-patterns in the serial stream to control functionality. After the generic designs, we move on to coding what will become the first cut at the `Deserializer` of our serdes class project.

Deliverables: Step 1: A correctly simulating deserializer which converts an incoming serial stream to registered (parallel) 32-bit words. Step 2: A modification of the Step 1 design which controls the deserialization process based on serial content. Step 3: A correctly simulating `Deserializer` which produces registered 32-bit words from a serial stream arriving in the packet format required by our serdes project.

Lab Procedure:

Step 1: Generic deserializer. Using the lecture material as desired, implement a deserializer which fulfills the generic description above (section 14.1.1.1) and realizes its `SerToParallel` block diagram's I/O's. The generic requirements are just to clock in the serial data until 32 bits have been acquired, and then to copy them onto the parallel bus, using the same clock (possibly on the opposite edge). Assume that the data are unframed and should be grabbed in 32-bit words so long as the `SerValid` flag is asserted.

Simulate the design at least for three deserialized, 32-bit words. Use `$random` for serial data (see figures 14-5 and 14-6). Do not spend time synthesizing this design.

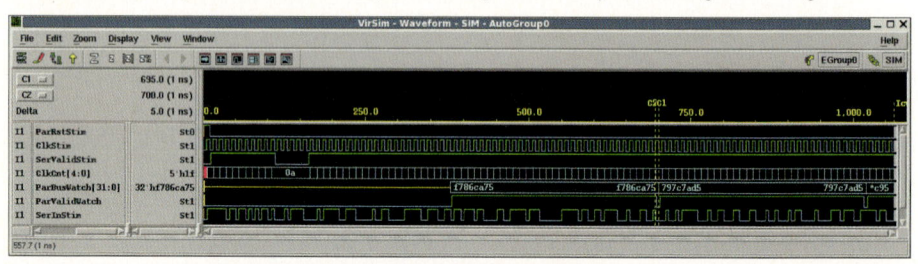

Fig. 14-5. The generic deserializer.

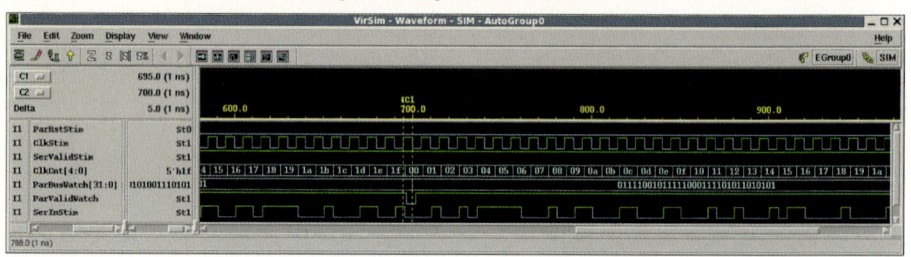

Fig. 14-6. Generic deserializer close-up of the serial bit counter wrap-around.

Step 2: Deserialization data-stream synchronization. Modify your generic deserializer from Step 1 so that whenever the serial stream contains 12 successive '0' bits in a row, those bits should be rejected, and deserialization should cease. After ceasing, with any incomplete word data (prior to the 12 0's) saved, the device then should wait until 12 '1' bits in a row arrive; after the 12th '1', the saved data should be restored and deserialization should resume with the shift register restarting at bit 13, where it left off.

One good approach to Step 2 would be to start by just shifting in a stream of bits and identifying the two 12-bit patterns by setting two flags, Found_stop and Found_start.

Continuing this approach, after you have simulated save/restore identification of the serial stream correctly, create a parallelizing register and add code to copy incoming bits to it until Found_stop is asserted. After Found_start subsequently is asserted, continue copying. The code asserting Found_stop could deassert Found_start, and vice-versa. Every time 32 bits have been parallelized, offload them to an output bus and set a parallel-bus-valid (ParValid) flag.

There are numerous different possible ways of implementing this design. Probably, there should be a shift-register which continues shifting while deserialization to the parallel bus is stopped.

This design also should be simulated (see figure 14-7) but not synthesized.

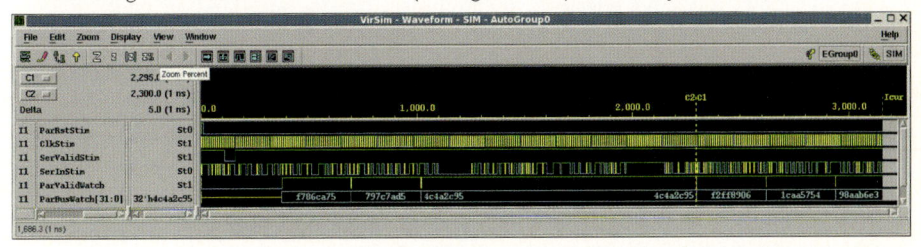

Fig. 14-7. Synchronizable, but unsynthesizable, generic deserializer, as described in the lab text.

Step 3: Deserialization Frame Decoder. We return now to our serdes project.

To establish data framing, we'll allow, optionally, a less rigorous criterion than when we locked in the PLL in *Lab 10*:

As a synchronization criterion, the current packet may be considered locked in on the received data byte immediately after a pattern of 8'b000_00_000 has been shifted in on SerIn. Deserialization may be performed throughout every period of synchronization. This criterion optionally might be stiffened to allow a lock-in only when all four pad patterns are in the shift register.

However, regardless of synchronization, the ParClk should be toggled on every 16th bit received on SerIn following the most recent synchronization, so that there will be 4 edges and thus two ParClk cycles for every 64 bits received, regardless of current synchronization state.

The synchronization loss criterion optionally may be that the embedded, 2-bit sequential pad number (..., 2'b11 -> 2'b10 -> 2'b01 -> 2'b00, ...) is found to miscount, or just that the first pad (pad_00) can't be located 64 bits after the previous one. Either way, synchronization loss should have no effect on ParClk: In this lab Step, your DesDecoder should include an oscillator which runs freely at exactly 1 MHz and is (re)synchronized whenever a packet has been identified. You may test your deserializer's resynchronization by using your testbench to vary slightly the serial input rate.

In our SerDes class project, eventually we will design a PLL which contains a free-running oscillator performing the same function as this oscillator, but with an adaptive frequency.

Deserialization should be halted whenever synchronization is lost; at this point, contrary to our previous work in this lab, incompletely deserialized data (fewer than 32 bits) should be discarded. Deserialization should be resumed with the first data byte following resynchronization. Resynchronization may be by either of the two criteria previously described—your choice: Either (*a*) because the sequence of four successive pad-byte counts is correct; or, (*b*) because a good pad_00 was detected 64 bits after the previous (good) pad_00. If you wish to create your own resynchronization criterion, this would be OK; but, if it is not effective, you might lose considerable lab time on it.

It is understood that resynchronization might cause a ParClk glitch. But, the DesDecoder can be designed not to malfunction because of such a glitch; and, for other design blocks on the digital side, that's one reason we have a FIFO in this deserializer.

In summary, the DesDecoder block will be supplied a digital data stream clocked in by an externally-generated signal, SerClk; however, this clock will not be made available to the deserializer. In this lab exercise, your testbench module should originate the SerClk, as well as a simulated SerIn data stream, using our 64-bit packet-framing protocol. See figure 14-8 below.

The conditions of this exercise may be obscure when compared with our previous ones; so, at the risk of pedantry, the conditions shall be restated just once more:

    1. There should be a testbench serial clock SerClk, but the deserialization should be done solely on a 1 MHz clock extracted from the serial data stream.

    2. Instead of a PLL-generated 32-MHz clock, for this Step you should provide your own 32 MHz clock (= SerClk) directly to the DesDecoder, and let the DesDecoder extract its own 1 MHz clock.

    3. If you want to serialize parallel data in your testbench, to supply the DesDecoder, just create a 32-MHz clock in your testbench and feed it to a counter to create a 1 MHz clock. This last, of course, then should be connected to be perfectly in phase with SerClk.

Digital VLSI Design with Verilog 299

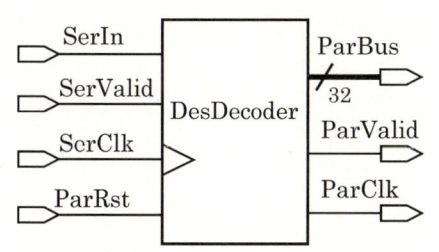

Fig. 14-8. Deserialization Decoder: First-cut block diagram.

So, implement the `DesDecoder` to extract a nominal 1 MHz clock from the serial data stream and use it to clock out the properly-framed, parallelized data in 32-bit words. Do not consider PLL functionality; just provide a 32 MHz serial clock input along with the stream of serial data. You should have: `SerIn` = serial data; `SerClk` = serial input clock; `ParClk` = output clock, extracted from the serial input stream; `ParBus` = 32-bit output parallel word, clocked by a two-cycles-per-MHz `ParClk`.

To complete this Step, use anything you wish from *Lab 10* or previous work, and implement the deserialization decoder block as shown above. Keep in mind that it may be simpler to start this design from scratch and only copy small fragments of your previous work. In particular, the shifting of the shift register can be much simpler here than for the Step 2 problem of this lab.

Consider breaking down the problem into several small `tasks`, setting up the calling of these `tasks`, and then implementing the bodies of the `tasks`.

Simulate your result; the simulation should resemble figures 14-9 and 14-10. Do not synthesize now; we shall synthesize this subblock of our serdes later in the course.

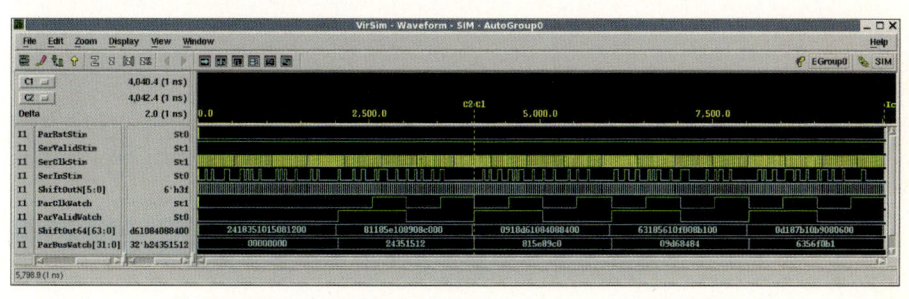

Fig. 14-9. The deserialization decoder (`DesDecoder`), not yet synthesizable.

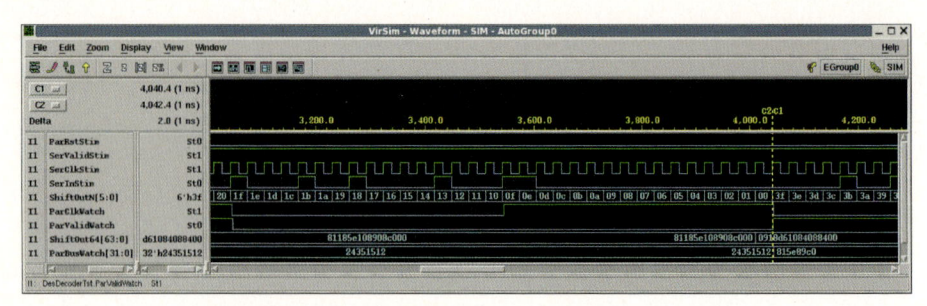

Fig. 14-10. The `DesDecoder`, zoomed in on the serial bit counter wrap-around.

14.3.1 Lab Postmortem

Think about problems in changing the packet width (to 128 bits, 36 data bits with parity), etc.

14.4 Additional Study

Reread this lab's instructions and review previous labs referenced in this one.

Chapter 15
Week 7 Class 2

15 Today's Agenda:

Lecture on UDP's, timing triplets, and switch-level models
Topics: User-defined primitives, timing triplets, switch-level primitives and nets, `trireg` nets.
Summary: We present the basics of verilog user-defined look-up table `primitives` (UDP's) and how to use them for simple component modelling. We then pass to a review of two- and-three-value delay specifications to introduce the second dimension of delay timing, the `min:typ:max` triplets, which may be used in place of single delay values anywhere. We then enumerate the switch-level primitives and explain their drive strength parameters. We end with the switch-level `trireg` net, which may be used to model decay transition time of a capacitor.

Lab
We model a combinational and sequential UDP first; then, we model several devices at switch level, including a `trireg` delay line.

Lab Postmortem
Q&A, and two little questions.

15.1 UDP's, Timing Triplets, and Switch-Level Models

We again put aside our serdes project to study some verilog in depth. This time, we shall look into the basics of the design of small devices at and below the gate level of complexity.

15.1.1 User-Defined Primitives (UDP's)

The UDP `primitive` is a verilog feature which gives a user a way to create SSI (Small-Scale Integrated) device models which simulate quickly and use little simulator memory. The target component for a UDP is any specialized kind of latch (flip-flop) or a complex combinational gate.

UDP's exist at the same design level as modules, and they have no functionality that a `module` can not have. Usually, several UDP models will be placed in one library file; they may coexist in such a file with models implemented as modules. However, often, these `primitives` are instantiated in a module wrapper to give them timing, multiple outputs, or other essential functionality.

UDP's are allowed to have only one output and any number of inputs. The verilog standards (IEEE Std 1364 section 8 and IEEE Std 1800 section 29) specify limits of at least 9 inputs for combinational UDP's and at least 10 for sequential ones. Every I/O must be one bit wide. The functionality is by a look-up table.

UDP's are not synthesizable.

The verilog structural organization of a UDP is: `primitive` keyword, UDP name, I/O declarations, a `reg` declaration if the device is sequential, (traditionally) an initialization block, and finally a `table` defining the logic. If an ANSI header is used, initialization may be done in the header; however, in our examples here we shall initialize only in the `initial` block. The `table` columns are somewhat different for combinational *vs.* sequential UDP's.

In exchange for simplicity and speed during simulation, a UDP may not include delays, 'z' states, or bidirectional (`inout`) ports. A 'z' on a UDP input is handled internally as an 'x'. Modern library components may be difficult to model as UDP's, because of the lack of delay or other technology-dependent functionality.

A typical `table` row for a combinational UDP has this organization:

$$(inputs \text{ in declared order}) \;:\; (output);$$

For example, three UDP `table` rows for a three-input *and* gate:

```
table
   ...
   1 0 0 : 0;
   1 1 1 : 1;
   1 1 x : x;
   ...
endtable
```

Of course, verilog already includes a primitive (= built-in) representing an n-input *and* gate; a more practical UDP implementing an *and-or* combination would be as follows:

```
primitive u2And1Or(output uZ, input uA1, uA2, uOr);
//
// Models uZ = (uA1 & uA2) | uOr.
//
table
// Output on right; inputs in declared order:
//   and'ed inputs     or'ed input
//    uA1 uA2           uOr           uZ
      0   0             0       :     0;
      1   0             0       :     0;
      0   1             0       :     0;
      1   1             ?       :     1;
      ?   ?             1       :     1;
endtable
endprimitive
```

Here, we have chosen the mnemonic pattern "u2And1Or" to represent u = UDP, 2And = two *and*s, and 1Or = one *or*. This saves some comment text and facilitates the understanding of a module or library file containing expressions with several UDP instances.

> Note on terminology: Compound combinational gates often are included in an ASIC library. Whether or not implemented as UDP's, they are named according to their functionality and number of logic-term inputs: "A" for *and*, "O" for *or*, "I" for inversion, etc. Consistent naming facilitates library maintenance. For example, a cell evaluating (A&&B)||C might be named, *AO21*. A cell evaluating !((A&&B)||(C&&D&&E)) might be named, *AOI23*.

For a sequential UDP, there are two columns in the table delimited by colons. The left column, surrounded by colons on both sides, represents the current state of the one, 1-bit, storage register allowed and maps to the output port. The rightmost column, following the second colon, represents the next state, given the current state and all inputs.

For example, here is a UDP which may used to implement a D flip-flop:

```
primitive uFF(output reg uQ, input uD, uClk, uRst);
//
initial uQ = 1'bx; // Not same as a module initial block.
//
table
// Output on right; inputs in declared order:
//                  current  next
// uD   uClk  uRst     uQ      uQ
    0   (01)   0  :    ?   :   0  ;  // Clock in 0
    1   (01)   0  :    ?   :   1  ;  // Clock in 1
    0   (0?)   0  :    0   :   0  ;  // Default to keep same 0
    1   (0?)   0  :    1   :   1  ;  // Default to keep same 1
// Unclocked:
    ?   (1?)   0  :    ?   :   -  ;  // Ignore negedge.
   (??)   ?    0  :    ?   :   -  ;  // Retain state.
// Reset asserted:
    ?     ?   (01):    ?   :   0  ;  // Posedge reset
    ?   (??)   1  :    ?   :   0  ;  // Ignore clock edge
   (??)   ?    1  :    ?   :   0  ;  // Ignore clock state
    ?     ?    1  :    ?   :   0  ;  // Ignore clock state
endtable
endprimitive
```

The table row entries in any UDP should be separated horizontally by whitespace for readability, but they need not be. UDP's were originated in the days in which it was a time-consuming calculation for a workstation computer to parse a few extra blank characters as spacers—and designers, used to punching and collating Hollerith cards to enter a program, couldn't read the input very well if they wanted to. Anyway, each row ends with a semicolon (';').

Edge sensitivity in a sequential UDP is described by the two states of an edge, written inside parentheses, with the initial state to the left; thus, "(01)" represents a rising edge, and "(10)" a falling one. Each `table` row is allowed only one edge. When an input change affects both a level and an edge column, the edge is evaluated first, then the level; this means that the level prevails in the event of a conflict. As in a `case` statement, '?' means a don't-care input. A hyphen ('-') means no change on an output.

A UDP `table` should completely cover all possible states of its inputs, because a state or edge definition in one row opens up possibilities for the simulator to misinterpret any other possible permutation of the input values. Thus, it is important to add a row for every possible foreseeable event. This becomes complicated for sequential UDP's, in view of the edge possibilities.

Several more rows would have to be added to the UDP above, if it became necessary to include 'x' output states selectively.

Like the verilog builtin primitive gates (`and`, `or`, etc.), UDP's may be instantiated without an instance name. An instantiation may include a delay expression.

To summarize what we have seen about UDP's:

UDP's are primitives defined by look-up tables. Keyword is **primitive**.

Structurally interchangeable with **module**s.

Advantages:
- May be instantiated without an instance name.
- Accept delays when instantiated.
- Fast compilation and simulation.

Limitations:
- Allowed only one, 1-bit output.
- Only 1-bit inputs (up to at least 9 of them).
- No timing or `parameter` declaration.
- No `specify` or other internal block.
- No 'z' or bidirectional functionality.

Not used in VLSI design.

Not synthesizable.

Sometimes used in verilog simulation library development.

15.1.2 Delay Pessimism

We move on, now, to a discussion of verilog delays. It is important to understand that *functionality* and *timing* are almost-orthogonal features of a simulation.

Functional edges. When a change results in assignment of a '1' level after any other level, including an 'x', the functionality is that of a rising edge, and an `always` event expression or any other edge expression will treat it functionally as a `posedge` event. Likewise, when a change results in a '0', it is functionally a falling edge, and functionally it means a `negedge` event.

Changes strictly between 'x' and 'z' are level-sensitive events but are not edge events.

Timing edges. When a (*rise, fall*) timing expression is associated with a change from '0' to '1', the rising-edge (`posedge`) delay will be given by the *rise* value. The `negedge` delay for a change from '1' to '0' will be given by the *fall* value. But, this correspondence with functional edges ceases when 'x' (or 'z') levels are involved.

When the change is ***to*** an 'x' level, neither the *rise* nor the *fall* in a (*rise, fall*) or (*rise, fall, to_z*) expression has a specific meaning. Instead, the smallest available delay value is used by the simulator. This quick change to an unknown value lengthens the duration of unknown states and thus is called "pessimistic" in regard to knowing the hardware value, which must be '1' or '0'.

The delay *from* an 'x' to another level is given by the delay to that level; pessimism does not apply. Unspecified delays involving 'z' are treated the same way as those involving 'x'.

15.1.3 Gate-Level Timing Triplets

In *Week 4 Class 1*, we saw how strength might be assigned to a gate output by putting one or two strength keywords in parentheses. For example, an NMOS *or* gate:

```
or (strong0, weak1) or_01(out_or, in1, in2, in3, in4);
```

We also have seen the same approach to assign delays, except that the delay values were preceded by a '#'. For example,

```
or #2 or_01(out_or, in1, in2, in3, in4);
```

Parentheses are optional around the single delay value and are omitted above. If both strength and delay are assigned, the strength specification is to the left of the delay specification:

```
or (strong0, weak1) #(2, 1) or_01(out_or, in1, in2, in3, in4);
```

As we have seen, delay values also may be assigned in multiples of two or three in parentheses. The interpretation of such values is as follows:

1 value: `#t inst_name(...);`	Every change on the output(s) is scheduled with this delay.
2 values: `#(tr,tf) inst_name(...);`	Every rise is scheduled with the first delay value *tr*; every fall with the second delay value *tf*.
3 values: `#(tr,tf,tz) inst_name (...);`	The first two are the same as for 2 values; the third delay is the delay to 'z' state for gates which are capable of it.

A Second Dimension of Delay. In the old days, when designs were mostly board-level assemblies of relatively small digital IC's and discrete passive components, logic simulators had to account for temperature, supply voltage, and fabrication differences across the board. Minimum and maximum delays were estimated separately for each chip. The resulting timing differences were simulated by assigning 'x' wherever the simultaneously-estimated minimum *vs.* maximum delays allowed it (=pessimistic). See figure 15-1.

However, most modern designs use VLSI IC's structured in blocks with latched (clock-synchronized) outputs, so this kind of pessimism has not been useful during simulation, even in recent deep-submicron designs. The chip itself tends to be more or less uniformly fast, typical, or slow; so, only one condition or extreme has been applicable in any one simulation evaluation. The min-max approach, however, usually is used during *static timing verification*, which we shall touch upon briefly later.

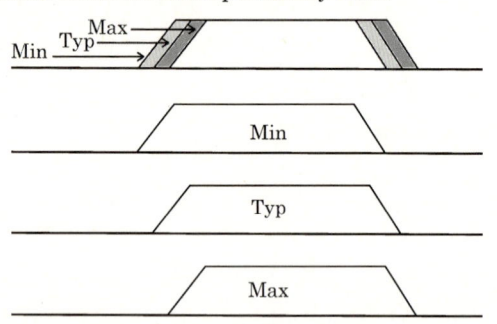

Fig. 15-1. Board-level simulation represents edge delay spreads as unknowns (top); IC-level simulations are run with no spread and each delay state separately (lower three). The sloping edges represent individual uncertainty intervals caused, for example, by skew or jitter.

In recent VLSI designs, with pitch down to and below 90 nm, the simulator has not been used to choose the condition as best, typical, or worst case; the designer has made this choice. Gates in libraries for logic synthesis still have to be characterized for temperature, supply voltage, and process variations, so there still is a need for a simulator to handle multiple possible delays on a gate. But now, devices being simulated almost always are assigned just one, single, specific, global range of delay, *minimum*, *typical*, or *maximum*, representing the expected global uniformity of a given, operating

Digital VLSI Design with Verilog

IC. The simulation then is repeated several times, each time with a different global alternative, *min*, *max*, or *typ*. VCS, as other verilog simulators, is told which global value to use by setting a command-line option when it is invoked. The default is *typical*.

There is some indication that verification of very large designs at 90 nm and below may require two, or all three, of each triplet to be used in a single simulation run in order to display local uncertainty of timing in a simulation waveform. This may become necessary, for one reason, because of the temperature differences which can develop across a chip as a function of operating time or mode of operation. The more that is put on a chip, the more likely that different functional blocks will be operating under different conditions. Also, a physically very large chip may be able to develop temperature differences simply because of the distance across it. Thus, the old board-level simulation displays, showing a range of unknowns around every edge, again may become common.

Regardless of simulator design or practice, to assign these triplet values to a gate output in verilog, they simply are substituted, separated by colons, for the single values which we have been using up to now. So, when indicated by characterization of the synthesis library, the table above may be modified to replace *t* with *t_min:t_typ:t_max*; and, each of *tr*, *tf*, and *tz* in that table above then will be triplicated correspondingly.

The first *or* gate example above then may be changed to,

```
or #(1:2:3) or_01(out_or, in1, in2, in3, in4);
```

Likewise, if a bufif1 were used to model a gate with a relatively slow turnoff, instantiation might require a delay specification given by,

```
bufif1 #(1:3:4, 1:2:4, 6:7:8) triBuf_2057(OutBit, InBit, CtlBit);
```

The simulator does not impose or require order in the values assigned to a triplet; in principle, one might find min:typ:max specified with min ≥ typ > max.

This isn't all there is to the calculus of delay in verilog; but, we shall put off internal delays in library components and other devices until later in the course.

15.1.4 Switch-Level Components

We introduced the different verilog strengths, and some of the different types of nets, in *Week 4 Class 1*. The net types and built-in gate types were studied further in *Week 5 Class 2*. We now shall go beyond this in modelling devices at the <u>switch level</u>, which is to say, at a level in which gates are treated as made up of individual substructures (transistors) which may be simulated as switching on and off.

To model at the switch level, we require switch-level primitives. These are supplied in verilog as the <u>MOS</u>, <u>CMOS</u>, <u>bidirectional</u>, and <u>source</u> switches.

Here is a list of all the available switch-level primitives:

> **MOS switches:**
>   nmos, rnmos (like bufif1)
>   pmos, rpmos (like bufif0)
>   cmos, rcmos
> **Bidirectional pass switches:**
>   tran, rtran
>   tranif1, rtranif1
>   tranif0, rtranif0
> **Switch-level net:**
>   trireg
> **Power sources:**
>   pullup, pulldown

Switch-level primitives, like UDP's, are not synthesizable.

MOS Switches. MOS stands for Metal-Oxide-Silicon, the main layers used to fabricate devices in this technology. It represents an advance in semiconductor technology over the more power-hungry bipolar (P-N junction; current-operated) semiconductors. MOS transistors have a source-drain potential which supplies energy for amplification, and a gate which turns source-drain current on or off electrostatically.

A P-channel transistor (pmos) conducts more current when its gate is more negative; and an N-channel transistor (nmos) conducts more current when its gate is more positive. Furthermore, P devices, which have hole majority carriers, are slower than N devices, which use electrons as majority carriers; so, in CMOS technology, the P side generally is made larger than the N side to make the response speeds more nearly equal. Also, CMOS P devices are fabricated to operate geometrically nearer the high supply voltage rail than ground, because, there they can be operated with delay comparable to that on the N side. Anyway, for a variety of reasons, P devices are fabricated near the *supply1* rail and N devices nearer the *supply0* rail (see the table below). Because of this device gravitation, the elementary logic gates in CMOS technology tend to be nand and not rather than and and buf.

The MOS primitives are named **nmos**, **pmos**, **rnmos**, and **rpmos**. The **r** means "resistive", and the **r*** primitives are meant to represent physically higher-impedance (smaller) devices than the others.

All the MOS devices individually are functionally identical to the bufif1 (nmos) or bufif0 (pmos) primitives we have used already in lab, except that the r* primitives have outputs which always reduce the strength applied at their inputs. For the r* devices, only ones of small-capacitor or highz strengths are not reduced in the verilog. See *Week 4 Class 1* for a table of verilog strengths.

Digital VLSI Design with Verilog

The rnmos and rpmos strength-changing rules are given on the left half of the following table; they may be found in section 10.2.4 of Thomas and Moorby (2002), or in IEEE Std 1364 section 7.12 or IEEE Std 1800 section 28.14:

| *Resistive* MOS strength rules ||||
| Strength || Strength keyword ||
In	Out	In	Out
supply	pull	supply0/1	pull0/1
strong	pull	strong0/1	pull0/1
pull	weak	pull0/1	weak0/1
large cap	medium cap	large0/1	medium0/1
weak	medium cap	weak0/1	medium0/1
medium cap	small cap	medium0/1	small0/1
small cap	small cap	small0/1	small0/1
highz	highz	highz0/1	highz0/1

The strengths on the right half of the table, with "0/1" (= *0 or *1) names, may be found in IEEE Std 1364 table 7.7 or IEEE Std 1800 table 28.7 and are used to resolve contention. The ***charge strengths***, described as large cap, medium cap, and small cap, are applied only to trireg net declarations, and they generally are written without a level, just large, medium, or small.

Tools should refuse to accept charge strength declarations such as "(strong1, **medium0**)" when the level is appended. This is because, according to the IEEE Std, the only legal use of charge strengths is for declaration of trireg nets. Incidentally, the default strength of a trireg net is medium.

CMOS Switches. The name stands for Complementary Metal-Oxide-Silicon. These devices are composed functionally of paired nmos and pmos transistors, just as are actual CMOS gates fabricated on a chip. A **cmos** switch even has two control inputs, just as a CMOS transistor on a chip would have two gates, each with one control. Of course, rcmos switches are composed functionally of paired rnmos and rpmos switches.

Verilog assumes a positive-logic regime ('1' = higher voltage; '0' = lower), so an N device turns on when its gate is at logic '1'; a P turns on when its gate is at logic '0'.

There are limits to the accuracy of a digital simulator at this level. For example, a cmos switch can be imagined to represent two MOS transistors in parallel. This does a fine job as a switch level model of a bufif1, as shown in figure 15-2.

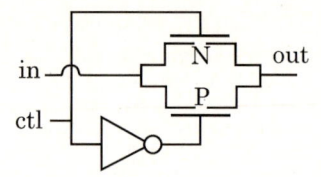

Fig. 15-2. CMOS `bufif1` modelled as `nmos` and `pmos` in parallel.

But, in verilog, the `nmos` and `pmos` primitives pull up their outputs (to '1') with the same strength as they pull down (to '0'). Real P *vs.* N devices on a chip differ in strength at the two logic levels (N pulls low stronger than high, and vice-versa for P).

A more reasonable, functionally correct, schematic representation would not include the inverter, as is shown in figure 15-3:

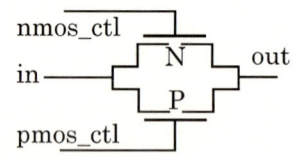

Fig. 15-3. CMOS `bufif1` modelled as `nmos` and `pmos` in parallel.

Viewed in terms of the analogue electronics, this would make a `cmos` a pass gate without any amplification and with some finite resistance. In digital logic, there is no such thing as resistance, and a `cmos` will drive its output in simulation with the same strength as its input.

When used to model a gate such as an inverter, the elements of a `cmos` should be rearranged so that the pass-gate "input" and "output" were connected between supply and ground, as shown in figure 15-3. In this arrangement, the inputs of the `pmos` and `nmos` switches would be tied to the power supply rails. The output of such an arrangement then always would be at *strong* strength, because, differing from the strength rules above, switches always reduce `supply` strength inputs to `strong` output strength.

Section 10.2.1 of Thomas and Moorby (2002) gives a nice switch-level model of a static RAM cell, and there is a model of a shift register in section 10.1, but there is some question as to what the purpose might be of such models, when a device of any comparable size can be modelled easily in SPICE, with extremely high accuracy.

Even so, CMOS devices are much closer to being symmetrical than individual PMOS or NMOS devices; for this reason, they are preferred in modern designs.

In this vein, consider the two different arrangements of transistor elements which could be used to implement an inverting buffer (verilog `not`) on a chip. Switch-level modelling does a fine job. By tying a `pmos` input to `supply1`, and an `nmos` input to `supply0`, it is possible to invert the control input, which is just what an inverting buffer does at the switch level.

This is shown in the next figure:

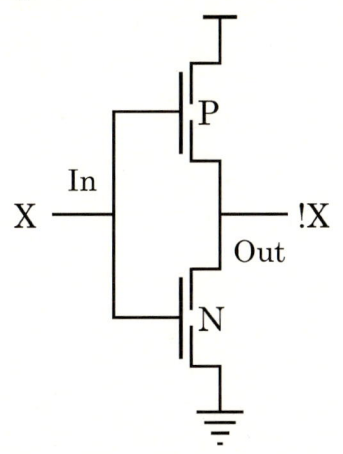

Fig. 15-4. CMOS not gate modelled as nmos and pmos in series.

Looking at the CMOS not gate model in figure 15-4, one might think that by reversing the N and P devices, the gate would become a noninverting buffer. This could be done, and the logic indeed would be noninverting; however, a P device operating near supply0 is extremely slow and inefficient; likewise an N device operating near supply1. Thus, for these technical reasons, simple noninverting buffers never are used for core-logic buffers in CMOS technology. If one requires a noninverting buffer in CMOS, the usual solution is to drive the input of a big inverting buffer with the output of a small one, the two inversions yielding a net noninversion.

The verilog cmos primitive approximates reality better than either pmos or nmos alone, in that a cmos also does not reduce the strength of an input (except that source becomes strong). Because a cmos has two control inputs, it has a different port definition than a bufif*x*. The ports declared for a (r)cmos are as follows:

      (r)cmos *optional_*inst_name (out, in, N_ctl, P_ctl)

In the event that one control is off and the other on, a cmos still will turn on; so, the second control presumably is intended to model requirements for on-chip connectivity, rather than to model functionality.

Bidirectional Pass Switches. These primitives model charge-transfer gates (pass transistors) and thus are called **tran** (figure 15-5), **tranif1**, and **tranif0**, with the corresponding high-impedance **rtran**, **rtranif1**, and **rtranif0**. They are not allowed to be assigned delays. Unlike trireg primitives, they don't store state. They merely pass a logic level applied at one I/O pin to the other I/O pin. They are used to model nets of transistors connected in arbitrary ways.

inout1 ⋈ inout2

Fig. 15-5. Schematic symbol of a `tran` switch.

Instantiation of a `tran` transfer gate follows the same format as `buf`, with two ports, both `inout`. A `tranif`*x*, where *x* means 1 or 0, has two `inout` ports and a third, control port; thus, it is analogous to a `bufif`*x*, except that it is bidirectional and can not include a time delay. More information on applications of transfer gates may be found in the layout-related references at the beginning of this Textbook and in the Additional Study suggestions below.

Source Switches. There are two of them, `pullup` and `pulldown`. Each accepts just one output net as argument. The `pullup` drives its net high with constant `source1` strength; the `pulldown` drives it with `source0`. Whereas the transistors being modelled in verilog at switch level usually are assumed to operate in enhancement mode (normally off), these would be assumed to be large transistors in depletion mode (normally on), or to be direct connections to an IC power or ground rail.

15.1.5 Switch-Level Net: The `trireg`

The functionality of the `trireg` net type is described in IEEE Std 1364 sections 7.13.2 and 7.14 or IEEE Std 1800 sections 6.3.2 and 6.6. In a word, a `trireg`, like a capacitor modelled as a time-limited `reg`, just stores a logic state.

A `trireg` is unique in that it makes use of the `small`, `medium`, and `large` strength values. For the `trireg`, such strengths are meant to represent the size of a capacitor which stores charge whenever the drive to the `trireg` enters the high-impedance ('z') state. However, a `trireg` driven at 'z' never enters the 'z' state; instead, it holds its last non-z state. If and only if the `trireg` has been assigned a delay, this last non-z state decays from '1' or '0' to 'x' as soon as the delay lapses.

The strength of a `trireg` is used to determine which of several `triregs` in contention will determine the delay to 'x'; this is the only use of the charge strengths. When strengths are equal, the rule of pessimistic simulation is used to determine `trireg` decay to 'x' just as it is to determine the result of other timing conflicts.

A delay value may be assigned to a `trireg` net when the net is declared. The delay format is different in one way from that of a three-state component: When three delay values are given for a `trireg` net, the first two refer to rise and fall, as for a three-state component; however, the third value refers to time to 'x', not time to 'z'. A `trireg` can not enter a 'z' state, although it could be initialized to one. A `trireg` with any delay value decays to 'x' after the given delay as soon as the net's last driver has turned off (to 'z'). A `generated` or `arrayed` structure including `trireg` nets thus could be used to model the refresh behavior of a dynamic RAM.

If a `trireg` net has no delay associated with it, it continues forever to drive its output(s) at the strength declared for it, at the logic level with which it last was being driven ('1', '0', or 'x'). This resembles procedural behavior and so explains the peculiar name, <u>trireg</u>, of this construct.

Example of a `trireg` declaration:

                        trireg (medium) #(3, 3, 10) medCap1;

Connectivity of `trireg` nets may be established by continuous assignment, or by port-mapping to switch-level component instances.

Here is an example of the use of `trireg` nets:

```
pullup(vdd);
trireg (small)   #(3,3,10)    TriS;
trireg (medium)  #(6,7,30)    TriM;
trireg (large)   #(15,16,50)  TriL;
// Pass transistor network:
tran (TriM, TriS); // left always wins.
tran (TriM, TriL); // right always wins.
// NMOS network:
rnmos #1 (TriM, TriS, vdd); // input has no effect.
rnmos #1 (TriS, TriL, vdd); // input controls output.
rnmos #1 (TriM, TriL, vdd); // Contention on output.
```

15.2 Component Lab 17

Topic and Location: UDP's, switch-level models, `trireg` models.

Do the work for this lab in the `Lab17` directory.

Preview: This lab does not involve any synthesis. We start by modelling a combinational UDP and then a sequential UDP. We model a switch-level inverter using n and p MOS switches. We examine a `cmos` switch. Next, we model a mux using pass transistor switches. We complete our study of switch-level components by modelling simple *nand* and *nor* gates. We simulate a delay line built on `trireg` nets.

Deliverables: Steps 1 and 2: Correctly simulating UDP component models. Step 3: A correctly simulating MOS-switch inverter component. Step 4: The Step 3 model with a `cmos` output added; a completed lab truth table for the `cmos`. Step 5: A mux built from pass transistor switches and simulated correctly both with `tranif` and `rtranif` primitives. Step 6: A module containing a correctly simulating switch-level model of a nand and a nor gate; a second version of this module using the Step 3 switch-level inverter to create and and or gates. Step 7 (optional): Two versions of a `trireg` pulse-delay line, one with `rnmos` primitive components, and the other with `rtran` primitive components.

Lab Procedure:

Keep in mind that many VLSI simulators will not simulate switch-level verilog correctly, because it does not synthesize. Silos generally will work, except for resistive primitives.

Step 1: Combinational UDP. Design a UDP in a module named `AndOr2Not4` which evaluates this logic function: X = (~a | ~b) & (~c | ~d), in which a through d are input names, and X is the output name.

SSI discrete component databooks or large ASIC libraries might include such a device.

Suggestion: Include a verilog comment line naming the table columns to help reduce entry errors.

Instantiate your component in a testbench module and simulate it to verify its functionality.

Step 2: Sequential UDP. Design a UDP named `AndLatch` which functions as a simple transparent latch but has two data inputs anded together before being latched.

Instantiate this latch in a module and simulate it to verify functionality.

Step 3: Switch-level model of an inverter. Create a module named `Nottingham`. Give it one 1-bit input and two 1-bit outputs. Combine a `pmos` and `nmos` primitive as described in the presentation above (figure 15-4) to model a `not`. This amounts to a CMOS inverter. Drive one of the two module outputs with this composed `not`.

Instantiate a verilog `not` gate and use it to drive the other `Nottingham` output. This will allow you to compare a verilog `not` with your `nmos`-`pmos` `not` in a simulation, both driven by the same input.

You could test such a design by feeding both outputs to the two inputs of an `xor` gate: A '1' on the `xor` output would indicate that the two gates were not identical logically.

Step 4: The `cmos` control inputs. Add a third output to the module in the previous Step and connect a `cmos` to drive it from the module input. Declare local control nets routed through new I/O's from your testbench.

Digital VLSI Design with Verilog   315

Simulate to fill in the first column of the cmos truth table:

cmos Truth table

out	in	n-ctl	p-ctl
	1	1	1
	1	1	0
	1	0	0
	1	0	1
	0	1	1
	0	1	0
	0	0	0
	0	0	1
	1	x	x
	1	x	0
	1	x	1
	1	1	unconn
	1	unconn	0

Can you predict what the output will be under each of these conditions? ("*unconn*" = 'z' state)

Step 5: Pass transistor mux model. The easiest demonstration of pass transistor logic is a multiplexer: The select input is decoded to turn on just one pass transistor; the turned-on logic level then is transferred to the output, where it wins the contention against the other output(s), which must be in 'z' state.

For a 2-input mux, all it takes is one tranif1 and one tranif0 with outputs tied together, and a one-bit select input to both control inputs:

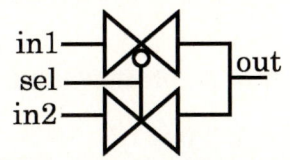

Fig. 15-6. Two-input mux modelled by tranif's.

The relevant verilog for the 2-input mux design shown in figure 15-6 would look something like this:

        tranif0 UpperTran(Out, In1, Sel);
        tranif1 LowerTran(Out, In2, Sel);

For this exercise, create a module named TranMux4 and design a 4-input mux for it, using nothing but pass transistors and nets. Simulate to verify your design.

After verifying your design, replace the tranifx switches with rtranifx switches and simulate again. (This may not work in Silos).

Step 6: nand and nor gates. Create a module named Nand_Nor. Give it three inputs and two outputs. The outputs should be named Nand and Nor. Use the schematic of figure 15-7 to enter a switch-level model of a 3-input *nand* gate and a 3-input *nor* gate.

Simulate to verify your design (see figure 15-8).

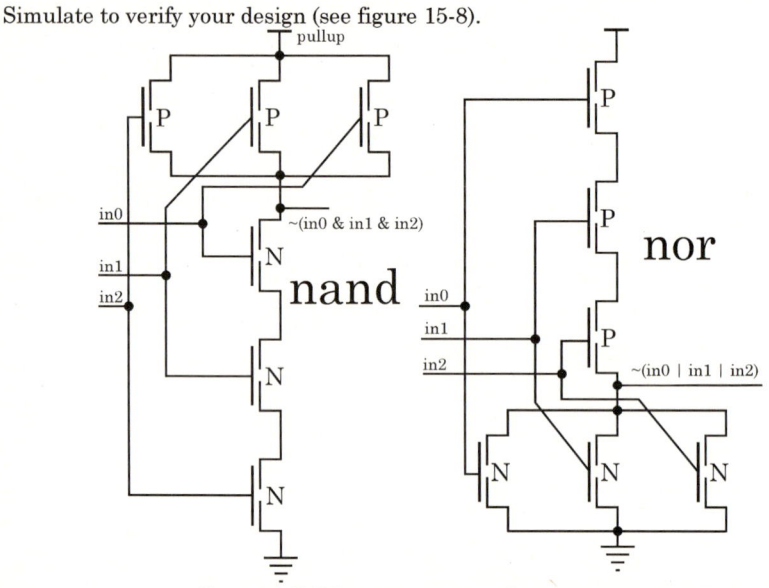

Fig. 15-7. A CMOS 3-input nand gate and a 3-input nor gate.

After this, add a CMOS inverter on each output to change the functions to *and* and *or*.

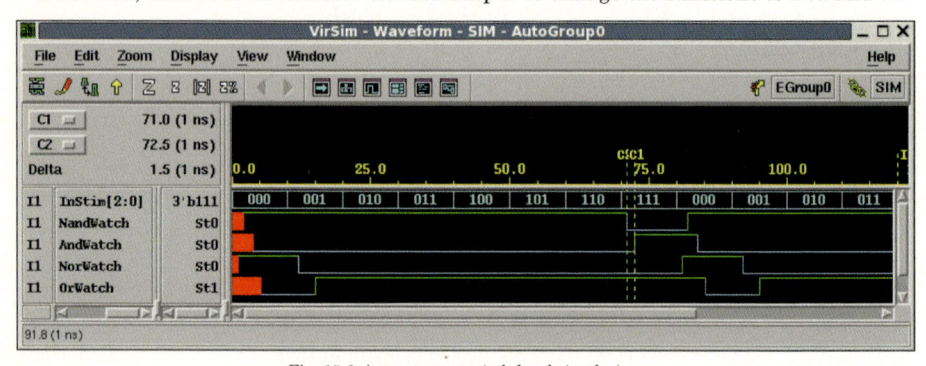

Fig. 15-8. A Nand_Nor switch-level simulation.

Digital VLSI Design with Verilog

Step 7 (optional): Trireg pulse filter. In this exercise, be aware that VCS and many other simulators probably will not simulate `triregs` correctly when connected to `rtrans`.

To prepare for simulation, use the schematic of figure 15-9 below. You will assign the delays as follows, naming the `triregs` Tri01 - Tri04, from left to right:

Name	Drive strength	Delay
Tri01	Large	(19, 23, 50)
Tri02	Large	(21, 29, 53)
Tri03	Medium	(9, 13, 17)
Tri04	Small	(5, 7, 11)

Create a new `module` named RCnet and use `trireg` switches as capacitors to implement the RC network, assigning a delay of 1 to each `rnmos`, as shown in figure 15-9. Use an enabled `rnmos` to represent each resistor shown, *in* on left; *out* on right:

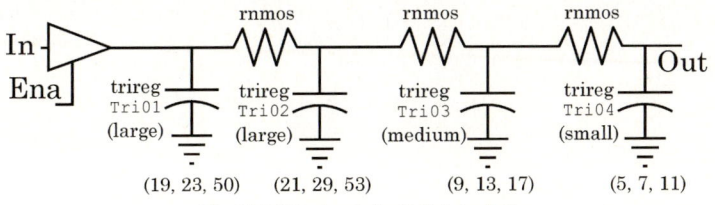

Fig. 15-9. RC network for digital simulation.

Simulate an input pulse to see how well these digital switches approximate analogue functionality. Next is a summary of what should be the result, with **rnmos** switches, each with delay of 1, and with Ena = 1:

If In goes from x to 0, Out will be 1'b0 after 23+1+29+1+13+1+7 = 75 ns.

If In then goes to 1, Out will be 1'b1 after 19+1+21+1+9+1+5 = 57 ns.

If In then goes to 0, Out will be 1'b0 after 23+1+29+1+13+1+7 = 75 ns.

   (a) If In then goes to x, Out will be 1'bx after 19+1+21+1+9+1+5 = 57 ns.

   (b) If In stays 0 and Ena goes to 0:

      Tri01 -> 1'bx@50 (*this is the one decay delay*)
          =>**Tri02**->1'bx@51+21=>**Tri03**->1'bx@73+9=>**Tri04**->1'bx@83+5.

      So, **Out** will be 1'bx after 88 ns.

Resistance happens to be of no importance here—just the delays. The delay of the input `bufif1`, if not zero, should be added to those above.

Repeat the above, the the same way, except with ***rtran*** substituted for ***rnmos***

Old VCS (*ca.* 2010) produces these results in figure 15-10:

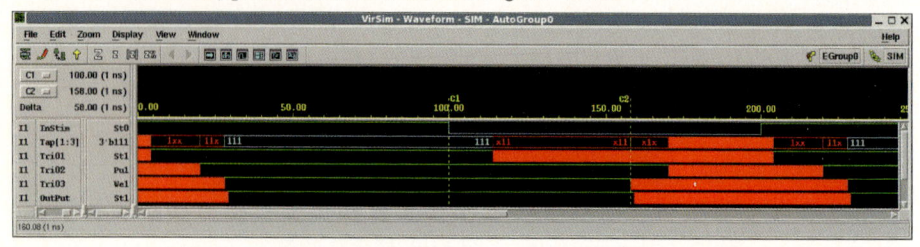

Fig. 15-10. Simulation of a delay line using enabled `rnmos` resistor models.

Replacing each `rnmos` in old VCS with an `rtran` to specify delays—but to display less interesting strength—produces figure 15-11:

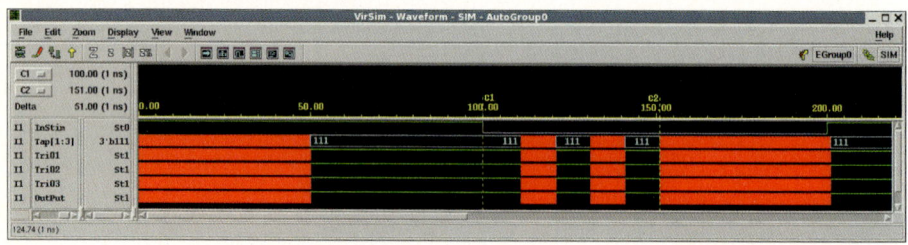

Fig. 15-11. Simulation of a delay line using enabled `rtran` resistor models.

The result of a repeated simulation in new VCS is similar to the one above and is shown below, followed by several *QuestaSim* simulation details. It should be emphasized that *QuestaSim* is an enhancement of an older tool (*ModelSim*) which was widely, and very successfully, in use long before synthesis became a major goal of gate-level design.

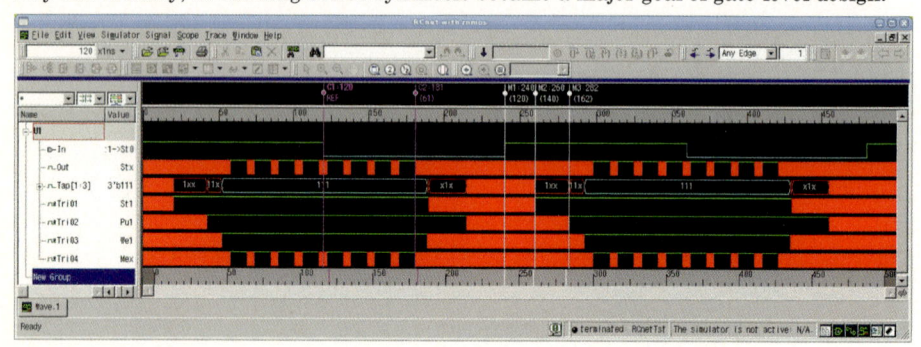

Fig. 15-12. Revised Fig. 15-10 = new VCS simulation of `RCnet`, with `rnmos` components (incorrect—output oscillates).

Digital VLSI Design with Verilog 319

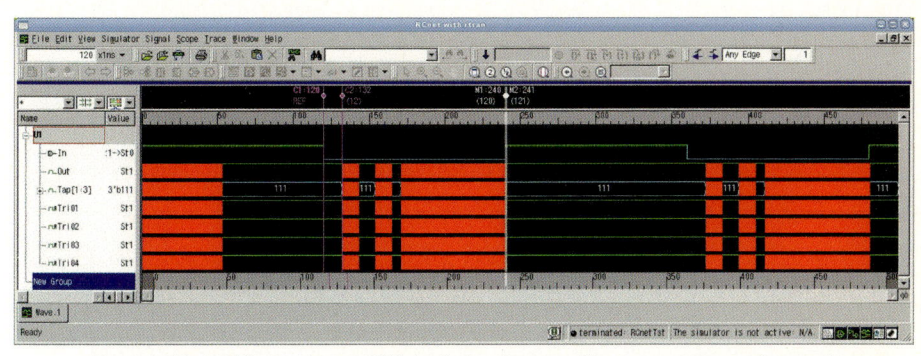

Fig. 15-13. Revised Fig. 15-11 = new VCS simulation of RCnet, with rtran components (incorrect).

The *QuestaSim* simulator (*v.* 6.5 of 2008) does a better job with switch-level constructs:

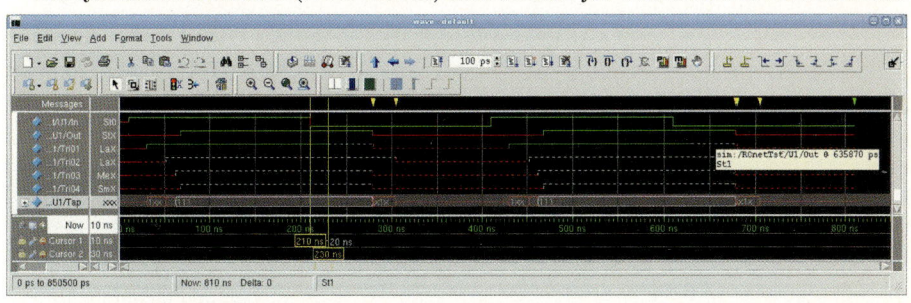

Fig. 15-14. Revised Fig. 15-10: *QuestaSim* simulation of RCnet, with rnmos components (overall view).

Further insight into the results based on *QuestaSim* simulation follow:

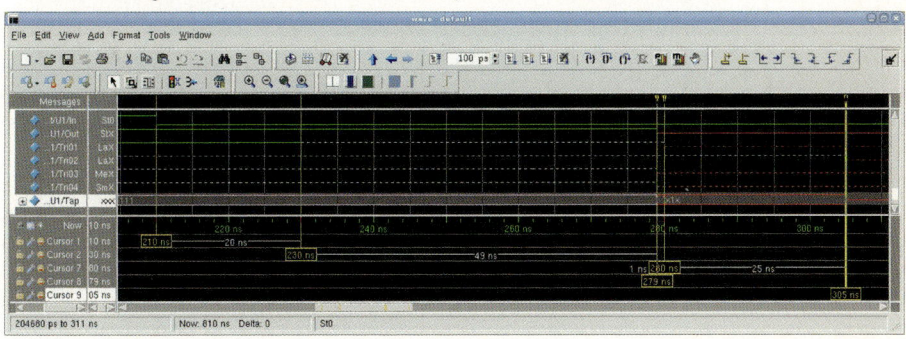

Fig. 15-15. Revised Fig. 15-10: *QuestaSim* simulation of RCnet, with rnmos components (buffer disable closeup).

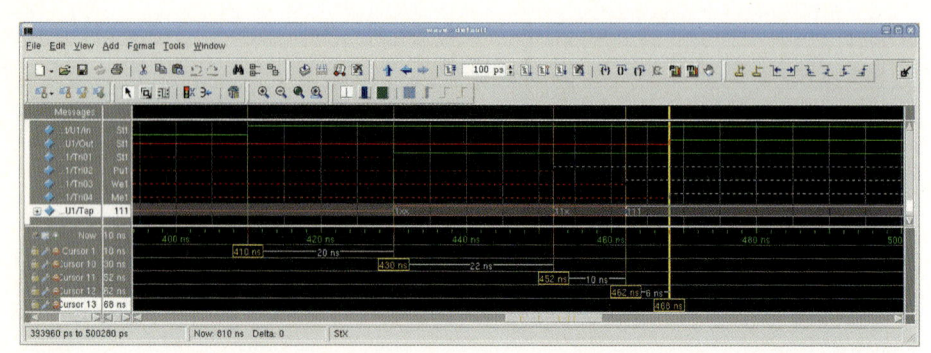

Fig. 15-16. Revised Fig. 15-10: *QuestaSim* simulation of RCnet, with rnmos components (buffer reenable closeup).

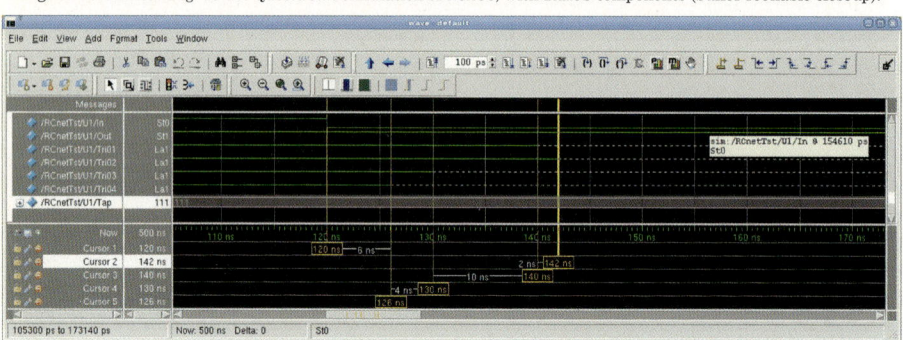

Fig. 15-17. Revised Fig. 15-11: *QuestaSim* simulation of RCnet, with rtran components (buffer disable closeup).

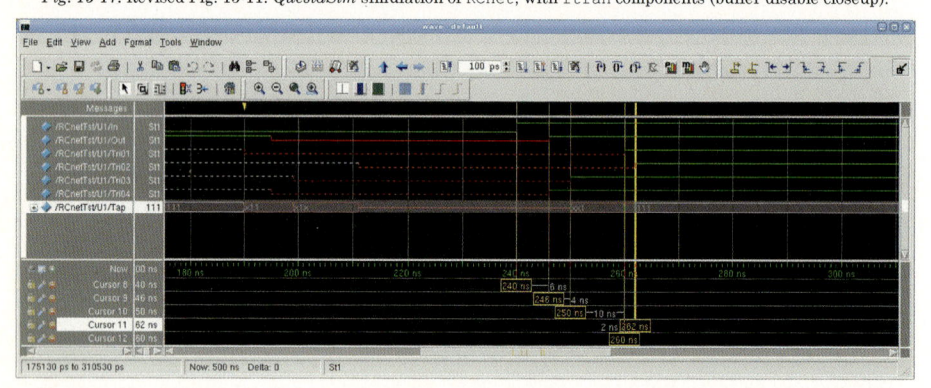

Fig. 15-18. Revised Fig. 15-11: *QuestaSim* simulation of RCnet, with rtran components (buffer reenable closeup).

15.2.1 Lab Postmortem

What do the three delay values mean when specifying delay on the output of a `trireg`?

Can an `rtran` be used to simulate a delay?

15.3 Additional Study

Read Thomas and Moorby (2002) section 6.5.2 on delay conflicts and pessimism.

Read Thomas and Moorby sections 6.5.3 and 6.5.4 on time units and timing triplets.

(Re)read Thomas and Moorby chapter 10 on switch-level modelling. However, ignore the "minisimulation" code.

(Optional) Read Thomas and Moorby chapter 9 on UDP's.

Optional readings in Palnitkar (2003):

Read chapter 5.2 on the basics of gate-level delays.

Read chapter 11 on switch-level modelling.

Read chapter 12 on UDP's. Notice especially section 12.5 which summarizes features of UDP's as contrasted with `modules`. Do section 12.7, problem 1, a 2-1 mux. Compare your result with the solution on the Palnitkar CD.

Study the section 11.2.3 model of a switch-level latch or flip-flop. There is a model of a flip-flop named `cff.v` on the Palnitkar CD; simulate it to see how it works.

Chapter 16
Week 8 Class 1

16 Today's Agenda:

This time we'll do some review of parameters and parameter passing. We'll also look into a few problems concerning design hierarchy.

Lecture on parameter types and module connection
Topics: Parameter types, port and parameter mapping, `defparams`
Summary: We examine in some detail the typing of `parameter` declarations, and ANSI and traditional port and parameter mapping. We mention `defparam`, a construct to be avoided, and `localparam`, a construct devised to prevent use of `defparam`.

Connection Lab
We exercise port and parameter declarations, as well as a hierarchical design comparing `` `defines `` with `parameters` to parameterize bus widths.

Lab Postmortem
We look into possible confusion of parameters with delays in component instantiation.

Lecture on hierarchical names and design partitions
Topics: Hierarchical names, verilog identifiers, design partitioning, and synchronization across clock domains.
Summary: We review and extend our coverage of verilog hierarchical names, generalizing to a discussion of the scope of identifiers in verilog. Then, we present various criteria which might be used to decide how a design should be partitioned into modules and module hierarchy. We dedicate some of the discussion to the problem of data which crosses independent clock domains and the related problem of how to synchronize the clocking of such data.

Hierarchy Lab
We experiment with hierarchical truncation and extension of mismatched bus widths. Then, we use a small design to see the benefit of latching module outputs with flip-flops.

Lab Postmortem
We point out a conflict between rules of thumb (*a*) to use continuous assigns to lump module simulation delays and (*b*) to latch all outputs.

16.1 Parameter Types and Module Connection

16.1.1 Summary of Parameter Characteristics

- Parameters are unsigned integer constants by default.

  parameter *Name* = *value*;

- May be declared signed (but some tools reject it).

  parameter signed *Name* = *value*;

- May be declared real (but not synthesizable and often rejected).

  parameter real *Name* = *float_value*;

- May be typed by vector index range.

  parameter[6:0] *Name* = *7_bits_of_value*;

- May be declarated anywhere in a module, but localparam preferred in body.

  localparam *Name* = *value*;

- Must be assigned when declared. Default width automatically is sized to be enough for the value assigned.

16.1.2 ANSI Header Declaration Format

module *ModuleName* #(parameter *Name1* = *value1*, ...) *port decs.*;

In the module header, we have advocated only the ANSI declaration of parameters, with pass of value by name. For example,

```
// Declaration sets defaults:
module ALU #(parameter DataHiBit=31, OutDelay=5, RegDelay=6)
            (output[DataHiBit:0] OutBus, ...);
  ...
// Instantiation:
ALU     #(.DataHiBit(63), .RegDelay(7))  // OutDelay gets the default.
  ALU1   (.OutBus(ResultWire), ...);
  ...
```

16.1.3 Traditional Header Declaration Format

```
module ModuleName (PortName1, PortName2, OutPortName1, ...);
  parameter Name1 = value1; ...
  direction[Name1:0] PortName1;   // direction = output, input, inout.
  direction[range] PortName2; ...
  reg[range] OutPortName1; ...    // output assigned procedurally.
  ...
```

The traditional module header ends with the first semicolon, just as does the ANSI header. However, the rest of the traditional header declaration does not end at any well-defined place in the module. This makes specification of the module interface error-prone and sometimes ambiguous.

16.1.4 Instantiation Formats

ANSI and traditional instantiation formats are identical:

16.1.4.1 Parameter Instance Override By Name

ModuleName #(.*ParamName1*(*value1*), .*ParamName2*(*value2*) ...)
        moduleInstName(.*PortName1*(*NetName1*), ...);

16.1.4.2 Parameter Instance Override By Position

ModuleName #(*value1*, *value2*, ...)
        moduleInstName(.*PortName1*(*NetName1*), ...);

Override by position is not a recommended practice.

16.1.5 Parameter Format Values

By default, a `parameter` is like an unsigned constant `reg`. When such a parameter is assigned to a variable, or used in an expression, it takes on the width and type of the destination.

An index range may be specified when the parameter is declared, making it an object of a certain width. For example, `parameter[7:0] CountInit = 8'hff;` declares a specific width, keeping the designer aware of what happens when the parameter is assigned to a variable of a given width.

Declaring a parameter with an index range is required if the parameter is to be used in a verilog *concatenation* expression: Concatenations must have an explicit width.

Also, a parameter, like a variable, may be declared `signed`; if so, arithmetic involving it may become signed arithmetic, if the other operand(s) also are signed types. A width or signedness specification in a module-header parameter declaration can not be overridden later in the module, although the value may be changed in an instantiation.

Examples:
```
// Note: 377 = 12'h179; -377 = 12'he87.
parameter signed[31:0] mul_coeff = -120*Pi; // Gets -376.9911 => -377.
//
// If the next were overridden by -120*Pi, it would get 32'hffff_fe87:
parameter signed[31:0] div_coeff = 32'h0000_0179;
```

Whether a module header has been declared in ANSI or traditional form, any `parameter` declared within a module can be overridden in instantiation, the same way as a parameter in the header. This is another good reason to use `localparam` declarations for parameters not intended to be overridden.

Overriding a header-assigned default during instantiation is the only recommended way of changing the declared value of a parameter.

Note: The Silos demo simulator which came with Thomas and Moorby (2002) or Palnitker (2003) may not recognize signed parameters.

16.1.6 ANSI Port and Parameter Options

Using the ANSI format, it is legal to pass values by position or by name, but it never is legal to pass a mixture of both. When passing by position (again, not recommended), all values to the left of each position must be provided.

Example (compare above):

```
module ALU #(parameter DataHiBit=31, OutDelay=5, RegDelay=6)
           (output[DataHiBit:0] OutBus, ...);
   ...
   ALU     #(31,5,8) // Must supply first two to change third one.
      ALU1  (.OutBus(ResultWire), ...);
   ...
   ALU     #(.DataHiBit(31),5,8) // Illegal.
      ALU2  (.OutBus(ResultWire[63:32]), ...);
   ...
endmodule
```

16.1.7 Traditional Module Header Format and Options

We briefly review here the traditional, *verilog-1995* module header format. This format is based on the old, pre-ANSI C language function header format ("K&R" C), invented by software pioneers Kernighan and Ritchie. While obsolescent, it still is generated by many automation tools such as netlist writers or file converters. The format is essential to understand because of these tools; manual editing of a preexisting verilog netlist is not advised but may be required to obtain a design which can be fabricated correctly.

In the old format, the header declaration lists the names, only, of the I/O's, if any. Immediately following the header, the parameters are declared, and the directions and widths of the I/O's are specified. After that, the types of the I/O's are specified; primarily, this means that outputs not driven internally by `wires` are redeclared to `reg` type. So, just as in ANSI format, parameter values may be used to assign widths to I/O's.

Example of a traditional declaration:

```
module ALU (OutBus, InBus, Clock); // Parens & contents optional.
parameter DataHiBit=31, OutDelay=5, RegDelay=6;
output[DataHiBit:0] OutBus;
input ...
reg[DataHiBit:0] OutBus;
...
```

Digital VLSI Design with Verilog

The more modern ANSI format avoids redundant name assignments and reduces the possibility of an entry error, so it should be preferred.

Override in instantiation is done the same way whether the header has been written in ANSI or traditional format.

16.1.8 `defparam`

There is a `defparam` construct in verilog which permits override of the value of any parameter which has been declared in the design in any module, however distant or unrelated. The syntax is just,

> `defparam` *hierarchical_path_to_parameter* = *new_value;*

This is a very risky construct, and it may be removed from the verilog language standard at a later date. It makes no sense within a given module, because within any one module, the parameter itself can be assigned just as easily (and under the same allowed circumstances) as it can be `defparam`'ed.

This construct has been retained in IEEE Stds 1364 and 1800, but its use has been discouraged in all Std versions through 1800-2012.

The `defparam` is the main reason for the existence of `localparam`'s: A `localparam` is identical to a `parameter`, except that it can not be overridden by `defparam`. A `localparam` is not allowed in an ANSI `module` header—because it also can't be overridden there.

It is recommended never to use `defparam` in a design. Like a *goto*, or like hierarchical references in general, it tends to introduce more of defects than it does of functionality.

16.2 Connection Lab 18

Topic and Location: Traditional and ANSI port mapping, instantiation of parameter overrides, hierarchical comparison of `` `define `` *vs.* parameter override.

Do this work in the `Lab18` directory.

Preview: We start with a simple exercise in reformatting a module header. Then, we do a few simple but perhaps unusual parameter overrides. After that, we view in VCS the hierarchy in a skeleton design with configurable bus widths, first using `` `define `` and then using `parameters`.

Deliverables: Step 1: A rewrite of a traditional-port module, followed by its compilation. Steps 2—4: Compiling versions of a design with various parameter overrides. Step 5: Compilation of the given design with macro bus widths in correct order. Answers for the two questions asked. Step 6: A compilable rewrite of the design of Step 5, but with parameter overrides in place of `` `defines ``.

Lab Procedure:

Step 1: Traditional port mapping. Type in and rewrite a copy of the following `module` as a new one named `nonANSItop`, with its header in traditional port mapping format.

The incomplete header is this one:

```
module ANSItop #(parameter A=1, B=3, parameter signed[4:1] List=4'b1010)
                (output[3:0] BusOut, output ClockOut
                , input[3:0] BusIn, input ClockIn
                , input[1:0] Select
                );
reg ClockOutReg;
assign #(2,3) ClockOut = ClockOutReg;
...
endmodule
```

Fill in some sort of module functionality—anything you want.

At the end of this Step, you should have two `module`s, in files `ANSItop.v` and `nonANSItop.v`, which will compile correctly in your simulator.

Step 2: Parameter overrides. Create a new verilog file named `ParamOver.v` and instantiate both `ANSItop` and `nonANSItop` from Step 1 twice each in a new `module` named `ParamOver` (you may use `ParamOver` for all of Steps 2—4).

For each of `ANSItop` and `nonANSItop`, override once by position and once by name as follows: Override `B` to be `20` and `List` to be `-2`.

Digital VLSI Design with Verilog 329

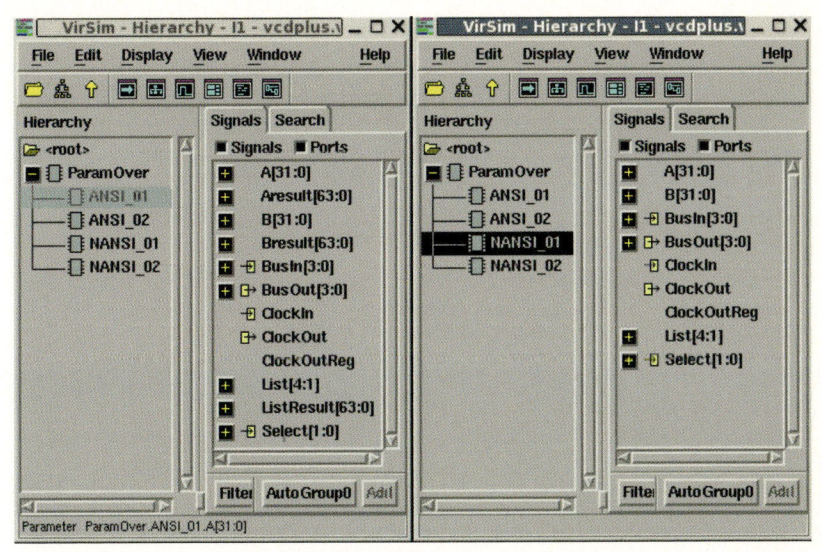

Fig. 16-1. `ParamOver` hierarchy views in old VCS: Branch panes for this one design are for `ANSI_01` (left window) and the nonANSI `NANSI_01` (right).

View the hierarchy in the simulator (see figure 16-1 for old VCS). QuestaSim may be invoked for a look at the parameter values, because old VCS doesn't display the compiled values of parameters (see below, if you are using new VCS). For best results, assign each parameter value to a 64-bit `reg` in an `initial` block in each module and use the simulator to display the `reg` value at time 0. With a width of 64 bits, the entire parameter value should be easily understood—aligned, of course, on the LSB of the `reg`. A `reg` used only in an `initial` block isn't doing anything in the design, so it will be removed during synthesis.

For the new VCS gui in figure 16-2 below, to bring up two hierarchy windows in order to match the old-VCS figures shown above, first set up the one shown on the left, more or less arranged the same way and displaying the same data. Then, open a second hierarchy window by opening `[Window]/[New Top Level Frame]/[Hierarchy + Data]` and double-picking `NANSI_01` in the hierarchy.

The result should resemble this:

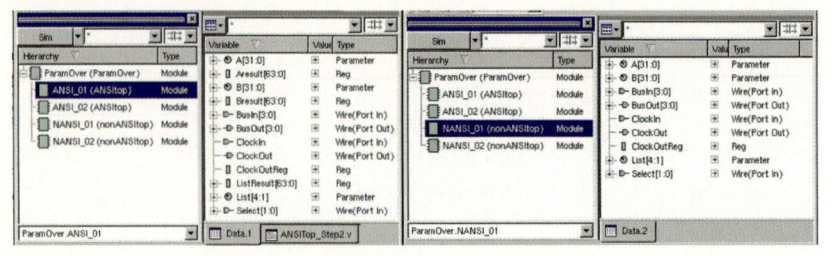

Fig. 16-2. `ParamOver` hierarchy views in new VCS: Branch panes on right are for `ANSI_01` (left window) and the nonANSI `NANSI_01` (right).

Also, as is suggested in this figure, the new VCS gui ***will*** display parameters and their values.

Side question: What was the default value of `List` in Step 1, expressed as a signed decimal number?

Step 3: Parameter width override. Instantiate `ANSItop`, overriding A by position to be 8'hab. Check the result by assigning the value to a 64 bit reg in an `initial` block and using the simulator.

Step 4: Parameter type override. Instantiate `ANSItop` again, this time declaring A and B `signed`, but otherwise with the same default values.

Override A by name to be 8'hab and B to be $-120*\pi$ (don't bother to write out π exactly). See figure 16-3.

Check the result by assigning the value to a 64 bit `reg` in an initial block and using the simulator. Change the display formats to 2's complement to see the signed integer values.

Digital VLSI Design with Verilog 331

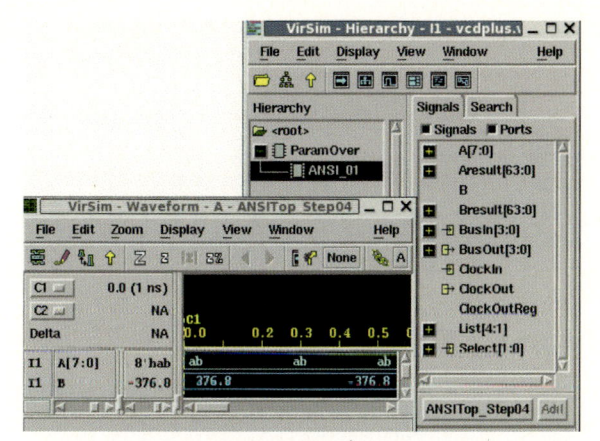

Fig. 16-3. Parameter real-valued expression in 2's complement wave display format.

Step 5: `` `define `` and ordering problems. The text of a skeleton `HierDefine` design is located in the `HierDefine` subdirectory of `Lab18`. Change to the `HierDefine` subdirectory to find the files, which are named `HierDefine.v`, `Level2.v`, `Level3.v`, and `Level4.v`. The bidirectional data bus is configured at each level so that its width is halved each time and it is distributed differently to each instance. This might represent a datapath design, for example part of a Fourier transformer.

A block diagram is given in figure 16-4 (see also the next Step):

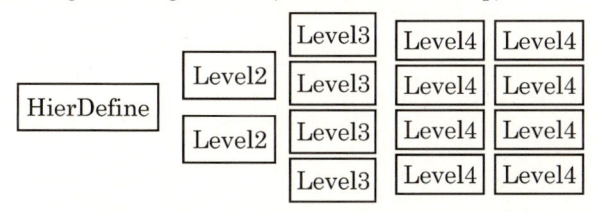

Fig. 16-4. Block diagram of the `HierDefine` instance hierarchy.

A. Set up a `.vcs` compilation file in correct order which would allow VCS to compile this design for simulation. The files contain "...", indicating omitted functionality; you will have to comment these to compile the (nonfunctional) design.

What happens if the order of two entries is reversed in the `.vcs` compilation file? What if `Level2` included a duplicated set of macro definitions, `` `define Wid 16 `` and `` `define ResWid 16 ``?

B. Suppose the design would work in different contexts if we doubled or halved all bus widths; how should `HierDefine` be modified, still using `` `define ``, to make such reuse convenient?

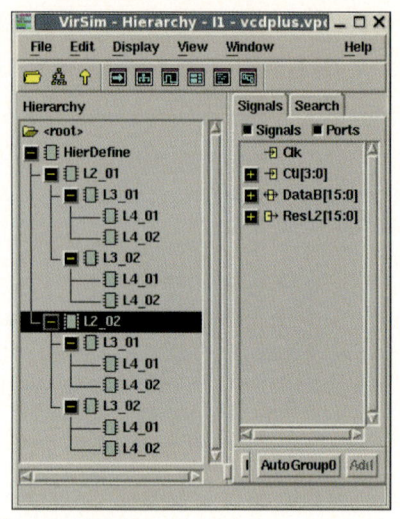

Fig. 16-5. VCS display of the `HierDefine` hierarchy.

Step 6: Parameter passing in depth. Assume a design `HierParam` with 4 levels of hierarchy, an exact copy of the `HierDefine` of the previous Step, except that parameters have been declared and there are no `` `define `` assignments.

The width of `DataB` must be halved at each level of hierarchy, as implied by the structural fanout at each level. This is shown in figure 16-6.

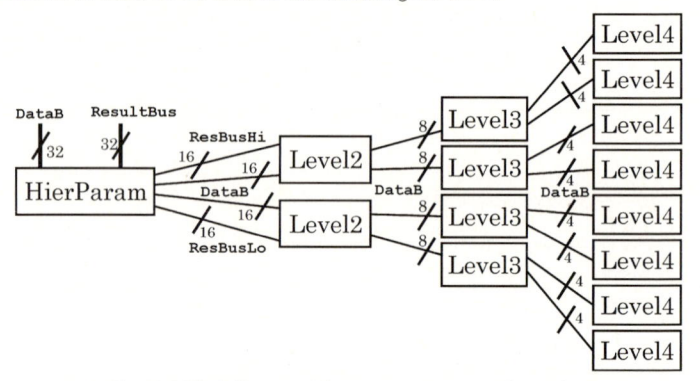

Fig. 16-6. Block diagram of the `HierParam` instance hierarchy.

All the module declarations have been collected into one file, `HierParam.v` in the `Lab18` directory, to make the problem more easily visible.

Digital VLSI Design with Verilog   333

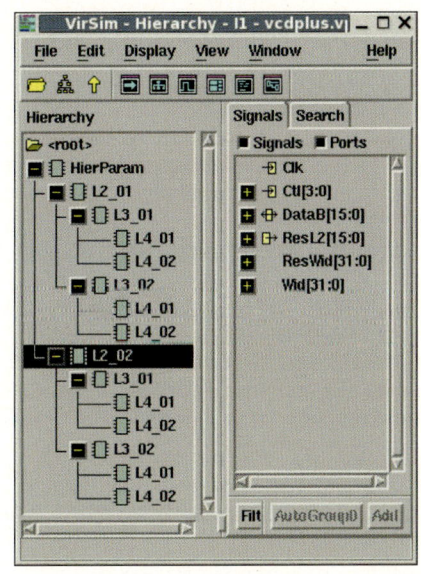

Fig. 16-7. VCS display of the modified `HierParam` hierarchy.

Modify `HierParam.v`, overriding the parameter default assignments, so that the value of the width of the `DataB` bus is changed consistently in every `module` by assigning just one `parameter` value at the top level. Don't change the default assignments declared in the file. Check your answer by compiling in a simulator and looking at the hierarchy (compare figure 16-7).

16.2.1 Connection Lab Postmortem

What if a designer wants to assign delays and pass parameters to the same instance?

16.3 Hierarchical Names and Design Partitions

16.3.1 Hierarchical Name References

We have studied hierarchy in several contexts, mainly `module` instance hierarchy and hierarchical name references in `generated` instances.

The hierarchy in verilog begins where the compiler (simulator or synthesizer) sees it, whether we are preparing for simulation or synthesis. If a submodule is being designed and tested, the names in it will be rooted at the top submodule. After this submodule is linked in to the larger design, and that design is loaded into the simulator or synthesizer, its name references will be rooted elsewhere.

For this reason, hierarchical references (A.B.C, ... etc.) should be restricted to the current module (as in a `generate` statement—recall *Week 6 Class 2*), or to the submodules of the module in which they appear. A hierarchical reference from the current module downward, toward the leaf cells of the design, can be controlled by the designer. This direction of reference is reasonably safe, because the current module depends on its instances for its functionality; changing the leaf cell (instantiation) structure means a redesign; in such a redesign, the designer is free to modify the hierarchical names as necessary.

On the other hand, upward references easily are broken beyond repair. Any `module` has some functionality as such; and, so, it may be reused not only in its specified location (if any is specified) but also elsewhere. If reused elsewhere, it is likely that the new instantiating module will differ from the previous one, probably invalidating hierarchical upward name references. The extreme case would be a verilog model of a library component; such a component should be designed for use anywhere; so, any hierarchical upward reference in it would render it limited or useless in functionality.

16.3.2 Scope of Declarations

Here's a review of the scope of identifiers of all sorts. We introduced some of these concepts in *Week 6 Classes 1* and *2*.

By ***scope***, or name scope, is meant that region, in the verilog, of visibility. The word "visibility" roughly means the same as "influence" or "effectiveness" in this context. Visibility may entail a name conflict when the visible object's identifier is redeclared or redefined.

Objects of wider scope conceal the names of those of narrower scope. For example, two different modules (wide scope) each may contain a different net locally declared and named `Clock`: The names of nets have narrower scope than names of modules, so the net names declared in one module are not visible in another module.

Inside a procedural block (including inside a `task` or `function` declaration), a name must be declared before its first use ("before" means above the use in the file). However, in a `module`, a *named block* declaration will be found by the compiler if it is anywhere in the `module`.

Here is a listing of name scopes which are reasonably distinguishable in verilog:

- `` `define `` macro identifiers: No real scope limitation or hierarchy. They are defined <u>everywhere</u> after the compiler encounters them, in compilation order (which may be unrelated to design structure).
- `module` names: Global scope, shared with UDP's (`primitives`) or `config`s only. One level of name scope; no hierarchy.
- `module` instance names. These may be extended to any number of levels of hierarchy. The top instance must be in a `module`; but, with this exception, nothing but instances may exist in a module instance hierarchy.

Digital VLSI Design with Verilog

- concurrent block names: These include every named `initial` or `always` block within a module. The `specify` blocks to be covered later also fall here. Concurrent blocks are allowed only one level of name scope, which means no hierarchy. However, `generate` blocks are in this scope and may contain a level of generated hierarchy after unrolling; such a hierarchy may include named procedural blocks other than `function` or `task` declarations.
- procedural block names: These include `tasks` and `functions`, as well as any named procedural `begin`-`end` block, such as one in a procedural `for` loop. These objects may be mixed with one another and extended to any number of levels of hierarchy.
- element names: Variables (`reg`, `wire`, etc.) and constants (`parameters`, `localparams`, `specparams`, etc.). These exist within a scope, but they define no scope of themselves.

Other language constructs such as expressions or literals (including delay values) have no name; and, so, scope is irrelevant to them

We shall look briefly at verilog `config` blocks later. A `config` is assigned an identifier (name) and specifies a collection of design objects, but it has no design functionality. A `config` may be located anywhere a `module` may be located. A `config` is more like a makefile, or a system disc file or directory, than a named block with a meaningful relationship to a hierarchy. The names in a `config` generally refer to objects in libraries external to the design.

16.3.3 Design Partitioning

Partitioning of a design may be done to reduce the complexity of the task of any one design engineer, to allow concurrency in the design work to speed the design to completion, or to achieve some design-tool related goal.

The most important single consideration in partitioning is that of the interfaces between the parts. Each different part has to have stand-alone, specific functionality differentiating it from all the others. When a part is to be used in several different places in a design, all uses have to be taken into consideration when writing design specifications for that part.

The interfaces have to be logical and intuitive, so that different designers, or the same designers at different times, recognize and easily understand the objectives of the partitioning. The interfaces also have to be well thought-out so that the nets crossing partition boundaries have completely specified types, widths, timing and sequential protocol. Any change in an interface must be considered in a system context and should be made with great reluctance, because it may imply redesign of several or all parts involved.

In view of the importance of interfaces, SystemVerilog, as an elaboration of verilog, includes a construct called an *interface* as one of its features. Interfaces are declared outside of modules and essentially predefine module headers; they also can add design

I/O functionality. An interface may be applied unmodified to several modules, thus guaranteeing I/O consistency among them. More information on interfaces is given in the SystemVerilog presentation of *Week 12 Class 2*.

In verilog (or SystemVerilog), partitioning always should be done on module boundaries. A module is the smallest verilog design unit which can be compiled separately. If a partition has to be redrawn, new modules separating its different parts should be declared, so that the parts might be compiled and tested separately. This allows the designer working on one partition to write a behavioral model or testbench representing the other(s), permitting effective debugging independent of other parts of the design.

16.3.3.1 Some Rules of Thumb Concerning Partitioning

Clock domains. Partitioning should be done to separate clock domains. Parts with different clock rates or sleep-mode states should be designed separately from one another. Synchronizers (see below) should be used wherever a foreign clock enters a new domain; this makes module inputs the logical place to insert such devices.

Voltage islands. Parts operating at different supply voltages should be separated; the gates generally have to come from different libraries and often different vending organizations. It is true that some CMOS libraries can be used in a range of voltages, for example 3 V to 5 V, but the power consumption or speed will be optimal only in one narrow voltage range. Separation also allows intelligent and efficient selection of voltage level-shifters.

Level-shifters usually are located in their own partition, or in the voltage partition of the level which is undergoing the shift.

Output latches. Each part of any substantial complexity should have clocked, latched outputs, making its internal timing separate from, and independent of, that of any other part. The work *latch* is being used here in its general sense, to refer to something that retains its value, something usually a flip-flop.

Control timing. When different parts pass control signals to one another such that the sequence of these controls is significant (for example, wait states on a read from a memory), the sequence may have to be enforced by latches on inputs or outputs, or by clocking on different cycles or opposite edges.

IP reuse. Large commercial IP ("IP, = Intellectual Property") blocks may be purchased separately and make up a large fraction of any modern VLSI IC. Microcontroller cores and memories are typical examples. Each of these blocks should be allocated its own partition in the design.

Test considerations. Design partitions usually should allow for internal scan insertion and possibly boundary scan. In any case, parts should allow observability of crucial intermediate results. Latches on outputs make ideal components to be replaced by scan flip-flops with little or no performance decrement; this often means that the same simulation test vectors may be used before and after scan insertion.

Digital VLSI Design with Verilog

Synthesis considerations. In general, logic optimization will work best on blocks of random logic in a certain size range, somewhere between about 300 to 50,000 transistors. This roughly would be about the same as 30 to 5000 instances, depending on the library, or 75 to 15,000 gate-equivalents. A block too large may take too long to optimize, lengthening the debug cycle; a block too small will not give the optimizer enough to work on, so that the designer's original input will not be improved. Partitioning should be planned to combine or split such logic, with the synthesizer's capabilities an important factor in the decision.

It's probably best for the designer to assume reliance on automation in the first approximation—the initial cycle of write, debug, and synthesize. After seeing the first working area and timing result, automated tuning followed by manual constraint tweaking would be a typical progression. After fully constrained synthesis, manual edits occasionally may be necessary. Manual editing of a synthesized netlist can produce results better than anything the tool can accomplish; but, on the other hand, optimization by the tool may be obstructed by too many "don't touch", manually-inserted structures.

A synthesizer usually allows for optional shifting of gates across sequential logic boundaries, thus combining random logic across the boundary and improving opportunities for speed of area improvements. In a flattened netlist, this kind of retiming may improve certain module instances more than others, providing refinements not available in a partitioning context. Except within a single module, this feature should be used only in the final design stages, because breach of the design partitioning is irreversible, and localization of defects or manual corrections may become very difficult after logic has been shifted in or out of what originally were separate verilog partitions.

However, keep in mind that a back-end tool such as a floorplanner can be made to reconstruct in the physical layout the original verilog source hierarchy, even from a flattened netlist. This generally is possible because the naming convention imposed during uniquification and flattening by the synthesizer or optimizer may be consistent with the original verilog module names.

16.3.4 Synchronization Across Clock Domains

A clock domain is defined by a clock which runs independently (by a different oscillator) of other clocks. A derived or generated clock created by a PLL or frequency divider may be said to be in a different domain from the original clock. However, we do not use this interpretation of the meaning of a domain in the present context. The reason is that derived or generated clocks are phase-locked to their origin clock, and problems of synchronization typically are limited merely to considerations of skew and jitter.

When data have to be transferred from one clock domain to another, problems arise which do not occur in a single-clock synchronous design. Specifically, when two clocks run at different rates, and are not derived from one another, any clock state can be simultaneous with any other at a given component input. This means that data (or control) from one clock domain can be sampled in an intermediate logic state in a different domain, far enough away from a '1' or '0' that a receiving gate's response (slew rate) may be very slow. Actually, intermediate-state sampling in general would be expected to occur frequently, and at random, given typical GHz clock rates in modern designs.

A simple mechanical analogy may help conceptualization of this problem: If a fencing foil is dropped a few inches at a random, almost vertical angle onto a hardened steel floor, there is almost no chance it will happen to balance upright on its tip for one full second before falling over. However, if such a drop was repeated a billion times, by simple chance several of the foils will have balanced upright for one second. Repeated enough of times, a foil or two will be found to have balanced for 10 seconds or more.

Likewise, on a fine enough time scale, an input always can be sampled so that it leaves the output undetermined, no matter how long the wait. See figure 16-8 for an illustration of oscilloscope-like, progressively increasing, time resolution.

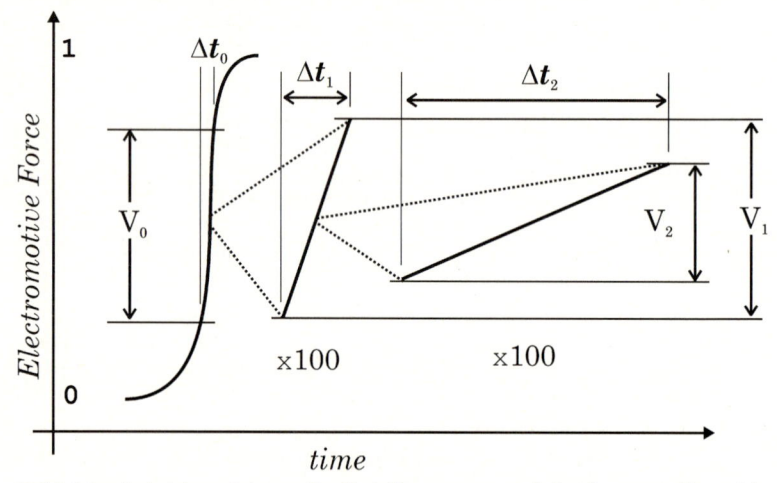

Fig. 16-8: Indeterminate intermediate sampling. Each V spans a range such that the output will go to '1' at the top and '0' at the bottom. Figure not to scale. If $\Delta t_0 = 100\Delta t_1$ and $\Delta t_1 = 100\Delta t_2$, then $V_1 \equiv V_0/100$; $V_2 \equiv V_1/100$, which implies approximately that the probability p that the gate output will be intermediate during the respective Δt will be $p_1 = 100p_0$ and $p_2 = 100p_1$.

This figure can be interpreted most simply as a representation of an oscilloscope display, showing a transition between '0' and '1' of a typical receiving gate. Increasing the horizontal sweep speed spreads out and flattens the voltage trace. No matter what the applied voltage, in a short-enough time interval, the voltage will have negligible effect in moving the potential of the gate from '0' to '1' or vice-versa. Thus, no transition will occur in a short-enough interval. If such a transition-less input is clocked (say, as the D input of a flip-flop), the receiving device's Q may not be able to change state quickly.

Digital VLSI Design with Verilog

We assume in this figure that an independently-running clock has provided an input data value of intermediate, indeterminate digital value. Viewed as a representation of a data input voltage near the input gate balancing point, for a high-enough clock frequency, no reasonable potential will be able to move the output either to a '0' or a '1' before the next clock.

There is an equivalent geometrical, statistical interpretation: At a low clock frequency, the input data will be sampled very rarely near the balancing point in any given interval of time. At low frequency, the slope of the transition curve, its rate of movement away from the balancing point, will be very high, as indicated in the leftmost Δt_0 interval. At a higher frequency, say 100 times higher, the slope at Δt_1 will be 100 times less, and at 10,000 times the original frequency, the slope will be approaching zero, as shown in the interval Δt_2. In effect, the voltage applied to clock in a distinct digital value will be reduced approximately proportionally to the frequency.

Thus, whatever the balanced-error rate at frequency f_0, this rate will grow with increasing frequency f approximately as f/f_0.

So, from time to time, probably many times per second for a GHz clock, a gate in one domain will sample an input voltage so exactly centered in its input range, that the gate will be almost perfectly balanced and can not switch to propagate either a '1' or a '0' before the next clock cycle in its own domain. As time passes after the balanced sampling, the gate eventually will switch one way or the other, but this may be too late for the correct logic level to be propagated.

This problem is overcome by latching inputs in the receiving clock domain. As shown below in figure 16-9, an output flip-flop or latch is triggered by the sending clock, the data are latched as usual, and the latched logic level then is sampled in a synchronizing flip-flop (perhaps the leftmost B-domain clock in figure 16-9) on a receiving-clock edge. However, as explained above, there is a remote chance that the synchronizing flip-flop will not have settled before its value is propagated, perhaps ambiguously, into the receiving logic. This chance is almost perfectly eliminated by using two synchronizing flip-flops in place of one, creating a little two-stage shift register.

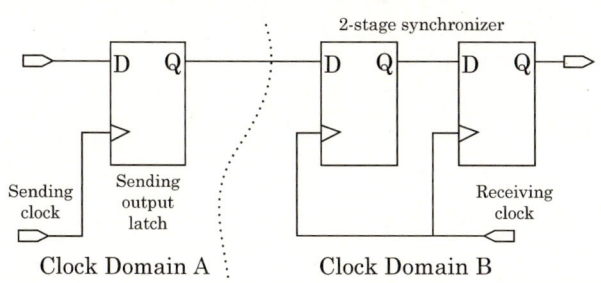

Fig. 16-9. Synchronization across two clock domains. Clocks A and B run at different and independent frequencies. Domain B must be prevented from clocking indeterminate data values from A.

If organized as shown in figure 16-9, the input balance point of the second flip-flop almost certainly will not be at the same voltage as the balanced output of the first. Then, if balanced, the old data in the first receiving flip-flop has almost an entire clock cycle to settle before being sampled by the second one, which is in a well-defined state anyway. The probability that the second flip-flop will be balanced on the balanced output of the first one is remote enough to be ignored or to be compensated, if necessary, by occasional ECC operations.

16.4 Hierarchy Lab 19

Topic and Location: Bus routing in a module hierarchy, synchronization across clock domains.

Do this work in the `Lab19` directory.

Preview: We create a skeleton hierarchy of module ports and see how bus-width mismatches are handled across hierarchical boundaries. We then write a testbench which clocks two 3-bit counters at different clock speeds. We examine glitching across the two clock domains and see how latched outputs ameliorate the glitching problem, even in a digital simulator which has no capability to represent different voltages.

Deliverables: Step 1: An empty hierarchy acceptable to the simulator compiler. Step 2–5: Answers to the questions, checked by the simulator. Step 6: A correctly simulating `ClockDomains` testbench.

Lab Procedure:

Step 1: Create a file named `MiscModules.v` and type in the following empty modules (headers only):

```
module Wide(output[95:0] OutWide, input[71:0] InWide);
endmodule
module Narrow(output[1:0] OutNarrow, input[1:0] InNarrow);
endmodule
module Bit(output Out, input In);
endmodule
```

This file will be used in the following Steps 2–5.

Step 2: Connecting a 1-bit signal to a 2-bit port. Using the file in Step 1, instantiate `Bit` in `Narrow`, and connect it to the `Narrow` I/O's as shown here:

```
module Narrow(output[1:0] OutNarrow, input[1:0] InNarrow);
    Bit Bit1( .Out(OutNarrow), .In(InNarrow) );
endmodule
```

Do not use a part-select. If you simulate, what do you think will happen? Only one bit in the `Narrow` busses can be connected: Which one?

Test your assumptions by implementing something trivial in `Bit`, for example,

```
assign Out = 1'b1;
```

Simulate `MiscModules` to see which bit is assigned.

Step 3: A hierarchy paradox. Without changing anything else in Step 2, instantiate `Narrow` in `Bit`; leave `Narrow` with no port map (I/O connection). Now you have two modules, each instantiated in the other. Do you think this is legal? Check it by compiling `MiscModules` for simulation.

New VCS comments: If you followed the instructions, the design now would have two top-level modules, `Wide` and `Narrow`, in the order shown in Step 1. Even though `Narrow` will have been made impossible in Step 3 as described just above, new *VCS* will issue no error and simply will ignore `Narrow` while compiling `Wide` correctly.

To make this example work and get a fatal error from new VCS, before performing Step 3, instantiate `Narrow` in `Wide`, mapping the `Narrow` output port to `OutWide` and the `Narrow` input port to `InWide`.

Step 4: Connecting a 2-bit signal to a 1-bit port. Before starting this Step, comment out the `Narrow` instance which you may have added in `Wide` according to the preceding Step 3 instructions.

Then, comment out the `Bit` instantiation in `Narrow` from Step 2, leave `Narrow` instantiated in `Bit`, and connect the `Narrow` ports analogously to the opposite connections in Step 2 (with no part-select or bit-select). This means, more or less, setting up `Bit` this way:

```
Narrow Narrow1(.OutNarrow(Out), .InNarrow(In) );
```

Again, try to predict which bit will drive the output. Add the following simple statement to `Narrow`, and simulate to test this:

```
assign OutNarrow = 2'b10;
```

Step 5: Repeat Step 4, using `Narrow` and `Wide` instead of `Bit` and `Narrow`. You should be able to predict the results. Optional: Simulate to see this result.

Step 6: Clock synchronization simulation. Set up a testbench module named `ClockDomains` which has two clock generators, one with half-period of 2.011 ns (the slow clock), and the other with half-period of 1.301 ns (the fast clock). These delay values are prime numbers and thus can not be matched by the periods of each other or of any smaller period. To resolve these periods in the simulator, set `` `timescale `` in the testbench to `1ns/1ps`.

As shown in the schematic of figure 16-10 below, write a model of a 3-bit RTL up-counter in a module, `Counter3`. Be sure to include a reset, and assign all three bits to the `Counter3` outputs. For this exercise, make the counter count clock edges, not cycles, using `always@(Clk)` rather than `@(posedge Clk)`. Instantiate this model twice in the testbench module. One instance will represent logic in the fast clock-domain, and the other, logic in the slower clock domain.

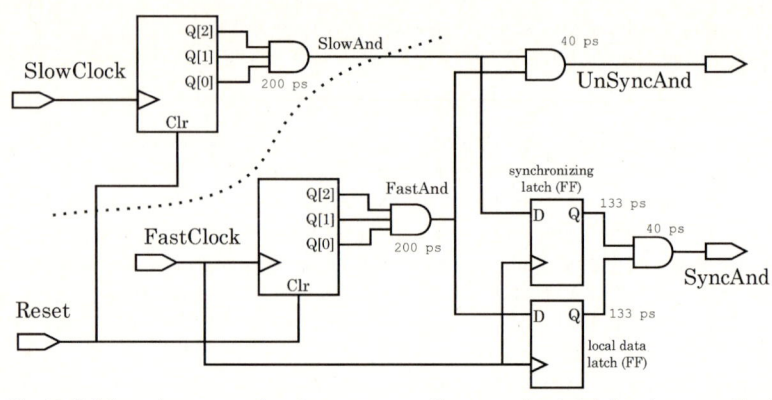

Fig. 16-10. Schematic representation of `ClockDomains`. Two uncorrelated clock domains are combined raw in `UnSyncAnd`. A synchronizing latch on `SyncAnd` makes slow-domain data more predictable when sampled in the faster domain. Delays are approximate.

Clock one counter instance with the fast clock, and the other with the slow clock. Use a continuous assignment and-reduction operator to assign the and of the three counter bits of each counter to its own net variable. Attach a small delay to these two continuous assignment statements, maybe 200 ps or so. These two nets are to be compared in the fast clock domain; when both are '1', the fast domain will have to do something (we leave the operation undefined in this example).

A. To see the raw, unsynchronized coincidence of these ands, just *and* them both in another continuous assignment onto a net called `UnSyncAnd`, as shown above. Attach another very small delay to this and statement, perhaps 1/5 of the other delay. Use decimal fractions (`1/5.0` rather than `1/5`) to force evaluation as reals and not integers. Simulate your model to verify it. You should see the `UnSyncAnd` occasionally going to '1' or glitching high. The different positive pulses will vary considerably in width because of the incompatibility of the clock frequencies (as already mentioned, both half-period values are prime numbers).

B. To add a synchronizing latch (= flipflop) in the fast domain, use a simple RTL statement to sample the 3-input and from the slow clock domain when the local (fast) clock is high. Give this component a slightly shorter delay than that of the ands.

We don't require a synchronizer for the fast clock and, but we do require a latch to hold its value; so, also write an RTL latch just as above for the fast-clock and value.

Digital VLSI Design with Verilog

Such a latching construct is suggested here:

```
localparam AndDelay = 0.200; // 200 ps 3-input and gate delay.
localparam LatchLagDelay = AndDelay/1.5;
...
reg HoldSlowAnd; // The synchronizing latch storage.
always@(posedge FastClock)
    #LatchLagDelay HoldSlowAnd = SlowAnd; // Sample the slow-domain and.
...
```

Then, complete the exercise by *and*ing both latched 3-input statements, one from each clock domain:

```
assign #(AndDelay/5.0) SyncAnd = HoldFastAnd & HoldSlowAnd;
```

Compare `SyncAnd` with `UnSyncAnd` in a simulation (see figures 16-11 and 16-12). You will find that the `SyncAnd` data are much cleaner and are glitch-free. Latching the fast side of the `UnSyncAnd` input makes little difference in the glitches and irregular pulses: Garbage in; garbage out.

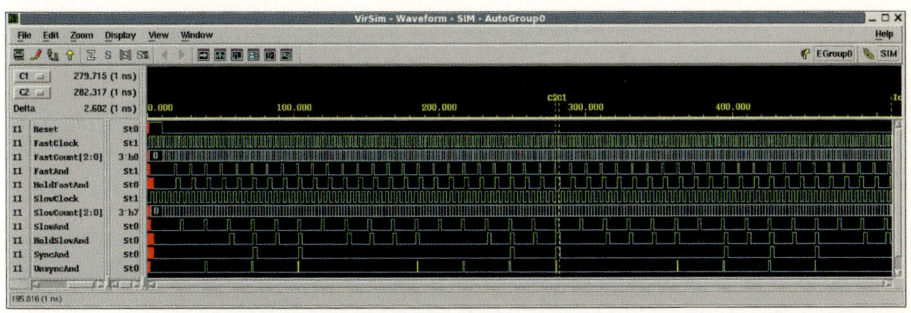

Fig. 16-11. The `ClockDomains` simulation, comparing the two *and*s.

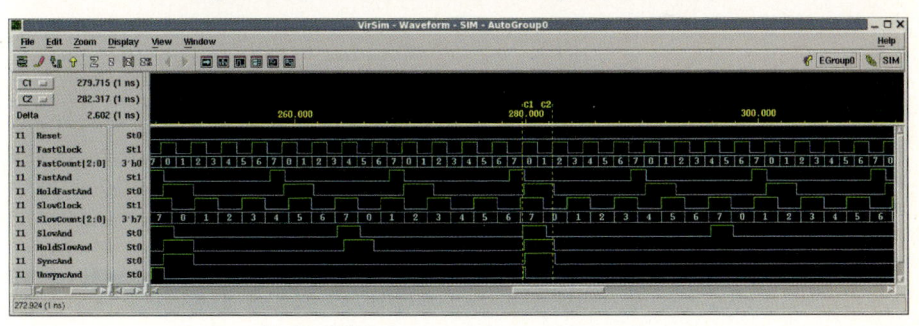

Fig. 16-12. Zoom in on `ClockDomains` narrow pulse.

Given that the clock frequencies are fixed, tuning the delays can introduce or eliminate potential glitches in the synchronized data. Also, some glitch-like pulses in the `UnSyncAnd` data can be admitted in the synchronized data by shifting or tuning the sampling windows. These are arbitrary issues when dealing with disjoint clocks, because synchrony is essentially arbitrary when the clocks are independent. It is the interface that defines synchrony, not the clock domains taken either separately or jointly.

16.4.1 Lab Postmortem

We have recommended keeping delays out of procedural blocks for synthesis reasons and, instead, putting the resultant delays in continuous assignments to module `output` or `inout` ports. How does this affect the partitioning practice of latching all data outputs?

16.5 Additional Study

Review parameters, hierarchical names, and connection rules in Thomas and Moorby (2002) sections 3.6 and 5.1–5.2.

Palnitkar uses traditional module header declarations extensively; Thomas and Moorby describes the differences briefly in Preface pages xvii - xix.

Chapter 17
Week 8 Class 2

17 Today's Agenda:

Lecture on verilog configurations
Topics: Libraries and `configs`.
Summary: We explain the relation of a library to a design and enumerate the major keywords for a `config`. There is no lab exercise or example available because of lack of vendor implementation.

Lecture on timing arcs and `specify` delays
Topics: Timing arcs within modules, `specify` blocks, path and full delays.
Summary: We build on the library concept to cover verilog delays within a module, especially a module for a library component. We elaborate on the use of lumped *vs.* distributed delays on a path. We present `specify` blocks and their delay statements, including `specparams`, leaving timing checks for a later class.

Lab
We study details of specify-block delays, concentrating on resolution of conflicts among overlapping paths. We write out an SDF file after a synthesis.

Lab Postmortem
We discuss simulation of conflicting delay specifications.

17.1 *Verilog configuration*s

The majority of designs begin with C models or behavioral (bus-transfer) models and, after verification of the partitioning and interfaces, proceed to further detail at RTL or gate level. This means that a module's implementation may change radically during the design process. Obviously, somewhere, there has to be a way of obtaining a complete list of the files comprising a design, or the design could not be used for anything. To substitute different versions of a module, designers have relied on makefiles or shell scripts. As an alternative to depending on the filing system or other nonverilog functionality, here we introduce a way of managing design versions and formats within the verilog language.

17.1.1 Libraries

The definition of a "library" of modules or technology-specific gates is vendor specific: Synopsys has its own definition, and other EDA vendors have theirs. Although file formats and byte-ordering make any compiled object nonportable, still, there has been some demand for a way within the verilog language for organizing a design. Verilog provides for a *library mapping file* as a way of mapping a library to the filing system. An example of this, for Synopsys synthesis, was given at the start of the generic DC compilation script provided at the beginning of this course. In that, the simplest case, the name of the library simply is paired with a directory containing the library contents; all subsequent references to the contents then are by the library name.

17.1.2 Verilog Configuration

The verilog *configuration*, new in *verilog-2001* (IEEE Std 1364 section 13; also in IEEE Std 1800-2012 section 33), is meant to make module versions easily interchangeable, and to make libraries more portable.

A configuration is bounded by the keywords `config` and `endconfig`. It describes a library comprised of everything in a design useful for compilation, simulation, regression testing or other design activity. It is stored in a design file at the same level as a `module`.

The format of a `config` is as follows:

```
config config_name;
   design design_top_name;
   default list_of_libraries;
   cell cell_name use library_name;
   instance inst_name [use] instance_liblist [.cell];
endconfig
```

Keywords are in bold above; the other words indicate design-specific identifiers or lists. Only the `design` and the `default` are required. Keywords and other reserved words or references used within a `config` are, of course, local and can not conflict or otherwise directly interact with anything outside of the `config`. Notice that the keyword **default** (`case` statement) is reused here in a very different context. The referents introduced by the keywords in a `config` are as follows:

- **config**. This introduces an identifier for this `config` statement. The intention here is that interchanging `configs` allows the design to be compiled or otherwise used with different collections of library elements or module versions, for different purposes. For example, a `config` early in a CPU design might be named `CPU_BusXfer`; later, a new `config`, with different references to the design library, might be named, `CPU_RTL`; then, later, `CPU_FloorPlanned`, and so forth. Any of the `configs` might be used to obtain information on the design at any `config`-identified stage.

  As another example, different `configs` could be used for simulation than for synthesis.

- **design**. This specifies the top-level module in the design assumed archived in the library involved. Verilog hierarchical names are used for the library locations. For example, if the library name was "IntroLib", then the top-level module might be identified by,

    design IntroLib.Intro_Top;

- **default**. This specifies the library, or libraries, in search order, from which the instances in the design shall be taken, unless otherwise specified in the **instance** statement. For example, if we used the Synopsys synthesis class library's class.db file for everything, we could complete this config for that exercise simply by stating,

    default class;

- **cell**. This specifies the library from which the named cell (module) shall be taken. The **use** clause is used to name the library cell itself, as shown above.

- **instance**. If more than one library was involved, and if a given module name was present in more than one, then the module or gate instances not to be found in the default library list are specified individually here. A **use** clause optionally may be used to pick a specific cell. The name starts with the name as given in the design statement. For example, recall that we used an *xor* expression (^) in the Lab 1 exercise. Suppose we had a special synthesis model of an xor gate, in library file special.db, which we wanted to be configured for our present purposes. Then, the xor in a previously synthesized gate-level netlist of our introductory lab exercise might be specified this way:

    instance IntroLib.Intro_Top.XorNor.xor_01 special;

Using the synthesis output library, the **instance** statement could be just,

    instance Intro_Top.XorNor.xor_01 special;

Library and design configuration maintenance is a very tool-dependent and specialized topic, and we shall not dwell more on it here.

The verilog *configuration* adds almost no functionality to the language, because modern designs always are configured already in the filing system. The language-based *configuration* thus may increase design failure by establishing conflicting multiple points of configuration control. No tool known to the author has implemented the verilog config, and so there is no lab exercise on the topic of verilog *configuration*s.

17.2 Timing Arcs and specify Delays

We move on to study timing arcs inside a module.

17.2.1 Arcs and Paths

Delays across a module in verilog are said to be distributed in space along timing arcs; the *arc* refers to a phasor angle swept out during a clock period. The rationale is that eventually the module will correspond to a geometric object on a chip, and the timing between distinct, identifiable places on that chip will be important.

A timing arc is equivalent to a delay between two points in a module. These points are viewed as structural, so they can not be represented solely by expressions or assignment statements; they have to be locations. Timing arcs represent scheduled verilog events and also a temporal distance between them. The distance may span just a single net, or it may span a network of multiple gates of combinational or even sequential logic. A timing arc may exist between a clock and a data pin. In a structural model, timing arcs may be defined between ports or internal pins of any module. At any level, each of these timing arcs may be assigned a ***path delay***, the delay of the timing arc mapped to a path.

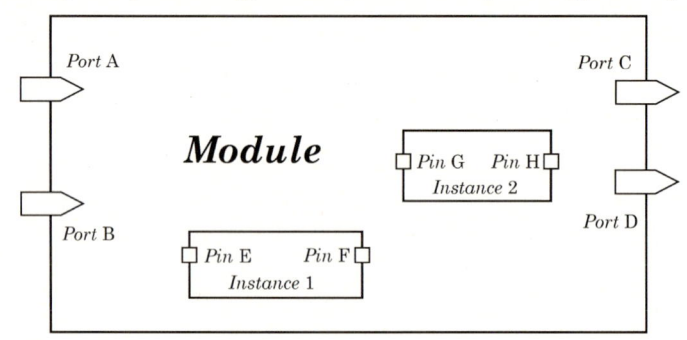

Fig. 17-1. Timing arcs are defined only between ports or pins. Assume A, B, E, and G are verilog inputs and the others are outputs.

In figure 17-1, a timing arc may be defined between module ports A or B and C or D. However, in general, if there is an arc between A and C, there is no single arc between A and both C and D. Timing arcs also may be defined between A or B and E or G, between F or H and C or D, and between F and G. No arc is visible between E and F or G and H in the module shown; however, such arcs may be defined in the modules which were instantiated as `Instance 1` or `Instance 2`. The delay on the paths between E and F or G and H may be calculated to determine arcs between, say, B and D and passing through one or both of the instances shown.

Although not shown in figure 17-1, any path must be represented by a causal connection, usually a simple wire or cloud of combinational logic—but possibly by layers of sequential logic, too.

The delay specifications we shall discuss here can be used by simulators or static timing analyzers.

17.2.2 Distributed and Lumped Delays

In our labs, we so far have assigned both distributed and lumped delays. A ***distributed delay*** is one which separately sums two or more delay values along a timing arc. Or, alternatively, a distributed delay is represented by a timing arc passing through two or more pins, each assigned a delay on a subarc. A ***lumped delay*** is one which is summed by the designer and assigned solely to the final point on a timing arc, the intermediate subarcs not being assigned any delay (including the simulation special delay of #0).

A simple comparison of distributed *vs.* lumped delays is as follows:

```
module DistributedDelay (output Z, input A, B);
  wire Node;
  assign #1 Z = Node;      // Output port delay.
  and #(2,3) (Node, A, B); // Pin delay.
endmodule
//
module LumpedDelay (output Z, input A, B);
  wire Node;
  assign #(3,4) Z = Node; // Total delay lumped on output port.
  and (Node, A, B);
endmodule
```

The distinction between distributed and lumped delays is made only in the context of modules with substructure; the distinction is not very relevant to simple behavioral or RTL models. To illustrate this distinction in a more elaborate example, consider a plain RTL model of a register composed of four three-state flip-flops.

We have:

```
`timescale 1ns/100ps
module FourFlopsRTL #(parameter DClk = 2, DBuf = 1)
                    (output[3:0] Q, input[3:0] D, input Ena, Clk);
  reg[3:0]  QReg;
  wire[3:0] Qwire;  // Not used yet.
  //
  always@(posedge Clk)
    #DClk QReg <= D;
  //
  assign #DBuf Q = (Ena==1'b1)? QReg: 'bz;
  //
endmodule
```

There is a delay from Clk to QReg, and another from QReg to Q, but the only timing arcs in this model are from D, Ena, or Clk to Q.

In this example, the (parameter) default delay on the D->Q path is 2+1 = 3 ns; the default delay on the Ena->Q path is 1 ns; and, the default delay on the Clk->Q path is 2+1 = 3 ns. Although the values can be summed, none of these defaults represent either distributed or lumped delay in any meaningful sense.

However, consider the more structural design below. This design, FourFlopsStruct, is functionally and timing equivalent to FourFlopsRTL; it contains some substructure implemented as arrayed instances.

We have:

```
module FourFlopsStruct #(parameter DClk = 2, DBuf = 1)
                       (output[3:0] Q, input[3:0] D, input Ena, Clk);
   wire[3:0] QWire;
   //
   DFF #(.DClk(DClk)) DReg[0:3](.Q(QWire), .D(D), .Clk(Clk));
   assign #DBuf Q = (Ena==1'b1)? QWire: 'bz;
endmodule // FourFlopsStruct.
// ----------------------------------------------------
module DFF #(parameter DClk = 2) (output Q, input D, Clk);
   reg   QReg;
   always@(posedge Clk) QReg <= D;
   assign #DClk Q = QReg;
endmodule // DFF.
```

The `FourFlopsStruct` model has the same timing as `FourFlopsRTL`, but the delays as calculated for `FourFlopsStruct` are distributed delays. The DReg instance output delay value `DClk` could be considered a lumped delay, viewed by itself.

Using what we know from previous models, `FourFlopsStruct` could be rewritten with lumped delays as follows:

```
module FourFlopsStructL #(parameter DClk = 2, DBuf = 1)
                        (output[3:0] Q, input[3:0] D, input Ena, Clk);
   wire[3:0] QWire;
   localparam DTot = DBuf + DClk;
   //
   DFF DReg[3:0] (.Q(QWire), .D(D), .Clk(Clk));
   assign #DTot Q = (Ena==1'b1)? QWire: 'bz;
endmodule // FourFlopsStructL.
// ----------------------------------------------------
module DFF(output Q, input D, Clk);
   reg   QReg;
   always@(posedge Clk)
      QReg <= D;
   assign Q = QReg;
endmodule // DFF.
```

Of course, as we might recall, the lumped delay `DTot` could be written to express rise and fall transitions separately, this way: (DTotR, DTotF). Furthermore, technology considerations might be taken into account to provide the verilog simulator with minimum, typical, and maximum delay estimates; from previous study, we know this might be written, (DTotRmin:DTotRtyp:DTotRmax, DTotFmin:DTotFtyp:DTotFmax). But, this is as far as we can go with what we learned up to now.

<u>17.2.2.1 Net Delays</u>

It is possible to distribute delays on nets, too, by assigning a delay when each net is declared. This is analogous to the way we have seen a strength assigned to a net. For

Digital VLSI Design with Verilog   351

example, "wire #3 DataOut;" schedules delayed changes in DataOut the same as though DataOut was being assigned from a temporary variable by a continuous assignment statement with the given delay.

The author never has seen net delays used in a design, although, in layouts in today's deep submicron pitches, the net capacitances have come to account for a significant fraction of the delay in a design, the remainder coming from the internal gate delays. Gates have locations; so, the general practice has been to assign timing in the verilog only to arcs between gates. The net delays in practice are accounted for when simulating with back-annotation from a floorplanned or placed-and-routed netlist.

17.2.3 specify Blocks

A specify block begins with the keyword **specify** and ends with the keyword **endspecify**. A specify block is at the same concurrent level in a module as an always or initial block. A specify block has no functionality of its own, but it may be used to determine the details of timing required for (a) an accurate gate-level verilog library model or (b) any other module with precisely known, and precisely required, timing.

Timing in a specify block can be more precisely and flexibly controlled than timing added to declarations, statements, or instantiations. A specify block makes it possible to model timing without knowing or defining any functionality; functionality can be added to the module later, with no further timing editing at all. A specify block may be used in a static timing model of an encrypted IP block without revealing any functionality.

We shall study timing checks later in the course; but, for now, it's important just to know that a specify block is the only place in a verilog model where a timing check may be stated. Possibly to prevent confusion over the '$' which begins the name of every timing check, system tasks, which also start with '$' ($display, $monitor, $dumpvars, etc.), are forbidden in a specify block.

A specify block may contain specparam definitions, timing checks, certain pulse filtering conditions, and module path delay specifications. We shall concentrate on the specparams and module path delays for now.

Names of ports declared in a module are visible within a specify block in that module; also, a net name, parameter value, or localparam value declared in the same module may be used in a specify block. Nothing in a specify block is allowed to reference declared reg variables or the port pins of components instantiated in that module.

In summary, specify blocks are at the same level in a module as always, initial, or generate blocks and are delimited by the keywords **specify** ... **endspecify**.

A `specify` block may contain:
- `parameter` or `localparam` references.
- module port or net references.
- `specparam` definitions.
- module delay specifications.
- timing checks.

A `specify` block may not contain:
- any other block.
- any declaration other than of a `specparam`.
- any assignment to a `reg` or net.
- any instance or other design structure.
- any `task` or `function` call, including system tasks and functions.

17.2.4 specparams

A `specparam` is a verilog parameter which usually is used only in a `specify` block. Although many tools will not allow it, a `specparam` also may be declared outside of a `specify` block.

This special kind of parameter exists for convenience of parsers and other tools which extract timing information from a model. There is only one unique feature to a `specparam`: It may be assigned multiple numerical values, for example in a declaration of the form,

> specparam *Name* = (x, y, z);

in which x, y, and z are numbers or timing triplets representing delays.

When creating a timing specification for a module, good practice is not to assign verilog literals to the delays; instead, the literals or other constants should be assigned to `specparams` in one or more `specify` blocks. Such assigned values then may be referenced in timing expressions later in their `specify` block. The `specparams` may be given mnemonic names and, of course, used in any number of timing expressions within that `specify` block. The reason that this is a good practice is that `specify` blocks are used in models of small devices; the `specparams` then would be named for specified performance parameters on the datasheet for such a device.

For example, we introduce our first path delay:

```
module ALU (output[31:0] Result, input[31:0] ArgA, ArgB, input Clk);
  ...
  specify
    specparam tRise = 5, tFall = 4;
    ...
    (Clk *> Result) = (tRise, tFall); // A simple full-path delay.
  endspecify
  ...
endmodule
```

A `specparam` also may be assigned a timing triplet for simulator min-typ-max alternatives. For example,

```
specify
  specparam tRise = 2:3:4, tFall = 1:3:5;
    ...
  (other stuff; maybe complicated)
    ...
  (Clk *> Q,Qn) = (tRise, tFall);
endspecify
```

Although a `parameter` may be referenced to define the value of a `specparam`, it is illegal in verilog to reference a `specparam` to define the value of a `parameter`.

Also, although a `specparam` may be assigned a multivalued delay of the form "`(x,y,z)`" in VCS, the IEEE Std 1364 section 4.10.3 or IEEE Std 1800 section 6.20.5 merely states that a `specparam` may be assigned "any constant expression"; these standards give only multiple single-valued examples such as

```
        specparam tRise_q = 50, tFall_q = 55;
```

So, before defining a `specparam` with multivalued delays, check your tool for the implementation.

17.2.5 Parallel vs. Full Path Delays

There are two main kinds of path delay statement, full-path ("`*>`") and parallel-path ("`=>`").

A <u>full-path delay</u> applies to all possible arcs between all bits of the ports or nets named. A full-path delay may imply a timing-arc fanout (as in the `Clk *> Result` example above) or a fanin. Bit-select or part-select delays usually would be specified by full path.

A <u>parallel-path delay</u> describes an arc, or a parallel set of them, which only can exist between endpoints with equal numbers of bits; no fanin or fanout of the arc delay is allowed. This kind of path is most usual between scalar (single-bit) ports. When the ports are vectors, the bit delays are mapped one-to-one, in parallel and in declared order.

Path delay specifications of any kind are illegal when fanned-in logic such as a `wor` or `wand` net directly drives a port; such fanin's must be replaced by logically equivalent gates with single outputs if path delays are to be assigned.

Examples of path delay specifications:

```
module FullPath (output[2:0] QBus, output Z, input A, B, C, Clock);
   ... (functionality omitted) ...
   specify
     specparam tAll=10, tR=20, tF=21;
     (A,B,C *> QBus) = tAll;
     (Clock *> QBus) = (tR, tF);
   endspecify
endmodule
// ---------------------------------------------------
module ParallelPath (output Z, input A, B, C, Clock);
   ... (functionality omitted) ...
   specify
     specparam tAll=10, tR=20, tF=21;
     (Clock => Z) = tAll;
     (A => Z)     = (tR, tF);
     (B => Z)     = tAll;
   endspecify
endmodule
```

An interesting feature of `specify` delay values is that there may be as many as six delay values for a path capable of turnoff. As stated in Thomas and Moorby (2002) section 6.6, a `specify` block assignment may assign six different delays to a path in the order, (0_1, 1_0, 0_z, z_1, 1_z, z_0). Because `specify` blocks often are used in simulation models of library components, models which never are synthesized, six or more delays on a path occasionally may be required. For paths on which delay pessimism must be overridden, up to twelve different delays are allowed, counting to- and from-`x` as well as to- and from-`z`. The special level of simulation timing implied by numerous conditional delays occasionally might be useful in switch-level modelling, but it primarily would be applied in high precision gate-level models or models of small IP blocks.

Levels of precision exceeding three delay values rarely are useful in modern design specifications of individual `modules`, because synthesizer or floorplanner back-annotation generally will be more accurate and less effortful than this level of designer guesswork in the source verilog.

17.2.6 Conditional and Edge-Dependent Delays

Any delay may be assigned to a path conditional on the nonfalsity of an expression. The usual verilog operators may be used in this expression, and the evaluation is nonfalse if it resolves to '1', 'x', or 'z'. If the expression is on a vector object, the (non)falsity depends only on the LSB. However, only simple statements are allowed; no *else*, *case*, or other constructs with alternatives are allowed. The keywords posedge and negedge have their usual meaning.

For example,

```
// output[3:0] Z, input[3:0] A, input Clk, Clear are declared ports.
...
specify
  specparam ClkR=2, ClkF=3, ClearRF=1, AThruR=4, AThruF=5;
  //
  if (Clk && !Clear)(A => Z)        = (AThruR, AThruF);
  if (A[0] && A[3]) (A[1], A[2] *> Z) = AThruR; // Lists are OK
  if (A[1] && A[2]) (A[0], A[3] *> Z) = AThruF;
  if (!Clear)       (negedge Clk *> Z) = ClearRF;
  if (!Clear)       (posedge Clk)*> Z) = (ClkR, ClkF);
                    (posedge Clear *> Z) = ClearRF;
endspecify
```

Although individual statements which are conditional are not allowed to include logical alternatives, a specify <u>block</u> default may be assigned for paths in a module for which the stated conditions and edges do not apply. This default is introduced by the **ifnone** condition, which may be read to mean, "<u>if none</u> of the preceding, do this".

For example,

```
specify
  if (!Clr) (posedge Clk *> Q,Qn) = (3, 4);
  if (!Clr)  (A, B *> Q, Qn)      = (5, 5);
  ...
  ifnone     (A, B *> Q, Qn)      = (1, 2); // For Clr asserted.
endspecify
```

It is illegal in verilog to assign an ifnone condition to a path also covered by an unconditional delay statement.

A path destination may be assigned a polarity, so that the delay is associated with a certain edge direction at the related port. This allows the data path to be used to determine the delay, as well as the output change.

For example,

```
// output Q, Qn, input D, Clk: Q1 and Q2 are logically equivalent.
...
specify
  specparam tR_Q = 5, tF_Q = 6.5;
  ...
  ( posedge Clk => (Q1 +:D) ) = (tR_Q, tF_Q);
  ( posedge Clk => (Q2 -:D) ) = (tR_Q, tF_Q);
endspecify
```

In the first statement above, `Clk` clocks in `D` to `Q1`; if `Q1` was '0' and `D` is '1', then `tR_Q` is the delay; if `Q1` was '1' and `D` is '0', then `tF_Q` is the delay.

In the second statement, the '-' means that the `D` polarity is inverted, so the effect of `D`, when compared with the explanation of the first statement, is to choose the other delay. So, if `Q2` was '0' and `D` is '1', then `tF_Q` is the delay; if `Q2` was '1' and `D` is '0', then `tR_Q` is the delay.

Similarly, both parallel and full path delays may be assigned with polarity dependence. The '+' means noninversion; the '-' means inversion. So, "-=>" refers to a change on the left which is in the opposite direction from the resultant, parallel-path delayed, change on the right. The same principle applies to full path statements, using "-*>" and "+*>".

For example,

```
...
specify
   ...
   (clk -*> Q1, Q2) = t_ClkTog; // Delay when Q1 | Q2 goes opposite clk.
   (clk +*> Q1, Q2) = t_ClkReg; // Delay when Q1 | Q2 goes same way as clk.
endspecify
```

The preceding special features might be useful in switch-level modelling.

Delays may be given explicitly for transitions to and from 'x' or 'z' as well as the other logic levels. As mentioned above, this leads to timing specifications not just of the two value (*rise, fall*) or three value (*rise, fall, turnoff*) variety, which we have seen and used, but also of six values (all transitions with 'z') or twelve values (all transitions with 'x' or 'z'). See Thomas and Moorby (2002) appendix G.8.1 - G.8.4 for the formal syntax of these specifications.

Keep in mind that overriding delays to 'x' or 'z' prevents the simulator from applying the principle of delay pessimism previously introduced (*Week 7, Class 2*).

17.2.7 Conflicts of `specify` with Other Delays

When a `specify` block contains a delay assigned to a path also delayed outside the `specify` block (in a delayed assignment or delayed primitive instance statement), ***the greater of the two delays*** will be the one simulated.

This said, keep in mind that individual simulators are likely to be equipped with invocation or configuration options to select among the different possible sources of delay, overriding the default specified by the verilog language.

17.2.8 Conflicts Among `specify` Delays

The verilog 1364 Std section 14.3.3 or IEEE Std 1800 section 30.5.3 say that when several active inputs change which individually would schedule the same event at different delays, the ***shortest*** delay is the one used.

Digital VLSI Design with Verilog   357

17.2.9 Conflicts with SDF Delays

It should be emphasized that an SDF delay, if applicable, supersedes and replaces any of the other kinds of delay described here.

17.3 Timing Lab 20

Topic and Location: Path delays and conflicts; SDF for a netlist.

Do this work in the Lab20 directory, using the verilog provided.

Preview: We use a small hierarchical design to study assignment of delays to component internal paths. We especially show how delays on multiple paths, potentially conflicting, are resolved in simulation. We synthesize a netlist and write out SDF to see the effect of the specify delays.

Deliverables: Step 1: A testbench counter for the SpecIt design which compiles in the simulator. Step 2: Correctly simulating distributed and lumped delays in SpecIt. Step 3: Various correctly simulating delays using a specify block in SpecIt. Step 4: A correctly simulating full *vs.* parallel path conflict, using specparams in a specify block. Step 5: Simulation and synthesis delays, and SDF delays, for a SpecIt netlist.

Lab Procedure:

The verilog for this exercise has been provided, minus timing, in the Lab20/SpecIt subdirectory. Before beginning this lab, copy every file in Lab20/SpecIt up one level to the Lab20 directory. The top-level schematic for the SpecIt design is given below in figure 17-2; the two lower-level schematics are given in figures 17-3 and 17-4. As previously discussed, module names and instance names are in different name spaces, so naming an instance exactly the same as its module, while not often recommended, is allowed.

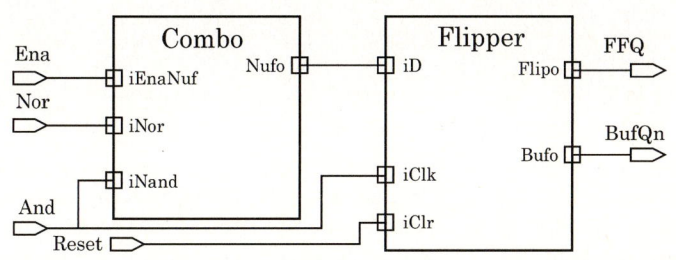

Fig. 17-2. SpecIt top-level schematic for timing path exercises. Port names and block instance names (= module names) are shown.

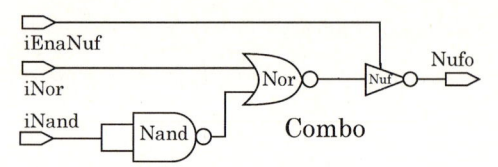

Fig. 17-3. The Combo schematic, showing gate instance and port names. Inputs begin with 'i'; outputs end with 'o'.

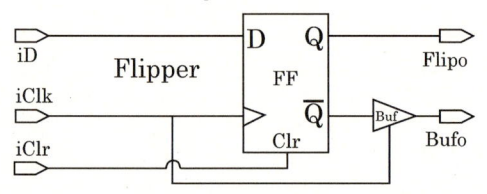

Fig. 17-4. The Flipper schematic, showing gate instance and port names. Inputs begin with 'i'; outputs end with 'o'.

Step 1: Instantiate the SpecIt module in a testbench and supply test vectors by means of a 4-bit up-counter. Use the count bits as stimuli on the four SpecIt inputs in this order: $\{MSB, \ldots, LSB\}$ = { Reset, Ena, Nor, And }. Clock the counter with a 100 ns (10 MHz) clock. Verify correct syntax by loading the design in the simulator (see figure 17-5).

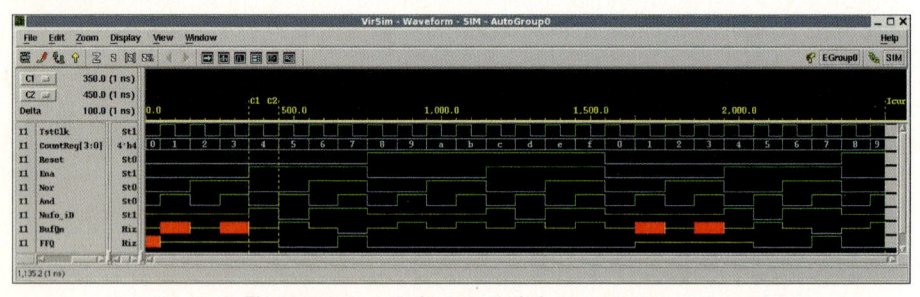

Fig. 17-5. SpecIt simulation to verify design correctness.

Step 2: Lumped and distributed delays. In the Nand, Nor, and Nuf gate instantiations in Combo, assign output delays of 3, 5, and 7 ns, respectively. This will establish a distributed delay totalling 12 or 15 ns between iNand and Nufo. Check this by simulating briefly (figure 17-6).

Digital VLSI Design with Verilog

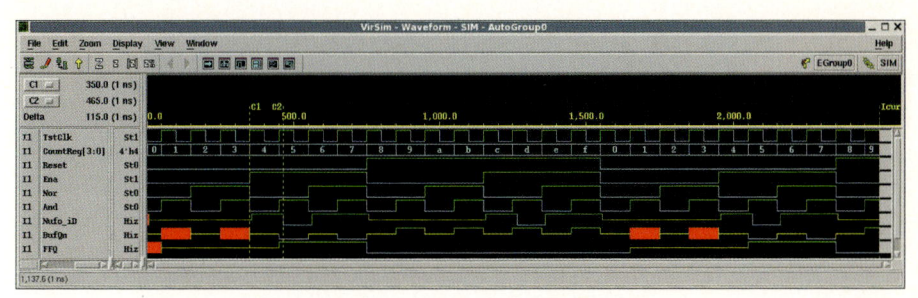

Fig. 17-6. SpecIt simulation with distributed delays.

What happens when two inputs change at the same time and are distributed different delays?

Try setting a lumped delay of, say, #10, on the Combo instance in SpecIt. Combo has just one output, so this delay should be unambiguous. What happens in this potential conflict?

Step 3: Specified path and distributed delay conflict. Make a new copy of Combo.v (with the distributed delays of Step 2) in a file named ComboSpec.v, but don't change the module name. Change your SpecIt file list so ComboSpec.v is read by the simulator instead of Combo.v.

A. In ComboSpec.v, add a specify block to the Combo module which assigns a full-path delay of 10 ns from any input to the output. Use a specparam for the time value. Simulate to see what happens.

B. Change the delay in the specify block to 20 ns rise, 21 ns fall, 22 ns turnoff. What happens?

According to the verilog standard, when delays conflict this way, they must be resolved pessimistically: As previously stated, in a change between '1' and '0', the longest delay is used. And, as usual, in a change to 'x', the shortest delay is used. All the new specify delays are longer than the distributed delays on the Combo instances.

C. Leaving the Step 3B specify values alone, define a localparam in the Combo module named InstTime and assign it a value of 25. Use InstTime to assign each of the distributed delays on the gate instances to 25 ns. Simulate. What happens?

D. Leaving the Step 3C distributed delays as-is, define a new localparam tZ = 30, and use it to assign the value to your specparam for turnoff time. Simulate. This shows another benefit of using parameters to change the timing of a model.

Note: Your simulator may refuse to use a generic named constant (parameter or localparam) to define a specparam; this is incorrect and a minor nuisance. A workaround might be to avoid specparams and use localparams instead.

E. In the Step 3D model, try changing the full-path assignment to a parallel-path assignment without changing anything else. What happens? Your simulator should issue an error or at least a warning when you do this.

Step 4: Specified parallel and full delay conflict. Copy Flipper.v to a new file FlipperSpec.v, keeping the module names the same. Use your original Combo.v (no instance delay) in combination with FlipperSpec.v for simulation in this Step.

A. Use specparams in a specify block in the Flipper module (not the DFFC module) to assign a parallel-path delay to both outputs of Flipper: Use a sum in the specify block to represent the conditions that Bufo adds a delay of 1 ns to any change on its input or control, and that clocked delays are 5 ns to a rise on Flipo and 8 ns to a fall on Flipo.

For example,

```
specparam tClkQR=5, tClkQF=8, dBuf=1, tClr=2;
    ...
    (posedge iClk => Bufo)   = (tClkQR+dBuf, tClkQF+dBuf);
    ...
```

B. Now assume that a negative clock edge turns off Bufo. To do this, add a turnoff delay of 1 ns to the specify block. Use only rise and fall delays, not posedge or negedge expressions.

C. Also assign a parallel-path delay of 2 ns to a clear of Flipo, appropriately adding delay to Bufo for that clear. Simulate to see the result.

D. With the changes in **A - C** imposed, add a full-path delay of 30 ns in Flipper from iClk to both output ports (using a list). Simulate. The 30 ns is large enough that you easily should be able to see a 30-ns shift in the waveforms, if one occurs.

E. Replace your 30 ns full-path assignment in **D** with an assignment to 0 ns. Simulate to see the effect. Question: Suppose there is a conflict between different delays to the same port pin; if so, which delay is used by the simulator, the shorter or the longer one?

Step 5: Using your Step 4E design for Flipper, reinstall the delays into Combo that you had at the start of Step 2, and simulate. You should be able to see waveforms the same as in figure 17-7.

Digital VLSI Design with Verilog

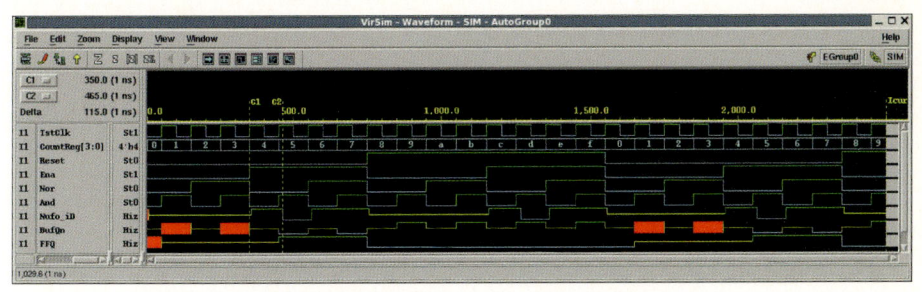

[screenshot of VirSim Waveform window]

Fig. 17-7. A `SpecIt` Step 5 verilog source simulation.

Synthesize `SpecIt`. Be sure to compile the correct files. Read the timing report; what happened to the timing in the modules?

Write out an SDF file (recall *Week 1 Class 1*) and look at it briefly in a text editor. Notice that there are timing triplets everywhere. We shall study this kind of file later in the course. If you should simulate the resulting netlist, you would see waves about the same as in figure 17-8.

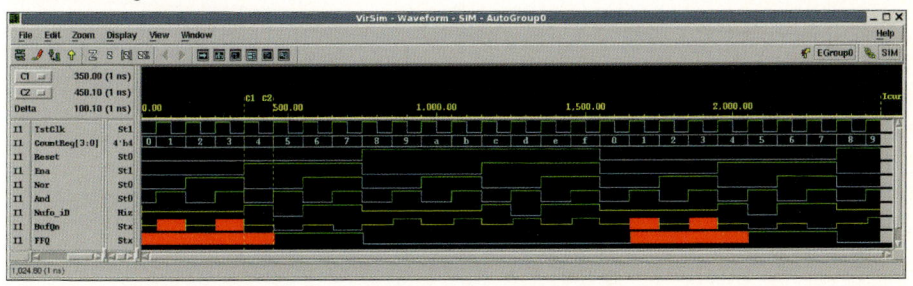

[screenshot of VirSim Waveform window]

Fig. 17-8. The `SpecIt` Step 5 verilog netlist simulation with SDF back-annotated timing.

17.3.1 Lab Postmortem

How does the simulator resolve conflicting delay scheduling times when the delays differ but the final state is the same?

What about when the final states differ?

17.4 Additional Study

Read Thomas and Moorby (2002) section 6.6 on `specify` block delays.

Optional readings in Palnitkar (2003):

Read sections 5.2 and 6.2 on gate-level and assignment-statement timing.

Read chapter 10 on path and other delays.

Try section 10.6, problems 1–3.

Chapter 18
Week 9 Class 1

18 Today's Agenda:

Lecture on timing checks and pulse controls
Topics: Relation to assertions, the twelve verilog timing checks, pulse filtering and delay pessimism.
Summary: This will complete our study of timing issues in verilog, as well as of the allowed contents of a `specify` block. We start by discussing the relationships among timing checks, assertions, and system tasks. After introducing the time-stamp/time-check rationale, we present the 12 timing checks and their default arguments. After describing conditioned events and timing-check notifiers, we explain verilog simulator pulse handling, inertial delay, the `PATHPULSE` task, and pessimism reduction in the `specify` block.

Lab on timing checks
We exercise the timing checks and pulse-filtering features of `specify` blocks.

Lab Postmortem
We entertain a Q&A session only, this time.

18.1 Timing Checks

18.1.1 Timing Checks and Assertions

A verilog timing check is of the form $name(arg1, arg2, ..., argN)$ and can check only things located in the module containing the check. There exist exactly twelve different *name*s; of course, this implies that there exist exactly twelve different timing checks. The timing check argument list, from two to three or more in length, identifies those running variables in the containing module which are to be monitored during simulation for timing violations.

Timing checks have the appearance of system tasks (both begin with '$'), but they are not system tasks (IEEE Std 1364 section 15 and IEEE Std 1800 section 31.2). System tasks are simulated in procedural code; timing checks are concurrent and are allowed only in `specify` blocks. However, some system tasks, such as `$display`, can be used to create assertions which resemble the result of a timing-check violation.

As we saw a while ago in *Week 4 Class 2*, simple system tasks can be used to construct assertion checks in verilog. While ***assertion*** often is taken technically to refer to something specialized, and while in some languages such as VHDL or SystemVerilog there is a builtin assertion mechanism, the functionality is just to check some condition on variables or other design objects during a simulation, and to create a warning message, and perhaps an error condition, when the assertion is not fulfilled. An error condition may stop the simulation.

Assertions generally are simulation tools, although, in principle, a synthesizer or static timing analyzer could be provided with an assertion mechanism to issue a message or stop the program when some structure or other condition was encountered during a traversal of the verilog input or during creation of the output.

Actually, the *Liberty* library format used for netlist synthesis includes its own timing checks very much the same as those used in verilog simulation. However, these checks can not be run by a verilog simulator; they are used (*a*) during synthesis and optimization to avoid violation of constraints and (*b*) during static timing verification.

The difference between assertions, debugging, and code coverage is that <u>assertions</u> represent a designer's insight into what might go wrong; <u>debugging</u> proceeds after a flaw has been uncovered; and <u>code coverage</u> estimates the need for debugging. In this context, a timing check can be viewed as a builtin, specialized assertion mechanism. All a timing check does is issue a message to the computer console when some design constraint has been violated, or when some device operating parameter has gone out of range. Simulators typically include an option to copy timing check messages to a log file.

A timing check itself normally does not change simulator waveforms or the values assigned to variables during simulation. However, there does exist a ***notifier*** mechanism in all verilog timing checks which may be used by the designer to change the course of the simulation as a result of a timing violation.

As already mentioned, timing checks are allowed only in verilog `specify` blocks; they can not be put in `always` blocks or in procedural code such as `tasks` or `functions`. Timing checks sit in their `specify` blocks, wired into the rest of the design, and they issue messages when the simulation goes wrong in some subtle, timing-related, way. The problem conditions which timing checks report often would go unnoticed by someone looking for violations in the simulator output waveforms.

To list the features of timing checks:
- They all are of the form, "$name_of_timing_check(argument_list);".
- Functionally, they are predefined assertions.
- They differ from procedural assertions:
    * They are allowed only in `specify` blocks.
    * They include builtin triggering logic.
    * They run concurrently.
    * They are part of the simulator and so entail little runtime overhead.

18.1.2 Timing Check Rationale

Timing checks usually are the best ways of enforcing device hardware operating specification limits during simulation. In terms of the technical functionality of the design being simulated, all timing checks are based on a ***reference event*** and a ***data event***. These events may be scheduled on one, or on more than one, variable in the simulation. There is no connection between use of the word *data* here, and its usage in a design context to distinguish something from "control" or "clock". A timing check imposes certain conditions on the reference event and the data event; and, so long as the conditions (equivalent to a logical expression) are true, the timing check does nothing. Generally, the reference event is conceived of as fixed in time, and the data event is conceived of as varying within or beyond a violation limit.

Two related technical terms are ***timestamp*** and ***timecheck***. In a timing check, whenever the first one of the above events is scheduled in simulation time (it may be either the reference or the data event), a *timestamp* is recorded. If and when the other event is executed, a *timecheck* is done to check the conditions on the two times.

As we shall discuss in detail below, the time limit in these checks often defines an open interval in simulation time the endpoints of which do not trigger a violation. In other words, assigning a time limit of 0 provides a convenient way to disable any timing check, with the sole exception of the skew-related checks.

By default, a timing check triggers only once per timestamp event, even at a time when more than one violating timecheck event has been simulated. Some of the timing checks can be enabled optionally to print multiple messages for repeated violations, for example violations on different bits of a *timecheck*ed bus.

The limits controlling timing checks must be constants and usually would be defined by specparams. The design variables may be vectors; if so, any bit change(s) in the vector mean(s) the same as that change in a one-bit variable; so, only one violation can be triggered (by default) for each such vector change.

Parameters and other inputs may be passed to a timing check by position, only.

18.1.3 The Twelve Verilog Timing Checks

Here is a complete list, grouped according to expected applicability; however, these checks work the same way regardless of the design functionality of the variable(s) checked:

Clock-Clock Checks	Clock-Data Checks	Clock-Control Checks	Data Checks
$skew	$setup	$recovery	$width
$timeskew	$hold	$removal	$period
$fullskew	**$setuphold***	**$recrem***	$nochange

* Avoid these, if possible.

Each of these timing checks is detailed below. Only the required inputs are listed below, except for $width, which includes an optional glitch parameter. Including that one exception, all optional inputs such as notifiers follow the listed ones. The notifier is explained below, but the reader should refer to an IEEE standard document for full details of features available in timing checks, such as edge specifiers or remain-active flags.

In QuestaSim, timing checks are treated as assertions, and verilog warning assertions must be enabled explicitly for timing checks to be run.

Clock-Clock Checks:

$skew. This check triggers on an excessively long delay between events on two variables. The delay value must be nonnegative. Typically, the reference (timestamp) event is a clock, and so is the data (timecheck) event. If the data event never occurs, there is no timing violation.

For example,

```
              ref. event           data event         limit expression
    $skew(posedge Clock, posedge GatedClock,    MaxDly);
```

$timeskew. This differs from $skew only in that, by default, if the specified time lapses, a violation occurs whether or not there ever is a data event.

$fullskew. This check is the same as $timeskew, except that it allows two nonnegative delays. The first delay specifies a limit when the data event follows the reference event; the second, when the data event precedes the reference event.

For example,

```
              ref. event R        data event D     R-D limit    D-R limit
    $fullskew(posedge Clock, posedge GatedClock,    MaxRD,      MaxDR);
```

Clock-Data Checks:

$setup. This check triggers a violation when the reference event, usually a clock, occurs too soon after the data event. The time limit must be nonnegative. The trigger-time window begins with the data event, which is the timestamp event for this check.

For example,

```
            data event    ref. event   time limit
    $setup(    D,      posedge Clk,  MinD_Clk );
```

$hold. This check triggers a violation when the data event occurs too soon after the reference event, which latter usually is a clock. The time limit must be nonnegative. The trigger-time window begins with the reference event, which thus is the timestamp event for this check.

For example,

```
            ref. event     data event   time limit
    $hold( posedge Clk,       D,       MinClk_D );
```

Digital VLSI Design with Verilog

$setuphold. This check requires two time limits, either of which may be negative; otherwise, the syntax follows that of $hold, with the setup limit first. It combines the functionality of a $setup and a $hold check when both limits are nonnegative. Because of error-prone complexity and thus likely false alarms, routine use of this timing check is not recommended. See the explanation below, in section 18.1.1.4, on negative time limits.

Clock-Asynchronous Control Checks:

Conceptually, these checks depend on internal delay characteristics of components to which several signals are applied; typically a clock train and an asynchronous control signal are assumed. The timestamp event is the reference or data event, whichever occurs first in the simulation. Refer to the waves in figure 18-1 below.

To clarify the difference, $recovery pertains to an asynchronous control event and the next clock edge following it; the timestamp is on the asynchronous control. Contrariwise, $removal pertains to the clock edge immediately preceding an asynchronous control event; the timestamp is on the clock.

$recovery. Recovery refers to recovery time from deassertion of an asynchronous control such as a set or clear, to effective occurrence of a clock edge. Given that the control has been deasserted, how much time must be provided to recover clock functionality? The time limit must be nonnegative.

For example,

```
            ref. event    data event       limit
$recovery(negedge Clr, posedge Clk, MinClr_Clk);
```

$removal. Removal refers to time of occurrence of an effective clock edge and deassertion of an asynchronous control. Given that a clock edge has occurred, how long after that edge does an asynchronous control have to remove itself in order for the clock edge to be ignored? The time limit must be nonnegative.

For example,

```
            ref. event    data event       limit
$removal(negedge Clr, posedge Clk, MinClk_Clr);
```

Note: $removal doesn't seem to work in the Silos demo version.

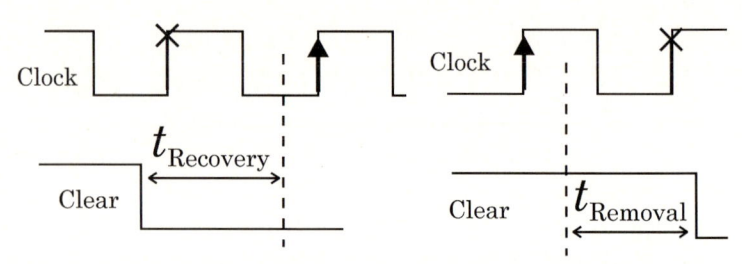

Fig. 18-1. Recovery and removal time limits. `Clear` is asserted high; violating edges are shown with 'x'; passing edges are shown with arrows.

$recrem. This check permits two time limits, <u>rec</u>overy limit first and then <u>rem</u>oval. When both limits are positive, it has the same effect as a $recovery and a $removal check, each with its respective limit. It allows for negative time limits. Because of error-prone complexity, routine use of this timing check is not recommended. See the discussion below, in section 18.1.1.4, on negative time limits.

Data Checks:

$width. This check verifies a minimum width of a pulse on a single variable. Whichever edge is provided is used as the timestamp event, and a timing violation is triggered unless enough time has lapsed before the opposite edge occurs. The width value must be nonnegative. A second timing parameter is optional, the glitch threshold, which suppresses the timing violation if the timestamped pulse is found to be narrower than specified by the glitch threshold.

For example,

```
            ref. edge       width      glitch thresh.
$width(posedge Reset, MinWid,    MinWid/10);
```

$period. This check triggers a timing violation if the same edge recurs on the specified variable within too short a time. The time value must be nonnegative.

For example,

```
            ref. edge       period
$period(posedge Clk,   MinCycle);
```

$nochange. This check requires an edge on the reference event (= timestamp) to define a level during which no data event (= timecheck) should occur. If a data event occurs during the reference level, a timing violation is triggered. Two offset times also are required; the first shifts the violation level start event (timestamp edge) and the second shifts the violation level end event (timecheck). The offsets may be negative; a positive value increases the duration of the violation window; a negative value decreases it.

nochange examples, showing `specparam` usage:

```
// Requires DBus constant during the entire positive phase:
           ref. edge    data event    lead shift    lag shift
$nochange(posedge Clk,     DBus,          0,            0);

// Violation starts 1 before negedge; ends 2 after posedge:
specparam MinSetup = 1, MinHold = 2;
$nochange(negedge Clk, EBus, MinSetup, MinHold);

// Shift the check 1 unit later:
specparam MinSetup = -1, MinHold = 1;
$nochange(negedge Clk, FBus, MinSetup, MinHold);
```

Note: $nochange apparently has been implemented recently only in the new VCS simulator.

18.1.4 Negative Time Limits

We have recommended against any use of the two timing checks, ***setuphold*** and ***recrem***, which permit specification both of positive and negative limits. This is because of complexity: A check on design timing should not be more complicated than the design; because, if it were so, an error in the writing of the check might cause the designer time lost over false violations or even perhaps an overlooked malfunction in the design.

However, when including a block of IP accompanied by a verilog model which lacks adequate internal timing checks, it may be necessary for the designer to add timing checks on variables at the boundary of the IP block. The internal variables may not be accessible. Under these circumstances, a skewed or even negative time limit may be necessary for the check.

To see why this might be so, as an example, compare the timing in regard to the reference event for simple setup and hold on an internal sequential element such as is shown in figure 18-2.

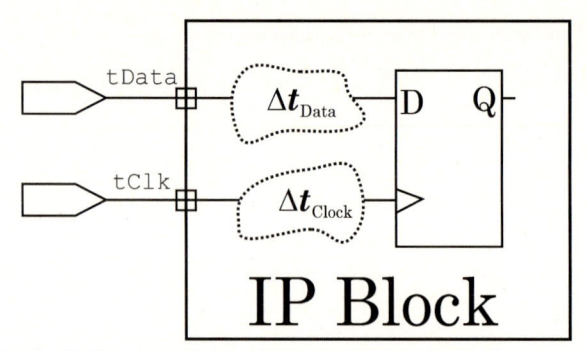

Fig. 18-2. An IP block with an inaccessible flip-flop. Delays within the block are shown as delta-delays. An accessible timing-check data event occurs on the block boundary at time `tData`; a reference event (clock) at `tClk`.

There are three different conditions possible in the figure 18-2 IP Block: (*a*) No differential delay through the IP logic up to the flip-flop; (*b*) additional delay on the flip-flop clock, perhaps because of a buffer tree; and, (*c*) additional delay on the flip-flop data, perhaps because of combinational processing.

The effects for mild skews are shown in figure 18-3.

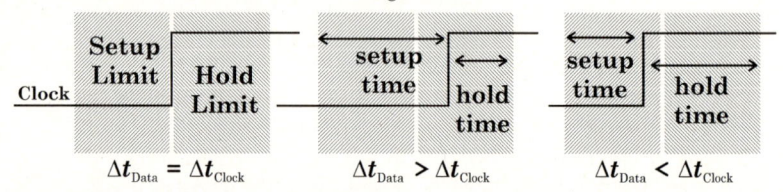

Fig. 18-3. Rationale for skewed time limits. Shaded regions represent fixed-width requirements of the sequential component. If the additional delays are equal, normal setup and hold timing checks have times equal to the limits. When the data are delayed more than the reference, the setup limit time must be increased and hold must be decreased; when reference is delayed more than data, the hold limit time must be increased and setup decreased.

When the additional IP delay exceeds a setup or hold limit, one of the times goes through zero and becomes negative. This is shown in figure 18-4.

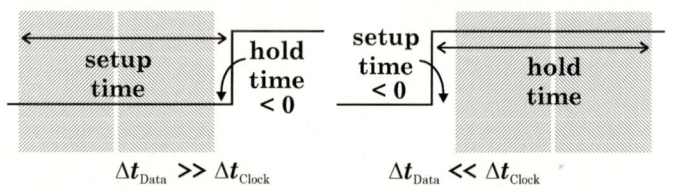

Fig. 18-4. Rationale for negative time limits. When the data or reference is delayed more than the setup or hold time limit for the isolated sequential component, the time at the IP boundary goes through zero and becomes negative.

Digital VLSI Design with Verilog   371

The present author argues against using $setuphold or $recrem, even though they permit direct entry of negative limits: When the internal delay differences are known, as they have to be to use $setuphold or $recrem negative values, the delay difference simply should be cancelled by delaying the timing-check data or reference event on a temporary net to cancel the difference. For example, if $\Delta t_\text{Data} \gg \Delta t_\text{Clock}$, then, delay the clock to the timing check by the difference, thus cancelling it, and use the databook requirement directly in the timing check.

Here is how to avoid negative time limits by adding an input delay:

```
wire ClockToHold;
assign #tCancel ClockToHold = Clock; // tCancel from IP vendor or experiment.
specify
$hold(posedge ClockToHold, DataIn, tHold);  // tHold from data book.
...
```

The tailored, delayed net (in the example, `ClockToHold`) will not be used elsewhere in the design and will be removed by the logic synthesizer (along with notifier `reg`'s, if any—see below). A delayed net may be used with the recommended timing checks above, even where it does not cancel the entire IP delay—just so long as the cancellation is enough to avoid the need for negative limits.

18.1.5 Timing Check Conditioned Events

Certain logical conditions are allowed with the variables named in a timing check. These are applied by the operator `&&&`, which logically is the same as `&&` in other contexts. The event must be "true" by `&&&` if the check is to be run. The operators, `==`, `!=`, `===`, `!==`, or `~` are allowed in the expression applied by `&&&`.

In a timing check, the conditioning expression RHS must be a scalar (1-bit) variable expression. A vector is allowed, but only the LSB value of such a vector will be used.

For example, here we don't want a violation if `Ena` is low while `D` changes:

    $setup(D&&&(Ena==1'b1), posedge Clk, MinD_Clk);

18.1.6 Timing Check Notifiers

It is possible to make a simulator's behavior depend upon a timing check. Perhaps, one would want the simulation to stop on a violation; or, maybe, in some cases a detailed message might be required, based on an assertion.

All timing checks allow at least one optional input in addition to those shown above, and the first optional input not given above always is the name of a ***notifier*** reg. This is a one-bit `reg` type declared visible to the timing check `specify` block, which is to say, in the `module` containing that `specify` block. This `reg`, of course, becomes part of the simulation model of the design. The net result is that anything possible on change of a `reg` value can be initiated by a timing violation through a `notifier`. Typically, a simulator system task such as $stop would be the only action a designer would want; or, perhaps an assertion failure might be programmed on notifier change.

When a timing violation is triggered by a timing check, the notifier value is toggled between '1' and '0'. If the notifier was 'x' when the violation occurred, it is toggled to '0'; if it was 'z', it remains 'z'. This last feature provides a way to disable notification without altering the timing check itself.

The reg passed to the timing check must have been declared, but it should not be used for anything but timing check related activities.

Example:
```
reg SetupNotify;
...
always@(SetupNotify) $stop;
  ...
specify
  ...
  $setup( D, posedge Clk, MinD_Clk, SetupNotify );
endspecify
```

Of course, each different timing check could have its own, different notification reg declared for it.

18.2 Pulse Filtering

Timing checks do nothing to alter the simulation, unless perhaps by the optional *notifier* feature. However, pulse filtering is an essential part of the simulator's behavior.

In normal, default inertial delay, pulses shorter than a gate delay are removed from the schedule of events on the gate input. In verilog, this default pulse filtering is viewed as occurring because two different time limit settings, the ***error limit*** and the ***rejection limit***, happen, by default, both to be set equal to the gate delay.

The verilog error limit always is greater than or equal to the rejection limit. If the rejection limit is different from (= less than) the error limit, three possible things may happen to an input pulse with an effect delayed on an output:

- (*a*) the pulse is wider than the error limit: The delay is imposed on the output, with the input width preserved;
- (*b*) the pulse width is between the error limit and the rejection limit: The leading-edge delay is imposed, and the output changes as though the input pulse had been narrowed to a width equal to that of the rejection limit; or,
- (*c*) the pulse width is below the rejection limit: It is cancelled, which is to say rejected, and it never affects the output.

These limits may be set by simulator invocation options or in an SDF file. We next shall show how they work by spending some time discussing the special specparam named *PATHPULSE*.

18.2.1 PATHPULSE Syntax

PATHPULSE is the only reserved word in verilog not in lower case; it changes the inertial delay behavior of the timing path involved.

In a peculiar twist of syntax, "PATHPULSE" is used in a `specify` block usually by allowing it to be prepended to the name of a timing path. The timing path is identified by expressing it in two $names: For example, "$In01$Out01" could represent the timing path between a module input named In01 and a module output named Out01. The full expression naming a PATHPULSE then would be PATHPULSE$In01$Out01.

The resulting expression then is used as the name of a `specparam`. For example, to apply PATHPULSE to the path between ports Ain and Bout, one might write,

```
specparam PATHPULSE$Ain$Bout = (R_limit, E_limit);
```

in which the designer-chosen term R_limit names the rejection limit value and E_limit names the error limit value.

It is possible to omit a path and write generically, "PATHPULSE$ = (R_limit, E_limit);" or "PATHPULSE$ = R_limit;". This assigns the limit(s) to every path in the entire module. When both this and a PATHPULSE naming a path are present, the one naming the path overrides the generic one on that path.

An example is shown in figure 18-5:

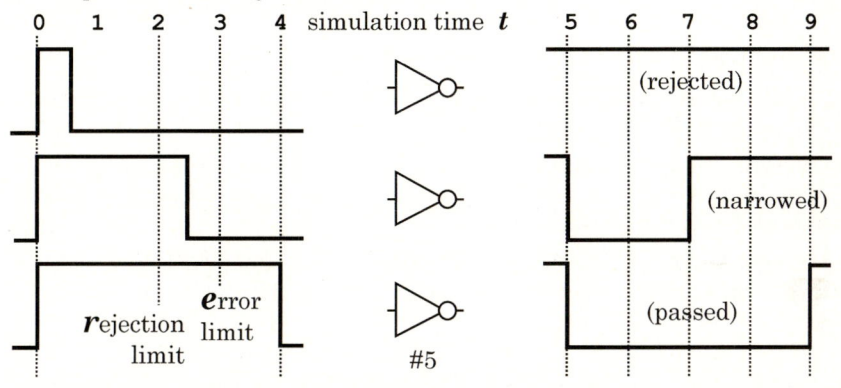

Fig. 18-5. Effect of PATHPULSE limits on pulse filtering. Assume PATHPULSE$ = (2, 3). With no PATHPULSE specification, all pulses would be rejected by simple inertial delay because of the gate delay of 5.

Notice that PATHPULSE, like other `specparams`, may be assigned more than one numerical value. Specifying the error limit is optional; assignment of one value in the line of code above would assign the rejection limit value only, the error limit value being taken to be equal to the gate delay involved. In PATHPULSE$*name1*$*name2*, the names in the path must be declared names and may not be bit-selects or part-selects.

Example:
```
module NewInertia #(parameter r_limit = 3, e_limit = 4:5:6)
                  (output Z, input A, B, C);
  ...
  specify
    ... (delays; timing checks) ...
    specparam PATHPULSE$     =   r_limit ;  // Module default.
    specparam PATHPULSE$B$Z = ( r_limit, e_limit );
  endspecify
endmodule
```

A `specify` block may contain full path delay statements, omitted from the example code above, which name multiple paths; when this is so, the "wildcarded" `PATHPULSE$` is applied only to the first path (first input to first output) in any such delay statement. Attempting to apply `PATHPULSE` selectively to any other path in such a delay statement, other than the first one, has no effect. Thus, `PATHPULSE` is most easily used for individual paths when those paths are one bit wide and when the timing specifications are parallel-path rather than full-path.

Note: As of 2011, `PATHPULSE` does not seem to work in Silos or QuestaSim, and it produces 'x' outputs instead of no change in VCS, when pulses are narrower than the rejection limit and the `+pathpulse` invocation option in force. Under these conditions, the VCS 'x' outputs are accompanied by warning messages.

18.2.2 `specparam` Improved Pessimism

Recall the discussion of delay pessimism in *Week 7 Class 2*. Pessimism improves chances of success, but it can be wasteful.

In addition to `PATHPULSE`, there are four other relevant reserved `specparam` types: `pulsestyle_onevent`, `pulsestyle_ondetect`, `showcancelled`; and, the negation of this last, `noshowcancelled`.

Unlike `PATHPULSE`, these are used by declaring lists of output variables assigned to them which are to be simulated with improved pessimism, which is to say, with increased early use of 'x' scheduling. They must be declared in the `specify` block before any path which is assigned a delay the expression of which includes one of the variables listed.

`pulsestyle_onevent`. This is the default behavior: When a contention yielding an 'x' exists, the contending events are scheduled normally; and, at the simulation time at which the contention first occurs, an 'x' is scheduled.

`pulsestyle_ondetect`. This is more pessimistic: As soon as the simulator has computed the contention, an 'x' is scheduled. This shifts the leading edge of the 'x' level to some time earlier than when it would have been had the default had been in force. Essentially, the output of a gate goes to 'x' as soon as an input event arrives which would cause that 'x'. The on-detect edge of the 'x' tells the designer how early a possibly

Digital VLSI Design with Verilog

uncontrolled state has occurred in the simulated design. In a big design, triggering a $stop assertion on detection, rather than on event occurrence, could save considerable debugging time.

showcancelled. Sometimes, different rise and fall delays may cause the leading edge of an 'x' event to occur at the same time as the lagging edge, or even before, creating a zero or negative duration of the 'x' level. By default, such events are cancelled silently by the simulator. Assigning an output pin to this specparam means that the simulator will schedule a zero or negative-width output pulse of 'x' at the original leading-edge output time; this pulse will be of the input-pulse width. If, in addition, this pin has been assigned to pulsestyle_ondetect, the output leading edge of the 'x' will be advanced to the detection time, widening the output 'x' pulse.

noshowcancelled. For debugging purposes, it is possible to select variables in a showcancelled list to be restored to default behavior by assigning them to the noshowcancelled specparam.

Note: None of these pessimism improvements seems to be implemented in any commonly-used simulator, although example usage is shown in the SystemVerilog (2012) section 30.7.4.2 on "Negative pulse detection". However, most simulators have invocation or interactive options providing functionality similar to the pessimism improvements listed here.

18.3 Miscellaneous time-Related Types

The verilog standard includes data types intended to be used in specify blocks. In addition to specparams, which are commonly used, these are the time and realtime types.

A **time** is an unsigned reg type of a predetermined width which is guaranteed to be at least 64 bits. It is meant to be used in testbenches or in conjunction with long-duration timing checks. Assigning the value of a time reg to a wide wire type yields a vector value which may be referenced in a specify block.

Of course, a user easily could declare a module-level reg[255:0] Time; if desired, increasing the width, in effect, without need for a possibly self-limited time declaration.

A **realtime** reg is identical to a real in all respects except name. This condition is reminiscent of that of the tri net, which is identical to a wire except in name.

Both time and realtime may be used anywhere in a module where a reg type is allowed. However, use of these types in design should be considered carefully. They add nothing to functionality and make the syntax a little more complicated. Instead of invoking a type with time-related associations, it is better to name the declared variables so that every use of their identifier recalls that they are time-related. Also, these types, having no unique functionality, may not be implemented in all design tools.

18.4 Timing Check Lab 21

Topic and Location: Timing checks and pulse filtering.

Work in the Lab21 directory. A subdirectory named PLLsync has been prepared for you there; it contains the entire PLLsync design from Lab10 (*Week 4 Class 1*).

Preview: We start with our old, unsynthesizable PLL design, not to continue the serdes project, but to use it as an example for the lab. After seeing how it runs for a long simulation, we add a specify block to the design so we can install timing checks. We use the PLL to study timing checks for the very common setup and hold constraints, as well as recovery, removal, width and period. We then switch to a simpler D flip-flop design, add a specify block to it, and systematically study pulse filtering and all the timing checks in the language runnable by our simulators.

Deliverables: Steps 1 - 4: specparam-controlled $setup, $hold, $recovery, and $removal checks simulated in PLLsync. Step 5: $width and $period checks simulated in PLLsync. Step 6: Various simulations of a D flip-flop (DFFC), with PATHPULSE filtering controls. Steps 7 - 10: Various timing check simulations, using DFFC.

Lab Procedure:

Be sure that your simulator is enabled to process PATHPULSE assignments and timing checks. These features are invocation options in VCS.

Recall that the old PLLsync design has the following component blocks: At the top level, PLLsync and Counter4; both of these are in .v files named for the modules. There is a testbench in PLLsync.v named PLLsyncTst. In a subdirectory named PLL, the PLL resides in five files: These are an include file and four others, the latter ones containing modules named PLLTop, ClockComparator, MultiCounter, and VFO.

The VFO in this model will run with its delay-adjusting functionality even if the applied clock frequency is constant; we shall use this to demonstrate timing checks.

Step 1: Preliminary simulation. Change to the PLLsync subdirectory. Using the testbench provided, run the PLLsync simulation for a very long time, say 25,000 ns. Notice that with a `VFO_MaxDelta of 2 (in the PLL include file), the VFO period oscillates within a 4 ns range. In your simulation, if you display the VFO_Delay variable value in VFO, you will see it switch in increments of 1 ns between 14 and 18 ns; this value should average 15.625 ns for a PLL clock with frequency of 32 MHz. This means that the counter counts with a delay of between 28 (= 32-4) and 36 (= 32+4) ns.

Step 2: Setup check. Suppose that we wish to use a positive edge on the Sample command to the VFO to sample the count value in the counter. Also suppose that we want to be sure to allow a setup time of 5 ns between these signals. Our recently renamed module, PLLsync, originates the Sample command as SyncPLL, and it also receives the counter count output as Behavioral.

So: Install a $setup check in a specify block in PLLsync which issues a violation at 5 ns. Make this limit depend on a specparam value. Simulate and watch the simulator console output window. Set the limit to 0 ns to disable this check.

Step 3: Hold check. Add a hold check, requiring 8 ns, to your specify block of the Step 2 timing problem. Simulate. After seeing the violations, set the limit to 0.

Step 4: Recovery and removal. Modify your testbench to apply two successive ClearIn pulses to PLLsync; they should be separated by 50 ns, and each should last 500 ns. The separation should be centered approximately on a clock edge around a simulation time of 1000 ns or so.

Add a recovery check of 100 ns for ClockIn so that it will cause a recovery violation from ClearIn on the falling edge of ClearIn; add a removal check of 100 ns on these edges. These values should trigger both timing violations. Simulate. Then, shorten the times so no recovery or removal violation is reported.

Step 5: Width and period checks. In PLLsync, add a width check to issue a violation when ClearIn is low for less than 100 ns. Simulate to see the result; then, disable this check by setting the time to 0. Add a period check to be sure that the PLL clock output period is not less than 30 ns, based on a positive edge. Disable this check after viewing the result.

That's all for the PLLsync design for now. For the rest of this lab, copy your DFFC.v model (D flip-flop with clear) from the Lab08 exercises (*Week 3 Day 2*) to the Lab21 directory.

Do the next Steps in order; they depend on one another.

Step 6: Pulse filtering. What happens when the input changes too rapidly?

To answer this, first change the DFFC model:

1. Remove the `timescale setting from the DFFC file; it's not necessary, because there will be one (see 3 below) in the testbench file. Change the old continuous assignment delays of #3 on Q and Qn to negligible but nonzero ones of #0.001 (1 ps).

2. Using mnemonically-named specparams, add a specify block after the module header declaration in DFFC, setting delays as follows:

Path delays from clock posedge: (*a*) To Q = 1000 ps rise and 800 ps fall; (*b*) to Qn = 1100 ps rise and 850 ps fall. Also add a full delay from clear to Q or Qn = 700 ps rise and 900 ps fall.

3. Instantiate the DFFC as a toggle flip-flop in the testbench (DFFC_Tst) which has been provided for you in the Lab21 directory. To make a toggle flip-flop, just wire the DFFC's Qn output to its D input.

The testbench in `DFFC_Tst` gradually decreases the clock period from 10 ns to 10 ps. It also provides a regular clear of 50 ns period and 50% duty cycle. It sets `` `timescale 1ns/1ps `` to resolve events on a 1 ps granularity.

A. Visual check. Simulate, being mindful that the float calculations will make this take a noticeable time, and examine the waveform. What is the clock frequency at which your D flip-flop's functionality becomes unreliable? Hint: `Q` and `Qn` both must work. How does this relate to the delays in the model? See figures 18-6 and 18-7 for one result.

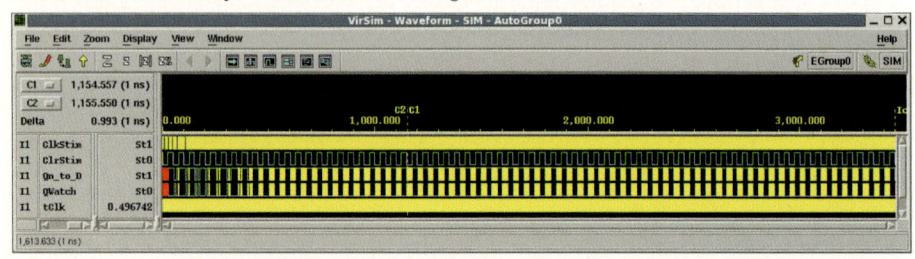

Fig. 18-6. Overview of the `DFFC` simulation with slowly increasing clock frequency.

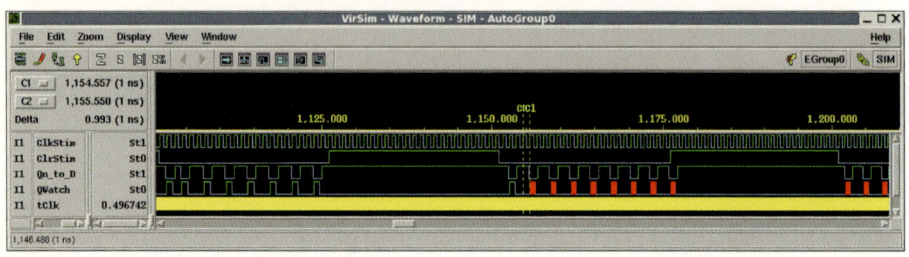

Fig. 18-7. Close-up of the location of the first timing failure in this `DFFC` simulation.

B. Reliability enforcement. Assume that we want an error limit of 1100 ps and a rejection limit of 500 ps for any pulse on any path to `Q` or `Qn` in our `DFFC` model. Add one or more `PATHPULSE` `specparams` to the `DFFC` `specify` block to accomplish this. Simulate to see the result; then, comment out the `PATHPULSE`(s).

Note: In this Step, new *VCS* will print a warning for each `PATHPULSE` error, as well as displaying an 'x' in the waveform. Also, a `$nochange` timing check has been added for this Step in the `Lab21_Ans` directory, just to show what the message looks like (assuming that your simulator can run `$nochange`).

Step 7: Width. We would like to be sure that `Q` and `Qn` stay high for at least 1 ns when they go high.

A. In DFFC, add `$width` timing checks for this. Simulate.

B. Change the minimum-width limit to 0.850 ns and simulate. This should produce a limited number of violations.

Digital VLSI Design with Verilog

C. Declare a new `reg` named `Notify` in `DFFC` and connect it to your `$width` timing checks as the fourth calling parameter; assign a glitch reject of 0 as a placeholder third calling parameter. Add the following `always` block below your `specify` block:

```
reg Notify;
...
specify
   ...
   $width(posedge Qn, twMinQQn, 0, Notify);
endspecify
//
always@(Notify) $stop;
```

Now run the simulation again. You can continue the simulation at your leisure, after each `$width` violation.

Change your minimum width `specparam` values to 0.500 ns to inhibit width violations, but leave in the width checks and `Notify` for now.

Step 8: Setup and hold. The pathological toggling at high clock frequency can be revealed several ways. One way is by adding setup and hold timing checks.

A. For the relationship between `Clk` and `D`, add a setup check for 1 ns and a hold check for 500 ps. Simulate (you may wish to stop early).

B. Connect the `$setup` and `$hold` notifiers to your `Notify` reg and simulate again. Decrease your setup and hold `specparam` values until all setup and hold violations vanish. Obviously, if the clock is going down to 10 ps, then it would be a good idea to start both limits here.

Note: the "10 ps" is in reference to the clock <u>period</u>, not to the half-period delay in the clock generator. What was the duration of the shortest one of each violation, and what was the earliest simulation time at which it occurred?

C. In the testbench, change the timescale to `10ns/1ns`. Run the simulation with setup and hold at the lowest limits which previously were causing violations. What happens? Notice how quickly (in wall-clock time) the simulation finishes with such a coarse time resolution.

Decrease the resolution limit to `10ns/100ps` and then to `10ns/10ps`, simulating each time. What is the effect of the resolution on timing violations and the way they are reported?

Restore the timescale to `1ns/1ps`, and disable the old `$width` check, as well as `$setup` and `$hold`, before continuing.

Step 9: Skew check. Add a `$skew` check between clock and clear, so that a violation will occur if clock ever goes high more than 49.99 ns after clear does. Simulate. Then, disable this check. As mentioned above, skew checks can't be disabled by assigning a 0; you will have to comment out the check or set the limit to 50 ns or more..

Step 10: Recovery and removal. These are perhaps the most easily misunderstood timing checks. They are designed to check on the relation of an asynchronous control, such as a set or clear, and a lower-priority synchronous control such as a clock.

A. Recovery check. Use `$recovery` to check when a `posedge Clk` has not been given at least 10 ps to recover after a `negedge` (deassertion) of `Clr`. Simulate. When does the first violation occur? Set the limit to 0 to disable this check.

B. Removal check. Use `$removal` to check that `Clr` has been removed (`negedge`) at least 10 ps before a subsequent `posedge Clk`. Simulate (see figure 18-8). Set the limit to disable the check when done.

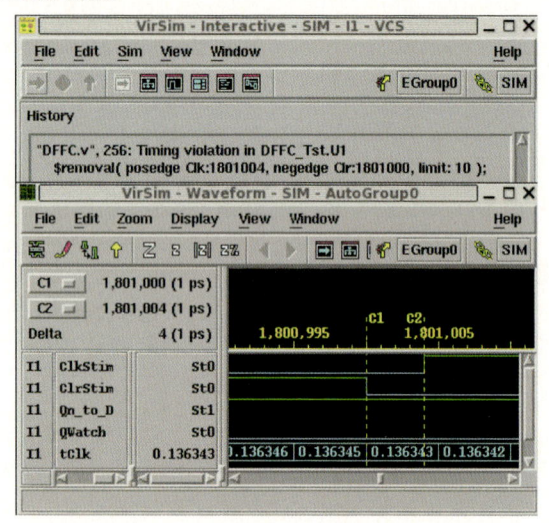

Fig. 18-8. A `$removal` timing-check violation, close-up.

18.5 Additional Study

Read Thomas and Moorby (2002) 8.1 and 8.4.4 for some insight into inertial delay.

Optional readings in Palnitkar (2003):

Read the section 10.3 on timing checks.

Do section 10.6, problems 6 - 8.

Chapter 19
Week 9 Class 2

19 Today's Agenda:

Lecture on sequential deserializer of our serdes
   Topics: Redesign of PLL for synthesis. Create blocks to prepare for the Deserializer.
   Summary: The "sequential" Deserializer is because our FIFO doesn't yet have a dual-port RAM; reads and writes have to be mutually exclusive (sequential; not concurrent). We review the status of our project and decide to set up the Deserialized now, but to complete it in the next lab. In the current lab, we prepare by redesigning the PLL to make it synthesizable and then setting up a new, permanent directory structure for the FIFO, PLL, and the Deserializer components (the deserializing decoder and serial receiver).

Lab on first-cut Deserializer
   Much of the work is in renaming and connecting blocks. However, we spend some effort in redesigning our PLL so it will operate about the same as before but will be synthesizable. We connect the PLL and our (single-port RAM) FIFO and then set up a divide-by-two clock for parallel data. We then connect the serial input in a testbench and run a simulation. In addition to redesigning the PLL, we make many small modifications to everything else. Finally, we synthesize for an idea of the design size and speed.

Lab Postmortem
   We focus on synthesis constraints and don't worry about the design functionality at this time.

19.1 The Sequential Deserializer

Let's review our serdes project and determine what remains to be done.

We introduced our serdes project in *Week 2 Class 2* and made significant progress on the SerDes PLL in that Week's PLL Clock Lab 6. Our progress included a serialization frame encoder in Step 8. We added serial frame synchronization in PLL Lock-In Lab 10 (*Week 4 Class 1*). We also completed a FIFO in FIFO Lab 11; however, this FIFO exhibits functionally incorrect synthesis because of unusual sensitivity lists and the resulting erroneous latch inference. Nevertheless, the FIFO did simulate correctly; and, because of this, we were able to refine the design of our deserialization decoder, DesDecoder, in Serial-Parallel Lab 16 (*Week 7 Class 1*).

In the present lab, we shall redesign the PLL for correct synthesis; we shall postpone revision of the FIFO for later labs. We also shall assemble and simulate the entire Deserializer by the end of the present lab.

So far, figure 19-1 shows where we are in the overall design, mostly from a data flow perspective.

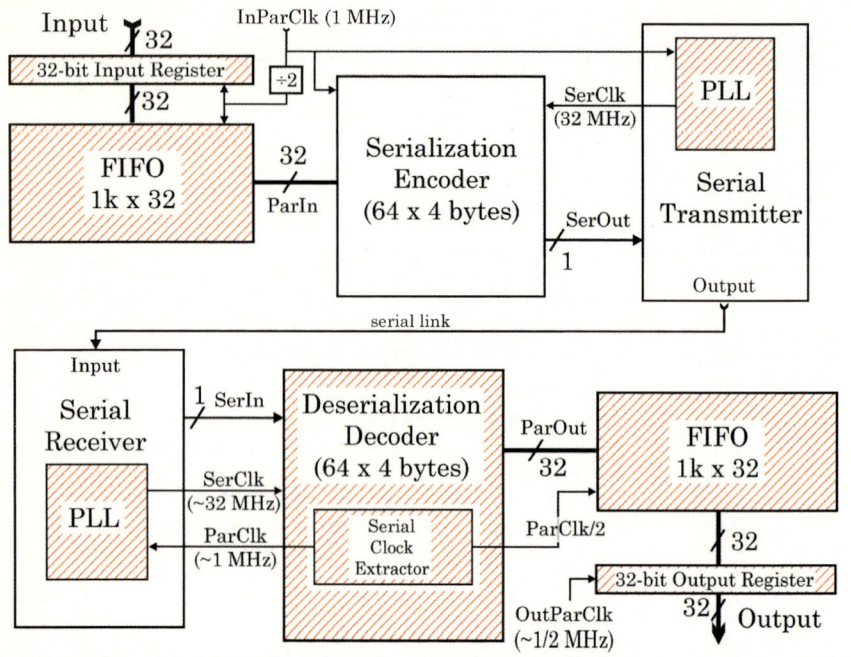

Fig. 19-1. Data flow and clock distribution in the serdes project. The upper half is the `Serializer`; the lower half, the `Deserializer`. Overall design updated as in Lab 16. Hatched blocks were completed separately in previous labs. None of the PLL, FIFO or DesDecoder will synthesize correctly, but all will simulate usably.

We'll first fix the PLL so that it will synthesize correctly; then, we can complete the `Deserializer` for simulation and almost-correct synthesis. After the PLL, most of the remaining tasks of today's lab will be organizational.

19.2 PLL Redesign

The main problem with the current PLL is in the VFO: It depends on programmable verilog delays in an oscillator implemented by a delayed nonblocking assignment statement. To get a predictable (but wrong) netlist out of this design, we added a preprocessor macro switch which forced the synthesizer to see a delayed blocking assignment instead of a delayed nonblocking one.

There is no really good way to design an analogue device such as a PLL in a digital language such as verilog. We can not create a variable capacitor, something that charges gradually on each clock, to control a VCO (variable-capacitance oscillator) for our PLL. To create such a thing, we would have to use a currently unsynthesizable, nonverilog

Digital VLSI Design with Verilog

language such as VerilogA/MS (see *Week 12 Class 2*). However, we can create a fast counter which adapts its count gradually on each clock.

19.2.1 Improved VFO Clock Sampler

Recall that we used a `Sample` pulse input, originating external to the PLL, in our 32x `Lab06`, `Lab08`, and `Lab10` PLL designs to reduce the rate at which the `VFO` frequency was adjusted. We did not use edge-averaging or any other technique for smoothing or refining the `ClockComparator`'s output. The sampling pulse idea, shown in figure 19-2, was not essential, but it did prevent the `VFO` from changing its frequency because of every tiny little clock misalignment.

The design looked like this:

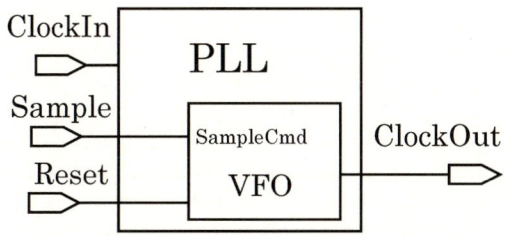

Fig. 19-2. The old `Lab06` VFO comparator-sampling command.

Keep in mind that our VFO both (*a*) receives an external 1 MHz `ParClk` input and (*b*) generates its own internal ~32 MHz `ClockOut`. To make our PLL self-contained, we shall not any more use an external sampling pulse for the VFO: We can build into the PLL itself a sampling clock which triggers a `Comparator`-based adjustment, once every cycle of one of the available, approximately 1-MHz clocks.

Looking ahead to synthesis, to generate a properly set-up sampling edge, we must allow our `Comparator` counters to settle at their current counts before each sampling. This set up may be provided easily by running the external input clock through a library delay cell (see figure 19-3) to retard the sampling edge for some reasonable, brief period. The counters then will be clocked by the other, VFO-generated, clock.

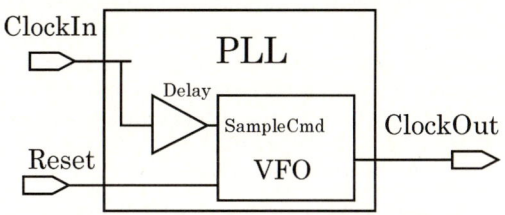

Fig. 19-3. The new VFO comparator-sampling command.

The verilog for the connection to such a library delay cell would be as follows:

```
module PLLTop (output ClockOut, input ClockIn, Reset);
  ...
  //                      delayed output     input
  Library_DelayCell DelayU1 (.Z(SampleWire), .I(ClockIn));
  //
  // (dont_touch DelayU1 synthesis directives)
  //
  VFO VFOU1 ( .ClockOut(MHz32), .AdjustFreq(AdjFreq)
            , .Sample(SampleWire), .Reset(Reset) );
```

19.2.2 Synthesizable Variable-Frequency Oscillator

Attempting to synthesize the old, Lab06 PLL will produce nothing usable. As already pointed out, we used a verilog macro to prevent the synthesizer from seeing the delayed nonblocking assignments and to substitute a nonoscillating delayed blocking assignment.

The netlist from the synthesized Lab06 VFO did have an oscillator, but it couldn't modify its frequency, and it rans at an unusably high frequency, depending on the synthesis constraints. Typical constraints produced a *nor* gate from our 90 nm library oscillating at a constant 6 GHz. The delayed blocking assignment workaround in the old PLL VFO amounted to an erroneously inferred latch and was synthesized to logic corresponding to the schematic of figure 19-4.

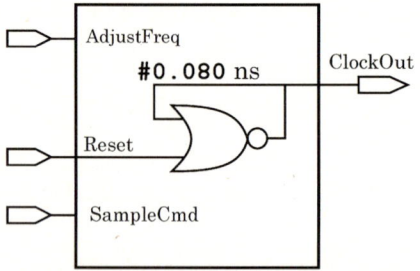

Fig. 19-4. The old Lab06 synthesized VFO netlist.

For a redesign which will guarantee synthesis, we shall use a high speed component available in our 90-nm synthesis target library: The simple, small, noninverting delay-cell buffer DEL005. We can implement a very fast oscillator based on a chain of DEL005 library delay cells. There will not be any verilog delay involved, except (*a*) as characterized from the library for the delay cells, and (*b*) as also characterized for the one additional inverting gate required to cause the oscillation.

To control the fast-oscillator frequency and use our redesign as the VFO, we can clock a fast counter and use the counter overflow as the PLL output clock. By varying the count at counter overflow, we vary the VFO frequency. Of course, this works because we require a frequency considerably lower than 6 GHz for a reliable fast counter.

Digital VLSI Design with Verilog 385

Fig. 19-5. The new, synthesizable VFO internal FastClock.

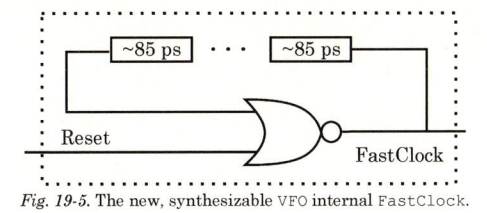

The frequency of our new fast counter oscillator is controlled by a delay line, which can be created of any configurable length by means of a verilog generate statement. Using several calibrated library delays of about 80 or 90 ps each will tend to reduce the delay variation in fabrication of the required inverter; different fabricated cells of the same type will tend to vary slightly in timing.

The oscillator output would be used internally by the VFO and is named FastClock in the schematic of figure 19-5. In that figure, we have replaced a simple inverter with an inverting *nor* gate, because the *nor* permits a simpler clock phase Reset initialization. The *nor* will be created by the synthesizer to implement an inverting continuous assignment statement as shown below.

The verilog for this oscillator follows:

```
reg FastClock;
wire WireToDelay, WireFromDelay;
assign WireToDelay = ~FastClock;  // oscillation here.
// ----------------------------------------------------
// A nor always block will allow initialization:
always@(WireFromDelay, Reset)
   if (Reset==1'b1)
        FastClock <= 1'b0;
   else FastClock <= WireFromDelay;
// ----------------------------------------------------
// The delays control the (fixed) fast oscillator speed:
LibraryDelayCell Delay0( .Out(Wire1), .In(WireToDelay) );
LibraryDelayCell Delay1( .Out(Wire2), .In(Wire1) );
    ...
LibraryDelayCell DelayN( .Out(WireFromDelay), .In(WireN) );
//(synthesizer dont_touch on all Delay* instances)
```

The delay line is shown unrolled in the code fragment above; it must be flagged with a synthesizer don't-touch to prevent its removal during synthesizer optimization. Because the delay line is an integral part of the VFO, the don't-touch is better done by a comment in the verilog rather than by a command in the .sct file. This verilog is entirely synthesizable either as unrolled in the code shown above or as generated.

We can make a VFO based on the new oscillator by tapping the FastClock and using it to clock a fast, programmable counter. Given this, we can use a selected, programmable count value chosen by AdjustFreq to define the VFO clock edges; this is shown in figure 19-6.

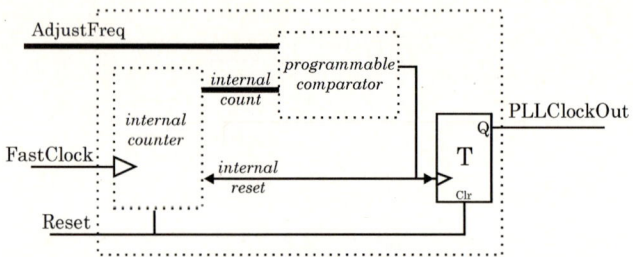

Fig. 19-6. The new VFO is based on an internal fast oscillator (FastClock) and a programmable counter.

The VFO internal comparator (not to be confused with the PLL ClockComparator module) stores the latched value of a programmable limit which is used to reset the VFO internal counter. An AdjustFreq adjustment is allowed to increment or decrement the programmable limit. The counter reset also toggles a flip-flop which creates the actual VFO output clock for the PLL.

The verilog would be somewhat like this:

```
// Assume a AdjustFreq vector declared which
//                   sets the VFO frequency:
reg[HiBit:0] Count;
reg PLLClockOut;
always@(posedge FastClock, posedge Reset)
  if (Reset==1'b1)
    begin
    PLLClockOut <= 1'b0;
    Count       <= 'b0;
    end
else begin
    if (Count>=AdjustFreq) // Programmable limit.
       begin
       PLLClockOut <= ~PLLClockOut;
       Count       <= 'b0;
       end
    else Count <= Count + 1;
    end
```

The AdjustFreq vector, of course, will be controlled by the AdjustFreq output of the ClockComparator block.

For our 32x PLL, we need a library fast enough to oscillate and count up to some small integer value with enough precision to vary the frequency of PLLClockOut reasonably around 32 MHz. If we assume a 16 ns half-period and 1-ns precision, we need a counter which can count to about 20 in 20 ns, which implies a 5-bit counter clocked at around 1 GHz. This speed is attainable easily in 130-nm or lower library technology.

We shall look at a synthesizable 1x PLL later.

19.2.3 Synthesizable Frequency Comparator

We also require a comparator which does not have a strange, latching sensitivity list. The current 32x PLL is clocked by the input, approximately 1 MHz, clock and samples the VFO (PLL output) clock independently in a change-sensitive block.

This was achieved as shown:

```
// OLD, unsynthesizable VFO:
always@(ClockIn, Reset)            // The input system clock.
  if (Reset==1'b1)
      ... (stuff) ...
  else if (CounterClock==1'b1)   // The PLL MultiCounter output clock.
          VarClockCount = VarClockCount + 2'b01;
      else begin
          case (VarClockCount) // The comparator object.
              2'b00: AdjustFreq = 2'b11;
              2'b01: AdjustFreq = 2'b01;
            default: AdjustFreq = 2'b00;
          endcase
          VarClockCount = 2'b00;
          end
```

Although this defined a PLL which simulated correctly, it is a classical case of latch inference which the synthesizer can not interpret meaningfully.

We can avoid latch inference entirely by using edge sensitivity. A reasonable improvement of the above, then, might be to write verilog for two very small counters, say, two or three bits each, one clocked by the external PLL 1 MHz input clock and the other by the PLL internal, approximately 1 MHz, counter overflow clock. The running counts could be compared on the edge of either clock to estimate the relative speed of the clocks. Depending on some measured difference between the two clock counts, we could decide whether to command an adjustment of the VFO frequency.

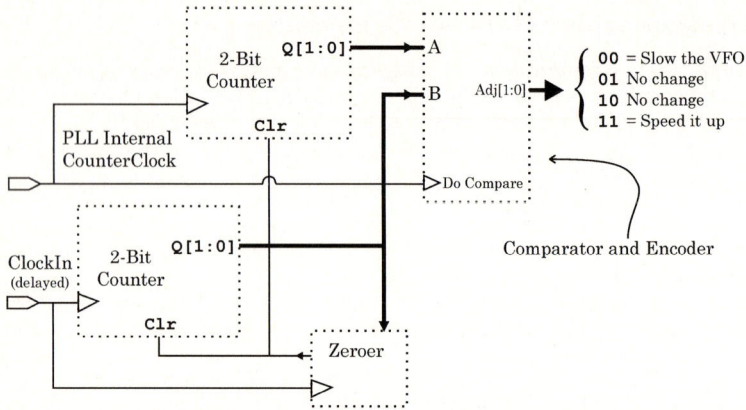

Fig. 19-7. The synthesizable, edge-sensitive `ClockComparator`. Dotted boundaries indicate functional blocks within the one module. The external `Reset` distributed to all sequential logic is omitted for simplicity.

In our implementation, two 2-bit counters, with possible output values of 0, 1, 2, or 3, count continuously (*a*) the positive edges of the 32x PLL `MultiCounter` output clock (about 1 MHz) and (*b*) edges of the input 1 MHz clock which is used as a stimulus for the PLL. On every positive edge of the PLL clock, a comparison is made of the relative edge-counts of the two counters, generating a continuously-updated adjustment decision to be used by the VFO. The `Zeroer` monitors the `ClockIn` count and reinitializes both edge counts simultaneously to 0, every time it receives a count of 3.

The VFO adjustment decision logic may be implemented with nested `case` statements located in an `always` block clocked by the PLL `MultiCounter` overflow clock.

This shows how:

```
case (ClockIn_EdgeCount) // The external ClockIn edges.
   2'b00: case (VFO_EdgeCount) // The MultiCounter overflow edges.
          2'b00: AdjustFreq = 2'b01; // No change.
          default: AdjustFreq = 2'b00; // Slow the counter.
          endcase
   2'b01: case (VFO_EdgeCount)
          2'b00: AdjustFreq = 2'b11; // Speed up the counter.
          2'b01: AdjustFreq = 2'b01; // No change.
          default: AdjustFreq = 2'b00; // Slow the counter.
          endcase
   2'b10: case (VFO_EdgeCount)
          2'b10: AdjustFreq = 2'b10; // No change.
          2'b11: AdjustFreq = 2'b00; // Slow the counter.
          default: AdjustFreq = 2'b11; // Speed up the counter.
          endcase
 default: case (VFO_EdgeCount) // Includes 2'b11 for initialization:
          2'b11: AdjustFreq = 2'b10; // No change.
          default: AdjustFreq = 2'b11; // Speed up the counter.
          endcase
endcase
```

19.2.4 Modifications for a 400 MHz 1x PLL

All the above verilog is meant for the 1 MHz, 32:1 PLL of our serdes project. But, what about the 400 MHz, unsynthesizable 1x PLL we discussed in *Week 2, Class 2* before Lab06?

The special concern is with the higher PLL output frequency. The internal counters described above will work unchanged, except that the VFO FastClock must be speeded up by limiting its delay chain to fewer delay elements. Also, the fastest available 90-nm library inverter should be instantiated structurally as part of the VFO oscillator.

In verilog, with `NumElems set to 3 instead of 5, the new FastClock generator becomes this:

```
reg FastClock;
wire[`NumElems:0] WireD;
//
generate
  genvar i;
  for (i=0; i<`NumElems; i = i+1)
    begin : DelayLine
    DEL005 Delay85ps ( .Z(WireD[i+1]), .I(WireD[i]) );
    end
endgenerate
//
always@(Reset, WireD)
  begin : FastClockGen
  if (Reset==1'b1)
      FastClock = 1'b0;
  else // The free-running clock gets the output of the delay line:
      FastClock = WireD[`NumElems];
  end
// The instantiated inverter:
INVD0 Inv75ps ( .ZN(WireD[0]), .I(FastClock) );
```

Our TSMC library DEL005 component provides a delay of about 85 ps. The new inverter, INVD0, lacks decision logic but is slightly faster. The required don't-touch directives are omitted for brevity.

With the sampling setup delay element in the containing PLL module, the synthesizable VFO adjustment is as above:

```
always@(posedge ClockIn, posedge Reset) //ClockIn delayed for setup.
  begin : FreqAdj
  if (Reset==1'b1)
      DivideFactor <= `InitialCount;
  else begin
      case (AdjustFreq)
        2'b00: // Adjust f down (delay up):
             if (DivideFactor < DivideHiLim)
                 DivideFactor <= DivideFactor + `VFO_Delta;
        2'b11: // Adjust f up (delay down):
             if (DivideFactor > DivideLoLim)
                 DivideFactor <= DivideFactor - `VFO_Delta;
      endcase // Default: leave DivideFactor alone.
      end
  end // FreqAdj.
```

The `InitialCount is set to half of the maximum count in the FastClock-created programmable counter.

In the faster 1x design, the Comparator decision logic can be biased to be more sensitive to real changes:

```
always@(posedge PLLClock, posedge Reset) //The delayed PLL ClockIn.
  begin : EdgeComparator
  if (Reset==1'b1) AdjustFreq = 2'b01;
  else
  case (ClockInN) // Count from the PLL external input clock.
    2'b00: case (PLLClockN) // Count from the VFO output clock.
            2'b00: AdjustFreq = 2'b01; // No change.
            2'b01: AdjustFreq = 2'b00; // Slow the counter.
            2'b10: AdjustFreq = 2'b00; // Slow the counter.
            2'b11: AdjustFreq = 2'b00; // Slow the counter.
            default: AdjustFreq = 2'b01; // No change.
           endcase
    2'b01: case (PLLClockN)
            2'b00: AdjustFreq = 2'b11; // Speed up the counter.
            2'b01: AdjustFreq = 2'b01; // No change.
            2'b10: AdjustFreq = 2'b00; // Slow the counter.
            2'b11: AdjustFreq = 2'b00; // Slow the counter.
            default: AdjustFreq = 2'b01; // No change.
           endcase
    2'b10: case (PLLClockN)
            2'b00: AdjustFreq = 2'b11; // Speed up the counter.
            2'b01: AdjustFreq = 2'b11; // Speed up the counter.
            2'b10: AdjustFreq = 2'b10; // No change.
            2'b11: AdjustFreq = 2'b00; // Slow the counter.
            default: AdjustFreq = 2'b10; // No change.
           endcase
    2'b11: case (PLLClockN)
            2'b00: AdjustFreq = 2'b11; // Speed up the counter.
            2'b01: AdjustFreq = 2'b11; // Speed up the counter.
            2'b10: AdjustFreq = 2'b11; // Speed up the counter.
            2'b11: AdjustFreq = 2'b10; // No change.
            default: AdjustFreq = 2'b10; // No change.
           endcase
    default: AdjustFreq = 2'b10; // No change; allows initialization.
  endcase
  end
```

In your Lab22_Ans directory, there is a subdirectory named PLL_1x_Demo containing the synthesizable version of the optimized 400 MHz, 1x PLL just described. This model does not, however do any decision averaging, and so it can not resolve delay times below that of one of its oscillator delay cells, (about 80 ps). This means that it never can lock in to a drifting clock, but it can come close.

Some 400 MHz waveform results are given in figures 19-8 to 19-11.

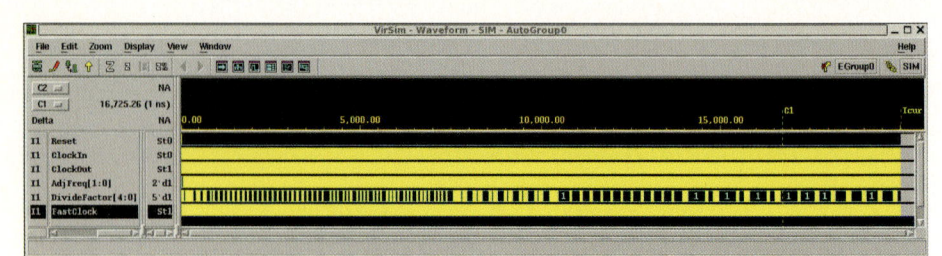

Fig. 19-8. The source PLL 1x verilog model: Entire 20 us simulation.

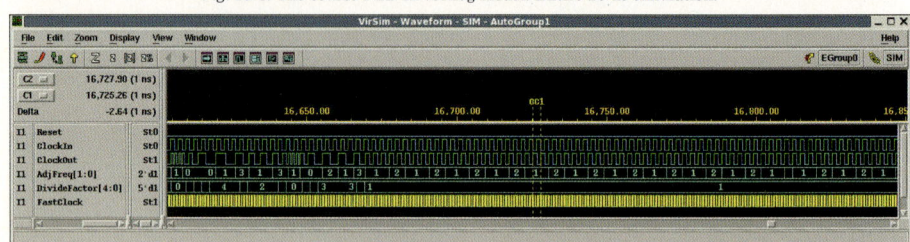

Fig. 19-9. The source PLL 1x verilog model: Lock-in detail near end of 20 us simulation.

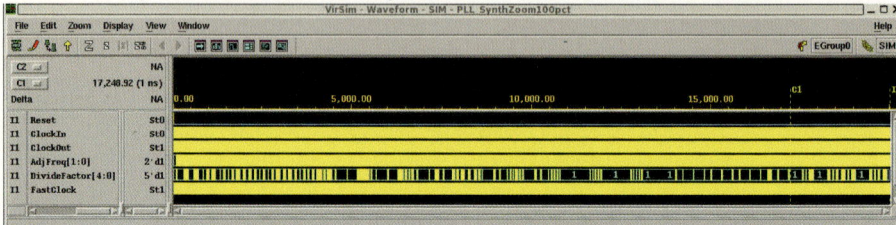

Fig. 19-10. The synthesized PLL 1x verilog netlist: Entire 20 us simulation.

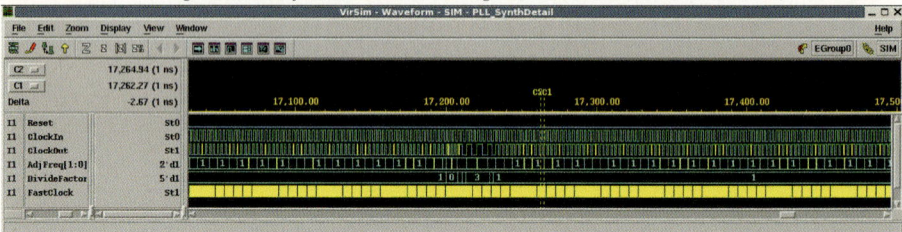

Fig. 19-11. The synthesized PLL 1x verilog netlist: Lock-in detail near end of 20 us simulation.

19.2.5 Wrapper Modules for Portability

As a small digression, before starting this lab, it should be mentioned that module I/O names will be very important to keep straight for the rest of our project. In your work as a designer, it may be necessary to adapt I/O names arbitrarily. If it is complicated or inconvenient to rename your own I/O's to match the required ones, a simple workaround is to instantiate your module in a "wrapper" module with the correct I/O names. Keep the file name the same, but add the wrapper module declaration above the functional one. Just connect all ports together, and the wrapper names will be the only ones visible in the rest of the design.

For example, suppose you have implemented and tested a module declared this way:

```
module MyModule(output[31:0] OutBus, ..., input ClockIn);
```

But, suppose the project required that this output port be named "DataBus" and the clock be named "Clock". To solve this problem, just rename your module slightly and put a wrapper in your MyModule.v file.

This is shown next:

```
// ---------------------------------------------
// This wrapper renames the MyModule ports as required:
module MyModule (output[31:0] DataBus
                , ...
                , input Clock
                );
  MyModule_ WrapperU1 ( .OutBus(DataBus), ... (1-to-1 wiring) ...
                      , .ClockIn(Clock) );
endmodule
//
// ----------------------------------------
// Begin original MyModule design (notice the underscore in the name):
//
module MyModule_ (output[31:0] OutBus, ..., input ClockIn);
  ... (valuable, tested functionality) ...
endmodule
```

In a large design, a wrapper is one of the few exceptions to the rule of declaring the contents of no more than one module in any one verilog source file.

19.3 Sequential Deserializer I Lab 22

Topic and Location: Completion of a first-cut Deserializer; synthesis.

Work in the Lab22 directory.

Preview: We reorganize and update our PLL design to make it synthesizable. We copy in our old PLL, FIFO, and old Deserializer files and organize them so they may be used easily throughout the rest of the serdes project. We install parameters for block reuse. We do considerable port renaming and other, minor modifications to connect all our previous blocks. We simulate when the `Deserializer` has been assembled fully; and, finally, we do some serious synthesis, attending to design rules and constraints.

This is a long and complex lab, realistic in what it does but more verbose than the usual commercial design process.

Deliverables: Step 1: A reorganized and renamed PLL design in its own subdirectory, `Lab22/PLL`. Step 2: A synthesized-netlist simulation attempt proving that the old PLL is not correctly synthesizable. Step 3: A redesigned PLL which simulates correctly. Step 4: A netlist simulation verifying that the new PLL synthesizes correctly.

Step 5: New `Deserializer.v` and `DeserializerTst.v` files in `Lab22`, with sketched-out ports and parameters as described.

Step 6: A parameterized and correctly simulating FIFO module in `Lab22/FIFO`, with a testbench in a separate file.

The next three Steps may be skipped to avoid time-consuming details. See the comments following the Step 6 instructions below.

Step 7: Two new subdirectories, `Lab22/DesDecoder` and `Lab22/SerialRx`, with contents and instantiations described, capable of being loaded into the simulator.

Step 8: A divided-by-two parallel-output clock generator in `Deserializer`.

Step 9: Completion of `SerialRx` by instantiation of the PLL; connectivity checked by loading into the simulator.

Step 10: A correctly simulating `DesDecoder`, slightly modified from a previous lab and with a testbench in a separate file.

Step 11: An expanded `DeserializerTst` testbench in `Lab22` which allows simulation, however imperfect, of a completed `Deserializer` instance.

Step 12: A `Deserializer` which occasionally correctly simulates storage and retrieval of data in the FIFO, and which can pass on serial data for parallelization.

Step 13: Two synthesized netlists of this first-cut `Deserializer`, one optimized for area and the other for speed.

Lab Procedure:

Step 1: Reorganize the `Lab21` version of the PLL.

Your `Lab22` directory contains an answer subdirectory, and a symlink to a file containing verilog models for the technology library we are using in this course.

Change to your `Lab22` directory; from the old `Lab21/Lab21_Ans` directory provided, copy the `PLLsync` subdirectory and all its contents (`cp -pr` in linux) to your new `Lab22` directory. The instructions in the present lab are somewhat complicated, so it will be best to use the answers provided rather than your own previous work.

The `Lab21` file organization was as shown in figure 19-12.

Digital VLSI Design with Verilog

```
        Lab21_Ans
            |
        PLLSync    PLLSync.v
            |      Counter4.v
            |
          PLL      PLLTop.v
                   PLLTop.inc
                   (rest of PLL)
```
Fig. 19-12. The old `Lab21` PLL file layout.

In the copied files, move everything in the new `Lab22/PLLsync` subdirectory up one level, into `Lab22`. In `Lab22`, delete `Counter4.v`, which won't be used any more, and delete the now-empty `PLLsync` subdirectory.

Rename `PLLsync.v` to `PLLTopTst.v`; rename `PLLsync.vcs` to `PLLTopTst.vcs`; also, move `PLLTop.inc` up to `Lab22`.

The `Lab22` reorganized files should be as shown in figure 19-13.

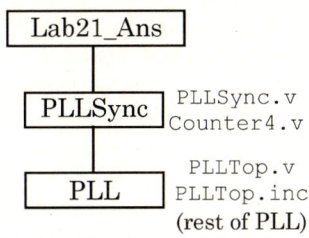

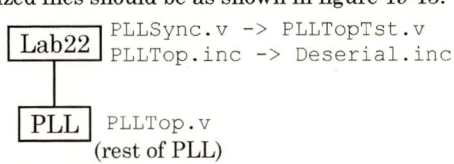

```
        Lab22    PLLSync.v -> PLLTopTst.v
            |    PLLTop.inc -> Deserial.inc
            |
          PLL    PLLTop.v
                 (rest of PLL)
```
Fig. 19-13. `Lab22` PLL reorganized layout.

Looking ahead, the design files in our `Lab22/PLL` directory will become those in the PLL subdirectory of our deserializer design; so, "Lab22" mostly will become "Deserializer". For this reason, now rename `PLLTop.inc` to `Deserial.inc`, and change the `` `include `` reference in `PLL/VFO.v` from `PLLTop.inc` to `Deserial.inc`. For simulation in VCS, don't put a path in `VFO.v`, because we plan to run VCS and DC from the `Lab22` directory, so `Deserial.inc` will be in current context whenever we compile `VFO.v`.

In the `Lab22` directory, edit `PLLTopTst.v` as follows:

- Delete all "`Step*`" defines, and change the `` `include `` to,
  `` `include "Deserial.inc" ``
- Delete the old `PLLsync` module, but keep its `Sample`-pulse generating `always` block, which now should be moved into the testbench module.
- Rename the testbench module to `PLLTopTst.v`.
- In the testbench file, the device under test now should be changed to an instance of `PLLTop`, and it should have an output pin named `.ClockOut`. The 1-bit wire mapped to `.ClockOut` should be named `PLLClockWatch`.

- The `PLLTop` instance should have a reset pin named `.Reset` and a `.Sample` pin driven by the `Sample`-pulse generating `always` block now located in the testbench. We keep this pin for now, although we already have decided to remove the `Sample` input of the old VFO.

Edit the `PLLTopTst.vcs` file so that it contains just the file name `PLLTopTst.v` and the (path)names of the verilog files in the `PLL` subdirectory.

You now should have a complete copy of the PLL design in your `Lab22/PLL` directory, with a PLL testbench in a separate file in `Lab22`. The `includes for the PLL should be located in `Lab22/Deserial.inc`.

Verify briefly that you can simulate the entire PLL by invoking and running the simulator of your choice in the `Lab22` directory.

Digital VLSI Design with Verilog

At the end of Step 1, your completed testbench should be somewhat like the one in the Lab22_Ans subdirectory for Step 2:

```verilog
`include "Deserial.inc"
module PLLTopTst;
reg   ClockStimIn, ClearStim, SamplePLLr;
wire  PLLClockWatch;
//
always@(ClockStimIn)
    #`HalfPeriod32BitBus ClockStimIn <= ~ClockStimIn;
//
always@(posedge ClockStimIn, posedge ClearStim)
  begin : Sampler
  if (ClearStim==1'b1)
        SamplePLLr = 1'b0;
   else begin
        #50   SamplePLLr = 1'b1;
        #100  SamplePLLr = 1'b0;
        end
  end
//
initial
  begin
  #0      ClearStim    = 1'b0; // #0 only required if 1'b1 init.
  #0      ClockStimIn  = 1'b0; // ditto.
  #1      ClearStim    = 1'b1;
  #9      ClearStim    = 1'b0;
          ClockStimIn  = 1'b1;  // Forces a phase change.
  #465    ClearStim    = 1'b1;
  #500    ClearStim    = 1'b0;
  #50     ClearStim    = 1'b1;
  #500    ClearStim    = 1'b0;
  #25000 $finish;
  end
//
PLLTop PLL_U1( .ClockOut(PLLClockOut), .ClockIn(ClockStimIn)
             , .Sample(SamplePLLr), .Reset(ClearStim) );
//
endmodule // PLLTopTst.
```

Step 2: Verify that the PLL is unsynthesizable to a correct netlist.

Change to your PLL subdirectory. Copy one of your .sct files from an earlier lab, say Lab15, into the Lab22/PLL directory. Modify the copy so that you can use it to synthesize the PLL. The PLL design at this point should consist of PLLTop.v, MultiCounter.v, ClockComparator.v, and VFO.v. Rename your .sct file to PLLTop.sct and edit it so it saves its synthesized verilog netlist into a file named PLLTopNetlist.v. Synthesize. The synthesis run should succeed; but, of course, the netlist can not be simulated correctly.

Change back to your `Lab22` directory to verify that the synthesized netlist does not simulate correctly. To do this, make a new simulator file named `PLLTopNetlist.vcs` in the `Lab22` directory with these contents:

```
PLLTopTst.v
PLL/PLLTopNetlist.v
-v verilog_library_2001.v
```

The `-v` library file, *verilog_library_2001.v*, is a symbolic link to our nonproprietary substitute of the library vendor's verilog models; you may have to copy it from the Extras `misc` directory, if the symlink is not valid. This library has been modified for these labs to have nominally realistic delays. The `-v` option causes the `-v` file to be processed by VCS as a library to resolve design netlist references; this means that the simulator will compile only those library modules used in the design.

VCS will be able to compile the resulting netlist; but, as we know, the synthesized VFO netlist can't adapt, because it will include a useless VFO more or less as in figure 19-14.

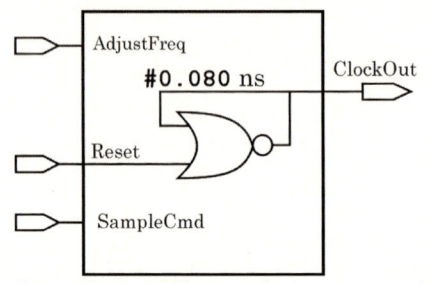

Fig. 19-14. (Same as figure 19-4) The reason for the incorrect synthesis of the copied VFO with verilog delayed blocking assignment. The incomplete sensitivity lists causes `AdjustFreq` and `SampleCmd` simply to remain unconnected.

Step 3: Rewrite the PLL so it becomes synthesizable.

Use the previous discussion in this chapter as a guide. You should rewrite substantially the `ClockComparator`, as described, and the `VFO`.

The `Sample` input port to the `PLL` should be removed, and a `Sample` pulse should be derived in `PLLTop` by passing `ClockIn` through a manually-instantiated delay cell and thence directly to the `VFO`, as shown above in figure 19-3. You should choose a verilog library delay cell simulation model in the verilog technology library file. Use an embedded synthesis script with don't-touch directives to preserve the delay cell instance(s), and their connections, from optimization.

The `AdjustFreq` code from `ClockComparator` to `VFO` should be modified so that 2'b00 requests a slowing down, 2'b11 a speedup, and other values no change; this will add some inertia to the new VFO's frequency adjustments.

Digital VLSI Design with Verilog

Some things to be alert for and avoid are (*a*) mixed edge and change sensitivity lists, and (*b*) `task` calls which have to be implemented with change sensitivity, only. The synthesizer will report these kinds of violation very explicitly.

> If you expect this exercise to be consuming more than an hour or two, and you are working in limited lab time, consider just copying the files in the `PLL` subdirectory from `Lab22_Ans` to your new `Lab22/PLL` subdirectory.

Your manually instantiated delay cell will have a delay of less than 100 ps; it would be advisable to set `` `timescale `` in your `.inc` file to `1ns/1ps`, for accurate simulation timing. The higher time resolution will slow the simulation somewhat; however, resolution of the exact value of the delay-cell delay is required for good verification.

A complete schematic of the synthesizable PLL might be helpful here:

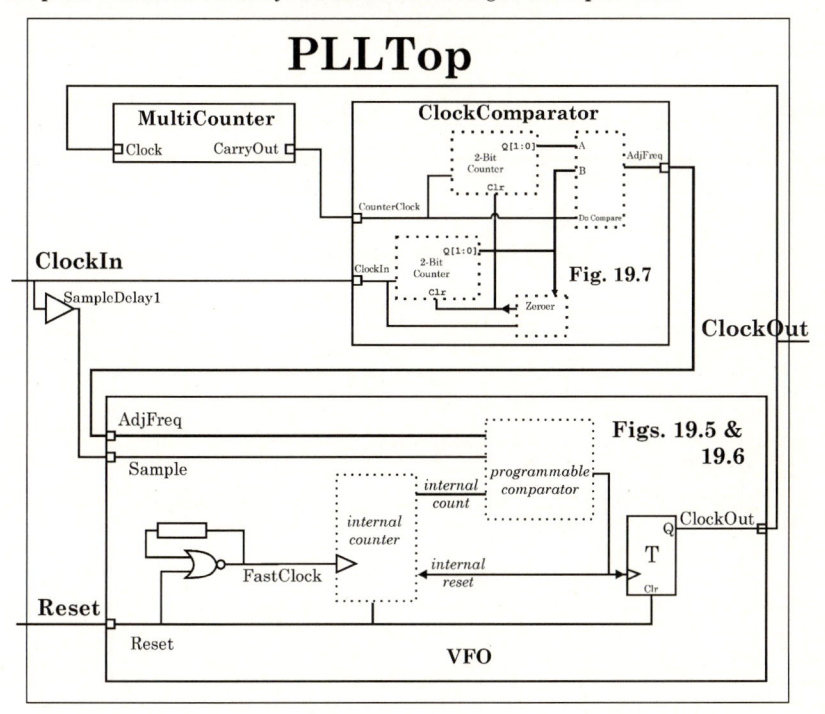

Fig. 19-15. Block-level schematic of the synthesizable PLL, with connection details. Some `Reset` lines omitted.

Step 4: Verify that the PLL now synthesizes correctly.

To do this, just synthesize `PLLTop` and simulate the resultant netlist by invoking VCS in `Lab22` again. If you have succeeded (see figure 19-16), you should find that the `PLL` netlist will generate a clock output about as good as did the verilog source design.

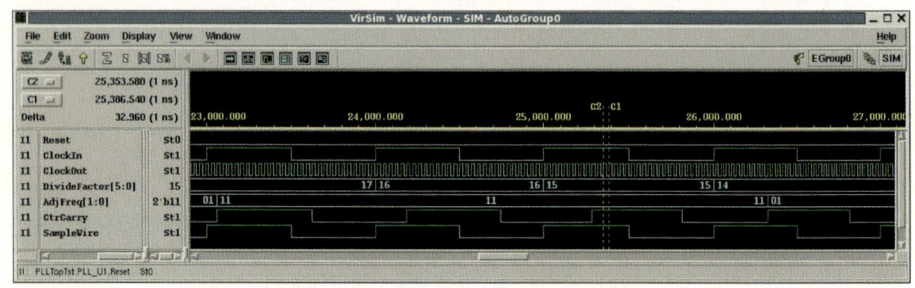

Fig. 19-16. Simulation of synthesized PLL netlist.

Step 5: Organization of the `Deserializer` block. Our goal in the next few Steps will be to implement a first-cut `Deserializer` as shown in figure 19-17.

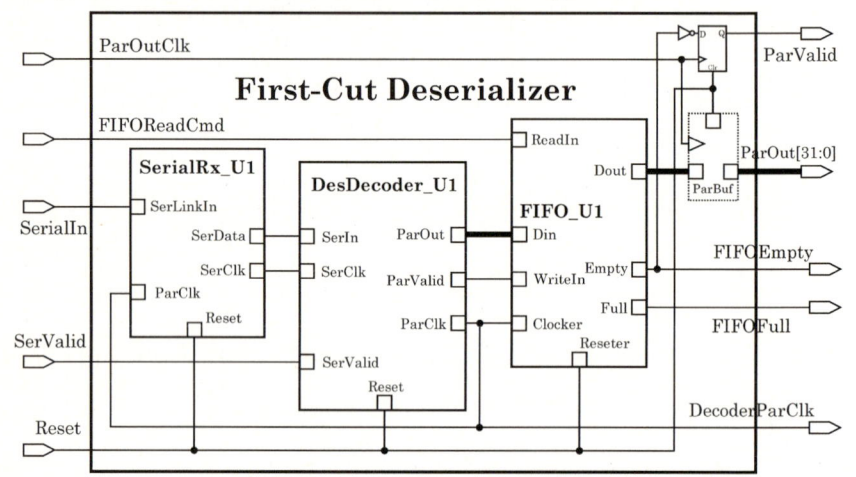

Fig. 19-17. Schematic of our first complete `Deserializer`.

The `Deserializer` of this step was introduced in figures 14-4 and 19-1.

Change to the `Lab22` directory. The `PLLTop*` files there are no longer useful and should be moved to the `PLL` subdirectory or deleted. In the `Lab22` directory, create a new module named `Deserializer` in its own file. Prepare an empty testbench, `DeserializerTst`, in a separate file.

In the `Deserializer` module, create empty-port instantiations of the FIFO (`FIFO_Top`), the `DesDecoder` (of Lab16), and a new serial receiver module named `SerialRx`.

The resulting top-level file organization should be as shown in figure 19-18.

```
                   Deserial.inc
      Lab22        Deserializer.v
        |          DeserializerTst.v
        |
       PLL         PLL_Top.v
                   (rest of PLL)
```

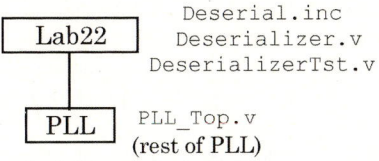

Fig. 19-18. Initial top-level file reorganization for the `Deserializer` implementation.

From the data flow figure 19-1 and the schematic figure 19-17, `Deserializer` should have one input named `SerialIn`, another input named `ParOutClk`, and a 32-bit output bus from the decoder to the `FIFO_Top` named `ParOut`. Create these ports.

Create an output port for the PLL's decoded incoming clock, which originates at the `DesDecoder` instance's `.ParClk` pin and goes to the `FIFO_Top` instance's `.Clocker` pin. To avoid confusion of this sending-domain 1 MHz clock with the input approximately 1 MHz parallel clock from the receiving clock domain, name the `Deserializer` connecting `wire` between these pins, `DecoderParClk` ("decoder's parallel-bus clock").

In view of the FIFO, we should require two more `Deserializer` outputs, `FIFOFull` and `FIFOEmpty`; these will not be useful to us in this lab, but they would be important in any system of which our serdes was part. Add these outputs.

We should add an external `SerValid` input in `Deserializer`, routed to the `DesDecoder` module, for test purposes. This would flag valid serial data from the sender. We also should have some sort of checking to generate a `ParValid` output from `Deserializer`. Finally, a `Reset` input should be provided to `Deserializer` so that the testbench can reset all the submodules.

Prepare to set the depth of the FIFO in `Deserializer` by declaring a `Deserializer` header parameter, `AWid` (= "Address Width" = width of register-file address bus). Assign a default value of 5 (for 2^5 = 32 words).

Set the width of the `Deserializer` output bus by another new header parameter, `DWid` ("Data Width"), with default value 32. This parameter then should be used everywhere in the design where the width of the parallel output bus appears. There is no special reason to have a 32-bit output bus as opposed to a 16-bit or 64-bit one, so, although we shall not alter the width in this course, our design will be parameterized to allow `DWid` to be changed to any other power of 2. We could, in principle, use a different value for our future `Serializer` input width, making the two ends of our serial lane independent in width. A very, very nice feature of serial connectivity . . . .

The `DeserializerTst.v` testbench file should include `Deserial.inc` for global timing definitions; add such an include directive to the testbench file.

Step 6: The FIFO. Create a subdirectory `FIFO` in the `Lab22` directory, and copy in the complete FIFO design from `Lab11/Lab11_Ans`; this should consist of `FIFO_Top.v`, `FIFOStateM.v`, and `Mem1kx32.v`. The `Mem1kx32` should be the one with separate input and output ports. Remove the timescale definitions from all these files.

In `FIFO_Top`, add `AWid` and `DWid` parameters as in `Deserializer` of the previous Step. The `DWid` parameter may be used anywhere in the FIFO design to determine the width of a register or of a parallel bus. Declare these parameters and replace every width and depth in the FIFO and its submodules (`FIFOStateM` and `Mem1kx32`) so that these parameters control the values.

The parameters `AWid` and `DWid` not only simplify a great deal of the code in the FIFO modules, but they also make it more convenient to change the size of the various FIFO elements; for example, they make it easier to change the FIFO size to one of 8 or 16 words instead of 32—or to change the FIFO word size from 32 to 16 or 64 bits.

Add 1-bit `Full` and `Empty` output ports to `FIFO_Top`, too; these will be wired to the corresponding new `Deserializer` ports of Step 5, as soon as the FIFO can be instantiated. Also, to simplify the `FIFO_Top` wiring, remove the old `E_FIFO` and `F_FIFO` nets; instead, substitute the new output port names (`Full` and `Empty`) into their continuous assignment expressions and the instance port maps. Then, delete the three nets, `ReadWire`, `WriteWire`, and `ResetFIFO` and replace them with the net names already used in `FIFO_Top` to assign to them. If your `Mem1kx32` instance is clocked with an inversion operator ("..., .ClockIn(~Clocker), ..."), remove the inversion operator ('~' or '!'), so that the memory responds to a rising-edge clock.

The `FIFO_Top.v` file includes a testbench. Parameterize this testbench with `AWid` and `DWid` as above. Be sure that the `FIFO_Top` instantiation has `.Full` and `.Empty` output pins, properly mapped in the FIFO testbench module. The testbench should have a `` `include `` of `Deserial.inc`. After these edits, separate the testbench into its own file, `FIFO_TopTst.v`, for possible future debugging.

You may have other testbenches commented out in `Mem1kx32.v` and in `FIFOStateM.v`; these testbenches should be deleted, because, if necessary, the `FIFO_TopTst` testbench may be used to exercise these submodules. If you find a bug in one of the FIFO submodules which requires a regression to a single submodule, you can recopy a local testbench from the earlier labs.

In `Mem1kx32`, add a `Reset` port and code which initializes all memory locations to 0. This will suppress startup parity errors when we use the memory later in simulation.

Finally, simulate with `FIFO_TopTst` briefly to verify that FIFO is complete and that the include file is referenced correctly.

Digital VLSI Design with Verilog

|--|

<u>Concerning Steps 7 through 9</u>: These next steps include some extremely detailed instructions. In part, especially in Step 7, they suggest a general approach to the rearrangement, duplication, and renaming of multiple module instances. This approach may be useful whenever redesign of a system including large, multiple instances becomes necessary.

> For the rest of this course, we shall be providing detailed instructions which, for literal and complete implementation, would be quite time-consuming. These instructions are intended primarily to provide important reading and coding experience for users with limited prior practice in writing verilog or SystemVerilog.

If the reader wishes to skip this part of the lab, it would be reasonable just to replace all `Lab22` data entirely by copying the `Lab22_Ans_Step10` contents to the working `Lab22` directory, saving only the `Lab22_Ans` directory and the `Deserializer.sct` and `tcbn90ghp_v2001.v` files already there. After this replacement, it would be a good idea to spend some time looking through and perhaps reading the new `.v`, `.vcs`, `.sct`, and `.inc` files before jumping ahead to Step 11.

|--|

Step 7: The `Deserializer` file structure. Create a subdirectory `DesDecoder` in the `Lab22` directory for the deserialization decoder, and another subdirectory, `SerialRx`, for the serial receiver.

Copy the `DesDecoder.v` from `Lab16_Ans` into the `DesDecoder` subdirectory. Remove the `` `timescale `` from `DesDecoder.v`, and pass the module header just one parameter, `DWid`. Change the name of the output bus from `ParBus` to `ParOut`, rename the `ParRst` port to `Reset`, and delete the testbench, if it still is present. Remove the unused `SHIFT` and `RESET` `localparams` if they are present in your copy.

Then, create a new file, `SerialRx.v`, in a new `Lab22` subdirectory named `SerialRx`. This file should contain just the following:

```
module SerialRx(output SerClk, SerData
               , input SerLinkIn, ParClk, Reset
               );
assign SerData = SerLinkIn;
//
PLLTop PLL_RxU1 (.ClockOut(SerClk), .ClockIn(ParClk), .Reset(Reset));
//
endmodule // SerialRx.
```

This finally shows where we shall instantiate our PLL: In the `SerialRx` module.

After all these changes, the resulting directory structure should put the `DesDecoder`, `FIFO`, `PLL`, and `SerialRx` blocks all directly under `Lab22`. This is shown in figure 19-19.

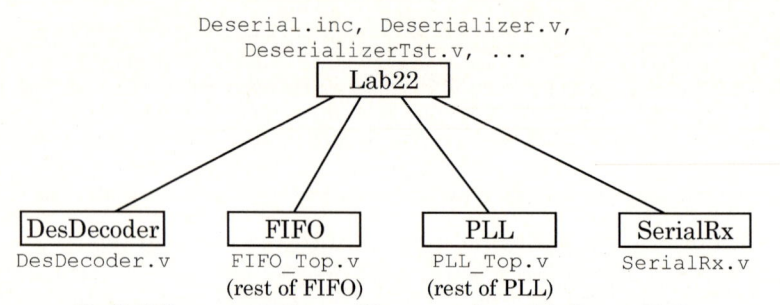

```
                    Deserial.inc, Deserializer.v,
                       DeserializerTst.v, ...
                         ┌─────────┐
                         │  Lab22  │
                         └─────────┘
            ┌───────────────┬────────┬───────────────┐
     ┌────────────┐    ┌────────┐ ┌───────┐    ┌──────────┐
     │ DesDecoder │    │  FIFO  │ │  PLL  │    │ SerialRx │
     └────────────┘    └────────┘ └───────┘    └──────────┘
     DesDecoder.v      FIFO_Top.v PLL_Top.v    SerialRx.v
                       (rest of FIFO) (rest of PLL)
```

Fig. 19-19. The new Lab22 directory. Files are named near their directory (block).

The Lab22 top level will become the Deserializer directory of our next lab exercise.

In the Deserializer module of Deserializer.v, instantiate the FIFO_Top module, and connect it with DesDecoder and SerialRx instances as well as possible for now, according to the design dataflow figure 19-1. Wire the SerialRx with port names as shown previously.

Because the width and name context of the DesDecoder and SerialRx instances strongly constrains their ports, create module templates for submodules of Deserializer.v first by wiring up the instance port maps, postponing module declarations for now. This way, you will have all connections immediately visible for your wiring. The result will look something like the next figure (before wiring):

Digital VLSI Design with Verilog

```
module Deserializer
    ...
// ------------------------------------------------
// Structure:
//
FIFO_Top #( .AWid(AWid), .DWid(DWid) )
FIFO_U1
   ( .Dout(), .Din()
   , .ReadIn(), .WriteIn()
   , .Full(), .Empty()
   , .Clocker(), .Reseter(Reset)
   );
//
DesDecoder #( .DWid(DWid) )
DesDecoder_U1
   ( .ParOut(), .ParValid()
   , .ParClk(), .SerClk()
   , .SerIn(), .SerValid(), .Reset(Reset)
   );
//
SerialRx
SerialRx_U1
   ( .SerClk(), .SerData()
   , .SerLinkIn(), .ParClk()
   , .Reset(Reset)
   );
//
endmodule // Deserializer
```

Then, compose the `module` wiring declarations by including them (illegally) in the instance port maps, using ANSI format. For example, in the code above, you might next change the above `FIFO_Top .Din()` to `output[Dwid-1:0] .Din(DecodeToFIFO)`, and the `DesDecoder .ParOut()` to `output[DWid-1:0] .ParOut(DecodeToFIFO)`.

Do this any way you want, but developing some sort of systematic way to define module ports and declarations top-down is an important skill in VLSI design.

One idea would be to start by declaring modules where instances will be wired. For example, to instantiate the FIFO in `Deserializer.v`, start by cut-and-pasting this into your `Deserializer.v` file:

```
module FIFO_Top #(parameter AWid = 5    // FIFO depth = 2**AWid.
                ,              DWid = 32 // Default width.
                )
     ( output[DWid-1:0] Dout(wire[DWid-1:0] FIFO_Out)
     , input[DWid-1:0]  Din(wire[DWid-1:0] DecodeToFIFO)
     , output Full(wire FIFOFull), Empty(wire FIFOEmpty)
        ...
     );
```

The port map wiring shown is illegal, because it includes width declarations copied directly from the ports. However, this is just an intermediate step.

Continuing this approach, you might next copy these combined instance-wirings plus declarations into their separate, proper .v files in the Lab22 subdirectories. After copying, the instance wirings in the submodules may be removed and the ports renamed or resized as necessary.

Finally, alter the code in the Deserializer module to change the module port declarations to simple instance pin-outs, leaving this in your Deserializer.v file:

```
FIFO_Top #( .AWid(AWid), .DWid(DWid) )
FIFO_Top_U1                    // Instance name.
  ( .Dout(FIFO_Out)            // pin-out (= port map).
  , .Din(DecodeToFIFO)
  , .Full(FIFOFull), .Empty(FIFOEmpty)
       ...
  );
```

Returning to the submodules, add reset inputs to all the submodule declarations, as well as any useful read, write, full or empty communication with the FIFO instance. Err on the side of adding ports; you always can remove unnecessary ports later.

Be sure to use parameters from the Deserializer declaration in all instance width declarations; pass the main submodules only DWid and/or the FIFO depth.

Finally, with only the FIFO and deserialization decoder actually defined, create a DeserializerTst.vcs file and load your mostly-unimplemented Deserializer.v structural design into the simulator, just to compile and check connectivity. Correct connection problems before continuing. When in doubt, add another port.

Step 8: The parallel output buffer (ParBuf) in Deserializer. For now, we shall implement the output register in figure 19-17 simply by declaring a reg named ParBuf in the top-level module, Deserializer.

Although our PLL generates a 1 MHz parallel-data clock from the serial stream, our packet format requires 64 bits per 32-bit word; this means that our serial line can deliver parallel-data words only at about 500 kb/s.

Therefore, we shall assume an external 1/2 MHz clock ParOutClk to clock data into this buffer this way,

```
always@(posedge ParOutClk, posedge Reset)
  begin : OutputBuffer
  if (Reset==1'b1)
      ParBuf <= 'b0; // To be wired to the ParOut port.
  else ParBuf <= FIFO_Out;
  end
```

The ParOutClk may be generated in the testbench module. The preceding OutputBuffer code says that when the FIFO is reset, only zeroes can be read from it. At

Digital VLSI Design with Verilog 407

this point, we also should provide a `ParValid` flag; this can be set in the same `always` block with `ParBuf`.

Step 9: The deserialization decoder. To begin, change to the `DesDecoder` subdirectory. The contents of your `DesDecoder` module should be copied from `Lab16/Lab16_Ans`, with the module I/O changes already made above (in this lab).

Make a second, new copy of the `Lab16_Ans/DesDecoder.v` file, renaming it to `DesDecoderTst.v`, and remove the design module; keep the testbench module, which originally was at the bottom of the file. In this new testbench file, change the timescale to 1ns/1ps, remove the `DC` compiler directives, and change the parameter references to `DWid`.

Also, in the new `DesDecoderTst.v` file, if the serial data to be shifted to the `DesDecoder` instance is shifted LSB first, reverse the order, for consistency, so that the data MSB goes out first. All data in this design should be shifted MSB first. Remove all delay expressions from the testbench, except those in the main `initial` block and in the serial clock generator.

Now, modify `DesDecoder.v` as follows:

- In case we might want a command from here to the FIFO, add a 1-bit output port named `WriteFIFO`.
- Remove all *#delay* expressions and rephrase or remove all comments concerning these delays.
- To put this module better under procedural control, change the runtime so that instead of four concurrent `always` blocks, there are just two: The `ClkGen` block, which should be sensitive to change in `SerClk` or `Reset`, and a new, second `always` block.

    The new `always` block should be done as follows:
    ```
    always@(negedge SerClock, posedge Reset)
      begin
      Shift1;
      Decode4;
      Unload32;
      end
    ```
    The sensitivity to the negative edge of `SerClock` is to avoid a race with the serial data shift.

- The `Unload32` task, as written for Lab16, includes an unnecessary edge-sensitive event control; replace this with a simple `if` test for `ParClk==1'b0`.
- The `ClkGen` task should be cleaned up a little more: Instead of having an `if` in the non`Reset` condition controlling nothing but serial clock gating, put everything, including the other `if`, under control of `if (SerValid==1'b1)`.

After these changes, make up a `DesDecoderTst.vcs` file and simulate `DesDecoder` alone, just to be sure that the new `DesDecoder` testbench connections are correct. Because of the complexity of this module, some further debugging may be useful at the unit level; but, this module should be almost usable without changes other than ones to permit synthesis, which will be made later in the lab. Additional modifications to make the `DesDecoder` synthesize correctly will be done later in the course.

Step 10: The `DeserializerTst` testbench. This may be based on the one from the `DesDecoder` of Lab16, although, if so, it will have to be modified to make the complete `Deserializer` work. The main problem with the Lab16 version is that there can be no serial clock in the `Deserializer`, whereas there was a testbench serial clock in the `DesDecoder` of Lab16. In the present lab, the `Deserializer` has to extract the parallel clock from the serial input stream and then to use the extracted clock to generate a serial clock synchronized with that same input serial stream.

So, we shall have to go easy on the `Deserializer`, at least at first. We shall use the testbench to feed `Deserializer` a serial stream at a clock rate only very slightly different from the base rate of the PLL. This will make extraction of the clock easy enough for us to validate the rest of the design. Later, we can see how far we can go in providing a serial stream farther off the deserializer's PLL base clock frequency.

With all this, the testbench *should* create a padded serial packet just as it did in Lab16; however, let us set the testbench serial clock half-period at 15.6 ns, which is only slightly off the `PLL` base rate of 500/32 ns = 15.625 ns = 15 ns after verilog truncation in integer division. The farther away from 15 ns we go, the more serial data the `Deserializer` will have to drop because of loss of synchronization, at least at this stage in the design.

Use the testbench to generate a separate 1/2 MHz clock to attempt reads from the `Deserializer` FIFO output register. Don't worry now about synchronization of `Deserializer` output reads with valid data in the `ParBuf` register. Then, instantiate `Deserializer` in this testbench and simulate it. It should be possible, as shown next in figures 19-20 and 19-21, to obtain some good data for the FIFO, although there will be many losses of synchronization.

Digital VLSI Design with Verilog

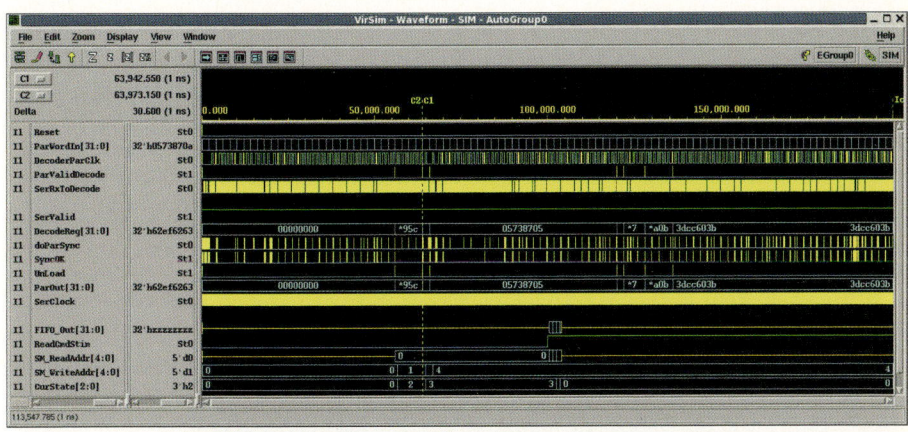

Fig. 19-20. Overview of source simulation of first working `Deserializer`.

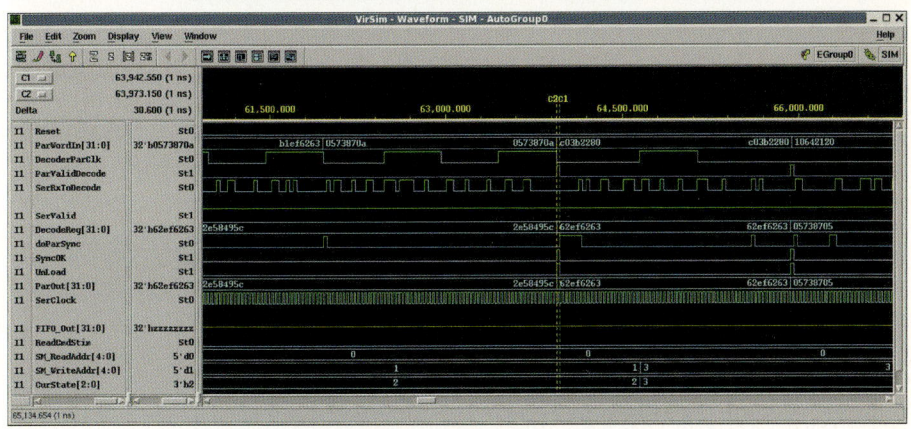

Fig. 19-21. Close-up of a parallel-bus unload of the first working `Deserializer`.

One further point: The new VCS may produce a first working source simulation with different output values than those shown here.

Also, the on-disc answers for this Step include some changes in the source verilog which have been made to prevent the new VCS from setting certain signals to 'x' where the old VCS did not. The main change is to comment out the FIFO state machine assignment of 1'bz to the `ReadCounter` when the FIFO is in its `empty` state.

Step 11: Tying up loose ends. When you have the `Deserializer` source verilog design more or less working (it should be generating a parallel clock and at least occasionally correctly parallelizing the serial data), refine the design as follows:

A. Note: This Step has been done already in the on-disc answer for Step 10. Anyway, remove the `DesDecoder`'s `WriteFIFO` output; this was completely superfluous. The serial-parallel input functionality will decide when to write to the FIFO, and it will be up to the external system to stop sending serial data if the FIFO should become full. Instead, in `Deserializer`, connect the `DesDecoder` instance's `ParValid` output to the `FIFO_Top` instance's `WriteIn` port.

B. Simulate. Your FIFO has no read requested, so it should fill up. If you used your own implementation, you may have to debug the FIFO to do this.

Run the simulation for enough time to fill the FIFO, and then to empty it at least once. To empty it, you will have to issue a read command in the testbench after the FIFO is full: Assert `FIFOReadCmd` to do this. See figure 19-22.

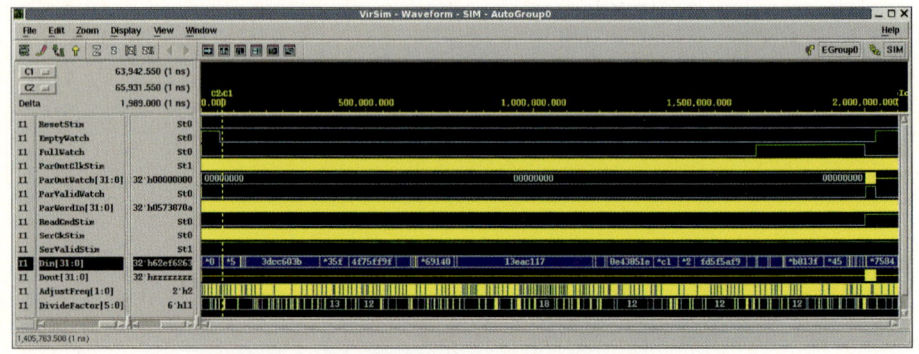

Fig. 19-22. Overview source simulation of a somewhat improved `Deserializer`.

C. Check the simulated transferred parallel data against the `Deserializer` testbench input for correctness, allowing for dropped serial words, if any. One implementation yielded the result shown in figure 19-23.

Digital VLSI Design with Verilog 411

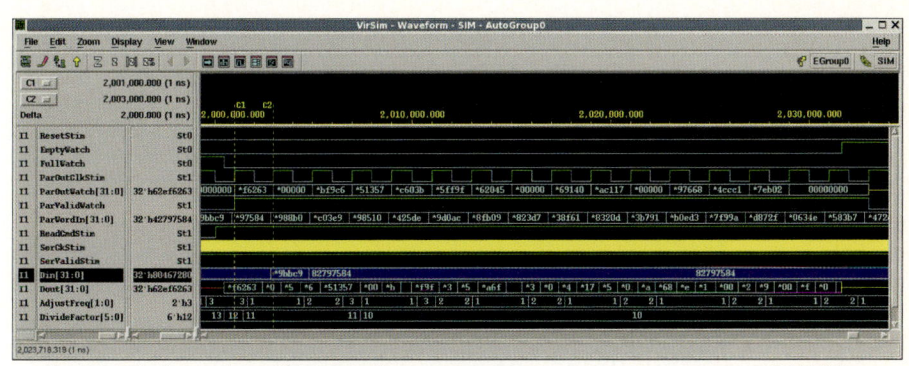

Fig. 19-23. The first word out of this improved `Deserializer` was `32'h62ef_6263` on `Din` at *t*=63,942,550 ns; however, it was copied to the `ParWordIn` bus, and stored in the FIFO, as `32'hb1ef_6263`. Clearly, there is more work to be done here.

Do not worry now about corner cases, but your FIFO should store data so that the value written to at least two different given addresses is the value later read out.

Recalling Step 8 above, typical code in `Deserializer` to read into the parallel output buffer would be,

```
always@(posedge ParOutClk)   // 1/2 MHz clock domain of the receiving system.
  if (Reset==1'b1 || F_Empty==1'b1)
     begin
        ParBuf      <= 'b0;    // Zero the output buffer.
        ParValidReg <= 1'b0;   // Flag its contents as invalid.
     end
  else begin  // Copy data from the FIFO:
        ParBuf      <= (FIFOReadCmd==1'b1)? FIFO_Out : 'b0;
        // Flag the buffer value validity:
        ParValidReg <=  FIFOReadCmd && (~FIFOEmpty);
     end
```

Two possible problems we shall address later: (*a*) Were FIFO addresses `0x00` and `0x1f` written and read? (*b*) Can the FIFO be read from and written to at the same time, so that the FIFO doesn't quickly fill up or go empty?

For new VCS, the specific data I/O values in the on-disc answers will differ from those above. For example, the first valid word out with new VCS might well be `32'h0573_8705`, as shown in the next figures, corresponding not-quite-correctly to the `ParWordIn` value `32'h0573_870a` at simulation time 61,947 ns.

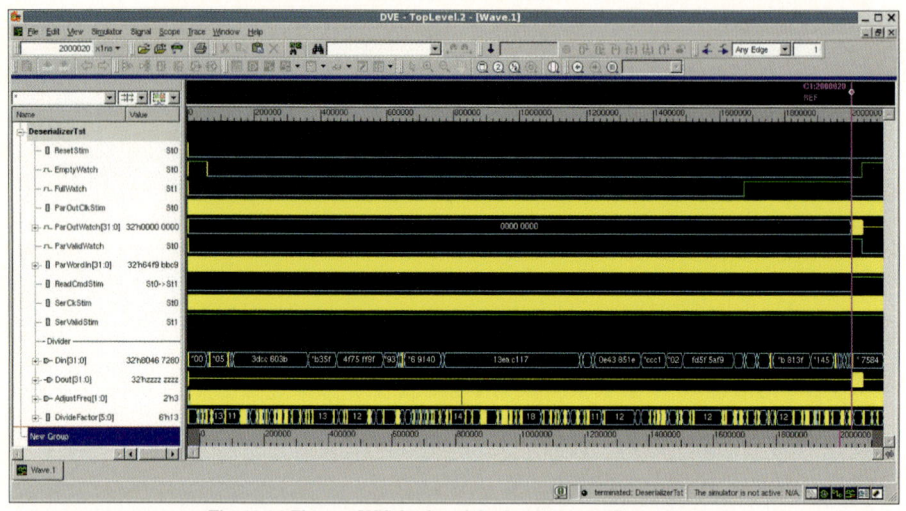

Fig. 19-24. The new VCS display of the first operating Deserializer.

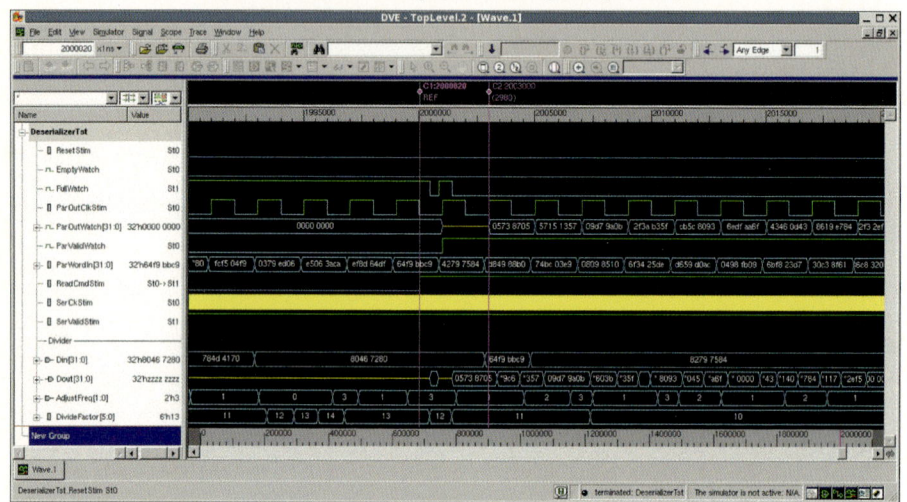

Fig. 19-25. The zoomed-in view of the new VCS Deserializer output (not quite correct).

Step 12: Synthesis. Attempt to synthesize the entire `Deserializer`. At this point, the goal is just to get some kind of netlist written out. You may have to modify certain parts of the design or conceal them from the synthesizer in `` `ifdef DC `` blocks. The `DesDecoder` may require several changes.

Digital VLSI Design with Verilog   413

Recall that the PLL now is synthesizable, but that the FIFO is not. The FIFO part of the netlist will be full of missing connections and dangling output pins.

To reduce the synthesis time, use a script which includes no design rule or speed constraint; just optimize for minimum area.

The resulting hierarchical netlist should total over 35,000 transistor-equivalents and may seem to be intact; however, it will not be functional, so don't bother trying to simulate it.

After this, use the same synthesis script again; but, just before the command to run the compile, insert a command to flatten the design. The synthesizer now will be able to remove unconnected logic across hierarchy boundaries, reducing the netlist size further.

So, we have a reasonable idea of the gate-size of our Deserializer, and we can see the extent to which flattening might reduce this size.

19.3.1 Lab Postmortem

How can one constrain synthesis for minimum area vs. minimum delay?

What is the difference between synthesis constraints and design rules?

Think about local delay or area minima and the multidimensional nature of the constraints.

What is incremental optimization?

19.4 Additional Study

Optional: Review parameter and instantiation features in Thomas and Moorby (2002) sections 5.1 and 5.2 (ignore defparam).

Chapter 20
Week 10 Class 1

20 Today's Agenda:

Lecture on concurrent deserializer lab
This brief lecture merely introduces the lab, which includes simulation and synthesis, the latter primarily as a verification tool.

Lab
We debug our first-draft Deserializer and make many small changes to modify the memory so it will operate as a dual-port. We change our FIFO state machine so it can take advantage of the dual-port memory.

Lab Postmortem
We stop for Q&A and discuss a few verilog debugging techniques.

20.1 The Concurrent Deserializer

In this chapter, we'll improve our deserializer implementation and shall assemble a complete serial receiver with imperfect (incomplete) functionality. The main change will be modification of the FIFO to permit concurrent (dual-port) read and write.

Figure 20-1 gives our working-sketch schematic of the top level:

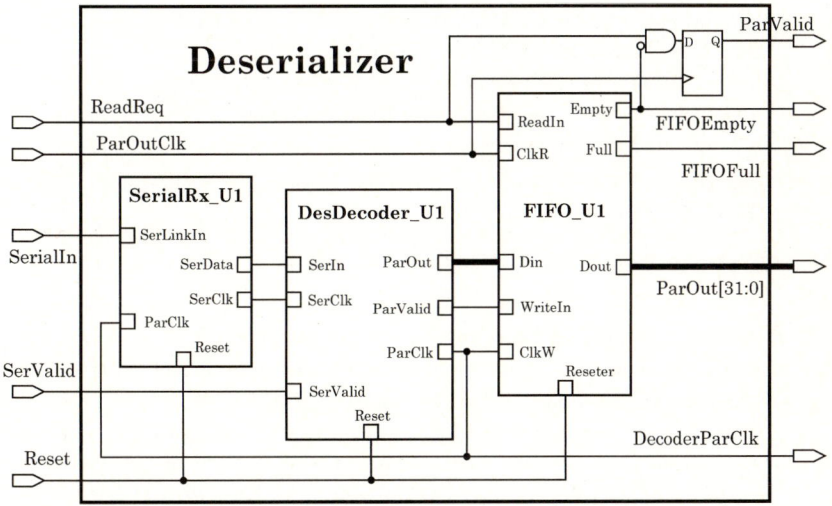

Fig. 20-1. Schematic of the concurrent Deserializer (with FIFO in two clock domains).

The schematic of figure 20-1 may be compared with that of the sequential Deserializer in the lower half of the schematic of figure 19-1. The PLL which is displayed in figure 19-1 as `PLL_RxU1` in `Deserializer.v/*SerialRx_U1` is especially noteworthy.

The clock speed for the FIFO's is 1/2 MHz, as was mentioned in this Textbook at several previous points and is detailed in (*Week 4 Class 1*). Two clocks at 1 MHz means (1 MHz)/2 = 1/2 MHz per FIFO clock.

To implement this part of our design, we shall modify our RAM to be dual-ported; also, we shall change the FIFO state machine (FIFO controller) so that it finds the difference between number of reads and writes on the current clock cycle and changes state according to the preponderance of memory usage on that clock. We read or write on one edge; we determine addresses and state transitions on the other. We'll also fix various things to make the `Deserializer` synthesizable—although perhaps not yet entirely functional.

20.1.1 Dual-Porting the Memory

This just requires enabling the memory to allow simultaneous (= same clock) read and write; of course, these will be to different addresses, as determined by the FIFO controller, not the memory. The FIFO may be in two different clock domains. We shall call the new memory, "DPMem1kx32", with DP for "**D**ual-**P**ort".

To upgrade our memory, we shall do the following in Step 4 of today's lab:
- Wire our chip enable (`Mem_Enable`) in `FIFO_Top` to a constant '1' so that the register file always is enabled. This is because of our use of the memory, not because of the new dual-port functionality.
- Install separate read and write address ports for the memory.
- Supply separate read and write clock ports for the memory.
- Make the memory read and write procedures independent of each other.

A dual-port memory need not have two different clocks, but it must provide simultaneously for two different memory addresses.

20.1.2 Dual-Clocking the FIFO State Machine

This is more complicated than dual-porting the memory and will be done in Steps 5 - 7 of the lab; it may be summarized this way:
- Supply independent read and write clocks (shared with the FIFO RAM).
- Permit nonexclusive read and write commands to the RAM.
- Derive a single state-transition FIFO controller clock from the input read and write clocks. This is because the FIFO read and write ports in general will be in different clock domains. Furthermore, it implies that an edge in either domain might be effective as a state-transition clock.
- Safeguard against the possibility that commands to the RAM might generate more than one read address or more than one write address per state-clock.

- Determine state transitions on each clock solely from the effect of read and write requests made during that clock cycle.

The rest is explained in the lab instructions.

20.1.3 Upgrading the FIFO for Synthesis

We will perform this upgrade in most of Step 8 of today's lab. The primary problems with synthesis of the FIFO when it has been modified as above for dual-port operation will be:

- The design still may include delay expressions; these are useless for synthesis.

  So, we shall remove all delay expressions.

- The incrRead and incrWrite tasks include edge-sensitive event controls but are called within a change-sensitive combinational block.

  So, we shall modify these tasks in two steps:
    (a) First, we shall prevent these tasks from running more than once per state-machine clock cycle, but we shall leave the change vs. edge event control problem alone;
    (b) then, to fix the change vs. edge event control problem, we shall remove the task event controls and break up the task functionality to put some of it in an external always block. The final result will synthesize correctly.

- The state machine clock generator includes an always block with a change-sensitive sensitivity list containing variables not used in the block; this causes dangling inputs to the synthesized (useless) clock-generator logic. This is just the usual synthesizer latch-inference problem.

  So, we shall derive the state machine clock directly from the mutually-independent FIFO read and write clocks.

- The read and write counters are controlled by combinational logic but are sequential elements just as much as is the state register. Thus, the logic in this particular machine never can be divided neatly into sequential logic for the state register and combinational logic everywhere else. The combinational block also causes incorrect synthesis of latches.

  So, we shall change the state-machine combinational block into a sequential block clocked on the opposite edge from the state-transition block.

20.1.4 Upgrading the Deserialization Decoder for Synthesis

We shall do this at the end of Step 8 and in Step 9 of the lab:
- We shall simplify the Shift1's always block trivially by removing its temporary registers.
- We shall rewrite the ClkGen task as a rising-edge sensitive always block; this will prevent latch inference. Also, we shall remove the SerClock gate from ClkGen and put it in a combinational block.

- We shall replace all clocked blocking assignments, which are relicts of the old DesDecoder tasks, with nonblocking assignments.
- We shall prolong the assertion of the *true* value of ParValid by adding a small counter (ParValidTimer) to the Unload32 always block. The duration of ParValid *true* will be guaranteed to be at least 8 serial-clock cycles.

20.2 Concurrent Deserializer II Lab 23

Topic and Location: Dual-ported RAM and concurrent FIFO read and write controls.

We'll do this work in the Lab23 directory.

Preview: We push down the Deserializer of our old Lab22 into a subdirectory for this lab, in anticipation of having a Serializer next-door later in the course. We check for obvious bugs in the first-cut Deserializer FIFO, and then we modify the FIFO memory for dual port operation. After the memory works for simultaneous read and write, we change the FIFO state machine so it can honor simultaneous read and write requests. We then modify the FIFO design further, and the DesDecoder, so that the Deserializer can be synthesized and be more or less correctly functional.

Deliverables: Step 1: Copy the Deserializer of our Lab22_Ans, checking it by a load into the simulator. Step 2: Correctly simulate the FIFO at corner cases. Step 3: Assign the ChipEna port a constant value of 1'b1. Step 4: Dual-port the RAM model (DPMem1kx32), and check it to load properly into the simulator. Step 5: Independent read and write for most of DPMem1kx32. Step 6: Some general renaming improvements. Step 7: Correctly simulating the empty-, full-, and normal-state FIFO controller, for simultaneous FIFO read and write. Step 8: A FIFO which simulates correctly, optionally both in source form and as a synthesized verilog netlist. Step 9: A correctly synthesizable DesDecoder. Step 10: An optionally synthesizable, complete Deserializer.

Lab Procedure:

Step 1: Duplicating the Deserialize of Lab22. Start by creating a new subdirectory, Deserializer, in the Lab23 directory. Copy the entire Lab22/Lab22_Ans/Lab22_Ans_Step12 contents into the new Deserializer subdirectory. Do not use your own previous work, unless you are willing to spend considerable effort in modifying the instructions to follow.

After the copy, delete the copied synthesized netlists and netlist log files.

This Deserializer should include a synthesizable PLL, a single-port RAM which prevents FIFO simultaneous read and write, and a FIFO which, as it happens, will not synthesize correctly. You may wish to recreate the symlink to the verilog simulation-model library (*LibraryName_v2001.v*) in the Deserializer directory, if it is not already present and usable. In Windows, a copy should be made.

Briefly simulate the copied `Deserializer`, using the copied `Lab22` testbench, to ensure a correct and complete duplication. The PLL synchronization will not be especially good, but the system should work.

Step 2: Deserializer corner cases and FIFO debugging. Change to the new `Deserializer/FIFO` subdirectory. You should have there the FIFO design, a FIFO testbench, and a simulation `.vcs` file.

Run the FIFO simulation, modifying it if necessary, so that long intervals of writes alternate with long intervals of reads. Verify that corner-case FIFO memory addresses `0x00` and `0x1f` both were read and written; don't bother about the data values, if any. Do not worry now about transition addresses `0x01` or `0x1e`. If the `0x00` and `0x1f` addresses were not accessed both for read and write, fix the `FIFOStateM` module so that they are.

Resimulate the FIFO again briefly, this time modifying, if necessary, the testbench so that both read and write are requested at the same time. Then, exit the simulator.

We shall be changing the FIFO design in this lab. To prepare for a thorough redesign, start by removing all *#delay* expressions from all FIFO design modules. Also, to distinguish the new FIFO from the old, rename all the files to remove the underscores, and change the module names accordingly. The result in the `Deserializer/FIFO` directory should be,

```
default.cfg or *.tcl     // Optional.
FIFOTop.v
FIFOTopTst.v
FIFOTopTst.vcs
FIFOStateM.v
Mem1kx32.v
```

Recall that all address parameters were renamed to `AWid` as part of the `Lab22` exercise. The `default.cfg` file (or newer `.tcl` file) is a VCS side file and is not essential to the design.

Step 3: How to enable memory simultaneous read and write. This Step mostly is preparatory discussion.

In the `Deserializer` simulation, you might have seen that write requests for the incoming serial stream were ignored when the FIFO was full; write requests also were ignored when there was a read request from the parallel-bus `Deserializer` output. If you did not notice this last, resimulate the `FIFO` briefly.

The original `Mem1kx32` which we adapted to use with our FIFO has only one address bus and thus can not execute a FIFO read and write simultaneously. The single address made the RAM easier to implement and debug, but now it prevents good FIFO operation.

It is the FIFO control logic that first of all enforces the read precedence so that a read prevents a memory write. To prepare to remove this control, in `FIFOTop.v`, delete the declaration and assignment to the `Mem_Enable` net. Then, wire the `ChipEna` port of the

Mem1kx32 instance in FIFOTop to a constant 1'b1, which will enable the memory permanently. There is only one RAM in this FIFO, so we don't need a chip enable control, anyway.

Although now enabled at the chip level, the memory still can't perform useful read and write at the same time because it has only one address port, and because the memory internal logic itself makes read and write mutually exclusive by if-else (in the Lab22 version of Mem1kx32). Furthermore, in FIFOTop, there is a continuous assignment which prioritizes read address over write address in determining the one possible memory address; this can't be removed, given our current, one-address memory.

Clearly, we have to redesign the memory.

In the following, keep in mind that the Deserializer will generate a FIFO write request whenever ParValidDecode is asserted by the DesDecoder.

Step 4: Dual-porting the FIFO memory. The original module ports are shown in figure 20-2.

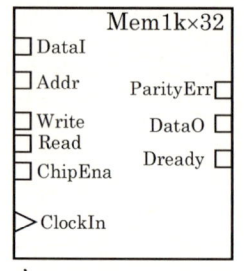

Fig. 20-2. I/O's of the original, single-port Mem1kx32.

Start by renaming the Mem1kx32.v to DPMem1kx32.v. Then, change the module name to match the file name. Modify FIFOTop.v and the .vcs file accordingly.

Next, edit DPMem1kx32 to add the second address port: Rename the existing address input port to AddrR, and add a second port named AddrW. Here, AddrR means "Read Address" and AddrW means "Write Address". Both should have widths determined by the AWid parameter value.

We will want to clock data in and out with separate clocks, so likewise rename the current, single clock to ClkR and add a second clock input port named ClkW. See the results in figure 20-3.

Digital VLSI Design with Verilog

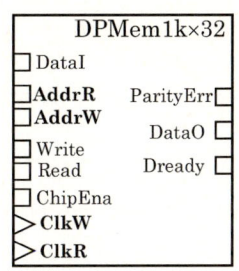

```
         ┌─────────────────┐
         │    DPMem1k×32   │
       ──┤ DataI           │
       ──┤ AddrR   ParityErr├──
       ──┤ AddrW           │
         │         DataO ├──
       ──┤ Write           │
       ──┤ Read    Dready ├──
       ──┤ ChipEna         │
       ▷─┤ ClkW            │
       ▷─┤ ClkR            │
         └─────────────────┘
```

Fig. 20-3. Mem1kx32 renamed for dual-port memory.

Internally, rename the old `ChipClock` to `ClockR`, and declare a new wire named `ClockW`.

Although we have disabled `ChipEna` in the `FIFO`, we still want the control inside the memory; this will make the memory more easily converted for use in a multiple-RAM design. So, put both clocks under control of the `ChipEna` input by wiring them to internal nets and muxing them to 1'b0 this way:

```
assign ClockR = (ChipEna==1'b1)? ClkR : 1'b0;
assign ClockW = (ChipEna==1'b1)? ClkW : 1'b0;
```

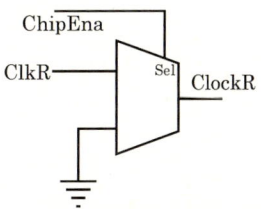

Fig. 20-4. ClkR clock gate logic.

The logic for the clock gating by the two preceding continuous assignments is as shown for the `ClkR` schematic of figure 20-4. With these changes, it is now possible to retain the gating of `Dready` and `DataO` by `ChipEna` without further alteration of the design.

We still have read and write logically dependent. To remove this dependency, put the read functionality and the write functionality into two separate `always` blocks as shown below. The continuous assignment gating is enough to control all activity, so we don't need a `ChipEna` either in the read or the write block. Because memory reads will be clocked externally from the `DPMem1kx32` output latches, reconvergence (*Week 1 Class 2*, section 3.1.3.5) of the internally gated clocks will not be a concern.

There is no special reason to initialize memory in the hardware, but to have fail-safe parity checking, we should generate an error if any memory bit ever goes to 'x' or 'z'. This means initialization of all memory locations on `Reset` to prevent false parity errors during the beginning of a simulation. `Reset` should be applied both to the read and the write blocks, but the memory initialization should be included only with the write code.

Be sure to avoid, ***always***, assigning to any variable in more than one `always` block; the risk of a race condition can not be overstated.

Also, it's about time we adopted a new `reg` naming convention. The synthesizer always renames design `reg` types, if they are preserved, to `*_reg` in the netlist; so, using our current convention of `*Reg` just creates `*Reg_reg` names in the netlist, which is redundant. So, in the `DPMem1x32` and all other FIFO modules, change all of our declared `reg` names which end in `*Reg` to `*r`. Thus, our new naming convention means that *anything*`Reg` will be renamed to *anything*`r`.

Thus, the separated read and write `always` blocks should be done this way:

```
...
always@(posedge ClockR, posedge Reset)
  begin : Reader
  if (Reset==1'b1)
       (init Parityr, Dreadyr, DataOr)
  else if (Read==1'b1)
              (do parity check and read)
  end // Reader.
//
always@(posedge ClockW, posedge Reset)
  begin : Writer
  if (Reset==1'b1) // Zero the memory:
       for (i=0; i<=MemHi; i=i+1) Storage[i] <= 'b0;
  else if (Write==1'b1)
              (do write)
  end // Writer.
...
```

As shown, name the newly separated blocks `Reader` and `Writer`. Because the `Reader` block has to do a parity check, it should initialize the parity error output to 0, as in the old `Mem1kx32` model. `Writer` probably should use its reset branch to do nothing but zero out the memory storage upon assertion of `Reset`. The `Writer` block should not assign `'z'` when it is inactive (this would be OK for a shared read-write bidirectional bus).

Finally, modify `FIFOTop` to remove the `SM_MemAddr` declaration and its assignment statement. Connect each of the two new `DPMem1x32` address input pins directly to its proper state machine output address pin. Also, connect the one available clock to both of the `DPMem1kx32` clock input pins. Then, verify that the FIFO simulates the same way with the new `DPMem1kx32` instance as it did with the original `Mem1kx32`.

After all these changes, the FIFO states will simulate the same way in both old and new VCS without modification; and, we can do no more at this time in `FIFOTop` or `DPMem1kx32`. To continue in a well-organized fashion, we have to modify the state machine design, which itself includes code assuming single-port memory operation.

Digital VLSI Design with Verilog   423

Step 5: Modifying the FIFO state machine I. The FIFO state machine is going to be more complicated than the preceding, so we shall do it in two stages. This is stage I.

First, change to the `Deserializer` directory and change all references to `"FIFO_Top"` to the newly spelled name, `"FIFOTop"`; likewise, change all references of the form `"Mem1kx32"` to `"DPMem1kx32"`. These references are, of course, the identifiers (`module` names) in the files in the `Deserializer` directory.

Then, simulate the `Deserializer`. It should be noted that the current read-address `task` is unsynthesizable mainly because of its `"@(posedge ClkR)"` control. You will notice that, during the memory reading, `ParValid` goes high, indicating a write from the incoming serial line. However, no write occurs, because the FIFO state machine has been designed to avoid "race conditions" by means of an `if-else` which always gives a read priority over a write.

However, we now have separate read and write addressing of the memory, which should prevent data corruption caused by a possible race. So, let us separate the state machine's read and write functions, allowing them to occur concurrently.

To do this, change to the `FIFO` subdirectory. In `FIFOStateM`, rename the `Clk` port to `ClkR` and add a second clock input port named `ClkW` (compare figure 20-3 above). After this, in the same module, modify both of the address increment tasks so that each one has a latching semaphore which will protect the address from being changed more than once per clock cycle.

For example, the read-address `task` may be modified as shown next:

```
reg LatchR, LatchW;    // Reset these to 1'b0 in SM comb. block.
...
task incrRead;   // Unsynthesizable!
  begin
  if (LatchR==1'b0)
    begin
    LatchR = 1'b1;
    @(posedge ClkR)
      ReadCount = ReadCount + 1;
    LatchR = 1'b0;
    end
  end
endtask
```

Revise both of the `tasks`, `incrRead` and `incrWrite`, in `FIFOStateM` to be like the one above.

We have a theoretical issue to address: A state machine can not be guaranteed deterministic unless it has a unique clock. But, now we have two clocks into `FIFOStateM`, so we must derive a single clock to determine the machine's current state.

This can be done by means of a simple, change-sensitive `always` block:

```
always@(ClkR, ClkW, Reset)
  if (Reset==1'b1)
      StateClock = 1'b0;
  else StateClock = !StateClock;
```

This construct is not synthesizable (`ClkR` and `ClkW` go nowhere), and we shall revise it later in the lab. For now, it will do as a quick simulation check. So, declare a `reg` named `StateClock` and add the above to `FIFOStateM`. Change the state transition clock name from `Clk` to `StateClock`. Also, change the state machine combinational block's enable to `StateClock==1'b0`. In the `incrRead` and `incrWrite` tasks, replace each `posedge` event control with one depending on `StateClock`.

Next, in the state machine combinational block, in the `empty` and `full` state code, remove all mutual dependence of read and write. Specifically, in the `empty` state, there is no reason for the `if` to test for a read request; in the `full` state, the `if` should test only for a read request.

Leave the `a_empty` and `a_full` state code alone for now.

In the `normal` state code, the counters are checked, so state transitions can be determined correctly on a simultaneous read and write in any order; therefore, in the `normal` state, just separate the read and write conditions each into its own `if` by removing all dependence of one upon the other request. Be sure to remove all statements which disable the opposite operation; for example, remove "`WriteCmdr = 1'b0;`" from the memory read statement.

Also, extract all transition logic from the two `if`'s, and collect it *below* the second `if`. Because the requests both may be processed now on one clock, we have to wait for the last one to lapse before deciding on the transition to the next state.

The resulting transition logic should look about like this:

```
(read & write command logic)
// Set the default:
 NextState = normal;
//
// Check for a_full (R == W+1):
tmpCount = WriteCount+1;
if (ReadCount==tmpCount)
    NextState = a_full;
//
// Check for a_empty (W == R+1):
tmpCount = ReadCount+1;
if (WriteCount==tmpCount)
    NextState = a_empty;
```

We now have independent read and write in the normal state. We are not finished yet, but you may wish to simulate briefly to check for gross errors. With the changes above, the FIFO register-file corner case addresses 0x00 and 0x1f again should be found to be read and written properly, as should the neighboring 0x01 and 0x1e.

Step 6: Routing the new read and write clocks. We need two clock inputs for the FIFO. In FIFOTop.v, rename the Clocker input to ClkR and add another input named ClkW. For the FIFO in the deserializer, ClkR will be for the receiving domain, and ClkW will be the parallel-data clock from the DesDecoder. In FIFOTop, connect the two clocks directly to the DPMem1kx32 and the FIFOStateM instances.

After this, in Deserializer.v, consistent with our FIFO naming convention, rename ParValidReg to ParValidr—in fact, now would be a good time to change all remaining design reg names from *Reg to *r.

Also, rename the FIFOReadCmd input port to ReadReq; we want to be sure to differentiate a request for a read (originating external to the FIFO) from a read command to the register file (within the FIFO). Modify the DeserializerTst testbench for the new port name.

In Deserializer, change the FIFOTop port map so that .Clocker is renamed to .ClkW, and a new .ClkR port is mapped to ParOutClk.

After this, simulate Deserializer briefly. The result should be changed, but the result still should be reasonable, if only intermittently good. The FIFO I/O ports now match those shown in the schematic of figure 20-1, but we haven't finished yet.

Step 7: Modifying the FIFO state machine II. When we have completed Steps 7 and 8, the organization of each state in the FIFO state machine combinational block will be as shown next, keeping in mind the transition limitations of the *empty* and *full* states:

```
always@(*)
...
case (CurState)
        ...
  stateX: begin
          1. Check for read & write requests.
             Update read & write addresses.
             Issue memory read & write commands.
          2. Check for updated addresses.
             Assign NextState.
          end
  stateY: begin
          ...
endcase
```

We should begin by modifying the a_empty and a_full states so that they will work properly for independent read and write requests. Looking at the FIFOStateM combinational block, the problem is that (*a*) as soon as one operation is triggered, the machine disables the opposite one; (*b*) the machine prioritizes one operation while mutually-excluding the other; and, (c) the machine does not check the counter values in these states and thus can not determine whether a read or write command, or both, has been issued on the current clock.

So, in `FIFOStateM.v`, for `a_empty` and `a_full` remove the disabling assignment to `ReadCmdr` in the writing blocks and to `WriteCmdr` in the reading blocks. The rationale here is the same as previously, for the `normal` state. However, leave the transition rules alone for the moment.

For `a_empty`, disentangle the transition logic from the memory command logic and collect the transition logic together at the end of the `a_empty` block, so that it can be made to depend on the counter values rather than on the memory command decision(s).

To ensure correct counter wrap-around, use an `AWid`-bit wide `tempCount reg` in `a_empty` as shown here:

```
// In this state, we know W == R+1; ...
// Memory command logic:
if (WriteReq==1'b1) ...
if (ReadReq==1'b1) ...
//
// Transition logic:
// Set default:
NextState = a_empty;
//
// Check for change:
tempCount = ReadCount+2;  // Destination determines wrap-around.
if (WriteCount==tempCount)
    NextState = normal;
else if (WriteCount==ReadCount)
    NextState = empty;
```

Do similarly for the `a_full` logic.

At this point, let's look again at the command logic in the combinational block. Consider the `normal` state. In that state, the command logic should be something like this:

```
// On a write:
if (WriteReq==1'b1)
  begin
  WriteCmdr  = 1'b1;
  incrWrite; // Call task, which blocks on posedge StateClock.
  end
// On a read:
if (ReadReq==1'b1)
  begin
  ReadCmdr  = 1'b1;
  incrRead; // Call task, which blocks on posedge StateClock.
  end
(transition logic ...)
```

With the code shown above, a write request would call incrWrite, which would block further execution until the next positive edge of the clock. This would cause loss of a read request on the low phase of the current clock.

A typical simulation solution to prevent this loss would be to put both the read and write requests together in a fork-join block; then, a simultaneous read and write both would be processed together, and both might possibly block on the positive edge, which, when it occurs, would unblock the command processing and allow the transition logic to update next_state and the address counters for the next positive edge. The result would be as shown next:

```
fork // unsynthesizable!   But do it for now.
// On a write:
if (WriteReq==1'b1)
  begin
  WriteCmdr  = 1'b1;
  incrWrite;
  end
// On a read:
if (ReadReq==1'b1)
  begin
  ReadCmdr   = 1'b1;
  incrRead;
  end
join
(transition logic ...)
```

Even though fork-join is not synthesizable, make these changes in your code for the a_empty, normal, and a_full command logic.

In our modified design, we already have made sure that both address-increment tasks will exit simultaneously on concurrent read and write commands; we did this by using only one posedge clock, StateClock, in those tasks.

From a previous Step, your tasks should have been rewritten this way:

```
task incrRead;   // Still incorrectly synthesized (but not done yet).
  begin
  if (LatchR==1'b0)
    begin
    LatchR = 1'b1;
    @(posedge StateClock)   // Read must not block longer than write.
      ReadCount = ReadCount + 1;
    LatchR = 1'b0;
    end
  end
endtask
```

The synthesizer doesn't use `fork-join` blocks (netlists are concurrent anyway) and, at least as of late 2011, rejects them; so, if we were going to synthesize at this point, each "`fork`" and each "`join`" in the combinational block would be surrounded by a `DC` macro test to prevent the synthesizer from reading it.

One last refinement: Let's change the state declarations from generic `localparams` to sized `localparams`. And, let's rename the states from *statename* to *statename*S, a final 'S' to make clear that this is a state identifier and not a common English word. These name changes can be done by a quick search-and-replace.

The result should be about like this:

```
localparam[2:0]
      emptyS = 3'b000, // all 0 = emptyS.
    a_emptyS = 3'b010, // LSB 0 = close to emptyS.
     normalS = 3'b011, // a_emptyS < normalS < a_fullS.
     a_fullS = 3'b101, // MSB 1 = close to fullS.
       fullS = 3'b111; // all 1 = fullS
```

Simulate your `Deserializer` again. In this simulation, be sure at some point first to disable reads until the FIFO goes full and then to read it long enough so that it goes empty. The design now should be much closer to being entirely correct than when we started this lab.

Step 8: Changing the FIFO state machine for correct synthesis. The current design will synthesize to a netlist, but the netlist still will be incorrect. The new PLL will be fine, but the FIFO remains a synthesis problem for two reasons:

1. The `StateClock` generator is an oscillator implemented as an `always` block with a change-sensitive event control containing nets not expressed in the block. This virtually ensures that the synthesizer will allow these nets to dangle.

2. The address-updating `tasks` include embedded edge-sensitive event controls. Although edge-sensitive, the `tasks` are not called on that edge; and, so, the embedded controls imply complex latches in the state machine combinational logic, which will not be synthesized correctly (as we know).

We shall resolve these problems in this Step.

Step 8A: Testbench for source and netlist. Before anything else, let's test to be sure that our FIFO testbench is adequate for unmodified use either with our FIFO source verilog or with a synthesized FIFO netlist.

At the start of the tests, we want pulsed write and read requests, as in previous FIFO testbench versions. After this, we want a run of back-to-back writes which fills the FIFO, followed by a similar run of reads which forces it empty. Finally, we want a run of mixed writes and reads which includes at least a few simultaneous read and write requests.

Modify your FIFO testbench to achieve these goals. Use two identical clock generators, one clocking the `FIFOTop ClkR` pin, and the other `ClkW`. If time is short in the lab,

Digital VLSI Design with Verilog

consider just copying all or part of the `FIFOTopTst.v` testbench in `Lab23_Ans/Lab23_AnsStep08A`. Don't bother about details of design functionality at this point.

Step 8B. Synthesizable `StateClock` generator. In `FIFOStateM`, we can replace the `always` block with one which is sensitive only to the two input clocks; this will be synthesizable.

For example,

```
reg     StateClockRaw;
wire    StateClock;   // See the next code example below.
//
always@(ClkR, ClkW)
  StateClockRaw = !(ClkR && ClkW);
```

An *xor* (`^`) would be more symmetrical than the *and* (`&&`), but it would not work if `ClkR` and `ClkW` were closely in phase; so, we use a *nand* expression. An *and* also would be fine, but in CMOS technology, *nand* usually is simpler and faster than *and*.

> Making the state clock generator synthesizable is a problem worthy of some further discussion:
>
> Using the receiving-domain clock as the state clock would work, but it would risk a variety of machine race conditions on some sending-domain clock phases.
>
> Using `"always@(ClkR, ClkW) StateClockRaw = !StateClockRaw;"` would not synthesize correctly, because neither `ClkR` nor `ClkW` are used in the statement. To avoid danglers, hypothetically one might try using any available edge on either domain, as in, `"always@(posedge ClkR, negedge ClkR, posedge ClkW, negedge ClkW)"`. However, event controls using both edges of the same variable are not synthesizable.
>
> The `"!(ClkR && ClkW)"` solution reduces the clocking problem to one of glitch filtering. Some experimentation shows that a plain *and*, `"(ClkR && ClkW)"`, actually works better than the book *nand*; so, this is used in the CD lab answers.
>
> However, any logical operation does imply that for certain very rare `ClkR` vs. `ClkW` phases, the state clock would stop—temporarily, assuming that at least one clock could drift in phase. Watchdog logic could avert this—but why complicate this FIFO further? It is, after all, a training exercise, and training priorities have to be taken into account.

The clock `reg` above is named `StateClockRaw`, and not `StateClock`, for the following reason:

A possible new problem is that there is no HDL delay here any more, and therefore there is no more inertial-delay filtering. This means that as the read and write clocks drift in phase independently, the simulator will be allowed to produce `StateClockRaw` glitches of arbitrarily short duration—and there will be no clear relationship to the performance in the synthesized netlist. This actually has been a simulation problem here, all along, ever since we deleted all the delay expressions in previous lab Steps.

The solution is to connect the StateClockRaw to a library component delay cell which will filter out the narrowest glitches; the delay cell then can be used to drive the StateClock. The same glitch filtering then will occur during simulation both in the source and in the synthesized netlist. For a delay cell, we might as well use the same library cell as we used in Lab22 for the synthesizable PLL.

We can preserve the delay cell from removal during synthesizer optimization by means of an embedded *TcL* script near the verilog instantiation.

The added code may be written this way:

```
// Glitch filter:
DEL005 SM_DeGlitcher1 ( .Z(StateClock), .I(StateClockRaw) );
//
//synopsys dc_tcl_script_begin
// set_dont_touch SM_DeGlitcher1
// set_dont_touch StateClock*
//synopsys dc_tcl_script_end
```

After commenting out the StateClock generator of previous Steps and deriving it from the read and write clocks as above, you will have to add a verilog simulation model of the delay cell to your .vcs file list to simulate the FIFO. Do this by referencing our library file, *LibraryName_v2001.v*, which is copied or linked in your Lab23 directory. Precede the file name with -v in your .vcs file, so that the simulator will search for and compile the one model, the "DEL005", instead of compiling the whole library.

Don't bother synthesizing yet; but, do try simulating the FIFO briefly. You may have to modify the testbench a little to get the FIFO to run all the way to full and to empty at least once during the simulation. Don't worry about simulation details; we shall be changing the FIFO, and thus perturbing the functionality, again.

Step 8C. Synthesizable FIFO address-generating tasks. We have to reexamine and change every state in our FIFO state machine combinational block.

We know that the event controls inside the old FIFO tasks will not synthesize correctly. So, let us separate the event controls from the task logic and put them in new always blocks. Instead of embedding an edge-sensitive event control in each task, we can use independently StateClock'ed always blocks, clocked on the positive edge, to update the address counters. An edge-sensitive always block can update a counter conditionally and be synthesized correctly.

With counter updates isolated this way, the tasks can be rewritten fully change-sensitive, meaning that they will synthesize correctly in the state machine combinational block.

Digital VLSI Design with Verilog

Our new `tasks` (shown below) will latch address state and issue read or write commands to the FIFO register file; the `always` blocks synchronously will change the addresses used by the register file.

With this new implementation, the `always` blocks also should be made to perform the `fullS` and `emptyS` address initializations on reset. Thus, all manipulation of register-file addresses can be encapsulated in the new `posedge always` block logic.

Now let's get down to implementation details:

Let's add input arguments to our `tasks` and adopt the convention that a `task` called with an input argument of `1'b1` will request an address increment, unless one is scheduled already on the current clock. A `task` called with a `1'b0` input argument will deassert the command controlled by that `task`. One way to write the required `always` blocks and their `tasks` would be this way:

First, we declare the following `reg`'s for use by the `tasks` and by the next-state logic:

```
reg[AWid-1:0] ReadAr, WriteAr          // Address counter regs.
            , OldReadAr, OldWriteAr    // Saved posedge values.
            , tempAr;                  // For address-wrap compares.
```

The `ReadAr` and `WriteAr` also should be used on the right sides of the usual continuous assignments to the corresponding `FIFOStateM` output ports, `ReadAddr` and `WriteAddr`. The `Old*` above are to retain state across calls, because we lost the sequencing capability which we had with the discarded (@ *edge-sensitive*) event controls.

After this, we may declare each new `always` block and its respective `task`. The `tasks` generally have to treat FIFO empty and full states specially. Our new `incrRead` and `incrWrite` blocks are developed first; their purpose is to replace the original and unsynthesizable `@(posedge Clk)` statements in the old `incrRead` and `incrWrite` of `FIFOStateM.v`.

For a register-file read, here is the `always` block:

```
always@(posedge StateClock, posedge Reset)
  begin : IncrReadBlock
  if (Reset==1'b1)
       ReadAr <= 'b0;
  else begin
       if (CurState==emptyS)
           ReadAr <= 'b0;
       else if (ReadCmdr==1'b1)
             ReadAr <= ReadAr + 1;
       end
  end
```

For a register-file read, here is the new `task`:

```
task incrRead(input ActionR);
  begin
  if (ActionR==1'b1)
       begin
       if (CurState==emptyS)
            begin
            ReadCmdr   = 1'b0;
            OldReadAr =   'b0;
            end
       else begin
            if (ReadAr==OldReadAr) // Schedule an incr.
                ReadCmdr = 1'b1;
            else begin // No incr; already changed:
                ReadCmdr   = 1'b0;
                OldReadAr = ReadAr;
                end
            end
       end
  else begin // ActionR is a reset:
       ReadCmdr   = 1'b0;
       OldReadAr =   'b0;
       end
  end
endtask
```

If the FIFO is empty, its first value can be written anywhere, so the `emptyS` starting value of `ReadAr` in principle need not be specified. However, the counter arithmetic we have adopted for state transitions requires that both counters start from a known value in the empty state. Therefore, the transition to `emptyS` must initialize both pointers to the same value, which might as well be 0. The write pointer must be initialized in its own `always` block, because we want to conform to the synthesis requirement that no `reg` may be assigned in more than one `always` block.

When the FIFO is full, its data occupies all storage locations, and the address of the first datum to be read is predetermined. Therefore, the writing `always` block, the only one allowed to modify the value of `WriteAr`, must be the one to initialize `WriteAr` to bring it from its undefined state to its one, correct value on the first read in `fullS`. That value equals the value of `ReadAr`, which, of course, will be incremented on the next `posedge` of the `StateClock`.

So, for a register-file write, the new `always` block would be:

```
always@(posedge StateClock, posedge Reset)
  begin : IncrWriteBlock
  if (Reset==1'b1)
       WriteAr <= 'b0;
  else begin
        case (CurState)
        emptyS: WriteAr <= 'b0;    // Set equal to read addr.
         fullS: WriteAr <= ReadAr; // Set equal to first valid addr.
        endcase
        if (CurState!=fullS && WriteCmdr==1'b1)
          WriteAr <= WriteAr + 1;
        end
  end
```

And, finally, for a register-file write, the new `task` would be:

```
task incrWrite(input ActionW);
  begin
  if (ActionW==1'b1)
       begin
       if (CurState==fullS)
             begin
             WriteCmdr  = 1'b0;
             OldWriteAr = ReadAr;
             end
         else begin
             if (WriteAr==OldWriteAr) // Schedule an incr.
                 WriteCmdr = 1'b1;
             else begin // No incr; already changed:
                 WriteCmdr  = 1'b0;
                 OldWriteAr = WriteAr;
                 end
             end
       end
  else begin  // ActionW is a reset.
       WriteCmdr  = 1'b0;
       OldWriteAr = 'b0;
       end
  end
endtask
```

All `task` calls in the `FIFOStateM` combinational block now have to be modified to pass a `1'b1` or `1'b0`. A call to both of `incrRead` and `incrWrite` should be added to the reset block in `emptyS`, and each should be passed `1'b0`; all other `task` calls should be passed `1'b1`. Our old `LatchR` and `LatchW` declarations and instances now should be removed from the design.

All reference in the combinational block to `ReadCount`, `WriteCount`, or `tmpCount` should be replaced by `ReadAr`, `WriteAr`, or `tempAr`, respectively, which latter also should be used in the continuous assignments to `FIFOStateM` output ports. None of these new identifiers should appear in the combinational block except in transition logic expressions.

We still have changes to make in the combinational block. First, why not delete all the `fork-join`s? Consider them a temporary simulation kludge, and remove them all, with their `` `DC `` controls (if any).

Because we are eliminating `fork-join` everywhere, we have to simulate a simultaneous read and write request in some other way. The individual `task` calls would be enough; but, we also have to deassert read and write commands on clocks not requesting either. We can do all this explicitly by using a `case` statement in each branch of the combinational block, as shown below.

The use of the new `task` blocks, minus their deleted `fork-join`'s, is shown next. Our new `task`s don't block anymore on simulation-scheduled events, so we can call them individually if there is just one of read or write requested—or, we can call them both (procedurally).

Here is how to do this in any of `a_emptyS`, `a_fullS`, or `normalS`:

```
...
case ({ReadReq,WriteReq})
  2'b01: begin
           incrWrite(1'b1); // On a write.
           ReadCmdr  = 1'b0;
           end
  2'b10: begin
           incrRead(1'b1);  // On a read.
           WriteCmdr = 1'b0;
           end
  2'b11: // On a read and write:
         begin
         incrRead(1'b1);
         incrWrite(1'b1);
         end
default: begin // No request pending, so deassert both:
           ReadCmdr  = 1'b0;
           WriteCmdr = 1'b0;
           end
endcase
...
```

In `emptyS` or `fullS`, there is only one operation possible, so we need call just the one `task` for those; but, we should deassert this command, too, when it is not requested.

Digital VLSI Design with Verilog 435

Finally, we have to break the rules to get the FIFO to synthesize. Normally, a verilog state machine is designed with a clocked block to determine state transitions and a combinational block to operate the transition logic and the rest of the machine. This usually works best. But, a FIFO is a strange animal.

Notice that we already have to use clocked logic to update the address counters, and that this is inconsistent with the rule of keeping everything but state register updates in a purely combinational block. Attempting synthesis with only the preceding changes still will result in pathological latch synthesis and incorrect functionality. We must prevent latches; we can do this by clocking what up to now was our "combinational" block.

To avoid race conditions, we modify the combinational block to be sensitive to the negedge of StateClock. This, incidentally, will allow us to associate an asynchronous reset with the erstwhile combinational block.

Your new "combinational" block now should resemble something like this:

```
...
always@(negedge StateClock, posedge Reset)
  begin
  if (Reset==1'b1)
     begin // Reset conditions:
     NextState  = emptyS;
     incrRead(1'b0);  // 0 -> reset counter.
     incrWrite(1'b0); // 0 -> reset counter.
     end
  else
  case (CurState)
  emptyS://  (The other previously combinational logic goes here)
  ...
  endcase
...
```

After completion of the above changes in this Step, we have a clocked "combinational" state machine block, but with blocking, not the usual nonblocking, assignments. This is so that the statements in this block still are updated <u>in order</u> as they are run, even if debugging delays should be added for some temporary reason. The usual setup and hold functionalities of nonblocking assignments in a simple clocked model are not relevant here, because we can be confident that everything will have settled to its final condition by the time the next posedge of StateClock arrives.

After these changes, simulate the FIFO again until it works well with your FIFO testbench.

> **Optional**: If time permits, **synthesize a netlist** from this design and
> simulate it. For correct synthesis, some setup and hold problems will have to be
> prevented; to see how to do this, examine the `FIFOTop.sct` file provided with the
> Extras in `Lab23/Lab23_Ans/Lab23_AnsStep08C/Deserializer/FIFO`. This
> synthesis will take some time, perhaps a half-hour.
>
> To reduce the time, let the synthesizer run until it has been in the "Area
> Recovery Phase" for about 5 minutes; then, press {control-C} <u>once</u> and use the
> text-based menu (which will appear after a little while) to abort optimization. The
> synthesis script will run to completion, leaving you with a suboptimal but
> functionally correct netlist.

The above optionally synthesized netlist of the FIFO should simulate more or less as well as the source FIFO. There may remain timing issues, but don't bother perfecting your FIFO testbench now; go on to the rest of the lab.

Concerning the optional synthesis just above, new DC (*ca.* 2011) will take only 15 minutes, but the netlist will not simulate correctly unless SDF timing back-annotation is used. The next two figures show the resulting SDF simulation waveforms.

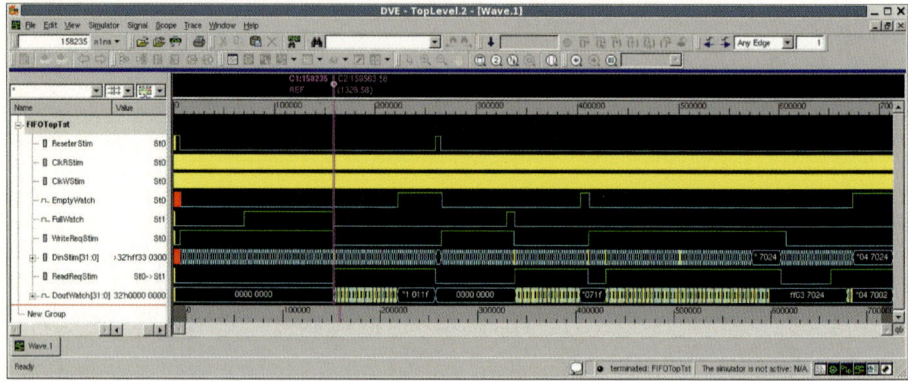

Fig. 20-5 Overall view of the new DC netlist simulation (with SDF timing back-annotation) of Step 8C.

Digital VLSI Design with Verilog 437

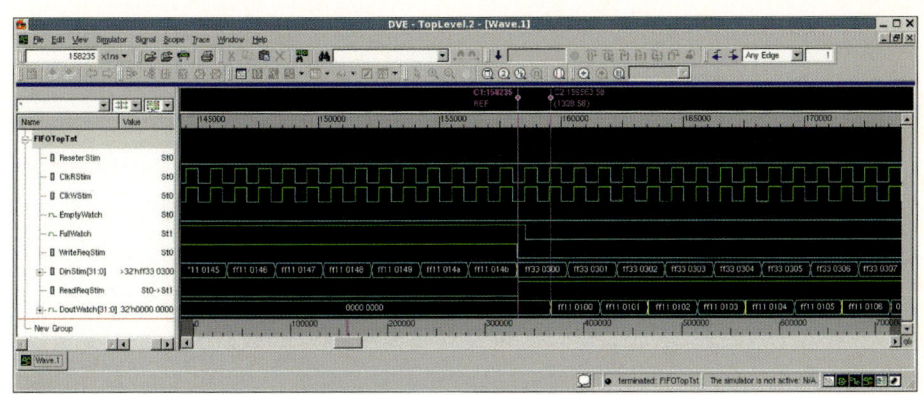

Fig. 20-6. Closeup of the Step 8C simulation.

The FIFOTop.sct file for the synthesis in this Step contains suppress_message* statements. These shorten the screen and .log report warning messages as we increase design and .sct complexity. Otherwise, eventually, the warnings would consume all the .sct runtime screen display, making error messages and all other messages disappear.

Step 8D (*Optional*). Verifying that the Deserializer from Step 8C does not yet synthesize correctly. If time permits, try the synthesis described in this Step; otherwise, just read the description here of the result.

Without any attempt to simulate the Deserializer from the top, change to the Lab23/Deserializer directory, run the synthesizer using the Lab23/Deserializer.sct script provided, and examine the waveforms resulting from an attempt to simulate the resulting netlist.

The netlist simulation will fail. You should be able to see that ParValid does not control FIFO writes correctly. The parallel-bus clock from the sending domain consists of approximately 160-ps wide pulses, repeated at about 2 ns intervals. The problem therefore is in the DesDecoder, which generates this parallel clock. Even though the main DesDecoder synthesis problem, edge-triggering task calls, was fixed, and the source verilog simulated correctly, the synthesized netlist is not correct.

To elaborate with a little more detail, depending on synthesis constraints, the DesDecoder ParClk may stay stuck at 1'b0. Also, whether the training-library (tcbn90ghp_v2001.v) or SDF delays are used will determine the exact way that the netlist simulation might fail.

The `Decode4` task, rewritten as a `negedge always` block, may be done this way:

```
always@(negedge SerClock, posedge Reset)
begin : Decode4
  if (Reset==YES)
       begin
       DecodeReg = 'b0;
       doParSync =  NO;
       SyncOK    =  NO;
       UnLoad    =  NO;
       end
  else begin : PacketFind   // Look for packet alignment:
       UnLoad    = NO;
       doParSync = NO;
       if ( FrameSR[7:0]==PAD0 )
          begin : FoundPAD0
          SyncOK = YES;
          if ( FrameSR[23:16]==PAD1 && FrameSR[39:32]==PAD2
               && FrameSR[55:48]==PAD3 )
               begin // All pads indicate all frames aligned:
               DecodeReg = { FrameSR[63:56], FrameSR[47:40]
                           , FrameSR[31:24], FrameSR[15:8] };
               UnLoad = YES;
               end
          else // Found a PAD0, but rest failed; so, synchronize:
               begin
               doParSync = YES;
               SyncOK    = NO;
               end
          end // FoundPAD0.
       end // PacketFind.
end
```

In the above `always` block we write <u>YES</u> (=1'b1) or <u>NO</u> (=1'b0). This is just for experimentation: See whether this adds anything to your general readability . . ..

The procedural complexity of this decoder requires blocking assignments. By generating the parallel clock on a `posedge` of the serial clock and using all `DesDecoder` results only on the `negedge`, race conditions easily are avoided.

Step 9: Synthesizing the `DesDecoder`. This Step is important; and, except for the "fancy" `localparam` expression after the `Shift1` discussion, it should not be considered optional.

Change to the `DesDecoder` subdirectory and read through the source code in `DesDecoder.v`. The `ClkGen task` definitely is a problem: It is called in a selectively change-sensitive `always` block and thus implies nonstandard latching. As usual, such a

Digital VLSI Design with Verilog

latch is too complex to expect the synthesizer to be able correctly to fulfill the design intent.

We have to rewrite the parallel-clock generator. We can start by cleaning up the rest of the `DesDecoder` module: Rename all `*Reg` to `*r`, if not done already; then, remove any remaining commented-out `task` remnants.

The clock-generator problem can be solved by rewriting the `ClkGen` task as an `always` block named `ClkGen`. This `always` block should be sensitive to the edge of `SerClock` which is opposite to the edge of the other `always` blocks in this module; so, `ClkGen` should be sensitive to the positive edge.

For example,

```
always@(posedge SerClock, posedge Reset)
  begin : ClkGen
  if (Reset==1'b1) // Respond to external reset.
      begin
      ParClkr <= 1'b0;
      Count32 <= 'b0;
      end
  else begin // If not a reset:
      if (SerValid==YES)
        begin
        // Resynchronize this one:
        if (doParSync==YES)
            begin
            ParClkr <= 1'b0; // Put low immediately.
            Count32 <= 'b0;
            end
        else begin
            Count32 <= Count32 + 1;
            if (Count32==5'h0) ParClkr <= ~ParClkr;
            end
        end // if SerValid.
      end // not a reset.
  end // ClkGen.
```

Edge sensitivity will eliminate complex latches, but further improvement is required. Start by removing the `SerValid` gate from the `ClkGen` block and putting it in a conditional continuous assignment.

Such an assignment easily is done this way:

```
assign SerClock = (SerValid==YES)? SerClk : 1'b0;
```

After this, if the `Shift1` still uses a temporary shift register, make it more compact by changing it to shift the `FrameSR` directly.

The following should work well:

```verilog
always@(negedge SerClock, posedge Reset)
  begin : Shift1
  // Respond to external reset:
  if (Reset==YES)
        FrameSR <= 'b0;
  else begin
        FrameSR    <= FrameSR<<1;
        FrameSR[0] <= SerIn;
        end
  end
```

Let's get a little fancy here, like someone designing configurable IP. In the simulation, you may have noticed the narrow, glitch-like pulses produced on the `ParValid` output. This net should be asserted for longer durations, several `SerClock` cycles, at least.

We cannot effectively program a pulse digitally; but, we can introduce a small, fast `SerClock` cycle-counting timer which can deassert `ParValid` after any convenient count.

So, add this to the beginning of the `DesDecoder` module:

```verilog
localparam ParValidMinCnt = 8;   // Minimum number of SerClocks
                                 //     to hold ParValid asserted.
localparam ParValidTWid = // Width of ParValidTimer reg.
                 (     2     > ParValidMinCnt )? 1
               : ( (1<<2) > ParValidMinCnt )? 2
               : ( (1<<3) > ParValidMinCnt )? 3
               : ( (1<<4) > ParValidMinCnt )? 4
               : ( (1<<5) > ParValidMinCnt )? 5
               : ( (1<<6) > ParValidMinCnt )? 6
               : 7; // Thus, width is declared automatically.
//
reg[ParValidTWid-1:0] ParValidTimer;
// ...
```

This approach will make it possible to adjust the width of the `ParValid` assertion, should system considerations require it.

Digital VLSI Design with Verilog

After this, the `Unload32` block may be rewritten along these lines:

```
always@(negedge SerClock, posedge Reset)
  begin : Unload32
  if (Reset==YES)
      begin
        ParValidr    <=  NO; // Lower the flag.
        ParOutr      <= 'b0; // Zero the output.
        ParValidTimer <= 'b0;
      end
  else begin
      if (UnLoad==YES)
          begin
            ParOutr       <= Decoder; // Move the data.
            ParValidr     <= YES;     // Set the flag.
            ParValidTimer <= 'b0;
          end
      else begin
          if (ParValidTimer<ParValidMinCnt)
            ParValidTimer <= ParValidTimer + 1;
          if (ParValidTimer==ParValidMinCnt && ParClk==1'b0)
            ParValidr <= NO; // Terminates assertion.
          end
      end // UnloadParData.
  end
```

This new `Unload32` block mostly differs from the original by including assignments to the new `ParValidTimer`. The `ParValidMinCnt` probably should be assigned a value of about 8 (= 1/4 of 32).

After simulating the rewritten `DesDecoder`, synthesize it and verify that the netlist simulates the same way as did the verilog source. Do not synthesize or simulate the entire `Deserializer` yet.

Step 10 (*Optional*). Synthesizing and simulating the `Deserializer` netlist. We complete this lab by demonstrating that the `Deserializer` from Step 9 now can be synthesized to a netlist which functions correctly at least occasionally.

Under proper constraints, with a 32-word FIFO, the `Deserializer` will synthesize as-is in about 15 hours to some 75,000 transistor-equivalents. However, this netlist probably will not simulate correctly. If you wish to run this synthesis, there is a .sct file available in the answer directory for your reference. It might be a good idea to start this synthesis just before leaving for the day.

A preliminary synthesis already has been done for you at this stage, producing a bad netlist in about an hour (ten or fifteen minutes with incorrect constraints); the result is in a `BadNetlist` subdirectory in your answer directory for Step 10. You also may want to see the Step 8C ending box above for a trick to get a quick and dirty netlist in 5 minutes or so)

The DC of early 2010 took about 45 minutes to synthesize a netlist which would simulate poorly but correctly without SDF back-annotation; however, some 2008 and 2009 DC versions would not succeed with any constraint set attempted by the author.

In your on-disc answers for this lab, there is a directory named OldDC_GoodNetlist which contains files which were synthesized by an older version of DC (2007). You can use new VCS to simulate the old netlist correctly: If you do this, you will be able to see that the ParOutWatch value 32'h3cf1_1979 was serialized from the input at about time 251,737 ns.

Because each Deserializer main subblock (FIFO, PLL, and DesDecoder) now synthesizes correctly, a mixed RTL-netlist simulation of the entire design is possible by putting together the synthesized subblock netlists instead of trying to synthesize a single netlist from the whole verilog design.

For a mixed simulation put together this way, you may leave Deserializer.v and SerialRx.v in source form and create a new DeserializerSubNets.vcs file containing the following list:

```
DeserializerTst.v
Deserializer.v
./DesDecoder/DesDecoderNetlist.v
./FIFO/FIFOTopNetlistSDF.v
./PLL/PLLTopNetlist.v
./SerialRx/SerialRx.v
-v tcbn90ghp_v2001.v
```

The answers for this Step include synthesis scripts to synthesize the subblocks. With the 2010 versions of DC, the SDF timing only is required for FIFO; the other modules will simulate without back-annotation; this is what is shown in the preceding VCS file.

In the .sct files for the answers to this Step, you will notice a considerably increased number of detailed assignment statements, along with an (expected) increased number of supress_message* statements. The increase in assignment complexity is normal as the design itself increases in complexity, which last causes correct synthesis to require more and more complex and detailed constraints.

However you approach it, there probably will be some debugging to do before the Deserializer will simulate reasonably even before synthesis. The main simulation problem probably will be synchronization of the serial clocks: The frequencies must stay within about 1/32 (close to 3%) before a complete 64-bit arriving packet will be aligned correctly and decoded; therefore, the testbench sending-domain serial clock should be set fairly close to the free-running Deserializer PLL base frequency. I have used a half-period delay of 15.5 ns successfully.

Digital VLSI Design with Verilog   443

Some things to consider:
- The `DesDecoder` synthesizable parallel clock generator now is called on just one edge, instead of on any change; therefore, it will run at an average of half of the previous speed. The receiving-domain clock frequency must be checked.
- The `DesDecoder` may call `doParSync` too often, thus perhaps causing the extracted parallel clock to fail because of frequent erroneous resynchronization on zero data rather than on `PAD0`.
- The PLL `` `VFO_MaxDelta `` possibly should be different for source simulation than for synthesis of a correctly-simulating netlist.
- The two VFO operating limits in `VFO.v` probably should be defined simply as `` `DivideFactor ± `VFO_MaxDelta ``, if not already so.
- The number of delay stages (`` `NumElems ``) in `VFO` should be set properly for correct source and netlist simulation. I have found 5 to be a good value here.
- In the top-level testbench, the total duration of the simulation and the sending-domain serial clock speed also may have to be tuned.

It is not important to have the `Deserializer` completed in this lab; there will be time for tuning later in the course. However, all the major submodules should simulate well by now on the unit level (`PLL`, `DesDecoder`, and `FIFO`), both as verilog source and as synthesized verilog netlist.

Success with the entire `Deserializer` system at this point would be indicated by just a couple of sporadic unloads of data from the FIFO onto the output parallel bus. Playing with the `DesDecoder` or the `PLL` probably will make more difference here than anything else. Examples of the waveforms are given in figures 20-7 through 20-9.

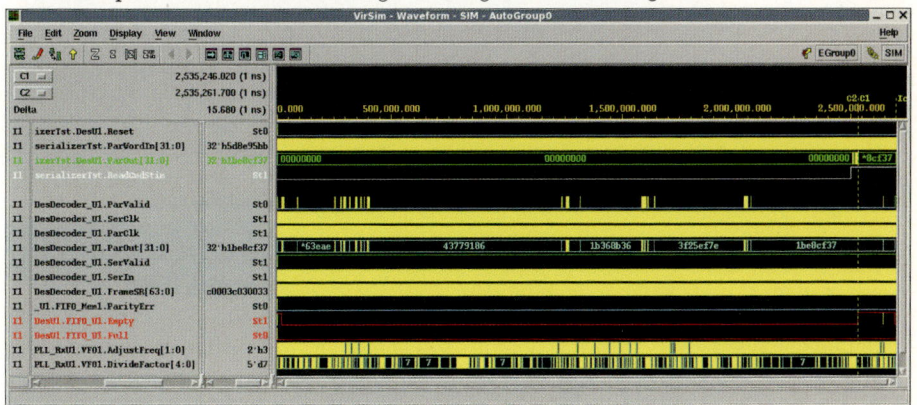

Fig. 20-7. First successful simulation result for synthesized `Lab23 Deserializer` netlist (not SDF back-annotated). The 32-word FIFO clearly goes from empty to a normal state; and, upon receiving-domain `ReadCmdStim`, the stored data are read out onto the parallel bus.

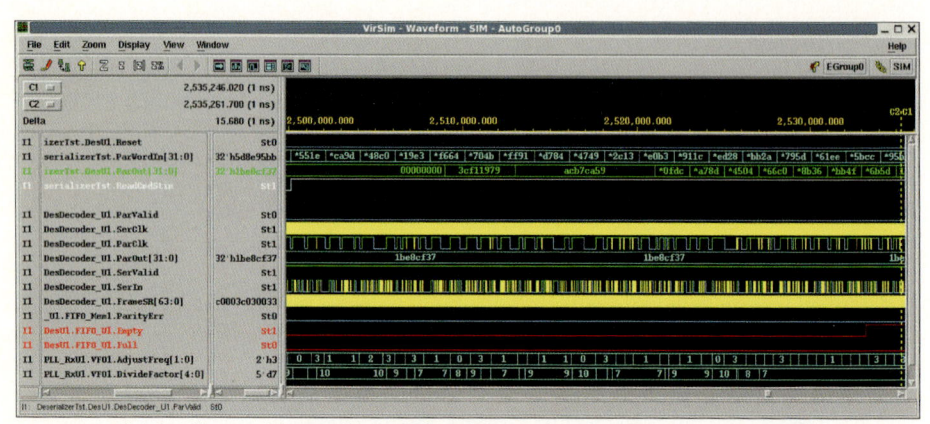

Fig. 20-8. Closeup of the stored data which are being read out from the `Deserializer` FIFO.

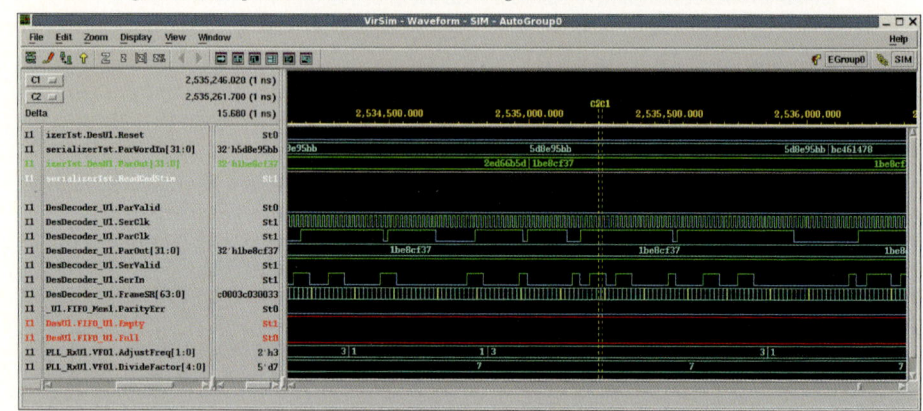

Fig. 20-9. Closeup of the arriving serial data during netlist simulation.

20.2.1 Lab Postmortem

Think about the possible kinds of verilog debugging technique.

20.3 *Additional Study*

Optional; Review `fork-join` functionality in Thomas and Moorby (2002) section 4.9.

Chapter 21
Week 10 Class 2

21 Today's Agenda:

Lecture on the Serializer and SerDes
Topics: Serializer submodules, assembly and checkout of the SerDes.
Summary: The brief lecture is used to outline the functionality of the `SerEncoder` and the `SerialTx`, and to point out the reorganization of the project files.

Lab
The `Serializer` model is completed, and it is attached to the `Lab23` `Deserializer` to form a complete serdes. We simulate and synthesize the result.

Lab Postmortem
We look into possible incompatibilities of our Ser and Des. We also question where assertions and timing checks should be installed.

21.1 The Serializer and the SerDes

Figure 21-1 below is an update of the SerDes block diagram which was presented at the start of *Week 9 Class 2*.

Notice the slight change in the location of the clock divide-by-two, which now is in the Serialization Encoder:

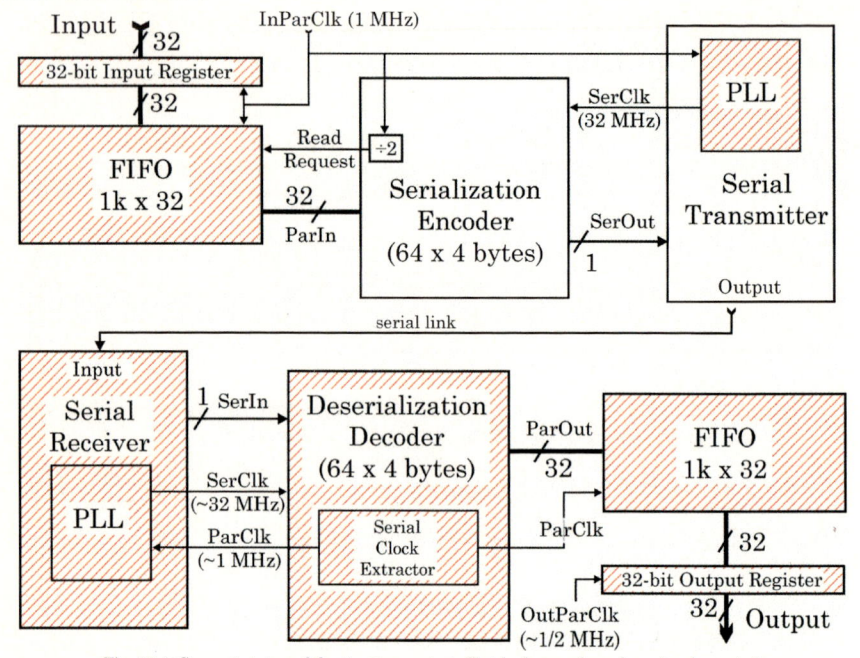

Fig. 21-1. Current status of the SerDes project. Hatched areas have been implemented.

Although the serializer looks almost as complex as the deserializer, it is much simpler. For one thing, there is a well-defined, independently generator 1 MHz input clock with no need for extraction from a stream. The serializer PLL only has to do a phase-locked 32x frequency multiplication to provide the serial output clock. Also, the serializer's FIFO is in a single clock domain.

The Serializer will use exactly the same PLL and FIFO modules as the Deserializer; however, we shall have to create new SerEncoder and SerialTx modules for the sending functionality.

21.1.1 The SerEncoder Module

This module, implementing the Serialization Encoder, reads from the serializer's FIFO to create serial packets.

It shifts out each packet serially to the SerialTx.

Digital VLSI Design with Verilog

21.1.2 The `SerialTx` Module

This module, implementing the Serial Transmitter, is what transmits the data on the serial line.

Because it contains the sending-side PLL, it is used to clock the `SerEncoder` shift register.

21.1.3 The SerDes

With a working `Serializer`, assembling the complete SerDes is trivial: One merely has to instantiate a `Serializer` and a `Deserializer` in a containing `SerDes` module, connect them with a wire, and supply a testbench for the `SerDes`.

This lab will complete the development side of our class project. We shall use the SerDes design later when we study design-for-test. We also shall make lab time available to improvements on the SerDes design during the remainder of the course.

21.2 SerDes Lab 24

Topic and Location: Design and simulation of a Serializer; assembly, simulation, and synthesis of a serdes.

Do this work in the `Lab24` directory.

Preview: We reorganize our files from previous labs so that the `PLL`, `FIFO`, `Deserializer`, and new `Serializer` are in the same working directory. Then, we design the `Serializer` to share packet pad definitions and otherwise to be compatible with our `Deserializer`. We connect the `Deserializer` and `Serializer` together with a single serial wire, and we demonstrate correctness by simulation.

Deliverables: Step 1: Copied previous `Lab23` answer files. Step 2: Reorganized design files tested by successful load into the simulator. Step 3: A correctly configured and assembled `Serializer`, to the extent of successful load into the simulator. Step 4: Completed connections for the `SerialTx`. Step 5: Completed `SerEncoder`. Step 6: A correctly simulating `Serializer`. Step 7: A correctly assembled and simulated, completed serdes project. Step 8: Synthesized, simulated unit netlists for the completed serdes' `PLL`, `FIFO`, `DesDecoder`, and `SerEncoder`. New and old VCS results. Step 9: Optional new and old SDF simulation.

Lab Procedure:

Step 1: Reuse the previous design. Start by creating a new `SerDes` subdirectory in the `Lab24` directory with a `Deserializer` subdirectory under it.

Make a new, complete copy of the provided solution from `Lab23/Lab23_Ans/Lab23_AnsStep10/Deserializer` into the new `Lab24/SerDes/Deserializer` directory. If you wish to debug your own `Lab23` work now, you may use your own files, but the details of the instructions below assume you are using the provided `Lab23` answers.

There is no reason to copy the `Lab23` unit-test files (the testbench files for the individual modules in `FIFO`, `PLL`, and `DesDecoder`) or the synthesis files at this time; if useful, they may be copied from the `Lab23` subdirectories later.

Step 2: Set up the `SerDes` files. Because the PLL and FIFO designs will be the same throughout the SerDes (but different instances may be parameterized differently), some reorganization is appropriate.

The directory structure should be changed as shown in figure 21-2.

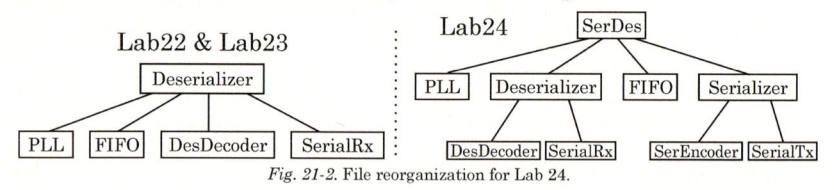

Fig. 21-2. File reorganization for Lab 24.

As shown in the figure, in the `Lab24` half, create a new `Serializer` subdirectory in the top-level `SerDes` directory at the same level as the `Deserializer` subdirectory. Also, to complete the modification of the previous `Lab23` setup shown, move the deserializer `PLL` and `FIFO` subdirectories up one level, so that `Serializer`, `Deserializer`, `FIFO`, and `PLL` all are in the same new, `Lab24` `SerDes` directory.

The new `SerEncoder` and `SerialTx` subdirectories shown in the figure will be created later.

Also, move `Deserial.inc` into the `SerDes` directory, renaming it `SerDes.inc`. This, with `SerDesFormats.inc` below, should be one of the two `.inc` file in the design, if we are going to simulate only the `SerDes`; but, if you have other `.inc` files, just leave them as-is.

Make up a simulator file list in the `SerDes` directory; name it `SerDes.vcs`, and verify the new arrangement by using `SerDes.vcs` to load the new design hierarchy into the simulator. Do not run simulation at this time.

Step 3: Prepare the `Serializer` module. Under `Serializer`, create a new `SerialTx` subdirectory; and, install a `SerialTx.v` file there, with an empty `SerialTx` module to instantiate the PLL. Use the header from `SerialRx.v` temporarily, if you want.

Likewise create a `SerEncoder` subdirectory in `Serializer`; it should contain a `SerEncoder` module in `SerEncoder.v`. Copying the `DesDecoder` header into `SerEncoder.v` may be a good way to reuse your previous work. A schematic of the resulting design is provided in figure 21-3.

Digital VLSI Design with Verilog

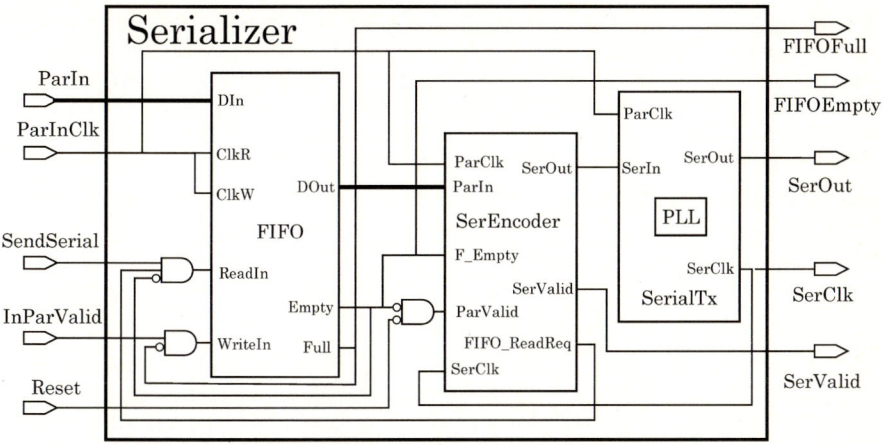

Fig. 21-3. Overall schematic of the completed `Serializer`. Not all `Reset` targets shown.

The serial packet format assumes the same framing on both ends of our `SerDes`; so, copy the pad `localparam` assignments and other shared information to a new include file, `SerDesFormats.inc`, located in the `SerDes` directory. In `Lab23`, these were in the `DesDecoder` module. Copy the comments explaining the format, too, in case debugging may become necessary. However, keep in mind that the specific format of the packets and their padding is parsed in detail from the `DesDecoder` verilog and is formed by detailed verilog coding in the `SerEncoder`, so this new include file has no portability function.

The file contents other than comments should be the following:

```
// SerDesFormats.inc:
localparam[0:0] YES  = 1'b1;       // Readability experiment.
localparam[0:0] NO   = 1'b0;       //      "                "
localparam[7:0] PAD3 = 8'b000_11_000;
localparam[7:0] PAD2 = 8'b000_10_000;
localparam[7:0] PAD1 = 8'b000_01_000;
localparam[7:0] PAD0 = 8'b000_00_000;
```

In this file, we again introduce various `localparam`s which in general provide a naming advantage for the `PAD`s. The `YES` and `NO` names may not be of much advantage for `1'b1` or `1'b0`—what do you think? The value of factoring out the pad format this way is that if the pad format should be changed in the future, it will be changed the same way both for encoding and decoding, thus avoiding possible maintenance problems. However, any pad format change will require detailed rewriting of content in both the Ser and the Des.

In the `Serializer` directory, declare a `Serializer` module in a file named `Serializer.v`, and instantiate in it `FIFOTop`, `SerEncoder`, and `SerialTx`.

Use the corresponding `Deserializer` modules as guides; you probably won't have to add new ports or connections, although names and directions will change. Be sure to retain parameters to size the FIFO and the address and data ports; the `Deserializer` parameters should be used directly for this. Use the same parameter names and defaults for the new serializer as you did for the `Deserializer`; the values always can be overridden differently, if desired, during instantiation.

Your `Serializer` ports should be:

> **Outputs**: `SerOut, SerValid, FIFOEmpty, FIFOFull, SerClk`.
> **Inputs**: `ParIn` (32 bits), `InParValid, ParInClk,`
> `SendSerial` (the request to send), `Reset`.

In the `Serializer`, you will want a control input to assert a request to <u>read</u> the (parallel) FIFO output into the serial encoder. Recall that in the `Deserializer`, you had instead a control input to <u>read</u> the FIFO output into the receiving system. By controlling FIFO read at both ends, the FIFO is most properly used as a buffer during continuous communication. Of course, on a FIFO-full or FIFO-empty condition, write may have to be controlled, too; but, we shall leave these problems to the protocol of the system containing our serdes.

The `Serializer` also should compose a FIFO write command to move new data into the FIFO from the parallel bus. Thus, the `Serializer` controls FIFO read and write requests, in addition to routing the 1-MHz parallel-data input clock (`ParInClk`) to the FIFO. `ParInClk`, of course, should be connected to both FIFO clock inputs.

The `Serializer` should provide a FIFO-valid control (`F_Valid`) for the `ParValid` input pin of the `SerEncoder`, to flag usable parallel data from the FIFO. This can be done very simply by a continuous assignment,

>   `assign F_Valid = !F_Empty && !Reset;`

The other controls just described may be provided this way:

>   `assign F_ReadReq = !F_Empty && SerEncReadReq && SendSerial;`
>   `assign F_WriteReq = !F_Full && InParValid;`

Here, the `F*` refer to FIFO pins; also, `SerEncReadReq` is `FIFO_ReadReq` from the `SerEncoder`; and, `InParValid` and `SendSerial` are `Serializer` input ports.

After completing the first cut at the `Serializer` as above, in the `Serializer` directory create a do-nothing testbench in a separate file named `SerializerTst.v`. Use it to simulate briefly to check your connections and file locations.

Also, create empty placeholder files named `SerDes.v` and `SerDesTst.v` in the `SerDes` directory.

The resulting file locations should be as shown in figure 21-4.

Digital VLSI Design with Verilog

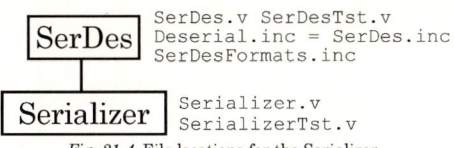

```
┌────────────┐   SerDes.v  SerDesTst.v
│   SerDes   │   Deserial.inc = SerDes.inc
└─────┬──────┘   SerDesFormats.inc
      │
┌─────┴──────┐   Serializer.v
│ Serializer │   SerializerTst.v
└────────────┘
```
Fig. 21-4. File locations for the Serializer.

Step 4: Complete the serial transmitter. This module, `SerialTx`, should be very simple, like the `SerialRx` module of the Deserializer. All it need contain is (*a*) a simple continuous assignment statement passing the serial data through from input port to output port and (*b*) a properly connected `PLL` instance.

Step 5: Complete the serialization encoder. Refer to figure 21-3 above, which gives a Serializer block diagram, and to the schematic figure 21-5 below, which gives connectivity details.

We shall design this module for synthesis.

The `SerEncoder` module has to use the serial and parallel clocks to frame the data and transfer it serially to the `SerialTx` module. The clocks are independent of this module's functionality and are guaranteed by the `SerialTx`'s PLL submodule to be phase-locked, so there is no need to extract phase or frequency information. However, the serialization has to be buffered so that it is not interrupted unless the FIFO, which provides the parallel-side input data, has become empty.

Recall that the 2:1 divided clock on the deserializer end of the SerDes (`Lab22`) could be provided by the `Deserializer` testbench; this was because the Des was operating at 1 MHz and was providing correct parallel data all by itself; it was up to the receiving domain to use the latched `ParOut` data properly. This is not feasible on the Ser end, because the Ser not only has to operate at 1 MHz, too, but it also has to determine the rate at which parallel data will be copied in and serialized. This means that a 1/2 MHz derived clock has to be part of the `Serializer` design. We shall derive this clock in the `SerEncoder`.

We can obtain the 1 MHz clock (`ParClk`) provided to the `Serializer` from an external source (initially, your `SerializerTst` verilog testbench). The `SerEncoder` therefore must contain a block, shown in figure 21-5, doing a simple frequency-halving of the `ParClk` input and using it to request FIFO reads.

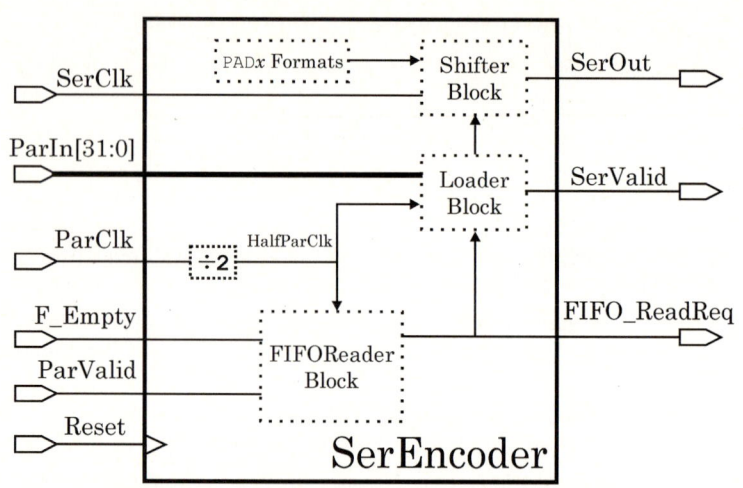

Fig. 21-5. The `SerEncoder`. Dotted outlines indicate blocks of code, not hierarchy.

In addition to this, we shall define three other functional blocks within the `SerEncoder` module, as shown in figure 21-5:

- `Loader` block. Read on every `posedge` of the half-speed `ParClk` (`HalfParClk`), this `always` block will do two things:

    1. Load the `SerEncoder`'s 32-bit buffer with a new word from the FIFO, or with all 0's if the FIFO was empty.

    2. Assert or deassert `SerValid`, depending on whether the FIFO was empty.

- `Shifter` block. This `always` block will be read on every `posedge` of `SerClk`, and exclusively will control the serial data values.

    `Shifter` simply lets its counter wrap at 64 and uses the count on each clock to determine (mux) whether a `PAD` bit, or a data bit from `SerEncoder`'s input buffer, is applied to the serial line out.

- `FIFO Reader` block. This block asserts a FIFO read request on every other `ParClk` to make available a new 32-bit data word for framing. It depends upon the 1/2 MHz clock, `HalfParClk`, derived from `ParClk`.

    There is no reason not to implement this block as simple combinational logic, for example as,

    ```
    assign FIFO_ReadReq
        = HalfParClkr && !F_Empty && ParValid && !Reset;
    ```

The complete block diagram of `SerEncoder` is what was shown above in figure 21-5.

Step 6: Use `SerializerTst` to simulate the `Serializer` until it operates correctly. One thing to check is that the FIFO should stop accepting writes when it is full. Also,

Digital VLSI Design with Verilog   453

parallel data should be flagged as invalid when the FIFO is empty. When the FIFO goes empty, the serial data should be flagged invalid as soon as the last valid packet has been shifted out.

Typical `Serializer` simulation results are shown in figures 21-6 through 21-8.

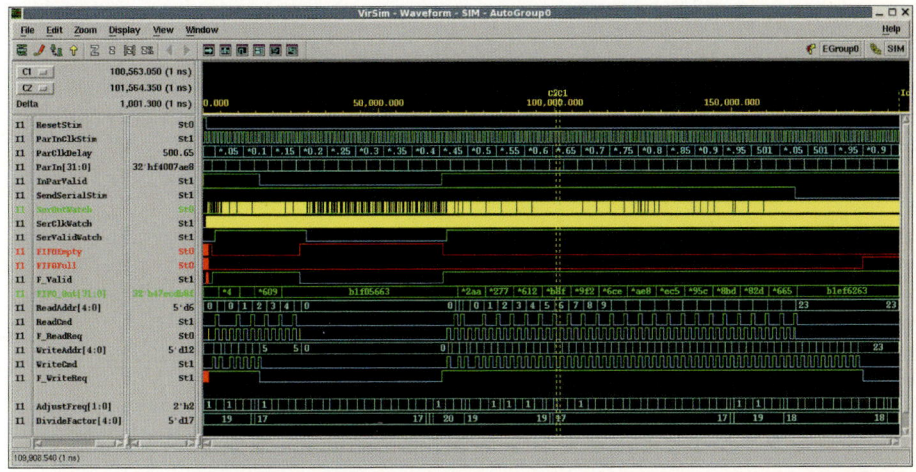

Fig. 21-6. Overview of the first good `Serializer` verilog source simulation. The FIFO clearly goes full and empty under reasonable conditions.

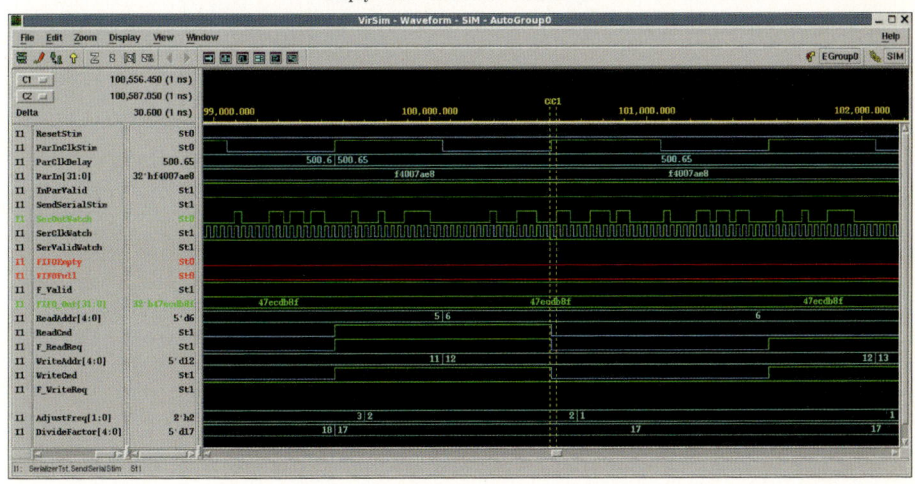

Fig. 21-7. Zoom in on first good `Serializer` source simulation, showing proper handling of the individual parallel-bus words.

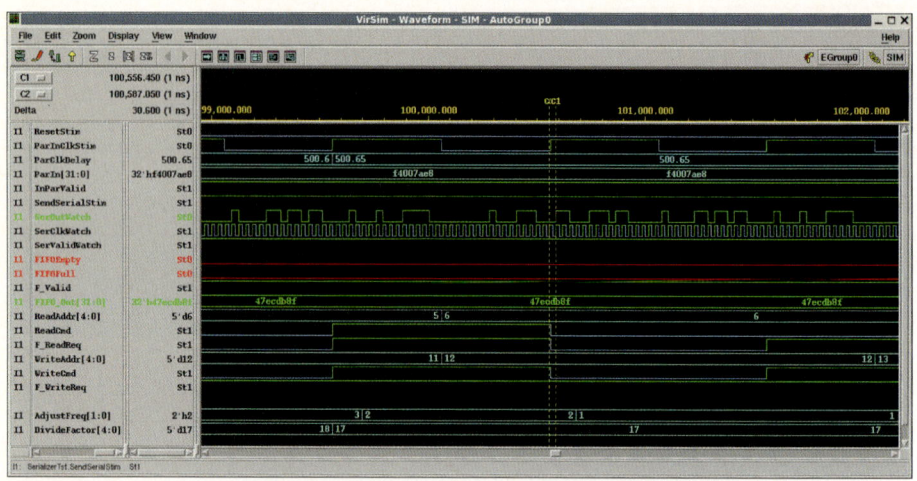

Fig. 21-8. Very high zoom in on the `Serializer` source simulation, showing individual serial-line clock cycles.

Step 7: SerDes simulation. In this Step, you will connect the `Serializer` serial output to the `Deserializer` serial input by instantiating both in a new SerDes module. You probably will have to rename some wires, for example to make collateral outputs from both halves distinct.

Before connecting the `Serializer` and `Deserializer` instances and simulating, use VCS to verify that the unconnected `Serializer` and `Deserializer` hierarchies, now residing in SerDes, are as expected. This is shown in the next figures, 21-9 and 21-10.

Digital VLSI Design with Verilog 455

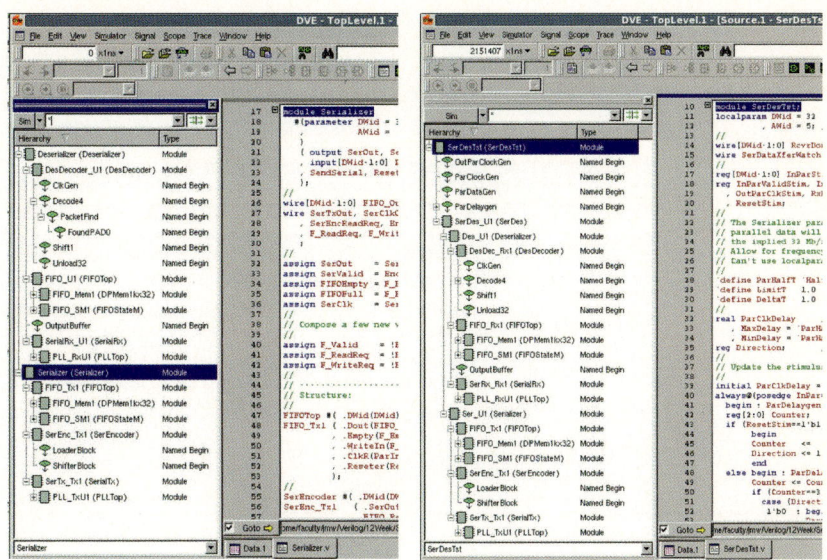

Fig. 21-9. The unconnected hierarchies. *Fig. 21-10.* The connected hierarchy.

Simulate, using at first your `SerializerTst` testbench (renamed to `SerDesTst`). Any serial transfer at all should be considered a complete success at this point. Depending on your FIFO state machine design, you may have to assert a reset twice to initialize everything at the beginning of the run.

Some typical `SerDes` simulation results are shown in figures 21-11 and 21-12. Both old and new VCS simulations are shown, just to emphasize that they are functionally interchangeable at this level, even though identical simulator regions are not displayed:

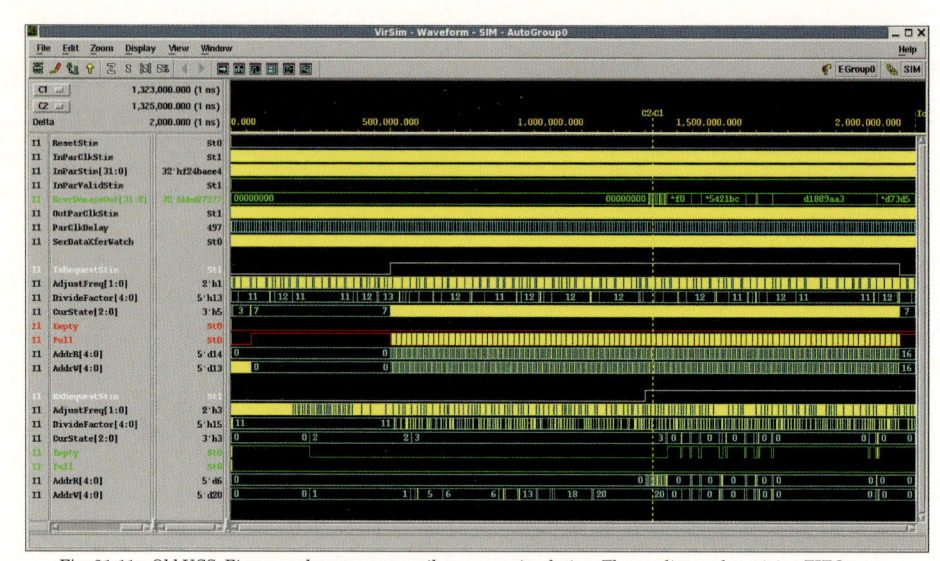

Fig. 21-11a. Old VCS: First complete `SerDes` verilog source simulation. The sending and receiving FIFOs seem to be operating reasonably.

Digital VLSI Design with Verilog 457

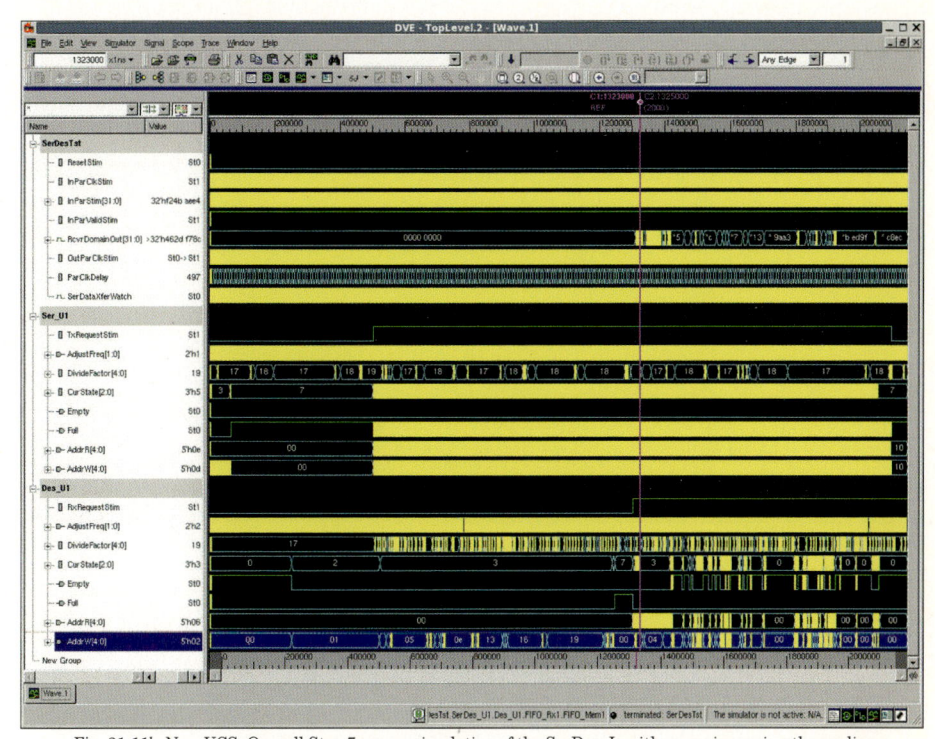

Fig. 21-11b. New VCS: Overall Step 7 source simulation of the SerDes. In either version, using the on-disc answers, the deserialized packet with data `32'h462d_f78c` appeared on the parallel-data input bus at time 28,598 ns.

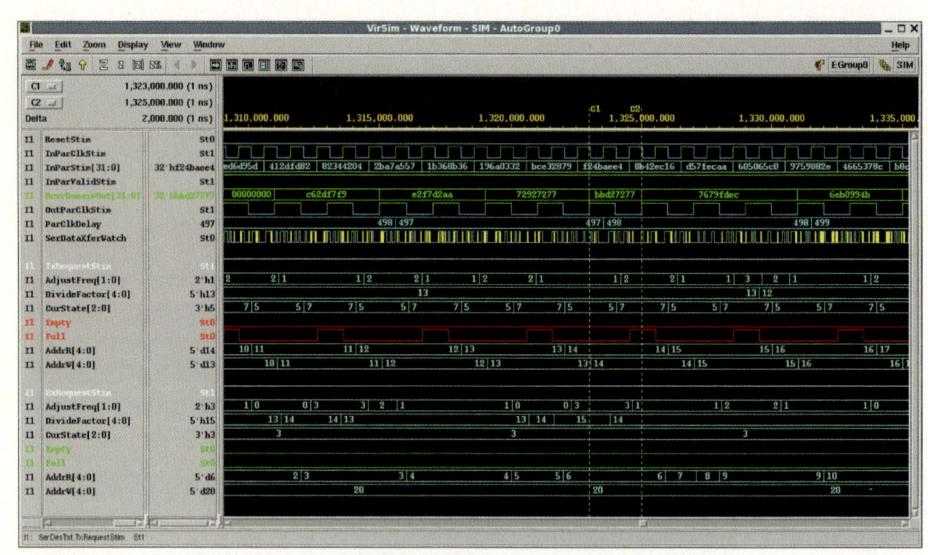

Fig. 21-12a. Old VCS: Closer zoom in on the first complete `SerDes` verilog source simulation, showing individual parallel-bus words in the receiving domain.

Digital VLSI Design with Verilog   459

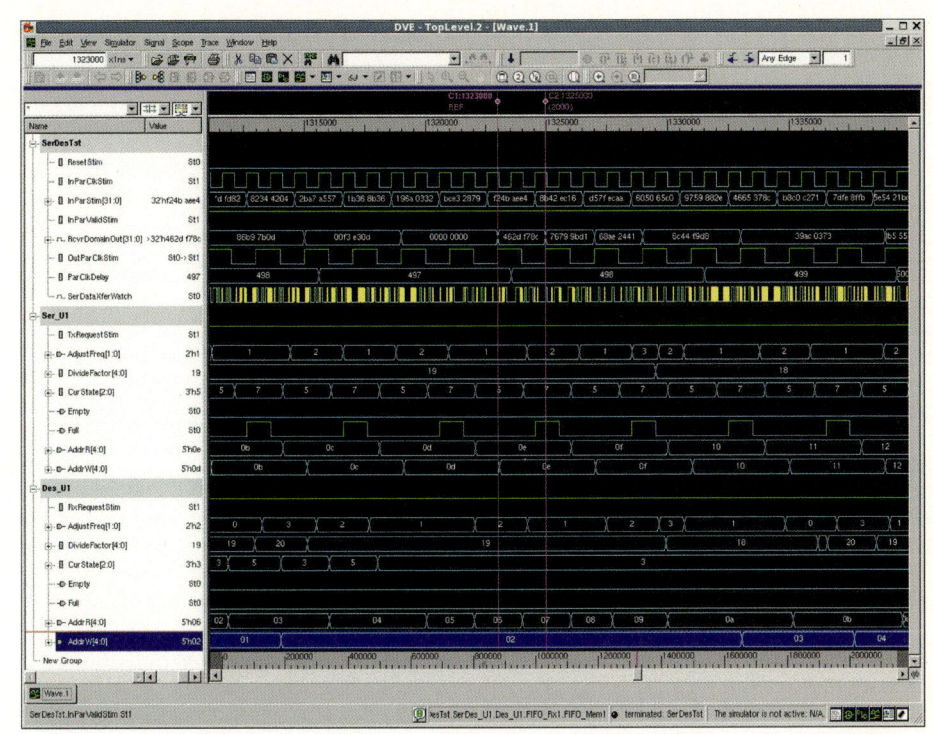

Fig. 21-12b. New VCS: Closeup of the Step 7 source simulation, showing the new VCS output just described.

Optional: After your SerDes is working, modify the Deserializer to have two different parallel data outputs, one clocked out as previously, on the receiving-domain clock, and a new one clocked out on the deserializing PLL embedded clock (ParClk), which is in a different domain.

Then, simulate. Notice the skew in the appearance of the 32-bit, clocked-out data. This is an illustration of the digital side of the clock-domain synchronization problem. Of course, a digital simulator can not display the more serious analogue problem of intermediate-value hangups which we discussed in *Week 8 Class 1*.

Step 8: SerDes unit-level synthesis. A synthesis of the full SerDes system will take too long for classroom scheduling at this point, because system timing refinements may require several full SerDes synthesis iterations to get the complete design working.

For this lab session, just verify that each major component of the SerDes can be completed on the unit level.

To do this, synthesize, under reasonable constraints, the following, each in its own, preexisting subdirectory:
- PLL
- FIFO
- Deserializer.DesDecoder
- Serializer.SerEncoder

The Deserializer.SerialRx and the Serializer.SerialTx are essentially empty wrappers, containing just a PLL instance, so there is no reason to synthesize them now: To synthesize these, you just could do PLLTop alone.

You may wish to look at the synthesis scripts in the answer subdirectories. Although the DesDecoder is a relatively small design, it requires tight setup and hold constraints for a perfect netlist, so the synthesizer will take hours of adjusting of the design rules to finish. Read the comments in the answer script provided to see how to sidestep this long wait.

Likewise, the FIFO, a large design for a single module, can take a very long time. Check the comments in the FIFO answer synthesis script, FIFO/FIFOTop_Hold_wDesRule.sct, for advice on obtaining a usable netlist early in the run. If hold-fixing is omitted, for example, the DesDecoder, SerEncoder, and PLL netlists should take less than 20 minutes each on a 1 GHz machine.

Synthesis in late-2011 DC of a good FIFO netlist took over 12 hours on a 1 GHz machine; the long run was because of the special need explicitly to fix hold times (as well as the usual setup times). By compiling with hold-fixing but without design rules, a good synthesis of the FIFO block should take less than a half-hour. In regard to this, see the Lab24_Ans_*.sct files for details.

After synthesis of the individual units listed above, use the same simulation testbench as you did for the source, to simulate each synthesized netlist. You should consider reusing simulation testbenches from previous labs as starting points. The unit-level netlists should simulate almost as well as, or better than, the source verilog for them.

Optional: During each unit synthesis, write out an SDF file; then, using your tcbn90ghp_v2001.v library, simulate the netlist. This library includes delays which will be overridden by SDF timing wherever the latter is available. The DC-generated SDF omits some delays, so a few paths in the netlist still require nonzero timing, which will be supplied by default by tcbn90ghp_v2001.v.

You may leave the synthesized netlists and side files in the individual subdirectories; those directories will not be written in higher-level simulation or synthesis runs.

Figures 21-13 through 21-21 show some typical netlist simulation results, using the approximated timing in the provided verilog-2001 library. SDF back-annotation makes for a slight difference in the simulation waveforms.

Digital VLSI Design with Verilog 461

The `PLL`:

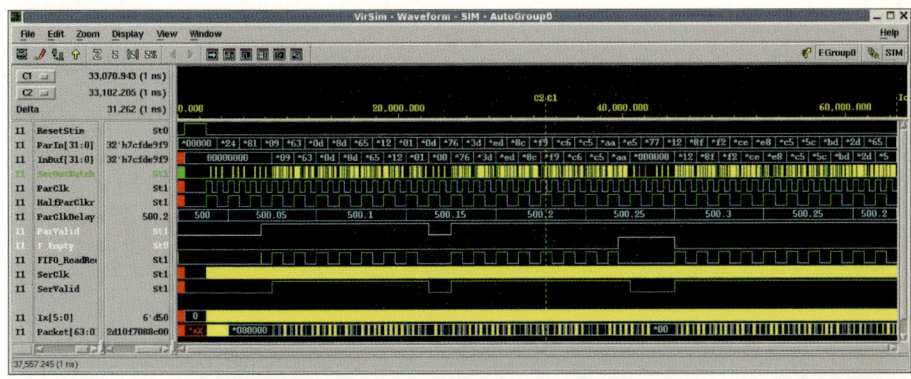

Fig. 21-13. Overall view of synthesized `PLL` netlist simulation.

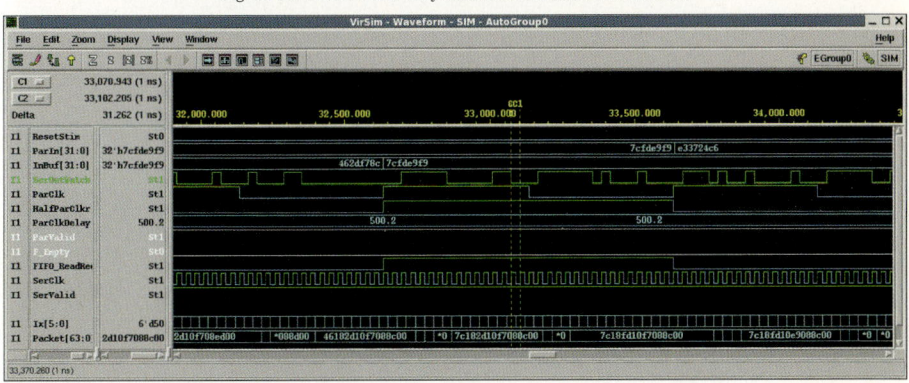

Fig. 21-14. Zoomed-in view of synthesized `PLL` netlist simulation, showing individual serial clock cycles.

The FIFO:

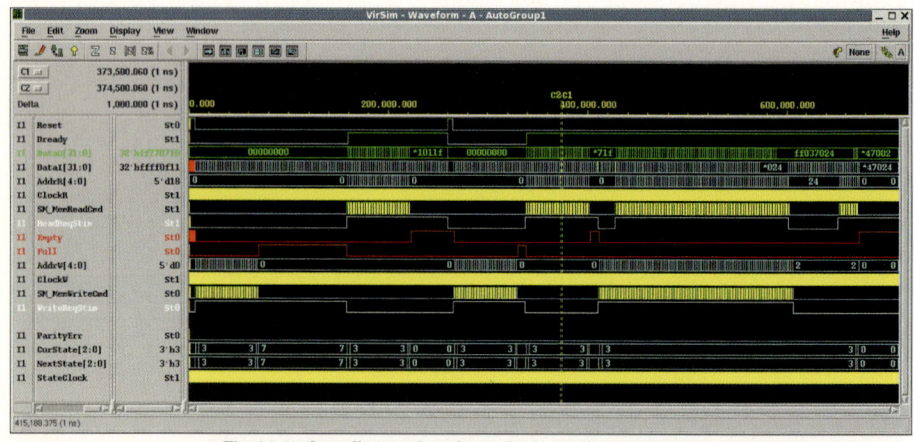

Fig. 21-15. Overall view of synthesized FIFO netlist simulation.

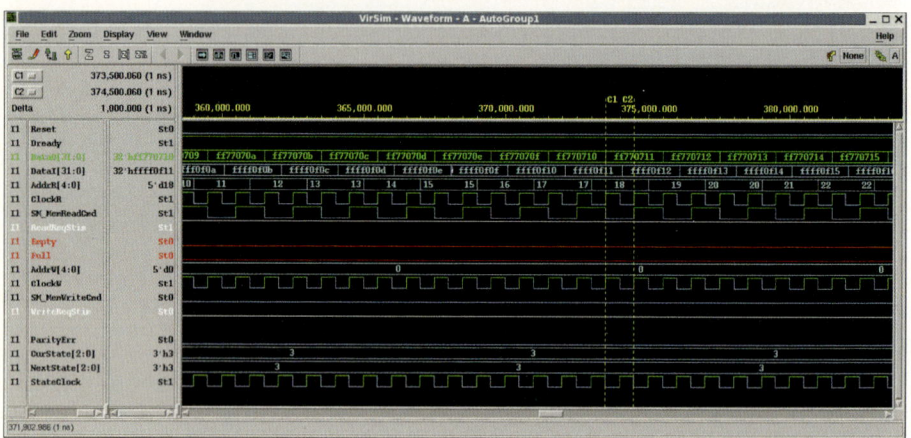

Fig. 21-16. Zoomed-in view of synthesized FIFO netlist simulation, showing individual 32-bit word transfers.

Digital VLSI Design with Verilog 463

The `DesDecoder`:

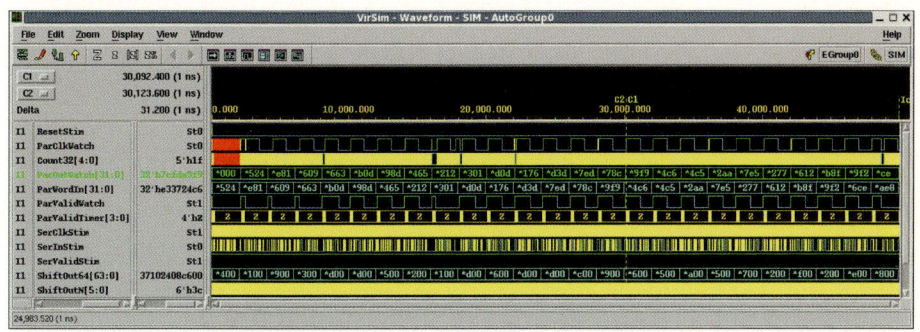

Fig. 21-17. Deserialization decoder (`DesDecoder`) overall view of synthesized netlist simulation.

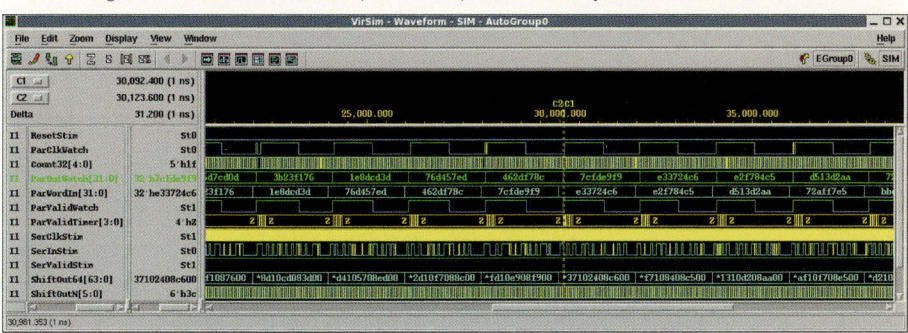

Fig. 21-18. `DesDecoder` netlist simulation, zoomed in to resolve individual parallel-bus words.

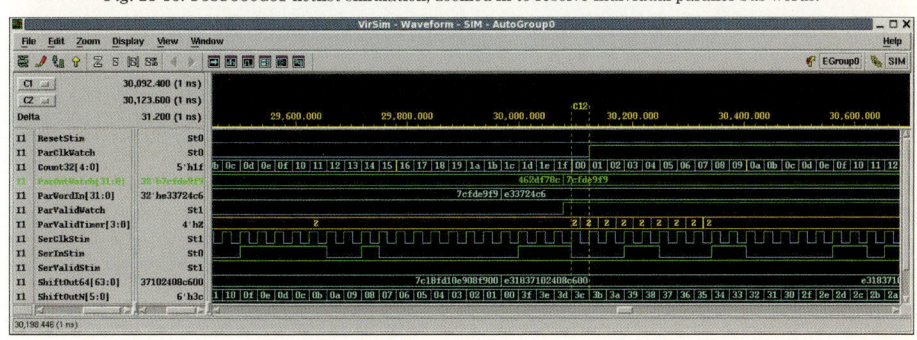

Fig. 21-19. `DesDecoder` netlist simulation, zoomed in to resolve individual serial clock cycles.

The `SerEncoder`:

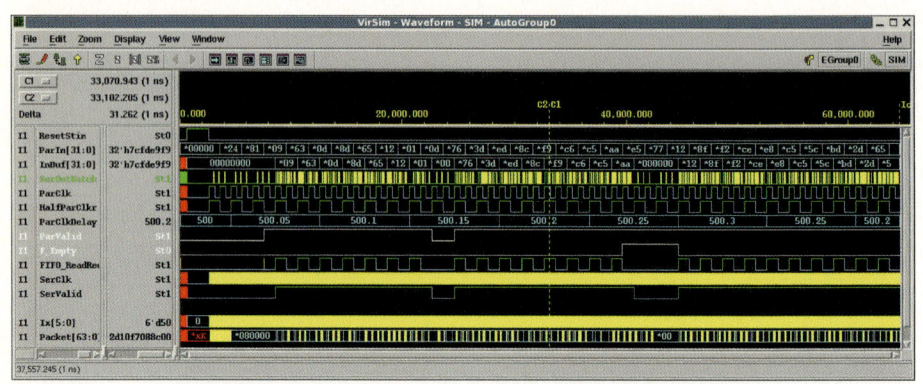

Fig. 21-20. The `SerEncoder`: Overall view of synthesized netlist simulation.

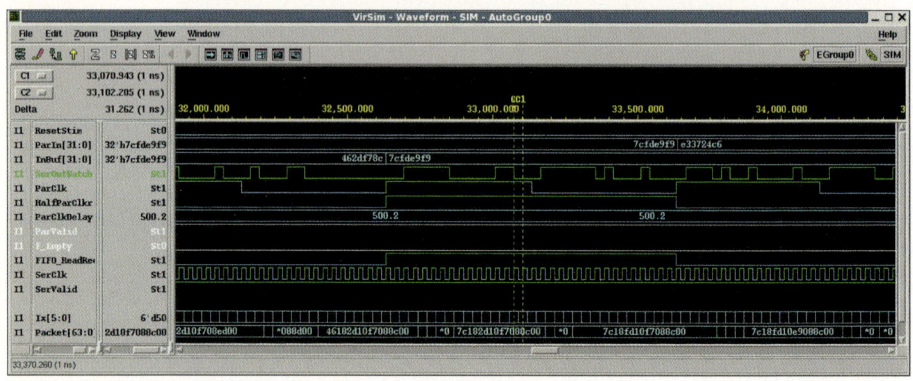

Fig. 21-21. Zoomed-in view of `SerEncoder` netlist simulation, showing individual serial clock cycles.

Step 9 (Optional): If you performed this optional part of the lab, by creating SDF files for each major component of the SerDes, you may quite easily simulate the entire `SerDes` hierarchy in a mixed RTL-netlist configuration. This would be how any large design would be synthesized and simulated during development.

Digital VLSI Design with Verilog

To do this, create a `SerDesSubNets.vcs` file with the following contents:

```
./SerDesTst.v
./SerDes.v
./Serializer/Serializer.v
./Serializer/SerEncoder/SerEncoderNetlistSDF.v
./Serializer/SerialTx/SerialTx.v
./Deserializer/Deserializer.v
./Deserializer/DesDecoder/DesDecoderNetlistSDF.v
./Deserializer/SerialRx/SerialRx.v
./FIFO/FIFOTopNetlistSDF.v
./PLL/PLLTopNetlistSDF.v
-v tcbn90ghp_v2001.v
```

The `*SDF.v` netlist files are the ones with `$sdf_annotate()`, which you probably used during the optional SDF-based simulation in Step 8. The other files are unsynthesized verilog source—they could be synthesized hierarchically to netlists and then edited for individual use, but this would be somewhat time-consuming and unnecessary, in view of the fact that they are simple enough that one may assume that the synthesizer easily would have created netlists which would simulate successfully.

During simulation of the design subnetted as above, the new VCS command window will report the SDF annotation as follows:

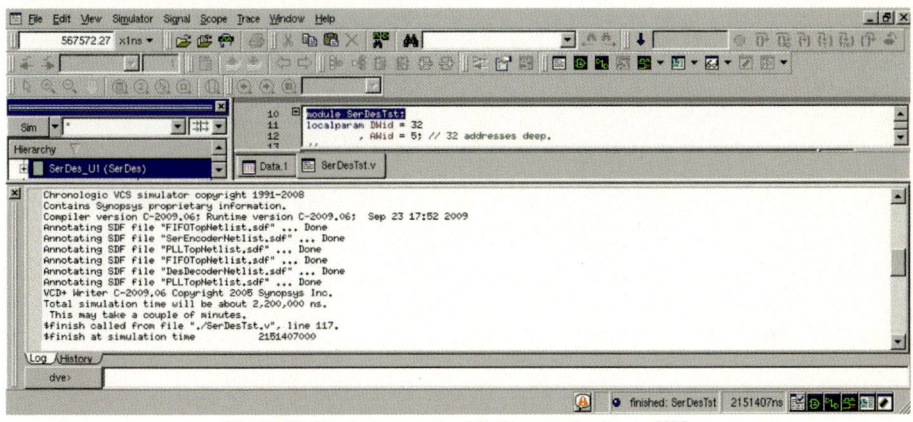

Fig. 21-22. Simulation reports of SDF annotation in new VCS.

Given this, an example of the overall `SerDes` waveform display would be

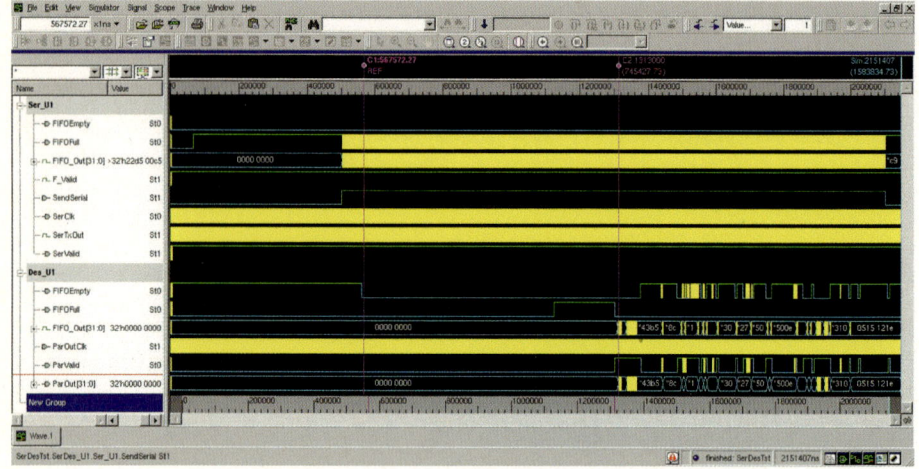

Fig. 21-23. SerDes simulation waveform display after SDF annotation in new VCS.

The two time markers visible above are set at 567,572.27 ns, when parallel data word 32'h22d500c5 was clocked out of the `Ser_U1` FIFO, and at 745,427.73 ns, when that word was clocked onto the `Des_U1 ParOut` bus.

Digital VLSI Design with Verilog

The next figure is a closeup of the time around which the `32'h22d500c5` word arrived on the Des output:

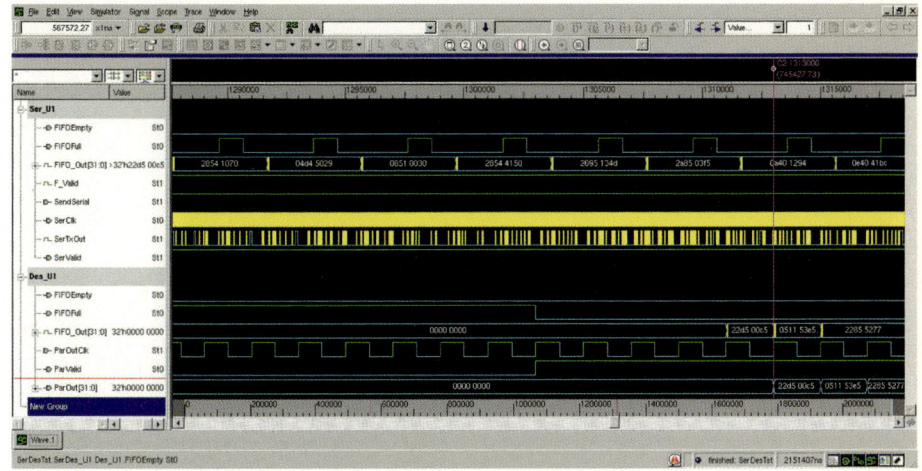

Fig. 21-24. Closeup simulation of the Des output after SDF annotation in new VCS.

There remain a few netlist glitches, but the `SerDes` clearly is operating more or less correctly.

21.2.1 Lab Postmortem

What might be possible incompatibilities of the Ser vs. Des devices?

What about assertions and timing checks for the SerDes?

21.3 Additional Study

Optional: Compare your shift register with the switch-level MOS shift register in Thomas and Moorby (2002) section 10.1. Why would you want to write a switch-level model if a logic synthesizer was available?

Chapter 22
Week 11 Class 1

22 Today's Agenda:

Lecture on DFT
Topics: Basics of DFT, observability, coverage, boundary scan, internal scan, and BIST.
Summary: After introducing DFT as a methodology, we relate it to assertions and then explain the meanings of observability and coverage. We then introduce boundary scan, internal scan, and BIST as the major DFT techniques.

Lab 25 on DFT
We recall an optional `Lab05` insertion of internal scan by the synthesizer. We show how to attach I/O pads in a verilog design for automated synthesizer insertion of boundary scan. We put a memory in a wrapper and attach a BIST module to it.

Lab Postmortem
We discuss the effect of the BIST on memory die size.

Lecture on full-duplex SerDes
Topics: Implementation parameters for connecting two of our SerDes into a full-duplex lane.
Summary: We give special FIFO depth requirements for our full-duplex system, and we discuss addition of DFT functions.

Lab 26 on a DFT, full-duplex SerDes
After putting together a working, full-duplex serdes, we add assertions, timing checks, and automated internal scan.

Lab Postmortem
We discuss the effect of the scan elements on SerDes size.

22.1 Design for Test (DFT)

22.1.1 Design for Test Introduction

Design for Test (DFT) is more an orientation, or perhaps a methodology, than a design technique. DFT means that the hardware will be testable, which is to say that, after implementation, its functionality will be verifiable. Testability usually is planned on the assumption of two major sources of malfunction, design errors and hardware failures.

<u>Design errors</u> can be found during testing only by observing a failure of functionality. Most of these errors will be found and fixed during simulation or synthesis; finding and fixing them means testing both (*a*) of the design source and (*b*) of the back-annotated netlist following floorplanning or completed layout. Errors of logic or of timing can be found and corrected before tape-out for mask creation.

Hardware failures, of course, can be discovered only after implementation and production of the hardware. The most serious of these are fabrication quality or physical design errors in which one or more individual gates become stuck in a nonfunctional state in every unit produced. This kind of highly localized failure cannot be detected unless it has an observed effect during testing of a design prototype or of the hardware units affected. Less serious are random fabrication errors which affect occasional units in a production run. Defective units must be eliminated before delivery to a customer; but, for this to be done, the defect in each unit must be observed during hardware testing.

Other hardware errors may occur after bonding and packaging, in the form of short- or open-circuit defects; these latter are serious but not difficult to discover, because they are on the boundary of the IC and thus can be observed easily.

There also are "soft" errors caused temporarily by thermal or mechanical stress, unexpected external electric or magnetic fields, or exposure to ionizing radiation. An error is "soft" if it spontaneously recovers itself permanently, or vanishes, after a reset or other change of device state. An intermittent hardware failure at a specific gate does not represent a soft error but rather a hard defect which has degraded that gate. In this course, we shall not be concerned further with soft errors.

22.1.2 Assertions and Constraints

The first line of defense against hardware failure is good software. This means not just good design specifications, but also meticulous attention to warning messages from the simulator, synthesizer, static timing verifier, or other tool. In addition, good software means good design insight into possible problems, by use of assertions and well-chosen simulation unit-test vectors. The functional specifications must be fully validated before attempting to create a hardware representation. Also very important are the timing checks in the verilog or synthesis library which monitor fulfillment of the library-level design constraints during simulation.

We have discussed the software side already; now let us look more closely at the hardware.

22.1.3 Observability

This refers to the capability to measure functioning of logical transformations and data transfers in the hardware. A chip would be 100% observable if every internal state could be measured externally. Ignoring sequences of tests to be applied, this reduces to measurement of every internal storage device Q (flip-flop; latch; memory cell) and every input pin I in the chip. The number of inputs then may be written as N_I and the number of internal storage devices as N_Q. Each test measurement in a digital device amounts to one bit of information, which makes for a factor of 2 in the number of states; so, the total number N of possible storage and input states to test then must be,

$$N = 2^{N_Q + N_I}.$$

Equipping a device with a simple thing such as an 8 kbyte cache adds as much as $2^{8*8*1024}$ = about $10^{20,000}$ new states to observe. This factor then is multiplied by the number of states calculated without the cache. It is easy to see why testability usually is expressed in terms of the number of pins and registers (a logarithm), rather than number of states.

A small board-level design can be made highly observable by providing an array of electrical test points on the traces on the board. Such test points are set up like nail-holes on a board; so, a "bed of nails" automated tester then can bring its probes in contact with these test points, stimulate the board, and observe the effects. Any failure ideally can be detected and localized for manual rework, or discard, of a board found to be defective.

Such an approach is impractical for large digital IC's, which have to be manufactured protected against electrical influences except through their I/O pads. Even probing the I/O pads without risking damage is mechanically difficult when the pads may number in the thousands for a modern ball-grid chip package.

Thus, observability has to be designed into the chip itself; it can not be left as an afterthought for someone to worry about after they receive the untested package or wafer in its final form.

22.1.4 Coverage

Coverage is a metric (statistic) which describes the quality of a sequence of test vectors during simulation or hardware testing. Coverage is related to observability. Given the total number of points observable under the test assumptions (I/O's only; or, all registers; or, whatever), coverage describes the fraction of them exercised by the given sequence of test vectors.

Of course, all points are observable in a software simulation. For coverage purposes, the test protocol may include the restriction that only I/O's will be monitored; this permits the software test vectors to be reused by a hardware tester. In such reuse, the simulated results would be compared with the hardware tester measurements in order to validate the hardware functionality.

The idea behind coverage metrics is to verify the absence of failed design pins, ports, or gates; failures of these usually can be described as "stuck-at" faults: A failing element gets stuck at a value of "1" or "0". If the test vectors can toggle every observable point, then coverage by those vectors is 100%. A *fault simulator* is a specialized logic simulator which checks for toggling of observable points.

In hardware testing, a sequence of test vectors may be more or less efficient at achieving a given coverage. Reducing ("collapsing") the number of vectors to achieve a given coverage is a desirable goal in testing, because it reduces the time required for the hardware tester to achieve the coverage. Time is money during manufacture.

Coverage can be used to describe a set of test vectors with reference to all points, not just to those observable under the given assumptions; in this case, the maximum achievable hardware coverage usually will be less than 100%.

Be aware through all this, that ECC in the hardware can correct many defects which are missed in testing; however, this only works for stored data, not for control logic. The goal of testing is to detect and eliminate all defects possible.

Coverage summary:

Coverage is a metric representing thoroughness of testing of observable states:
- Coverage is a measure on a set of test vectors.
- 100% coverage means all observable states have been tested.
- Low coverage means new or better vectors have to be used.
- Coverage usually is considered poor until it reaches 95% or more.

Coverage can refer to software (simulation) vectors:
- Represents lines of code or statements executed.
- Highly order-dependent. Non-Markovian.
- Independent of functional verification.

Coverage can refer to hardware vectors:
- Represents random defects checked.
- Generally only hardware stuck-at fault detection.
- Represents hardware functional verification.

22.1.5 Corner-Case vs. Exhaustive Testing

Hardware testing is a sophisticated field, and we shall touch on just some of its principles here. An instructive article on setup and hold corners in a netlist is the one by Bansal, *et al*, cited in the References near the beginning of this Textbook.

Recalling the 8-kB cache computation above, it is completely unrealistic to assume 100% hardware coverage of an entire chip of any practical size. To test all combinations of states possible, looking for random defects and assuming a test vector of length 1024 bits, and a vector application rate of 1 GHz, the 8 kB cache example above would take about $10^{19,980}$ years to complete. This is calculated from 10^7 seconds/year = 10^{9+7} vectors/year = $10^{9+7+3} = 10^{19}$ bits/year. The age of the known universe is only about 10^{10} years.

Happily, one can assume that not all states must be observed to be sure of the hardware functionality. For example, interaction between adjacent storage cells in a memory array on chip is very possible; so, a defect might be observed in one cell only when adjacent cells were in a certain state. However, such interaction is very unlikely among cells separated even by one other cell. For this reason, applying an alternating pattern, for example 'b010101... vs. 'b101010..., to a cell vector (register) generally reveals any single-bit defect or defect dependent on any two adjacent cells. Combining this with a "walking 1" and "walking 0" test (shifting a single, isolated '1' through the register and then shifting an isolated '0'), one can assume at a certain level of confidence that all states which would reveal a defect in a hardware register had been observed.

Let's look at the 8 kB cache example again, assuming that the memory is organized (in at least two dimensions on the die, remember) so that only adjacent bits in a stored byte can interact. Then, alternating patterns as above, twice applied each, require 4 vectors per byte. Walking '1' and '0' require, say, 8 vectors each, for a total of 20 8-bit vectors to test one byte. Thus, 8 kB would require 8 x 1024 x 20 8-bit vectors, or a little over 10^6 bits of state. Assuming 1024-bit hardware test vectors, this is only 1.3 k vectors. At 1 MHz, the time to test one cache memory this way would take only a millisecond or two. In practice, a millisecond easily would be available to test individual 8 kB cache memory dice on a wafer; so, more elaborate test vectors, for example vectors validating adjacent pairs of bytes, could be accommodated.

The term ***corner case*** usually refers to the spatial or temporal boundary of a range of values. The last cache test vector example above in a sense was a corner case, because it was intended to test adjacency, whereas, only a small fraction of all memory bits can be adjacent to those being tested at any one time.

More usually, corner case testing refers to selection of isolated values to test. For example, in a design including a verilog `for` loop that iterated through `i=0` to `i=127`, one would pay special attention to the values `0, 1, 126,` and `127` as the software corner cases of the loop. And, watch out for `-1` and `128`!

Spatially, physical design problems should be sought especially on the boundaries of voltage islands or clock domains. And, in defining hardware test vectors, vectors of all-'0', or all-'1' always should be included, because these represent externally-defined corner cases.

Temporally, within the tested device, states immediately following chip-enable, chip-disable, read, write, and so forth would be given greater coverage than others. Look for something to go wrong every time you turn a corner!

22.1.6 Boundary Scan

Boundary scan (and internal scan) were introduced briefly in *Week 2 Class 1* (section 4.1.8) and in `Lab05`.

Boundary scan provides for increased observability of the pin-outs of chips on a board; these are the <u>boundaries</u> of the chips. Rather than provide numerous direct electrical contact points for a bed-of-nails automated board tester, or for manual probing, in boundary scan dedicated traces on the board are connected to the chip I/O's; after this, the board I/O's themselves are used to apply stimuli and observe results. This makes possible the observation of inter-chip communication on wires or traces not normally routed to a board I/O.

For boundary scan, each scanned chip is equipped with a test port, the TAP (Test Access Port), for controlling the scan. The test logic of the TAP controller may be very simple, as in the internal scan exercise we did in *Week 2 Class 1*, or it may include a device-specific, programmable state machine which automates some of the test mode shifting to save time or computation by the external hardware tester.

Very much the same as the JTAG internal scan port of *Week 2 Class 1*, the TAP has five pins defined: TDI (Test Data In), TDO (Test Data Out), TCK (Test Clock), TMS (Test Mode Select), and TRST (Test controller Reset). Except TMS, the TAP pins physically may be pins having other chip functions; they may be control, address, or data pins, the test functionality being selected by asserting TMS.

A generic boundary-scan is shown in figure 22-1. In test mode, the scan latches (or flip-flops) are chained together, stimuli may be shifted in by TCK, and outputs may be shifted out. Each scan latch in a cell is connected to input and output muxes so it can be bypassed during normal chip operation. At any time during normal operation, TMS and TCK may be asserted to store the output states in the latches for subsequent shift-out and inspection.

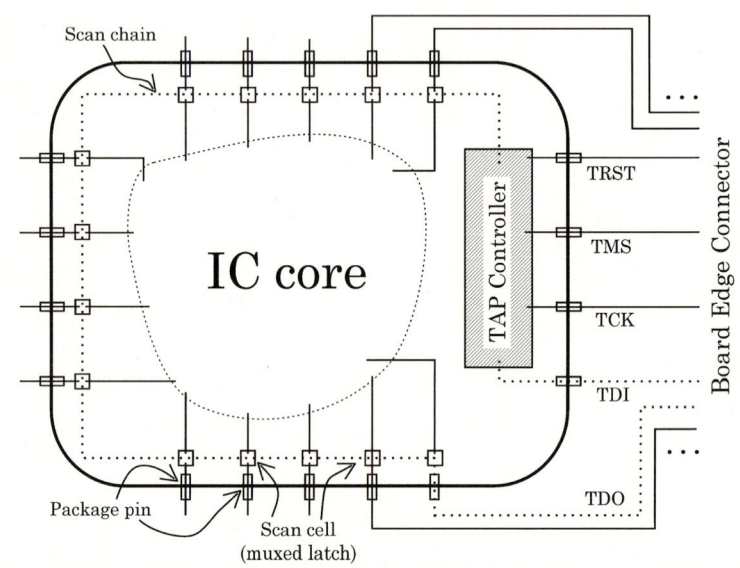

Fig. 22-1. Boundary scan logic on a board-mounted IC. Other IC's on the board are omitted, as are the TCK and TMS distributed to each of the scan cells. The IC is shown in test mode, with the scan chain the dotted line connecting the scan cells. Boundary scan makes points internal to the board, but not internal to the IC, observable.

Boundary scan may be combined with internal scan or BIST (or both) in the same chip.

Digital VLSI Design with Verilog

22.1.7 Internal Scan

We have presented the rationale for muxed flip-flop internal scan in *Week 2 Class 1*. Briefly, in internal scan, the registers in the IC are replaced by scan cells which can be configured in test mode as one or more chip-wide shift-registers. Several different ways of designing for internal scan are described in the readings recommended at the end of this chapter.

Sometimes the terms "full scan" and "internal scan" are used interchangeably. However, because it is possible to omit some registers (for example, an entire memory) from an internal-scan chain, internal scan does not imply full scan. Therefore, we prefer to use the two terms differently.

Full internal scan connects every register in the IC in the chain. Because inputs are observable already, this means that full internal scan in principle permits observability of all $2^{N_I+N_Q}$ possible internal states. This observability allows a level of confidence which may be required in certain life-critical systems, or systems in space or undersea vehicles which can not be maintained or repaired during use. Such systems often are simple enough to make 100% coverage feasible.

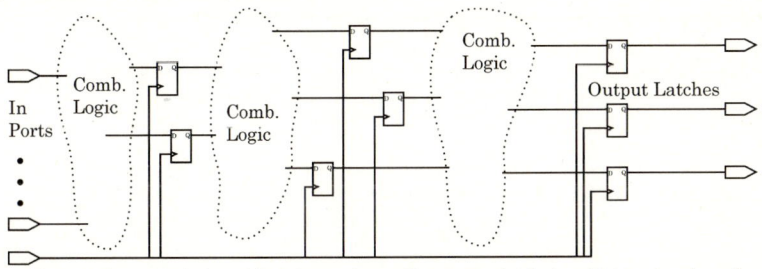

Fig. 22-2. A generic design without internal scan. Outputs are latched, and internal regions of combinational logic are separated by sequential elements. A clock is shown distributed to all the sequential logic.

Internal scan does not include pad cells, if they are present in the scanned module. Internal scan logic inserted in the design of figure 22-2 is shown in figure 22-3.

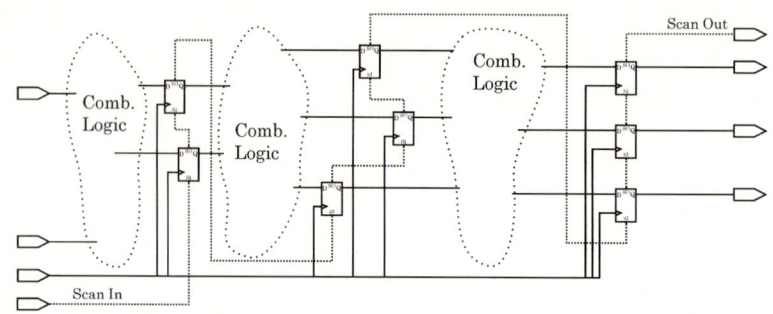

Fig. 22-3. The same design as above, after internal scan insertion. Every sequential element is replaced with a scan cell. The scan data chain is shown (each SI to SO). The clock shown may be considered the scan clock. Scan mode, as well as other controls, is omitted.

Recall that almost all IC's will have outputs latched for synchronization reasons. Outputs, then, will be in the scan chain. Other logic may exist which is not observable; but, with full internal scan, this unobservable logic can not affect functionality of the chip and thus usually may be ignored for test purposes. Such logic may be present because of design errors or oversights, or because of design or production work-arounds. A famous example of the latter was the 386-SX microprocessor: Whenever a manufacturing defect was found in the onboard cache of a 386 die, the cache was disabled by tying off a pin, and the chip was packaged and sold at a lower price as a perfectly usable, cacheless, 386-SX. These modified 386 processors had thousands of nonfunctional gates.

Unobservable logic, such as just discussed, may affect operating parameters of a chip, because unobservable gates still may draw clock current and may leak the same way as functional gates, causing additional power dissipation merely because they exist on the chip.

It should be mentioned again that internal scan and boundary scan are not mutually exclusive and can be combined in the same chip.

22.1.8 BIST

Built-In Self-Test is an idea not restricted to IC design. Every PC has a ROM-based memory check routine which is executed automatically when it is turned on; likewise, when a hard disc is formatted at low level, the software, often in a disc-controller ROM, automatically performs tests to verify read and write access to every bit.

One advantage of BIST is that it requires no external test apparatus; it does, however, require a control input and a result output of some kind. In almost all cases, these additional I/O requirements can be met by assigning multiple functions to preexisting design pins. Even more important, any number of BIST-equipped chips can be tested concurrently during manufacture or operation. So, BIST allows design size and complexity to grow, and to be verified, with little or no additional hardware production testing time.

Digital VLSI Design with Verilog

Thus, the cost of BIST in terms of chip I/O is negligible. However, the core silicon required for BIST may be substantial, because the self-test controller and program must be stored somewhere; also, the BIST usually must be executed upon the functional logic by means of additional, dedicated internal routing. If the design is complicated and includes many internal registers or modes of operation, the BIST must be very elaborate to generate vectors to achieve even minimal confidence that the hardware is fully functional; this implies significant design overhead. Furthermore, test results must be evaluated on-chip, which requires yet more storage and logic.

The process of BIST insertion is illustrated in figures 22-4 and 22-5.

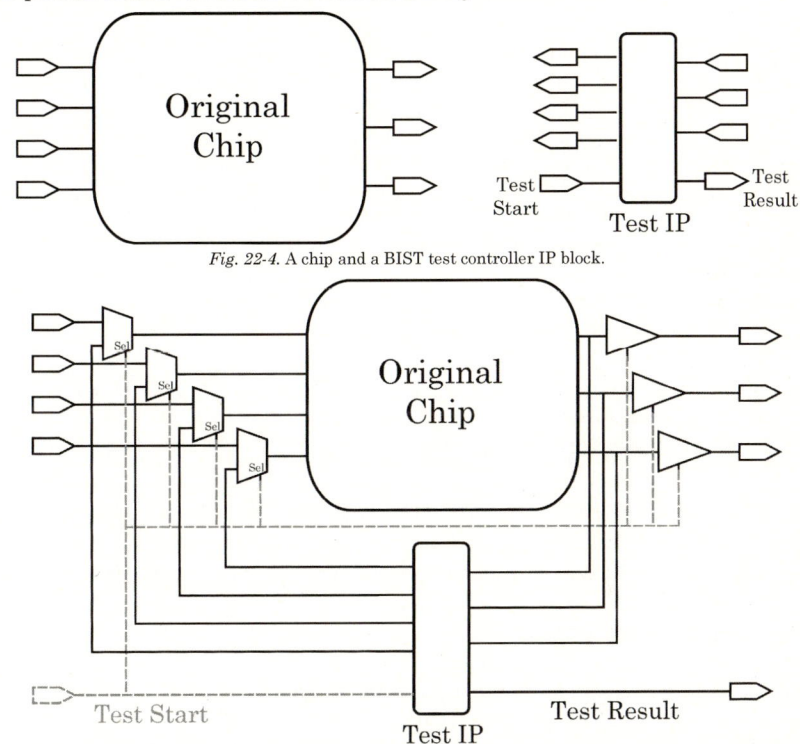

Fig. 22-4. A chip and a BIST test controller IP block.

Fig. 22-5. Typical BIST insertion.

Because of this, BIST is most efficient for IC's of large size and regular, repetitive structure—in other words, for memory IC's. If the memory is fabricated with spare cells or words, BIST can be used during manufacture to find defective cells and replace them with intact ones, increasing the production yield of the manufacturing process.

There is a variety of ways to implement BIST and BISR (Built-In Self Repair) of on-chip memories. For repair, the usual way is to manufacture the memory with one or two spare rows (addressable words) or columns (bit-offsets within all words), depending on the hard error rate expected. Then, BIST is run during manufacture to identify defective bits. Knowing the location of the defects, if any, the tester programs a within-chip or within-board register with the spares to be substituted. By this means, either the spare row(s) and column(s) are substituted by permanent "burned-in" logic, or on-chip logic is programmed dynamically by shifting in a substitution configuration (as in programming an FPGA).

Defects are rare enough so that almost all memories can be repaired by substitution of just one or two spare rows or columns.

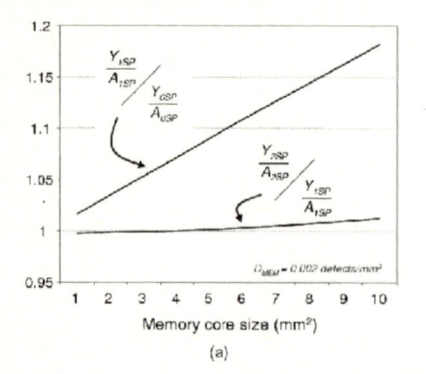

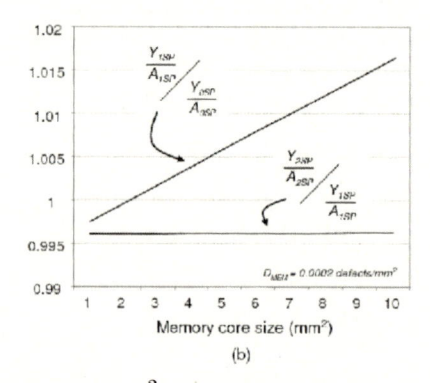

Fig. 22-6. The graph (a) is for a typical memory with 0.002 defects/mm^2; the graph (b) is for a defect rate 1/10 as great. The upper curves show the improvement in BIST pass rate gained by having one spare row or column available vs. none; the lower curves show the improvement by having two spare rows or columns vs. just one. Clearly, there is little expected benefit in providing for more than two spare rows or columns. The statistic is Yield/Area. After figure 7 of the Mentor *Memory Repair Primer* of November 2008. Reproduced with permission of Mentor Graphics, Inc.

BIST usually depends on a standard TAP controller interface. BIST may be combined with scan logic. The BIST may accept a test-mode input; as a result, the programmed test may

   a. put the chip into test mode,
   b. scan in preprogrammed patterns,
   c. scan out the latched states,
   d. evaluate the result; and,
   e. terminating with a go/no-go signal to the rest of the IC or to the containing system.

A random-logic state machine or a ROM may be involved. LFSR"s may be used to generate pseudorandom test patterns efficiently.

BIST often is used in systems which require verification without direct access—for example, in the space program or in underseas devices. The currently active Mars rovers are equipped with elaborate BIST, and associated redundancy, for error recovery. The

Digital VLSI Design with Verilog

high radiation exposure requires special accommodation to soft errors caused by the Solar wind and cosmic rays, against which Mars has no magnetic or atmospheric shielding.

22.1.9 Design for Test Summary

In summary, DFT is a methodology which implies use of a certain collection of design techniques. The methodology includes attention to corner cases and test coverage of points of probable failure. Attention must be paid during software development; during hardware design; and, after device manufacture, to the software and hardware itself. The techniques of DFT encompass simulation, fault simulation, insertion of special hardware devices for internal or boundary scan, inclusion of built-in self-test apparatus, and use of I/O's to provide observability of internal functionality.

22.2 Scan and BIST Lab 25

Topic and Location: Automated internal scan review; automated boundary scan insertion; manual BIST design, insertion, and synthesis.

Do these exercises in the Lab25 directory.

Preview: We recall our Lab05 optional automated scan insertion exercise and now complete it, if it was not done during Lab 5. We then manually add I/O pads to the old Intro_Top design in order automatically to insert boundary scan elements. After this, the lab is dedicated to attachment of a BIST module to our DPMem1kx32 RAM: We decide on the BIST test protocol and then detail how to create a wrapper module to enclose the RAM and BIST. We then implement the BIST by using a supplied "IP" BIST module and synthesize the result.

Deliverables: Step 1: Completed Step 9 of Lab05. Step 2: A modified Intro_Top design with boundary scan automatically synthesized. Step 3: A BIST for our DPMem1kx32.v RAM, synthesized. Step 4: Some BIST planning. Step 5: A wrapper module for the memory plus BIST, correctly connected as verified by load into the simulator. Step 6: A new BIST separated from the memory. Step 7: Implementation of BIST controls. Step 8: A correctly simulating, partly optional, completed memory with BIST. Step 9: A synthesized netlist of the memory with BIST, to be compared in size with the one from Step 3.

Lab Procedure:

Step 1: Internal scan exercise. If you have done the Lab05 (*Week 2 Class 1*) Step 9 scan synthesis exercise already, just proceed to Step 2.

If you have not yet done Step 9 of Lab05, create a subdirectory named IntScan in the Lab25 directory, copy in the purely-combinational design from Lab05 and do Lab05 Step 9: Add the FF's, renaming the design to Intro_TopFF; insert the scan cells using the synthesizer; examine the result; and, proceed to Step 2 of this lab.

Step 2: Synthesizer boundary scan insertion. Boundary scan is intended for entire chips on a board; like BIST, it is inefficient for a small design. In this lab, we shall use the synthesizer to insert boundary scan in the original Lab01 design just to see the result. We choose the Lab01 design because automatic insertion has certain requirements which would be unnecessarily complicated to meet in our larger SerDes design.

A. Start by creating a new directory BScan in the Lab25 directory. Copy in the original Lab01 Intro_Top design and simulate it briefly with its TestBench.v, to verify connectivity.

After this, copy in the verilog pad model wrapper file tpdn90g18_3PAD.v, which should be symlinked in the VCS directory to tpdn90g18tc_3PAD.v. Also copy in the synthesis script file BScan.sct from the Lab25 directory. All these files will be found named in your Lab25 directory. If the symlinks have not been created, you will have to create them now or copy the files from the Extras misc directory.

B. Add a test port. Automatic boundary scan requires a preexisting TAP, so add one new top-level output, ScanOut, to Intro_Top; also add four new inputs, ScanIn, ScanMode, ScanClk, and ScanRst. These are just the usual internal-scan JTAG port names for the TAP.

C. Instantiate pads. For the synthesizer to insert boundary scan, we must have pads in the design. We need three different kinds of pad: input, output, and three-state output. A three-state output pad is required for the TAP TDO (= ScanOut).

In a verilog design, pads are added inside the top-level module ports, as individual components within the top-level module. The module port names are signal names, not hardware pin contacts, in this context. Of course, a top-level wrapper module normally would be used; but, regardless, putting the pads inside the top design module permits the same testbench to be used for a design before and after pads have been added.

The TSMC pad library that matches our core library can not be used for synthesis, because those pads each are multipurpose and may be controlled by inputs to be input, output, or bidirectional. The TSMC models include switch-level elements and are too complicated both to be accurate in simulation and to be synthesizable. If you are licensed for the TSMC libraries, the data book for this pad library probably will be available in the course installation VCS directory, if you wish to examine it later.

For use in this lab, three pads have been selected from the pad library, reduced so that they are not proprietary copies, and provided with verilog wrapper modules for simulation, only. The wrappers may be seen in the verilog file, tc*padlibename*_3PAD.v, which should be available as a link in the Lab25 directory.

Digital VLSI Design with Verilog

> **Note**: The detailed instructions here are to provide some experience with inserting and connecting pads in a design; they may be skipped if such experience is unnecessary. At the end of this Step, you may simulate the design if you wish, using the `Lab25` answers provided for Step 2C. The modified design should function as the familiar `Intro_Top` of previous labs; however, the undriven `ScanOut` pad output will remain unknown.

The names of these wrappers, only, should be used to instantiate pads for the `Lab25` boundary scan exercises. The names are: `PDC0204CDG_18_Out` for output pads, `PDC0204CDG_18_In` for input pads, and `PDC0204CDG_18_Tri` for the one three-state output pad. The "0204" represents a drive strength (02 = 2 mA), and the "18" means 1.8V pad I/O (for 1.0V core logic).

The port declarations for these library components are typical for verilog:

```
module PDC0204CDG_18_Out  (output PAD, input I);
module PDC0204CDG_18_Tri  (output PAD, input I, OEN);
module PDC0204CDG_18_In   (output C, input PAD);
```

The three-state `OEN` (<u>o</u>utput <u>en</u>able) pin is asserted low in the library; however, the logic has been inverted to be asserted high for the three-state component in the verilog wrapper.

Remember that the ports of the `Intro_Top` module were named *X*, *Y*, and *Z* (outputs) and *A*, *B*, *C*, and *D* (inputs), and that each port pin in verilog is associated with an implied wire of the same name. After declaring a few new, obviously-named *to-* and *from-* wires, the resulting pad structure for `Intro_Top` should look something like the following:

```
PDC0204CDG_18_Out Xpad1( .PAD(X),    .I(toX) ); // X is port; toX is wire.
PDC0204CDG_18_Out Ypad1( .PAD(Y),    .I(toY) );
PDC0204CDG_18_Out Zpad1( .PAD(Z),    .I(toZ) );
PDC0204CDG_18_In  padA1( .C(fromA), .PAD(A) );
PDC0204CDG_18_In  padB1( .C(fromB), .PAD(B) );
PDC0204CDG_18_In  padC1( .C(fromC), .PAD(C) );
PDC0204CDG_18_In  padD1( .C(fromD), .PAD(D) );
```

The TAP port has to connect to its pad cells. The core I/Os of the pads should be left dangling, so that the synthesizer can determine how to connect to them:

```
PDC0204CDG_18_Tri TDOpad1( .PAD(ScanOut) /*, .I(), .OEN()*/ );
PDC0204CDG_18_In  padTMS1( /*.C(),*/ .PAD(ScanMode) );
PDC0204CDG_18_In  padTDI1( /*.C(),*/ .PAD(ScanIn)   );
PDC0204CDG_18_In  padTCK1( /*.C(),*/ .PAD(ScanClk)  );
PDC0204CDG_18_In  padTRST1( /*.C(),*/ .PAD(ScanRst) );
```

The pad instances will be treated as any other instances during synthesis and netlist optimization. To protect the pad wiring from being modified during synthesis, add a comment directive to `Intro_Top.v` which flags all pad instances as `dont_touch`. You may use a wildcard, for example, `"*pad*"`, for this.

The synthesis `.sct` script will be used to tell the synthesizer which ports (see Step 2 **B** in this lab) we want it to use for the TAP.

D. Synthesize the boundary scan logic. After instantiating and wiring in the pads, you may use the `BScan.sct` synthesis script in the `Lab25` directory to insert the boundary scan cells and TAP controller.

After compilation, you may write out the hierarchy and view it in `design_vision`. To do this, read in the file, `Intro_TopNetlistBSD.v`. You may wish to back-track in the schematic from the `tdo` pad to the boundary scan tap controller, for example.

VCS also can be used to view the netlist schematic: Select "`Topper01 (Intro_Top)`" in the control window hierarchy view, and then pick the "and gate" icon just above. This may take a minute, but the result will be better-arranged than the one from `design_vision`. Tool details are in your *DC_Synth_Lab01_Summary.pdf* handout.

This exercise merely showed how to handle TAP ports; the controller has not been programmed, so this netlist can not be simulated usefully. The TAP controller could be implemented by means of DC commands too time-consuming for this course. As implied in the previous paragraphs, the final netlist can be simulated without TAP; however, the scan-mode scanned output data will not be meaningful.

The remainder of this lab is a memory BIST design.

It will include some detailed file rewriting. If the user is not planning to work with BIST development and has sufficient experience with verilog, Steps 3 - 8 may be skipped to conclude with the Step 9 synthesis exercises. In any case, it would be instructive to read the `Lab25_Ans` answers provided for all the steps of this lab.

Step 3: Design a BIST for our `DPMem1kx32` RAM.

A. Set up a working location. Do this by creating a new subdirectory named `BIST`; copy into it the `DPMem1kx32.v` RAM model from any of the `FIFO` answer subdirectories of `Lab24`. Name the new copy `Mem.v` and rename the module correspondingly.

B. Generate a benchmark synthesis result for later use; then set up a simulation testbench. This preparation will reacquaint you with the (already simulated) memory functionality and thus speed development and lessen the likelihood of design errors.

First, synthesize the renamed RAM of **A**. To do this, size `Mem.v` for 32 words of 32 addressable bits each and run the synthesizer to optimize for area only. For this synthesis, impose simple logical-netlist design rules such as maximum fanout, etc. Make a copy of the synthesizer's area report in a text file for later reference. Also be sure you keep a copy of the synthesis constraints you used to obtain this area. ***Do not simulate***

Digital VLSI Design with Verilog

the synthesized netlist—we are just interested in the approximate size of it, to compare sizes when the BIST is added.

For simulation of the verilog (source) design, instantiate the renamed RAM of **A** in a new `MemBIST_Tst.v` file, and create a small testbench there that exercises the RAM to demonstrate both write and read. Use a common clock for read and write. Set parameters to size the memory to be 32 addressable bits wide and 32 words deep. Simulate briefly, just to verify basic functionality. Your setup should be the same as in figure 22-7.

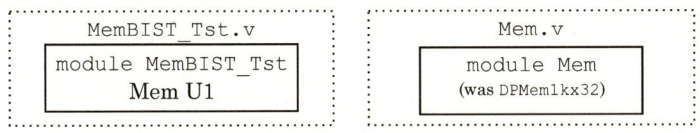

```
  MemBIST_Tst.v                    Mem.v
  ┌──────────────────────┐         ┌──────────────────────┐
  │ module MemBIST_Tst   │         │   module Mem         │
  │      Mem U1          │         │  (was DPMem1kx32)    │
  └──────────────────────┘         └──────────────────────┘
```

Fig. 22-7. **Lab 25, Step 3 B**. The two verilog files are one testbench file and one memory model file. Mem is instantiated in `MemBIST_Tst`.

C. Prepare a BIST plan which describes what the built-in self-test should do. Of course, it will be intended for random hardware defects, only. We shall assume that only isolated defects can occur; this is just for lab convenience; but, even so, in a real production run, we would catch the majority of hardware fabrication defects on this assumption alone.

We must be sure that our tests accommodate the memory parameters of width and depth, so we shall not assign numerical values to width or depth anywhere in the BIST module. Our memory is equipped with parity, so we should monitor parity throughout testing, in case of a single-bit failure during testing—even a soft one.

Here's what our plan for this lab says the BIST should do:

First test pattern: Validate the addressability of bits. Write a different value to each address word; then, read the values out to verify that the memory hardware can store the expected number of different data. This will detect permanently shorted or open address lines, or obvious address decoder failures.

We shall write a value which counts from 0 to the max address, this count being replicated at several offsets in each word and easily recognizable visually. An example of such a pattern sequence, counting from address 0 to address 31 (5'h0 to 5'h1f), would be,

    32'he0e0_e0e0, 32'he1e1_e1e1, ..., 32'hffff_ffff.

Second test pattern: Validate integrity of every bit. Write '1' to every bit at all memory addresses, then read out and check the value. Repeat with '0'. This will find any bit or bits stuck independently at either level and missed by the previous counting test.

We shall stop with these two patterns, although more efficient, more thorough, or more exotic tests could be devised. For example, alternating checkerboard patterns on adjacent words might be written and read, or walking '1' tests run. Testing of a large RAM can be very elaborate, and thus there is an advantage to programming the test into the RAM chip, so that all RAMs in a manufacturing run, or installed in a system, might be tested at once, with minimal need for external test apparatus.

Step 4: Plan the BIST interface. We shall put the BIST logic in a separate module. In an actual manufacturing run, the layout almost certainly will require that the memory core be a regular cell array in its own block; so, the BIST logic would have to be kept in a separate partition of the design.

Our BIST will have to address the memory by its address bus, and it will have to read and write on the memory data bus. The BIST and memory clocks can be shared. There will be a new input in the test-equipped memory for a test-start signal, and at least one new output to report test status; we shall keep these ports separate and not worry about sharing test functions on preexisting pins. During self-test, the Mem module must be unresponsive to external addressing or read or write requests, and it will have to disable its output drivers.

The best way to accomplish all this is to instantiate the Mem in a wrapper module which also instantiates a module containing the BIST logic. During self-test, the BIST-Mem system then can be isolated from the external world by the wrapper logic.

Step 5: Create the BIST wrapper: Create a new module named MemBIST_Top in a file named MemBIST_Top.v by making a copy of Mem.v. Give the new MemBIST_Top module the same I/O's as Mem, except for one new input, DoSelfTest, and two new outputs, Testing, and TestOK.

After modifying the MemBIST_Top header for the new I/O's as just described, instantiate Mem in the MemBIST_Top module, and, declaring explicit wires for every I/O, directly connect all Mem I/O's to the corresponding MemBIST_Top ones using continuous assignment statements. Explicit wires in this exercise are important and will simplify interconnection of BIST and Mem later in the lab. There is no area or timing overhead for such wires, because the synthesizer will convert implicit wires to explicit ones anyway.

Your setup should be as in figure 22-8:

```
┌─────────────────────────────┐    ┌─────────────────────────────┐
│      MemBIST_Top.v          │    │           Mem.v             │
│  ┌───────────────────────┐  │    │    ┌──────────────────┐     │
│  │ module MemBIST_Top    │  │    │    │    module Mem    │     │
│  │ Mem Mem_U1            │  │    │    └──────────────────┘     │
│  └───────────────────────┘  │    │                             │
└─────────────────────────────┘    └─────────────────────────────┘
```

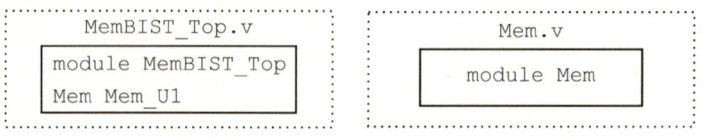

Fig. 22-8. **Lab25 Step 5.** The MemBIST_Top wrapper has almost the same module ports as Mem. On the left, Mem is shown instantiated inside MemBIST_Top.

We are done with MemBIST_Tst.v of Step 3. After instantiating Mem in MemBIST_Top as above, and completing the wiring, rename MemBIST_Tst.v to MemBIST_TopTst.v, changing the testbench module name accordingly. Change the Mem instantiation to

Digital VLSI Design with Verilog 485

MemBIST_Top. Then, simulate the new two-module memory (MemBIST_Top containing Mem) briefly to check connectivity before going further.

Step 6: Define the BIST interface. Create a new BIST.v file for the built-in self-test module, BIST, by making a copy of MemBIST_Top.v.

But, before changing anything in the new file, instantiate BIST in the original MemBIST_Top.v by making a copy of the existing Mem instance and editing it as will be described. In this design, because Mem already has a complete interface to guide us, it will be easiest to connect a BIST instance first, and then to edit the copied module file in BIST.v to work out declarations of the required BIST I/O's.

The setup for this Step thus is as shown in figure 22-9, and the work will be to modify the BIST instance so we will know how to change BIST.v.

```
┌─────────────────────────────┐   ┌─────────────────────────────┐
:       MemBIST_Top.v         :   :           Mem.v             :
: ┌─────────────────────────┐ :   : ┌─────────────────────────┐ :
: │ module MemBIST_Top      │ :   : │     module Mem          │ :
: │ Mem    Mem_U1           │ :   : └─────────────────────────┘ :
: │ BIST   BIST_U1          │ :   :          BIST.v             :
: │  (port map copied       │ :   : ┌─────────────────────────┐ :
: │   from Mem_U1)          │ :   : │     module BIST         │ :
: └─────────────────────────┘ :   : │ (copied from MemBIST_Top)│:
└─────────────────────────────┘   └─────────────────────────────┘
```

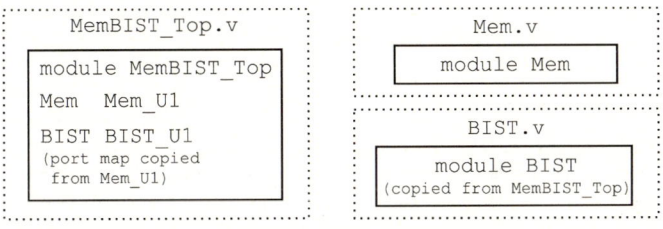

Fig. 22-9. **Lab25 Step 6**. Using wrapper MemBIST_Top to help determine the new BIST module ports.

So, looking at the new BIST instance in MemBIST_Top.v:

First, we shall have to pass the AdrHi and DWid parameters to BIST; so, we should retain these exactly as copied from Mem.

We can eliminate Dready and ChipEna from the BIST I/O's.

Second, we require control of the Mem data input and output, so we should retain these I/O's in BIST. Soon, we shall multiplex these Mem busses so they can be directed either to the MemBIST_Top ports, or to the BIST ports. We shall connect the DataO output port of the Mem instance to the DataI input of the BIST instance, and vice-versa. So, we retain the DataO and DataI ports in the BIST instance.

Third, we should add a BIST address output bus, as well as read request (ReadCmd) and write request (WriteCmd) outputs. These also will have to be multiplexed with Mem inputs. Because none of our planned tests involves simultaneous read and write, a single BIST address bus can be supplied both to the RAM read and write address input ports.

Fourth, we require a BIST clock input and a hardware Reset input. To implement these, we shall have our test procedures clocked by the RAM read clock (ClkR); so, the read clock is the one we shall supply for the BIST.

Finally, we should install the three special BIST I/O's (the DoSelfTest input, and the Testing and TestOK outputs), all of which will be routed directly to the

MemBIST_Top module ports. We also should monitor the RAM's ParityErr output during testing by using a corresponding BIST input.

At this point, after the above described changes had been made, the verilog code for your BIST instance in MemBIST_Top should look something like this:

```
wire ClkRw, Resetw, ...;   // Assigned from MemBIST_Top module inputs.
...
BIST  #( .AdrHi(AdrHi), .DWid(DWid) )
  BIST_U1
    ( .DataO(), .Addr(), .ReadCmd(), .WriteCmd() // outputs.
    , .Testing(), .TestOK()                       // outputs.
    , .DoSelfTest(), .ParityErr(), .DataI()       // inputs.
    , .Clk(ClkRw), .Reset(Resetw)                 // inputs.
    );
```

Note: In the verilog code, ClkRw and "Resetw" are named "*w" because they have been declared wires.

Step 7: Implement the BIST controls in MemBIST_Top. After adding port names as above, complete the wiring in MemBIST_Top to BIST_U1, including the multiplexed data and other busses. You can do the muxes most easily as continuous assignments with conditional operators; you'll have to declare new wires for the BIST instance to do this. If, initially in Step 5 above, you used continuous assignment statements to wire the connections between the MemBIST_Top header and the Mem instance, only the new wire declarations and the conditional expressions will have to be added.

Briefly, when the DoSelfTest input goes high, the edge will initiate the test. Muxes in MemBIST_Top will be put in test mode by the assertion of Testing, which will be kept at '1' while self-test is in progress; TestOK will be asserted after a self-test if the test found no defect (otherwise, TestOK will stay at '0').

After this, go to the new file named BIST.v, which you created above, remove everything but the header and parameters from your new BIST module, change the port names and directions to match the code fragment above, and simulate MemBIST_Top briefly, using MemBIST_TopTst, just to check connectivity.

Step 8 (partly optional): Implement the BIST functionality. Before anything else, the test mode control should be defined.

Digital VLSI Design with Verilog

Here is one way to do this:

```
...
reg AllDoner  // Flag completion of testing for internal BIST use.
  , Testingr; // Sets test mode for the BIST.
//
assign Testing = Testingr;  // Testing is a BIST output port.
//
always@(posedge Clk, posedge Reset)
  begin : TestSequencer
  if (Reset==1'b1)
       begin
       Testingr  <= 1'b0;
       AllDoner  <= 1'b0; // Normal level (noncommittal).
       ...
       end
  else // Must be a clock:
       begin : TestClocked
       if (DoSelfTest==1'b0)
            begin // Init, but leave TestOK alone:
            Testingr <= 1'b0;
            AllDoner <= 1'b0;
            ...
            end
       else Testingr <= 1'b1; // Entering test mode for this clock.
...         ...
```

All reg names end in 'r'. A rising level on Testingr then can be used to start the self-test; the test routines should determine when the testing is finished.

A clocked always block can be used as a simple state machine to sequence the BIST through its tests. Our tests in Step 3 above address all of memory for write and then for read. It seems reasonable to implement each test as a separate always block containing an address counter and clocked on the BIST input Clk . The always for each test would be run selectively because of a flag set in the test-sequencing block. Actually, each always block used to write memory could be different from the one verifying the stored results. The test sequencer could check test status on every clock; when a test is complete, the current always block could set a flag reporting results and telling the test sequencer it is done.

Implementation of this BIST makes a very good ***optional project***; but, it is very time-consuming and probably is at least a day's work. Therefore, a complete implementation is provided for you in BIST_Done.v in the Lab25/Lab25_Ans directory.

To continue this lab, copy BIST_Done.v into your BIST directory. Copy your own BIST.v to a different name to save it. After looking through BIST_Done.v, copy or link it to BIST.v to replace your empty interface model. Simulate briefly to check connections.

If you want to spend some lab time on the BIST module, it might be interesting to rewrite part of the answer provided as a verilog *task*: If you do this, your task may be called under as many as six different test conditions, using the test number as the (single) task input. The test sequencer code for phases 1 through 6 is very repetitive; this suggests replacing it with six calls to a task with a single number as input.

Simulate to validate your changes: A good simulation might exercise the memory a little; then, run a self-test; then, exercise it a little more. See figure 22-10.

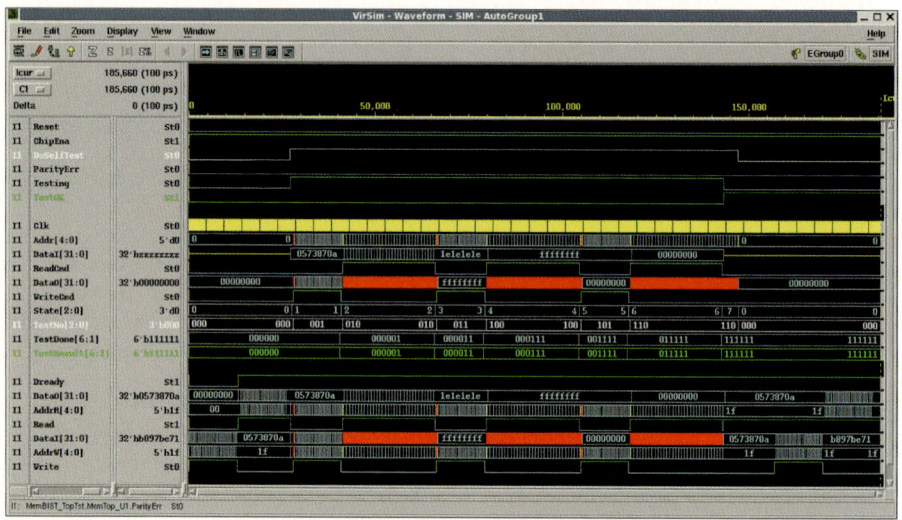

Fig. 22-10. Typical MemBIST verilog source simulation.

Step 9: Synthesize the completed MemBIST_Top design, optimizing it for area with the same constraints as you used in **Step 3 B**. The synthesis will take about 10 minutes with mild constraints. With a quick netlist, don't bother with simulation; the netlist may not simulate because of lack of clock constraints, but its size will be about right. Compare the size of the design, as synthesized area, with and without the BIST.

Optional: After comparing sizes by areas, you may add clock definitions and a maximum output delay constraint to your synthesis script, set constraints to fix hold violations, and resynthesize (see the answer directory for specific values). This netlist will take a half-hour or more to synthesize, but the result, as in figure 22-11, will simulate correctly.

Digital VLSI Design with Verilog

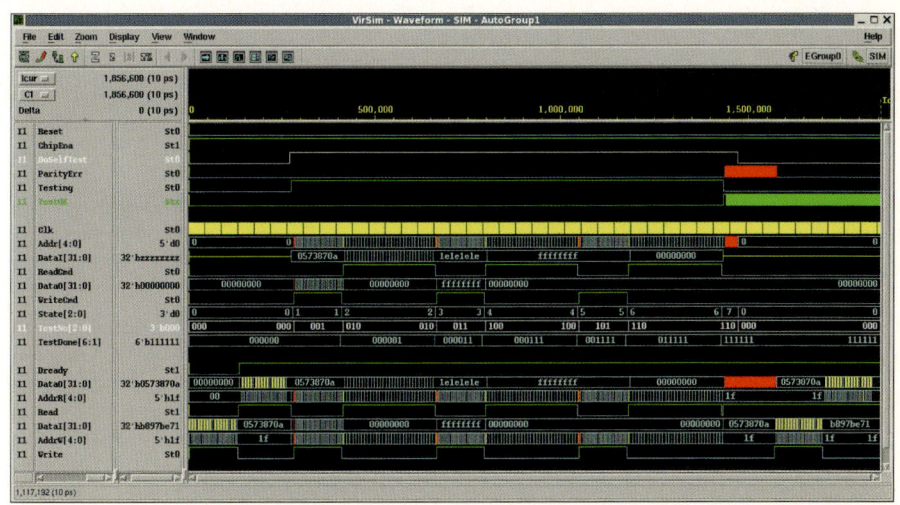

[Figure: screenshot of VirSim Waveform window showing MemBIST simulation signals including Reset, ChipEna, DoSelfTest, ParityErr, Testing, TestOK, Clk, Addr, DataI, ReadCmd, DataO, WriteCmd, State, TestNo, TestDone, Dready, DataO, AddrR, Read, DataI, AddrW, Write]

Fig. 22-11. Simulation of the synthesized `MemBIST` verilog netlist, using the same testbench as for the source.

Note: Using the latest 2011 version of DC, synthesis of the memory with BIST and with timing constraints, but without `set_fix_hold`, took about 15 minutes on a 1 GHz machine. The resulting netlist simulated correctly, just using the approximated delays given in `tcbn90ghp_v2001.v`.

22.2.1 Lab Postmortem

How big was the `BIST` netlist, compared with the one for our bare `DPMem1kx32` RAM?

Would you expect the use or arrangement of the BIST `tasks` to make any difference in synthesis?

22.3 DFT for a Full-Duplex SerDes

We'll finish up today with a lab on test insertion for our `SerDes`. First, we'll modify our `Lab24 SerDes` design to have full-duplex functionality, as would a *PCI Express* lane. Then we'll add test logic.

22.3.1 Full-Duplex SerDes

A full-duplex lane is just a dual serial line, with one serdes sending in one direction, and a second serdes sending in the other. The two serial lines being independent, this lane can send and receive simultaneously in both directions.

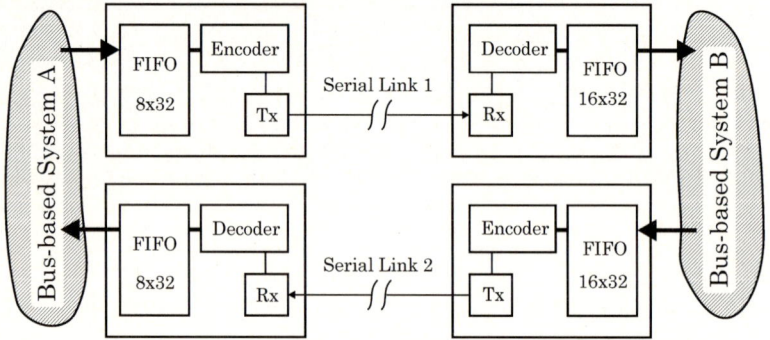

Fig. 22-12. Two instances of the `Lab24 SerDes` assembled into a full-duplex lane. A simulator testbench can represent the two communicating systems, *A* and *B*. The FIFO depths differ for the two systems.

For variety, we shall assume that system *A* can provide a FIFO only with 8 words, but that system *B* can provide one 16 words deep. See figure 22-12. Such a difference would not be unusual, assuming that our duplex serial lane was between different chips on a board. For example, the communicating chips might be from different suppliers, in different technologies, or fabricated on availability of different kinds of IP. Many systems would omit a sending-side FIFO on the assumption that sending was more controllable than receiving.

It is important to understand the organization of the full-duplex lane proposed. Each SerDes device model can be used to span the opposite sending and receiving ends of a single serial line; or, each SerDes could be interpreted as referring to one end (send and receive) of a full-duplex lane. The difference is illustrated in the top and bottom of figure 22-13.

Digital VLSI Design with Verilog

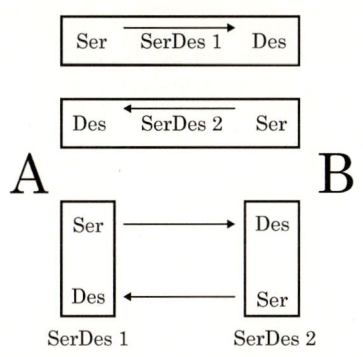

Fig. 22-13. Two ways of using a pair of `SerDes` devices between two systems, *A* and *B*.

Our choice is the upper one shown; it permits immediate instantiation of our `SerDes` design. We assume that our two serdes will reside in a system (maybe on a single chip) which will include a floorplan positioning each ser some distance away from its des, so that a serial data transfer between *A* and *B* would be useful.

22.3.2 Test Logic Questions

Before proceeding to the lab, some thought might be given to the following points:
- How should we equip our serdes for testability?
- How good is observability without test logic?
- Where should we add assertions?
- How much internal scan?
- Would boundary scan be useful?

22.4 Tested SerDes Lab 26

Topic and Location: A full-duplex serdes; addition of assertions and timing checks, synthesis of the netlist, and synthesizer insertion of internal scan.

Work in a new subdirectory of the `Lab26` directory for these exercises. The instructions will say how to name files and directories in it.

Preview: We first implement a full-duplex serdes by instantiating two of our previous unidirectional `SerDes`. We use parameters to size the `FIFO`s at the two ends of the lane differently, emulating different system characteristics. Once the instances are connected, we simplify their interfaces to accommodate the full-duplex lane, which we then synthesize. We make a new copy of our fully functional `SerDes`, and, after adding an assertion *pro forma*, we insert a few timing checks and experiment with them to tune them according to the desired `SerDes` performance. We then make a second new copy in which we use the synthesizer to insert internal scan.

Deliverables: Step 1 - 3: Modify and assemble the SerDes parts. Step 4: Simplify, connected, and parameterized full-duplex instances, as proven by loading into the simulator and examining the hierarchy. Step 5: A simulating copy of the full-duplex SerDes, with FIFO-warning assertions. Step 6: Four timing checks in DesDecoder, tuned properly by simulation as described in the lab procedure. Step 7: A synthesized, constrained netlist for the original SerDes (without assertions or timing checks). Step 8: A new SerDes netlist with internal scan inserted automatically by the synthesizer.

Lab Procedure:

> *NOTE*: This lab again includes considerable file rewriting, which may or may not be instructive to a user already skilled in verilog. Steps 1 - 4 may be skipped, using the provided Step04_FullDup answer to run the SerDes simulation described in the last two paragraphs of Step 4.
>
> After this simulation, if skilled in the use of verilog assertions and timing checks, the user may read the Step 5 and Step 6 instructions carefully and then simulate a new copy of the SerDes, with assertions and timing checks, the copy being made from the answer in Step06_FullDupChk.
>
> Steps 7 and 8 of this lab are new and important and should not be skipped.

Step 1: Gather the parts of the full-duplex serdes. Create a directory named FullDup in Lab26, and create in it a subdirectory named SerDes. Make a copy of the entire contents of the SerDes directory from Lab24/Lab24_Ans/Lab24_Step08 in it. For now, use the provided answer design; there will be opportunity later to return to your own SerDes implementation, if you should want to do so.

What we'll do now is to create place-holder files for our full-duplex serdes by modifying the files for the Lab24 unidirectional SerDes.

Leave SerDes.v in the new SerDes directory, but move the other files up one directory, into the new FullDup directory. These files should include: SerDes.inc, SerDesFormats.inc, SerDes.vcs, and SerDesTst.v. The only things remaining in the SerDes directory should be SerDes.v and the four subdirectories originally there.

Rename the moved design files by changing "SerDes" in their names to "FullDup". For example, you would have in the FullDup directory, FullDup.inc, and FullDupTst.v, among others. Don't change the contents of these files yet.

Step 2: Use a testbench copy as a template for FullDup.v. Copy the file you just renamed to FullDupTst.v, to a new file named FullDup.v in the FullDup directory.

Open FullDup.v in a text editor and delete everything in it except the old SerDes instance, with its parameter map.

Copy-and-paste a second, identical copy of this instance into FullDup.v. Name the upper instance in the file SerDes_U1 and the lower instance SerDes_U2.

Digital VLSI Design with Verilog

Having pasted the instances just described, create a starting copy of the `FullDup.v` module header as follows: Open the `SerDes.v` file in the `SerDes` subdirectory, and copy-and paste the `SerDes` module header declarations from it into the top of `FullDup.v`. Change the new module name to "FullDup".

`FullDup.v`, which once was a copy of `FullDupTst.v`, now should consist of the module port declarations from `SerDes.v`, and two instances of `SerDes` each with different instance names. The top module name in `FullDup.v` should be `FullDup`.

Next, we shall edit `FullDup.v` to make minor changes in the `SerDes` part of the design; then, we shall complete the full-duplex design of `FullDup`.

Step 3: Resize the `SerDes` FIFOs. It would be fun arbitrarily to vary both widths and depths of the FIFOs for the two systems, *A* and *B*, shown in figures 8-6 and 22-13. However, the frame encoding would have to be altered to change the width, so we shall leave all widths at 32 bits (33 including parity), and just modify the depths. Also, we have used exact counter wrap-arounds in the FIFO state machine, so each FIFO must have a depth in words equal to a power of two.

As shown in figure 22-12, instead of 32 words, each *A* FIFO should include 8 words, and each *B* FIFO 16 words. So, we need separate depth parameters for the *A* and *B* sides of each `SerDes`.

To accomplish this in the `FullDup.v` file, remove the old `AWid` FIFO depth parameter and replace it in the `FullDup` header with four new ones, one for each FIFO instance in the full duplex design. The result should look something like this:

```
module FullDup #(parameter DWid = 32
              , RxLogDepthA = 3, TxLogDepthA = 3   // 3 -> 8 words.
              , RxLogDepthB = 4, TxLogDepthB = 4   // 4 -> 16 words.
              )
...  (module port declarations)  ...
```

You should recall from basic communications hardware usage that "Rx" names a receiver and "Tx" names a transmitter. The new parameters should be passed to the `SerDes` instances; but, first we have to decide which instance in the verilog corresponds to which one in figure 22-12 above. Let's take the first `SerDes` instance in the file, which we have named `SerDes_U1`, as the top serial line in figure 22-12 ("Serial Link 1").

Then, postponing port-mapping details, the parameters should be passed this way:

```
...
SerDes #( .DWid(DWid)
        , .RxLogDepth(RxLogDepthB)
        , .TxLogDepth(TxLogDepthA)
        )
SerDes_U1 ( .ParDataOutRcvr ... // A sender; B receiver.
 ...
SerDes #( .DWid(DWid)
        , .RxLogDepth(RxLogDepthA)
        , .TxLogDepth(TxLogDepthB)
        )
SerDes_U2 ( .ParDataOutRcvr ... // B sender; A receiver.
 ...
```

At this point, the `SerDes` module can not use the new parameters, so we must modify it to accept them.

In our `FullDup` design, a `SerDes` module gets two FIFO depth parameter declarations, one for the receiving (Deserializer) FIFO, and one for the sending (Serializer) FIFO. Of course, at the `SerDes` level, there is no distinction between '*A*' and '*B*'. In the `SerDes` subdirectory, in `SerDes.v`, pass in the new parameters, and find every occurrence of `AWid` and rename it to `RxLogDepth` or `TxLogDepth` appropriately. As you do this, do not rename the FIFO parameter; just change the name of the value mapped to it. For example, in `SerDes.v`, "Serializer #(.DWid(DWid), .AWid(AWid)) ..." will become, "Serializer #(.DWid(DWid), .AWid(TxLogDepth)) ...". The Serializer's `.AWid` gets the Tx parameter; the Deserializer's `.AWid` gets the Rx parameter. Assign 5 (32 words) to be the module header default for both depth values; this value will be overridden to 3 (8 words) or 4 (16 words) during instantiation.

There is no need for further editing of parameters lower in the design; the `Serializer` and `Deserializer` will pass the proper values to their submodules as they did before.

In particular, each `FIFO_Top` instance now properly will pass on the depth values required to define FIFO addressing and number of memory words. Because of our previous planning, there is no reason to rename the parameter declared in the `FIFO_Top` module or to change anything in the FIFO or PLL parts of the design.

Step 4: Complete the connection of the full-duplex lane. Back again in `FullDup.v`, we have to decide what should be our module I/O's, and how they should connect to the `SerDes` instances.

The `SerDes` instances are independent; there should not be any communication between them except over the serial line, and this simplifies our decisions. What we shall do next, is decide what to call the nets connecting to the `SerDes` instances, and what should be the `FullDup` module I/O's. For most of the rest of this Step, we shall try to winnow away all but the minimum required `FullDup` I/O's.

Let's start with a simplification of the system relationships: `SerDes_U1` (link #1 in figure 22-12) transmits data from A to B; `SerDes_U2` transmits from B to A. Therefore, on the A side, the only `SerDes_U1` serializer output port would be the serial line (`SerLineXfer`); the A serializer inputs would be `ParDataIn`, `InParClk`, `InParValid`, `Reset`, and `TxRequest`. On the `SerDes_U1` B side, the B deserializer, the outputs would be `ParOutRxClk` (in the B clock domain) and `ParOutTxClk` (in the A clock domain). The B deserializer inputs would be `OutParClk`, `Reset`, and `RxRequest`.

The opposite I/O relations must hold for `SerDes_U2`.

That takes care of the instance pins. As for the nets, to avoid elementary errors or confusion, we shall start by renaming all of them. Initially, we'll prepend 'A' or 'B'; later, when decisions have been made, we shall rename the nets by moving the prepended letters to the end of the net name, following the example above of "RxLogDepthA", etc.

So, let's begin by adding an 'A' prefix to every `SerDes_U1` and `SerDes_U2` A net, and a 'B' prefix on every other net connected to the `SerDes` instances. For example, in `SerDes_U1`, we would have `AInParStim`, `AInParClkStim`, etc. At this point, the only net lacking an 'A' or 'B' would be `ResetStim`, which, as shown above, we shall share between A and B.

After this, we can go up to the `module` port declarations and make a <u>complete, duplicate copy of the current `FullDup` module ports</u>. Do this simply by copy-and-paste of the entire module header port declarations, just doubling the original number of `FullDup` I/O's.

Once the `FullDup` module ports are duplicated, doubling the number of I/O's in the `FullDup` header, prepend an 'A' to every net name in one copy and a 'B' to every net name in the other.

We could declare a few new nets and stop here, but let's simplify things a bit:

Simplification **A**. Eliminate one `FullDup` module input port by declaring just one common `Reset` in the module header to be routed (later) to both `SerDes` instances.

Simplification **B**. We want to transfer data between A and B, which are bussed systems; we don't really require the serial lines to be visible outside of `FullDup`. So, declare wires for the serial lines named `SerLine1` and `SerLine2`, and substitute them for the nets on the `SerLineXfer` pins of `SerDes_U1` and `SerDes_U2`, respectively. These output nets will remain otherwise unused, and this allows us to delete the two corresponding module output ports in the header of `FullDup`.

Simplification **C**. We may assume that A and B correspond two distinct clock domains. Therefore, the A clock for clocking in data to `SerDes_U1` should be the same clock as the A clock used to clock out `SerDes_U2` data. Declare a single module input named `ClockA`, and connect it to `SerDes_U1.InParClk` and `SerDes_U2.OutParClk`. Then, declare `ClockB`, connect it correspondingly, and delete the module input ports for all other clocks. This eliminates a total of another two `FullDup` module ports.

Simplification **D**. It was suggested optionally at the end of `Lab24` to create two parallel-bus output ports from the `SerDes`. Data from one would be clocked out by the receiving-domain's clock, and data from the other would be clocked out using the parallel clock embedded in the serial framing. These ports were named `ParDataOutTxClk` and `ParDataOutRxClk`. The purpose was to see the way the data delivery differed because of the different clock domains.

Thus, you may have two parallel output ports on each `SerDes`. We don't want these two ports in `FullDup`; we only want outputs clocked in the receiving domain. So, if you have them, delete both `ParDataOutTxClk` ports, even if they were commented out in your `Lab24 SerDes` instance. This leaves just one parallel-data output port in each `SerDes` instance and allows us to delete as many as two more `FullDup` module output ports.

Simplification **E**. Let's look at the `RxRequest` and `TxRequest` inputs. They were convenient for debugging our `SerDes` and demonstrating FIFO functionality; but, now they can be reduced. However, to preserve flexibility in case our reduction plan doesn't work out, let's keep the port connections and declare four *request* wires to assign them explicitly; this will be shown in a code example below. So, we'll tie `RxRequest` high permanently ("= `1'b1`"); we'll control everything with the input `ParValid` signals at both ends: If `ParValid` is asserted in either system, it also will assert `TxRequest` in that system. This allows us to remove four `FullDup` ports and simplifies the `SerDes` operational protocol.

Simplification **F**. Wrap-up. Except `ClockA`, `ClockB`, and `Reset`, rename the `FullDup` module ports so each remaining input begins with "`In`" and each output with "`Out`". Also, except `Reset`, rename all ports so that every *A* port ends with '`A`', and every *B* port with '`B`'. This moves the '`A`' and '`B`' prefixes to the end of each name, to become postfixes; the whole renaming sequence was an accounting technique to help keep the port renaming free of confusion or errors.

After all this, your `FullDup.v` should contain something close to the code shown below:

Digital VLSI Design with Verilog

```verilog
`include "FullDup.inc"   // timescale & period delays.
//
module FullDup
    #(parameter DWid = 32                      // 32 bits wide.
      , RxLogDepthA = 3, TxLogDepthA = 3       // 3 -> 8 words deep.
      , RxLogDepthB = 4, TxLogDepthB = 4       // 4 -> 16 words deep.
      )
      ( output[DWid-1:0] OutParDataA, OutParDataB
      , input[DWid-1:0]  InParDataA,  InParDataB
      , input  InParValidA, InParValidB
      , ClockA, ClockB, Reset
      );
//
wire SerLine1, SerLine2
   , RxRequestA, RxRequestB, TxRequestA, TxRequestB;
//
assign RxRequestA = 1'b1;
assign RxRequestB = 1'b1;
assign TxRequestA = InParValidA;
assign TxRequestB = InParValidB;
//
SerDes #( .DWid(DWid)                    // Parameter values.
        , .RxLogDepth(RxLogDepthB)
        , .TxLogDepth(TxLogDepthA)
        )
SerDes_U1 // Ports reordered:
    ( .ParOutRxClk(OutParDataB), .SerLineXfer(SerLine1)
    , .RxRequest(RxRequestB),    .ParDataIn(InParDataA)
    , .InParValid(InParValidA),  .TxRequest(TxRequestA)
    , .InParClk(ClockA),         .OutParClk(ClockB), .Reset(Reset)
    );
```
(*similarly for* SerDes_U2; *but, with* SerLine2, *and with* 'A' & 'B' *suffices reversed*)

Simulate briefly to verify connectivity, file names, and file locations. Remember that by default some simulators will attempt to open `include files on a path relative to their invocation context. Note that the SerLineXfer ports are just for simulation convenience; they are unnecessary to SerDes functionality at the FullDup level.

You may reuse your SerDesTst.v testbench (which was renamed to FullDupTst.v in Step 1) for this simulation after changing names and adding the new FIFO parameters. Or, to expedite the testbench, which should be quite elaborate for this design, you might consider just copying the FullDupTst.v testbench file provided for you in the Lab26_Ans/Step04_FullDup directory.

Use the simulator to check the hierarchy tree to be sure that both SerDes instances, and all FIFO instances, are present. Make sure all warnings about bus widths, assignment widths, and so forth are corrected or well-understood before proceeding. Figures 22-14 and 22-15 show some FullDup simulation results.

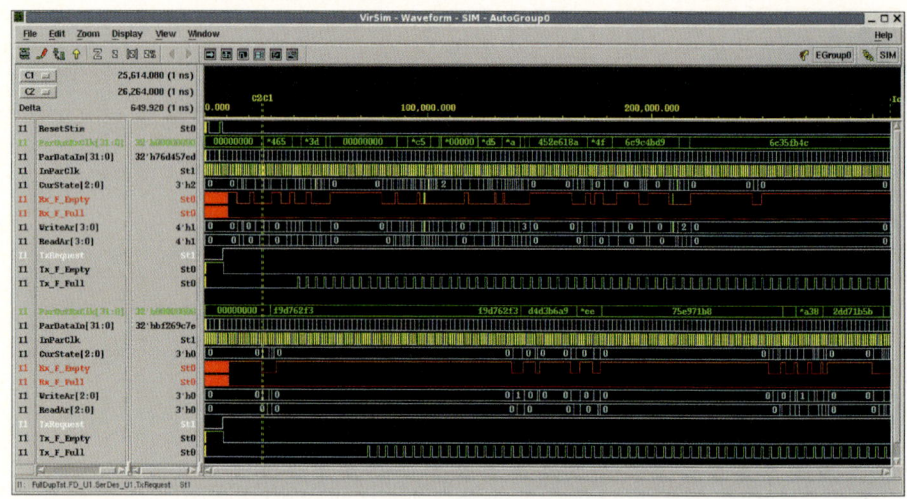

Fig. 22-14. The first full-duplex serdes (`FullDup`) simulation. SerDes U1 waves are on top; U2 below. Data are random and different in each direction; the *A* and *B* FIFO's are of different depth.

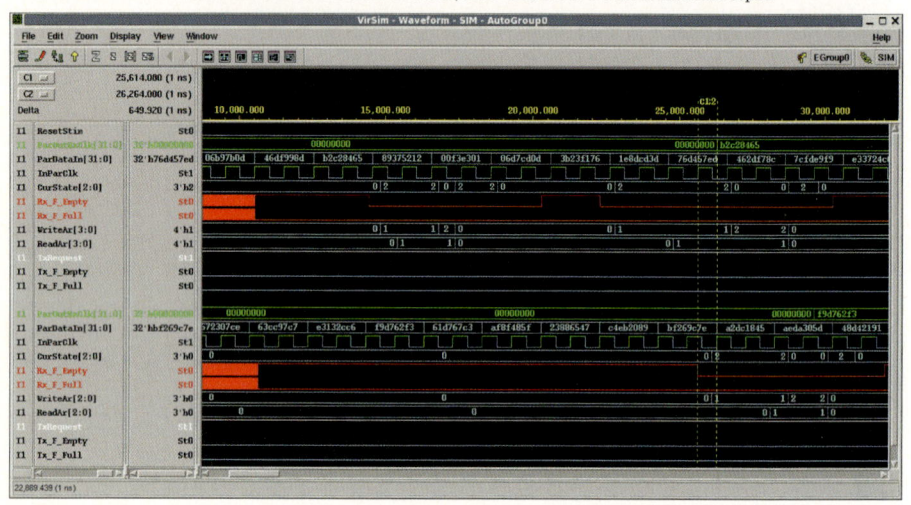

Fig. 22-15. Close-up showing clock skew between the two, independent parallel-bus domains.

Step 5: Add assertions. Create a new subdirectory `FullDupChk` in the `Lab26` directory and copy everything in `FullDup` into it. Do this Step in the new directory.

Digital VLSI Design with Verilog

We've ignored assertions and timing checks in our serdes design, except for one parity check assertion in the FIFO memory model. Let's correct this deficiency now.

No FIFO is visible at the `FullDup` level of the design, but we still would like to know when the FIFO's go full or empty. The first appearance of the full and empty flags in the design occurs at the `SerDes` level, in the `Serializer` and `Deserializer` instance port mappings.

Add four assertions to `SerDes`, two in `Serializer` and two in `Deserializer`, that check that the four FIFO flags are false. Each assertion should print a <u>warning</u> message to the simulator console, warning of a FIFO full or empty. The assertions may be your own, or they may be modelled after our generic assertion in *Week 4 Class 2* (Lab 11). Use `%m` so that the assertion will print the `module` instance name in which it was triggered.

Our design will not fail under these warning conditions, but they imply that possibly the containing *A* and *B* systems will have to resend lost data, the loss being determined from the data rather than from `FullDup` hardware. We could use these assertion warnings to decide whether the design word depth of the `FullDup` FIFO's should be changed.

Simulate. There should be at least a few FIFO-empty states to exercise your new assertions.

If you have not done so already, add testbench code to enable transmission (`Tx`) from both `Serializers`; your testbench also should make `ClockA` and `ClockB` independent. You might consider copying the testbench provided in `Lab26_Ans/Step04_FullDup`, to save time in lab. This testbench is fairly sophisticated and includes independent clock-frequency drift in both domains, as well as different random input data for each serdes.

Step 6: Add timing checks to monitor the `DesDecoder` control pulse widths. Perform this entire Step in `FullDupChk`. Using a simulation testbench from `Lab26_Ans`, or one very similar, display the signals in a `DesDecoder`. There is only one `SerDes` `DesDecoder` module, so both instances will implement your checks. Our simple approach to control in this module has allowed `ParValid`, `ParClk`, `doParSync`, and `SyncOK` to be asserted apparently as pulses of wildly varying width.

Let us study these pulse widths. We wish to determine the narrowest positive pulse that occurs during simulation. To do this, add `$width` timing checks in `DesDecoder` to report a violation whenever the width of a positive pulse on one of these signals is less than the value given by a `specparam`. Declare a different `specparam` for each timing check.

Vary the `specparam` values while repeatedly running simulation. For example, after setting up four `specparams` and four `$width` timing checks, set all the `specparams` but one to 0; set the other one to 100 (=100 ns); this will disable all but one. Then, run the `FullDup` simulation for a relatively long time, say 500k to 1M ns.

If there is no violation reported, increase the nonzero criterion `specparam` value by 50 or 100 ns, or more, and repeat the simulation until violations occur (you will know a violation message when you see one). Use the simulator [Stop] button if violations are too numerous. Going by the message numbers or guesses, then decrease the criterion and hunt until you have found the maximum criterion value at which no violation occurs. Leave the `specparam` there.

Displayed next, in figure 22-16, is the new VCS console output showing a `$width` timing violation at time 21,935 ns.

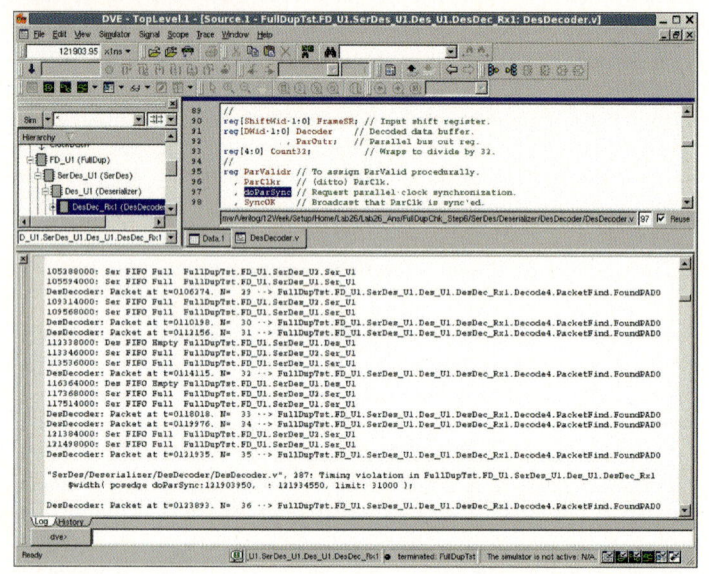

Fig. 22-16: The Step 6 `DoParSync` pulse is too narrow at time 21,935, causing a `$width` violation.

Even zoomed in, the violating pulse (obscured by the measurement cursors) does not draw attention in the waveforms, as shown in figure 22-17:

Digital VLSI Design with Verilog

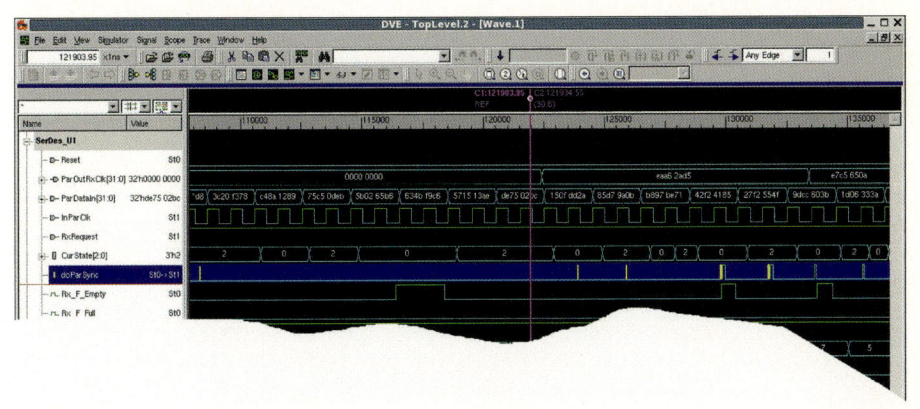

Fig. 22-17: The Step 6 `DoParSync` violating pulse in a new VCS waveform view.

Repeat the procedure for the other three `specparams`, separately or simultaneously, until all are at their maximum no-violation level.

Now, any unusual state causing a shorter pulse will be revealed by these checks. Record your timing-check widths in this table:

Variable	Maximum no-violation width (ns)
ParValid	
ParClk	
doParSync	
SyncOK	

It pays to be able to find unexpected brief pulses. As the old naval shipyard saying goes,

"Loose glitch sinks chips".

Step 7: For this Step, change out of `FullDupChk` and return to the `FullDup` directory. Make up a synthesis script file, `FullDup.sct`, using as a template, for example, a script in one of the subdirectories at `Lab24/Lab24_Ans`.

With this script, synthesize the design while imposing only zero-area and just one timing constraint, as follows: Set a stringent maximum delay limit on all outputs, one which the synthesizer can meet with only a little slack. Do not define a clock. Flatten the design before running `compile` with all defaults (no option); this will make the resulting netlist comparable with the scan-inserted netlist we shall create next. Be sure to log an area report into a text file for later reference.

The purpose of this Step is just to prove quickly that the design verilog is entirely in synthesizable format. Without any clocking, clearly it should not be expected to simulate correctly. However, with clocks and other timing constrained, and with hold violations fixed, this `FullDup` design can be made to synthesize to a correctly simulating netlist. See the Extras answers for various ways of applying workable synthesis constraints. For curiosity's sake, figures 22-18 and 22-19 give waveforms for a fully-constrained, ***completed class SerDes project***:

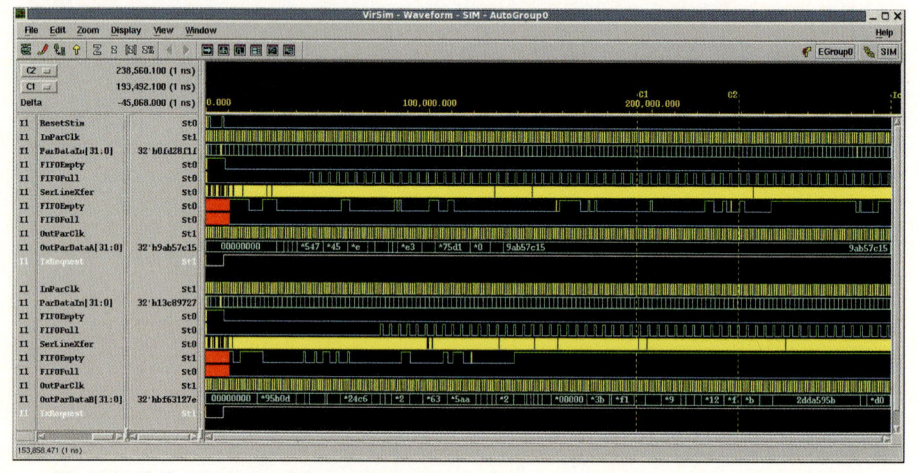

Fig. 22-18. Netlist simulation of the completed full-duplex SerDes, synthesized with hold-fixing for all clocks.

Digital VLSI Design with Verilog

Fig. 22-19. Close-up of the completed full-duplex SerDes netlist simulation, demonstrating that the *A* domain input parallel word `0x0fd2_8f1f` (time ~193 ms) is transferred correctly to the *B* domain output bus (time ~235 ms).

Some closing remarks: The `FullDup` source verilog simulation should be about the same as shown here, depending on precise coding. Concerning the synthesis answers in the `Lab26_Ans/Step7_FullDup` directory, the DC script for the quick proof of synthesizability is named `FullDup_Simple.sct`. Its only output netlist is named `FullDupNetlist_Simple.v`. In 2011 software, this netlist is not fully functional because of timing inconsistencies caused by lack of constraints. The latest 2011 version of DC synthesizes this `SerDes_U2` in a partially working state.

For one form of a correctly constrained, monolithic `FullDup` netlist, there is a `FullDup_TimingNoHold.sct` script in the answer directory for this Step. You can see the numerous constraints in the `.sct` file; these were arrived at by trial and error. This script produces a different netlist named `FullDupNetlist_Hier.v`. Using the 2011 version of DC, and the given constraints, the hierarchical netlist simulates mostly correctly without SDF timing; the output from the *B* direction being correct, but that from the *A* direction being only occasionally correct. Notice that `set_fix_hold` was **not** used for these netlists, but that fairly stringent max delay constraints were used instead, to avoid hold failures by speeding up all gates. Hold errors vanish for fast-enough gates. It is possible that these constraints could be improved without hold-fixing, but that was not attempted.

Hold-fixing, using `FullDup_TimingHold.sct`, improves the netlist so that the hierarchical netlist simulates correctly without SDF timing; however, the synthesis with hold-fixing took over 24 hours on a 1 GHz machine.

It took over a week of repeated synthesis attempts to arrive at the constraints used above. In a commercial project, perhaps a week to a month of development time would be allowed for refinement of the netlist synthesis of such a device. This illustrates the advantage of purchasing IP when gate constraints may be hard to meet.

There is a different set of exploratory synthesis results in the Step7_FullDup/Netlists/OLD_Netlists directory. These are explained in a *ReadMe* file. There is a correctly simulating netlist there which was created by aborting a synthesis run (with set_fix_hold) before it had completed. This exact result is chancy and perhaps can not be duplicated with new DC. In particular, the DesDecHold netlist may not simulate correctly in other versions with or without SDF.

Step 8: Insert scan logic. Create a new subdirectory named FullDupScan and copy the entire contents of the FullDup directory into it. Simulate briefly at the new location to verify file locations and connectivity. Do the rest of this lab in FullDupScan.

Use the synthesizer to insert internal scan as follows:

A. If you wish to reuse your copied synthesis script, begin by renaming it to something like FullDupScan.sct; edit it to change its output file names (netlist and log) so they will be identifiable as scanned outputs. A better plan might be to copy the FullDupScan.sct file from the Lab26_Ans, Step 8 directory.

In these .sct files, notice the large number of DC warning messages which have been suppressed; each one was studied individually and deemed unimportant. It is a good general idea not to suppress warnings unless they are understood well and either should be ignored or are so numerous as to interfere with other, perhaps unanticipated, synthesizer messages.

In the script file, you may have to delete the existing compile command and substitute in its place the scan-insertion commands used in Lab25. These may be found at Lab25_Ans/Step01_IntScan/DC_Scanned/Intro_TopFFScan.sct. You will have to change the Clk and Clr names and the scan clock period. You also will have to edit the FullDup module header in FullDup.v to add two new input ports, tms and tdi, and one output port, tdo, for the scan circuitry.

B. Run the synthesizer to insert scan elements. This will take only a little longer than did plain synthesis. Be sure to log an area report in a separate text file so you can compare it with the one in the previous Step. The purpose of the netlist comparison between the results of Lab26 Step 7 and Lab26 Step 8 is to get an idea of the netlist size difference caused by scan insertion.

The same constraints should not be difficult to meet with scan inserted, because previous compilation should have shown they could be met without scan. This might have been guessed by recalling that scan insertion replaces sequential elements with scan elements, and that the extra gate count from (effectively) one mux per flip-flop really can't amount to very much. Without dwelling on library issues, inspection of the library we are using for synthesis indeed shows that the area of a typical scan flip-flop is only about 30% greater than that of an ordinary flip-flop.

This exercise primarily was to show the effect of scan on netlist size, not speed. Do not bother with simulation at this point; the TAP controller will not be functional because the synthesis design constraints did not include a clock, preventing creation of a working scan test mode.

C. Browse through the scan-inserted netlist file in a text editor to see where the scan elements were substituted.

22.4.1 Lab Postmortem

Where in `FullDup` might other timing checks be added?

What was the area impact of scan insertion in this design?

How hard would it be to provide a hardware test engineer with a key to the scanned design? In other words, if you define a `FullDup` scan test vector to be a vector the length of the entire `FullDupScan` scan chain, how wide would this vector be? Then, how would one go about providing a key giving the location of each design bit in the test vector?

22.5 Additional Study

Read *The NASA ASIC Guide*, in the References at the start of this Textbook.

If you want a little more detail on LFSR pseudorandom number generation, try the Koeter article, also in the References to this Textbook.

Optional: Retrieve your `BIST.v` empty interface file of `Lab25`, Step 8, and complete the BIST functionality in your own way.

Chapter 23
Week 11 Class 2

23 Today's Agenda:

Lecture on SDF back-annotation
Topics: Back-annotation basics; SDF file use with a verilog design.
Summary: We briefly explain the use of back-annotation and describe some of the characteristics of an SDF file. We show how SDF fits into the design flow. We then give the syntax for the use of SDF with a verilog module.

SDF Lab
We synthesize the `Intro_Top` design and then simulate the netlist with back-annotated timing.

Lab Postmortem
Q&A, and how to back-annotate part of a design.
(rest of the day may be used for retrospection on recent, previous labs)

23.1 SDF Back-Annotation

Today, we'll look into SDF just enough to understand how it works.

23.1.1 Back-Annotation

Back-annotation refers to the reevaluation of an existing netlist because of newly assigned or updated attributes, usually delay times. A tool is run; and, the result is interpreted as going <u>back</u> to the netlist; whence, <u>back</u>-annotation. The netlist itself is not modified; the back-annotations instead are stored in a side file.

Although the synthesizer can estimate the effect of trace length on delay time while optimizing a netlist, these delays initially are in the synthesis library's wire-load model, and they do not appear as such in the synthesized netlist. The netlist merely contains component instances with certain timing and drive strengths chosen during optimization to be adequate to meet the timing constraints, given the expected trace lengths. These component choices are fairly inaccurate when compared with actual performance after placement and routing. An unusually good estimate of many of the layed-out delays based solely on the initially synthesized netlist might be off by 10%; a typical estimate would be off by more.

After a netlist has been used to implement a floorplan or placed-and-routed layout, more accurately estimated delay times can be associated with the placement of individual component instances. To associate the new delays, the verilog library propagation delays are ignored and back-annotated delays are used instead. Usually, this means that a back-annotation side file should contain a description of the timing of every net and every component instance in the netlist. The side file generally used is an SDF file.

23.1.2 SDF Files in Verilog Design Flow

The special file format for netlist delay back-annotation is called SDF (Standard Delay Format). It is specified in IEEE Std 1497; its use with a verilog netlist is described in section 16 of IEEE Std 1364 or in section 32 of IEEE Std 1800. We wrote out SDF files in Lab01 and in Step 5 of our *Week 8 Class 2* Lab20.

When a delay is programmed in the verilog (#*delay* or in a specify block), but no such path under the same conditions is annotated in the SDF, the verilog delay is simulated unchanged. In all other cases, the SDF delay replaces the verilog delay; the latter is ignored.

Various other back-annotation file formats, for example SPEF (Standard Parasitic Exchange Format), are used for a variety of other physical layout purposes not included in the present course. SDF contains delay times; SPEF contains capacitances. In any case, the flow for the back-annotation is as in figure 23-1.

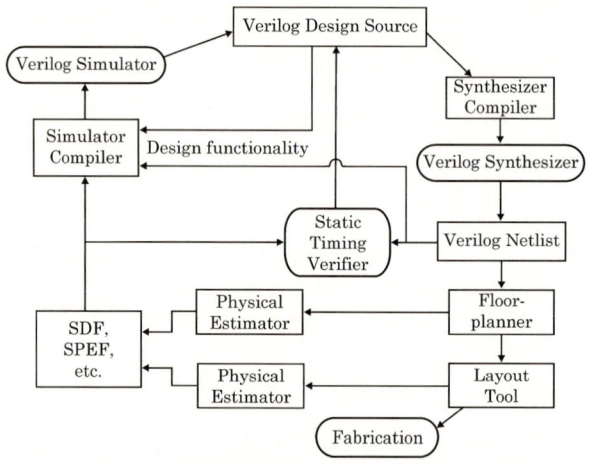

Fig. 23-1. Back-annotation in a typical design flow. The estimated delays or physical parameters are kept in a side file associated with the design netlist by the tools used to create them.

Timing in an SDF or other back-annotation file will override verilog library specify block path delays, including conditional delays. The net (interconnect) delays attributable to trace length also can be back-annotated in SDF, although in the verilog, such delays generally would be lumped to component outputs. Verilog path delays are associated with the IOPATH keyword in SDF; net delays are associated with the INTERCONNECT keyword. The keyword CELL is used for any component or module instance. Except for colons in timing triplets, the only SDF punctuation is by parentheses ('(' and ')') and quotes for literals; so, the language looks a bit like lisp or EDIF at first glance.

Digital VLSI Design with Verilog

In an SDF file, the SDF keywords are written in upper case but are not case-sensitive; references to netlist objects are copied literally from the netlist and are case-sensitive.

23.1.3 Verilog Simulation Back-Annotation

This process can be very simple and straight-forward. A verilog netlist is created somehow, for example by synthesis, and an SDF file is obtained somehow for that netlist. Typically, after the netlist has been floorplanned, an SDF file may be written out by the flooplanning tool. Another way to obtain an SDF file is simply to write one out from the synthesizer after saving the netlist in verilog format. This last is the approach which we shall use in this Textbook.

Once a verilog netlist and corresponding SDF file have been obtained, the verilog system task, $sdf_annotate("*sdf_file_name*"), is inserted into the netlist at whichever level of hierarchy the SDF file was written. This should be the only editing ever made to a netlist which is being back-annotated. When the simulator then is invoked, those modules which were back-annotated (usually, the whole design) will be simulated using the SDF timing instead of the verilog component library timing.

Complexity can arise if timing is to be back-annotated from several SDF files. If the files are nonoverlapping, the ABSOLUTE keyword (which appears everywhere in our SDF files in the lab below) may be used to make the last file read overwrite every common delay with its ABSOLUTE value. It is possible to use an INCREMENT keyword instead; in that case, the INCREMENT delay values are added to previous ones.

In summary:

SDF (Standard Delay Format) is defined in IEEE Std 1497. It is
- A netlist representation of timing triplets.
- Written by the synthesizer, static timing tool, or layout tool.
- Hierarchical; an SDF file may be rooted at any verilog module level.
- Associated with a design netlist by editing that netlist as follows:
    initial $sdf_annotate("*SDF_file_name*.sdf");

Our presentation here is meant only to show how SDF relates to verilog; we cover only a minimum of what an SDF file can do. In particular, the SDF language can include conditional and multivalue delays, PATHPULSE overrides, and timing checks.

SDF back-annotation may be used for netlist-based tasks such as simulation or static timing verification; it can not be used for synthesis.

23.2 SDF Lab 27

Topic and Location: Simulation and synthesis of Intro_Top; simulation of SDF back-annotated netlists.

Do this work in the Lab27 directory.

Preview: We simulate the old `Intro_Top` design and compare timing of the original verilog with that of the synthesized, back-annotated netlist. We then edit the SDF file manually to see how it controls the netlist simulation timing.

Deliverables: Step 1: A correctly simulating `Intro_Top` design in the `orig` directory. Step 2: A correctly simulating synthesized netlist of `Intro_Top` in the `orig` directory. Step 3: A correctly simulating netlist of `Intro_Top`, with written-out SDF back-annotated timing, in the `ba1` ("<u>b</u>ack-<u>a</u>nnotated #<u>1</u>") directory. Step 4: A correctly simulating netlist of `Intro_Top` in the `ba2` directory, with manually modified SDF back-annotated timing.

Lab Procedure:

<u>An operational reminder</u>: In new VCS, to display two different simulation waveforms at once, begin by simulating one of them as usual. After arranging the wave window nicely, save the session. Then, use the [x] icon in the upper left-hand corner of the ***console*** window to dismiss the console window; this will leave the wave window displayed and active. Bring up a second shell window, cd to the same directory, and run VCS again for the second simulation. Load the saved session to create a new wave window with the same file-lists and geometry as the first one.

Also: Older versions of VCS used *config* files (`.cfg`) to save window setups and wave-window signals; new VCS uses *session* files (`.tcl`), which are Tcl scripts.

Step 1: Simulate a design. Create three new subdirectories, `orig`, `ba1`, and `ba2`, in your `Lab27` directory. Change to the `orig` directory and copy in the verilog files for our old friend, the `Intro_Top` design from `Lab01`. Use the original files provided, including the original `TestBench.v` file. If you have edited those files, get a fresh copy from the Extras.

In `TestBench.v`, comment out the `` `include `` of `Extras.inc`, and change the `TestBench.v` timescale to `1ns/1ps` or `1ns/10ps` before simulating; this last will make small differences between the source and netlist simulations more easily comparable.

Then, invoke the simulator and simulate the design. After simulating, do not exit VCS; but, rather, as described above, leave up the waveform window displaying the simulation of the top-level primary inputs and outputs.

If you are using VCS and save your VCS configuration to a file, this file may be used, as described above, to invoke VCS for the rest of this lab with the same exact window sizes and waveforms displayed.

> NOTE: The simulator process ***must be left active***; otherwise, the wave window will not resize or redraw if moved. The easiest way to guarantee this is to do each step in this lab in a simulator instance invoked in a new terminal window. Once the simulator has appeared, it should be possible to delete (or at least minimize) the invoking terminal window.

Digital VLSI Design with Verilog

Step 2: Synthesize a netlist. After simulation in Step 1, replace your copied `Lab01 Intro_Top.sct` with the modified one provided in your `Lab27` directory. The design rules in the `Lab27` version have been changed so that the gate count in the netlist is minimized. You may wish to use a text editor to compare the original `Intro_Top.sct` with the new one to see how this was done.

After copying the new `Intro_Top.sct`, synthesize the design with it. The new script will write out the synthesized netlist and an SDF file automatically. The SDF file will be based solely on the synthesis library's generic wireload model, but its timing will be far more accurate than that of the original verilog design.

Make up a new simulator file list containing `TestBench.v`, the netlist file just created, and the *LibraryName*`_v2001.v` verilog component library file provided originally in your *misc* directory:

```
TestBench.v
Intro_TopNetlist.v
-v tcbn90ghp_v2001.v
```

Simulate the synthesized netlist and compare the waveforms with those of the original design in Step 1.

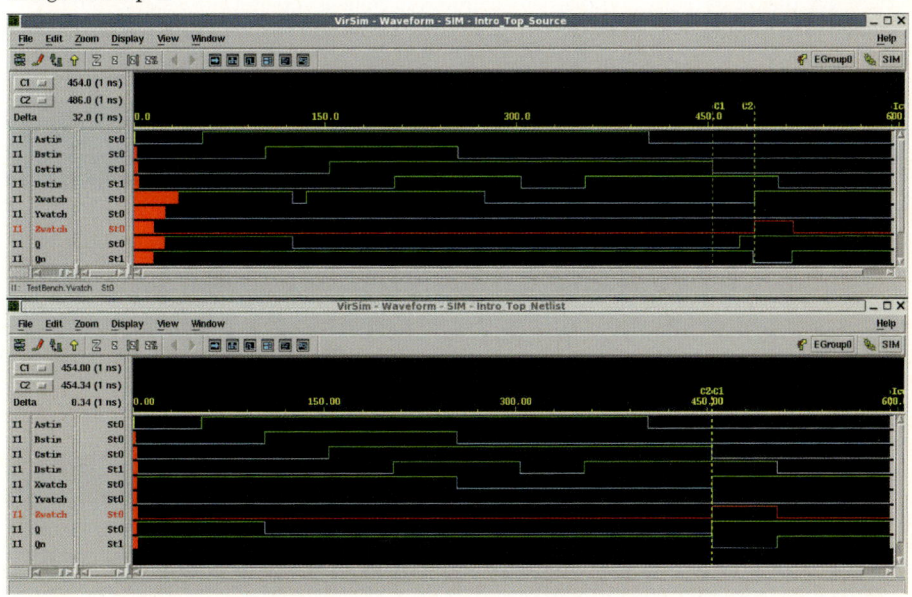

Fig. 23-2. Timing differences between the verilog source (above) and its synthesized netlist.

As shown in figure 23-2, the netlist-simulation wave shapes should be somewhat different from those of the source verilog, because the actual library gate delays are very short (approximated in the *LibraryName*_v2001.v file) compared with the delays in the source verilog. Most noticeably, a negative glitch in the Xwatch output wave will be missing in the netlist simulation.

After examining the simulation timing, leave up the netlist simulation waveform window which you just have examined; however, close the Step 1 wave window, or exit the simulation session, of the original design.

Step 3: Simulate the back-annotated netlist. Copy TestBench.v, the netlist file (Intro_TopNetlist.v), the simulator file list, and the SDF file of Step 2 into the ba1 directory you created above.

Open the netlist file in a text editor and find the module Intro_Top. Add an initial block in that module, below its header specifications, to reference the SDF file. Your initial block should be,

```
initial $sdf_annotate("Intro_TopNetlist.sdf");
```

Simulate again, and compare the SDF back-annotated waveforms with those of the original netlist of Step 2. All things considered, the shapes should be the same, and the timing of the outputs should be close to the same. There will be small timing differences (see figure 23-3), because the original netlist used timing from the verilog component library, *LibraryName*_v2001.v, which was hand-created with approximate timing, whereas the back-annotated netlist uses more accurate SDF delays created by the synthesizer from the original *LibraryName*.lib (compiled by Synopsys, if using its synthesizer, as the *LibraryName*.db) database.

Digital VLSI Design with Verilog

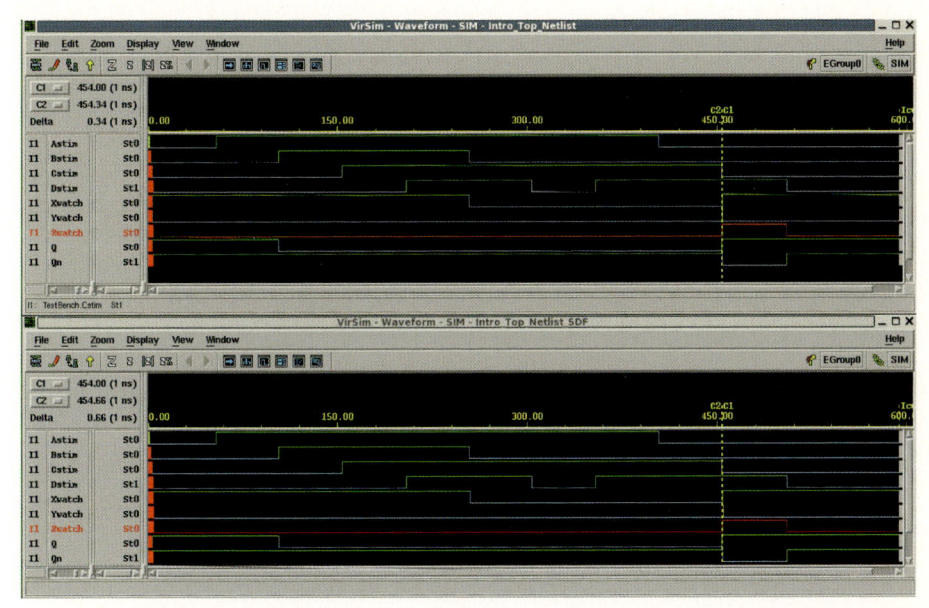

Fig. 23-3. Netlist timings for an approximated verilog library (above) and synthesizer SDF back-annotation.

When you have examined the delays, close the simulator waveform display window from Step 2 but leave up the waveform display of `Intro_TopNetlist.sdf` from this Step.

Step 4: Modify the SDF timing and simulate to see the difference. Copy all files from `ba1` of Step 3 to the `ba2` directory you created previously. Then, *cd* to the `ba2` directory.

Let's modify the timing of the .Z output port of `Intro_Top` to see how such a modification might be done during the back-annotation by a floorplanner or layout editor. This is just for instructional purposes: A designer never would edit manually an SDF file under normal conditions.

Open the verilog netlist file (`Intro_TopNetlist.v`) in a text editor, find the `Intro_Top` module, and locate the driver of the Z output port; this probably will be an inverter gate of some kind. Just find the last driver cell of the Z output and note the component type *ctype* ("INV*") and the instance name *inst_name* (maybe "U2") of that directly-connected driver cell.

Then, open the SDF file in a text editor and look for this statement,

```
(CELL
  (CELLTYPE "ctype")
  (INSTANCE inst_name)
```

Below the instance name, there should be a `DELAY` block which contains an `IOPATH` statement. It is the `IOPATH` statement which determines the delay on that path in *inst_name* when the netlist is simulated with the SDF back-annotation. The first triplet is rise delay; the second is fall.

The correct instance will be at the top of the design represented in the SDF file; it will **not** be preceded with a hierarchical `INSTANCE` path as may be, for example, U2 in "OutputCombo01/U2".

You might like to look in the *LibraryName*_v2001.v file to see how the hand-entered verilog delay differs from the synthesizer's delay for this component. This difference will account for our simulator display differences. However, in general, an SDF delay will be instance-specific; whereas the library delay can be only component (module) specific. If you are so licensed, the actual database used by the synthesizer to calculate the SDF delays can be seen in *LibraryName*.lib, in your Synopsys library installation directory area.

Here are some samples of other SDF statements (yours may differ):

```
// Wire delays are "INTERCONNECT":
(CELL
  (CELLTYPE "Intro_Top")
  (INSTANCE)
  (DELAY
    (ABSOLUTE
      (INTERCONNECT SRLatch01/U2/ZN U2/I (0.000:0.000:0.000))...
// Instance internal delays are "IOPATH":
(CELL
  (CELLTYPE "XOR2D1")
  (INSTANCE OutputCombo01/U3)
  (DELAY
    (ABSOLUTE
      (COND A2 == 1'b0 (IOPATH A1 Z (0.073:0.076:0.076) (0.079:0.082:0.082)))
```

These examples do not include the one to be edited next.

Anyway, to see how the SDF file controls timing, edit the `IOPATH` rise delay for the Z output at `Intro_Top` to be about 20 times the original value, and the fall delay to be about 10 times it. You may wish to comment out the original line; use a verilog "//" to do this. Then, save the file and repeat the simulation. The effect of the new delays should be fairly obvious: As is visible in figure 23-4, the final Z positive pulse in the simulation should be displaced to a greater time and should become narrower than it was with the unaltered SDF back-annotation.

Digital VLSI Design with Verilog

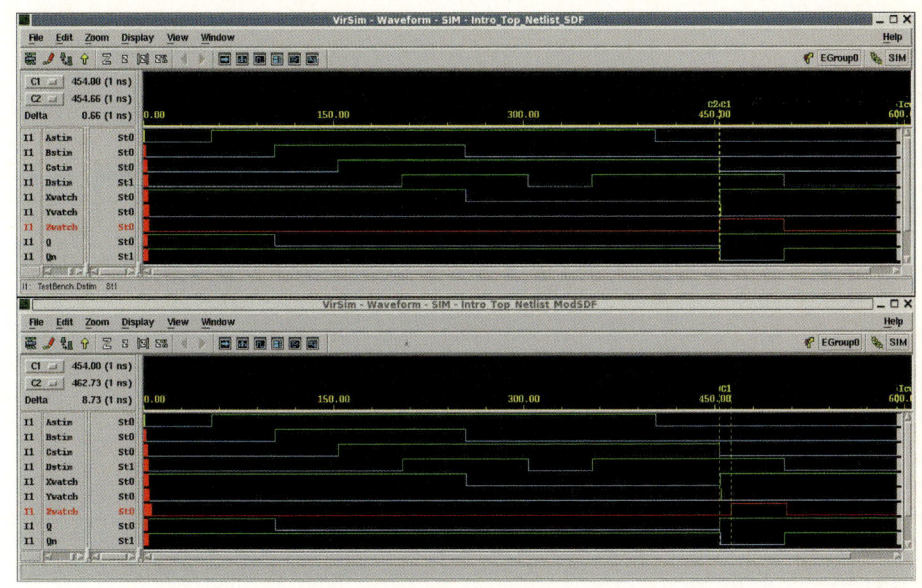

Fig. 23-4. Back-annotated netlist timing: Original SDF (above) vs. hand-modified SDF.

Optional. You may consider repeating Step 4 but entering different *min vs typ vs max* values in the SDF file. After this, invoking the simulator will cause it to use the changed sets of delays.

23.2.1 Lab Postmortem

What if the $sdf_annotate task call is located in the wrong module?

How might just part of a design netlist be back-annotated?

23.3 Additional Study

Optional. Recall Step 2 of today's lab. Explain in detail why a brief transition to 0 on the top-level X output occurred during simulation of the original design (figure 23-2, top) but not for the original synthesized netlist (figure 23-2, bottom).

Chapter 24
Week 12 Class 1

24 Today's Agenda:

Lecture wrapping up coverage of the verilog language
   Summary: We describe briefly the differences between *verilog-1995* and *verilog-2001* (or *verilog-2005*). We review the synthesizable subset of the language and list the constructs not explicitly covered in the course. We explain the relationship of the verilog PLI to the language. We mention a few wiring tricks which may be found in legacy code.

Continued Lab Work (Lab 23 or later)

24.1 Wrap-Up: The Verilog Language

24.1.1 Verilog-1995 *vs.* 2001 (or 2005) Differences

We have consistently exercised one major difference between *verilog-1995* and standards later than *verilog-2001*: The `module` header declaration format. Some other differences we used were: Implementation of `generate`, multidimensional arrays (*verilog-1995* allowed only 1 dimension), the `signed` arithmetic keyword (only `integers` or `reals` were signed in *verilog-1995*), and the `automatic` keyword for recursion in `tasks` or `functions`. In addition, *verilog-2001* included standardization of the use of SDF, as well as VCD dump file enhancements.

With the release of *verilog-2001* there were a host of other, perhaps less important changes, all explained in the Sutherland paper referenced in today's *Additional Study*. The further enhancements of verilog which are included in its successor standard, SystemVerilog, are discussed in the final class of this Textbook.

24.1.2 Verilog Synthesizable Subset Review

By now, all this may be well understood. However, here's a brief review:

The synthesizer can not synthesize:
- delay expressions,
- initializations (`initial` blocks),
- timing checks,
- verilog system tasks or functions;

or, generally, anything else which is transient in, or depends on, simulation time. The component delays used by the synthesizer for timing optimization and timing constraint violations come from its own library models and not from `specify` blocks or other verilog constructs.

The synthesizer may reject, or issue a severe warning about, procedural blocks containing even one nonblocking assignment with a delay. However, it will synthesize correctly any sequence either of blocking or nonblocking assignments which are not associated with delays. Delays written in the verilog never are used for synthesis.

The synthesizer will synthesize `generate` blocks, looping or conditional.

The synthesizer will enforce certain good coding practices by rejecting assignments to the same variable from different `always` blocks, and by rejecting mixed blocking and nonblocking assignments to the same variable or in the same `always` block.

The synthesizer has great difficulty synthesizing correctly any construct which is equivalent functionally to a complicated latch—which is to say, to a complicated level-sensitive or transparent latch. To avoid latch problems, (*a*) do not omit variables from an `always` block with a change-sensitive event control list; and (*b*) do not omit logic states from items assigned in a change-sensitive block, unless the code represents a simple latch with just one output bit.

Keep in mind that `fork-join` can not be synthesized (as of 2011) and should be avoided in any final implementation of simulation code—other than a testbench, of course.

24.1.3 Constructs Not Exercised in this Course

We have indeed covered the entire verilog language, including some constructs not useful in VLSI design; but, we have omitted a few details which differ unimportantly from the ones we have covered, or which are not frequently implemented in any tool. Here is a complete list:

- All the verilog language ***keywords*** are listed in IEEE Std 1364 appendix B or as part of IEEE Std 1800 annex B. Thomas and Moorby (2002) section B.5 also contains a table of them. A miscellaneous few of the verilog keywords have been ignored in the present work.

- File input/output manipulations. See the "File I/O" system tasks and functions listed below. For Windows users, note that the correct verilog filesystem name divider is '/', not '\'! Thomas and Moorby (2002) sections F.4 and F.8 present the available file-related system functions. There also is some relevant discussion of file I/O in Bhasker (2005), section 3.7.4.

- System tasks and functions. For the most part, consistent with the synthesis orientation of the course, we have called attention to these constructs only when necessary. The many builtin tasks and functions we have ignored are reasonably self-explanatory, given our practice in the few we have used. Some good explanation may be found in Bhasker (2005), section 10.3. There is a complete list of them below.

- Many compiler directives (`` `directive``). Those which we have not used also are quite easily understood, for the most part, from their names. See a list of them below.

- Attributes. These are boolean expressions in a scopeless block delimited by the two tokens, "(*" and "*)", the same tokens as used for comments in the Pascal language. For example, (* dont_touch U1.nand002 *). Verilog attributes are vendor- and tool-specific and are meant to control synthesis or simulation behavior. Current tools use comment directives such as "//synopsys ..." for this purpose and generally will not recognize verilog attributes.
- defparam, force-release, assign-deassign. As has been explained, these constructs should be avoided, although force-release sometimes might be useful for special testbench functionality or for debugging. As previously mentioned, defparam and procedural assign-deassign are likely to be removed from the language, according to the current *SystemVerilog* IEEE Std 1800-2012. Don't use them.
- Declaration of module output ports as **reg**. This is permitted in verilog-2001, as a compatibility carryover from verilog-1995. It is not recommended as a design practice, because internal wires no longer can be connected to such ports (bringing back the error-proneness of the 1995 module header format), and module delays on such ports are difficult to make visible. Also, the presence of mixed wire and reg ports requires perpetual attention to the mutually exclusive assignment statements used for addressing ports (assign vs. always), negating any small savings of time because of omission of the usual internal reg declarations paired with continuous assignments.
- The verilog PLI, which is discussed below.
- Three verilog wiring tricks, which you may encounter in preexisting code but which are not recommended for reasons of style:
    A. Multiple continuous assignments may be comma-separated. Example:
    assign WireX = A && B, WireY = C||(D^E), WireZ = WireY | A && D;
    B. A wire declaration with initialization, and optional delay, may be used instead of the declaration of the name followed by a continuous assignment:
    wire WireX = A && B;

    These two tricks are not especially risky but are not recommended because they complicate syntax and make searches for declarations or continuous assignments less predictable. They may break shell or Tcl scripts.

    C. A name used in a port map implies a wire and need not be declared:
    ALU U1 (.Result(Res), .Ain(A), .Bin(B), .Clk(**GatedClock**));
    ...
    bufif1 ClockGater (**GatedClock**, Clock, Ena);
    GatedClock is implied and need not be declared.

    Not recommended: (*a*) widths of implied wires always are 1, which is a minor limitation; (*b*) there may be lost time and confusion searching for the

declarations of the implied names; and (c) what if, near the top of the same module, an implied wire later is declared differently?

Closely related, omitted topics which are not part of verilog:

- Many synthesizer constraints have not been exercised. The command reference manual for the Synopsys DC synthesizer contains hundreds of commands, thousands of options, and exceeds 2,000 printed pages.
- The old Synopsys DC Shell ***dcsh*** constraint syntax of `dc_shell` or `design_vision` command mode. This is very similar to Tcl syntax, but it is now obsolete.
- The Synopsys Design Constraint language (SDC), which is an adaptation of Tcl and which is licensed free by Synopsys. SDC is meant to provide scripting and constraint portability among tools such as synthesizers, static timing verifiers, power estimators, coverage estimators, and formal verifiers. It can be used with VHDL or any other language such as SystemVerilog or SystemC, if the tool will accept it.

24.1.4 List of All Verilog System Tasks and Functions

These are documented in the IEEE Std 1364 section 17 and also may be found, among others, in the IEEE Std 1800 section 20. The verilog-specific ones fall into ten categories; a complete list is given in the next table. The functionality usually is described adequately just by the name, but the Std or other help documentation should be consulted for full explanations. Many of these system tasks are intended solely for testbench use.

No tool set known to the author has implemented all the system tasks and functions; however, all tools known to the author have implemented the subset used in this Textbook. The ones we have exercised are in bold in the table below.

In choosing a new system task or function for some special purpose, it is advisable to verify that all the working tools (simulator, synthesizer, floorplanner, etc.) likely to employ it have implemented it correctly or at least in exactly the same way.

Type	Task or function name
Display	**$display** $displayb $displayh $displayo **$monitor** $monitorb $monitorh $monitoro $monitoroff **$strobe** $strobeb $strobeh $strobeo $write $writeb $writeh $writeo $monitoron
Time	**$time** $stime $realtime
Sim. control	**$finish $stop**
File I/O	$fclose $fdisplay $fdisplayb $fdisplayh $fdisplayo $fgetc $fflush $fgets $fmonitor $fmonitorb $fmonitorh $fmonitoro $readmemb $swrite $swriteo $sformat $fscanf $fread $fseek $fopen $fstrobe $fstrobeb $fstrobeh $fstrobeo $ungetc $ferror $rewind $fwrite $fwriteb $fwriteh $fwriteo $readmemh $swriteb $swriteh **$sdf_annotate** $ssacf $ftell
Conversion	$bitstoreal $itor $signed $realtobits $rtoi $unsigned
Time scale	$printtimescale $timeformat
Command line	$test$plusargs $value$plusargs
Math	$clog2 $ln $log10 $exp $sqrt $pow $floor $ceil $sin $cos $tan $asin $acos $atan $atan2 $hypot $sinh $cosh $tanh $asinh $acosh $atanh
Probabilistic	$dist_chi_square $dist_exponential $dist_poisson $dist_uniform $dist_erlang $dist_normal $dist_t **$random**
Queue control	$q_initialize $q_remove $q_exam $q_add $q_full
VCD	**$dumpfile $dumpvars** $dumpoff $dumpon $dumpports $dumpportsoff $dumpportson $dumpall **$dumpportsall** $dumplimit $dumpportslimit $dumpflush $dumpportsflush $comment $end $date $enddefinitions $scope $timescale $upscope $var $version $vcdclose
PLA modelling	$async$and$array $async$nand$array $async$or$array $async$nor$array $sync$and$array $sync$nand$array $sync$or$array $sync$nor$array $async$and$plane $async$nand$plane $async$aor$plane $async$nor$plane $sync$and$plane $sync$nand$plane $sync$or$plane $sync$nor$plane

24.1.5 List of All Verilog Compiler Directives

A complete list follows below. No tool known to the author has implemented all these directives, but many tools which have not implemented a directive are likely to ignore it or just issue a warning. As suggested before, if feasible, use `undef at the end of any file in which a macro name has been `defined.

The Compiler Directives:

`begin_keywords `celldefine `**default_nettype** `**define** `**else** `**elsif** `endcelldefine `**endif** `end_keywords `**ifdef** `**ifndef** `**include** `line `nounconnected_drive `pragma `resetall `**timescale** `unconnected_drive `**undef**.

24.1.6 Verilog PLI

PLI stands for "Programming Language Interface". The PLI, strictly speaking, is not part of the verilog language, so we have not discussed it previously in this course.

However, the PLI is specified in the same IEEE standard document (Std 1364 or 1800) which specifies the verilog language.

The PLI language is C, and the PLI is a library of routines callable from a C program. The PLI is a way of extending the capability of the verilog simulator which allows users to design and run their own system tasks and even to create new verilog-related applications to be run by the simulator. Examples of such applications would be fault-simulators, timing calculators, or netlist report-generators. Like the builtin system tasks such as `$display`, the user-defined ones also must be named beginning with '`$`'.

The PLI generally is not especially useful for a designer, but it may be valuable to someone creating tools for a design team. Another use might be to implement certain simulator features not present in the current release available from a particular tool vendor.

Examining the messages printed by the VCS simulator when it compiles a new module, one can see that it simply is a specialized compiler which compiles and links an executable C program. The resulting VCS GUI is a different, precompiled program which optionally creates an interprocess communications link with the PLI and other executables when the simulator runs; the executable's I/O then is formatted and displayed by the GUI. To save run time and memory, VCS can be invoked in a text mode, with no GUI; the VCS help menu shows how. QuestaSim, Silos, Aldec and other verilog simulators work the same way as VCS, although any of those GUIs and simulators may be more or less tightly coupled.

In the past, the verilog simulation executable was *interpreted* (like a BASIC program or shell script) rather than being *compiled*. Except for running slower than a compiled executable (but being modifiable faster), the interpreted version worked essentially the same way as the compiled version.

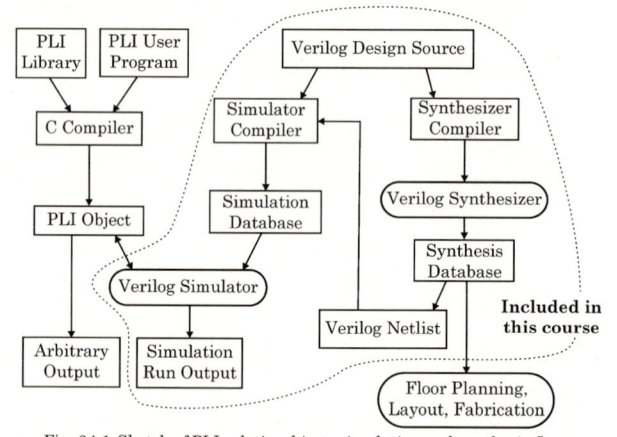

Fig. 24-1. Sketch of PLI relationship to simulation and synthesis flows.

Digital VLSI Design with Verilog

The PLI operates on the simulator itself (actually, on its internal data structures and operational runtime code), as shown in figure 24-1. Two independently accessible databases representing the verilog design are (a) the verilog design source code, which may be accessed by shell scripts or by visual inspection in a text editor; and, (b) the synthesized netlist, which, if written out in verilog format, also may be accessible by shell script or text editor.

The simulator's internal data structures generally are in a proprietary binary format; but, if a design is simulatable, the internal data must consist of objects representing the design structure, component models (including primitives), nets, and attributes of the corresponding verilog objects. Busses and ports must be represented independently for every bit, and, for netlist simulation, all loops must be represented unrolled. Over and above the objects in the design, the simulator database includes state, delay, and simulation time information.

As of the IEEE 1364 2005 Std, the PLI consists solely of the VPI or **_verilog procedural interface_** routines. Past versions of the PLI included two other collections of routines, TF or **_task-function_** routines and ACC or **_access_** routines, which now are absorbed into the VPI. All three collections are overviewed in chapter 13 of Palnitkar (2003), and we won't go into their properties further here.

24.2 Continued Lab Work (Lab 23 or Later)

24.3 Additional Study

On the enhancements introduced in *verilog-2001*, read Stuart Sutherland's excellent summary, "The IEEE Verilog 1364-2001 Standard: What's New, and Why You Need it", in the reference section at the beginning of this Textbook.

Read the summary of *verilog-2001* enhancements in Thomas and Moorby (2002) Preface pp. xvii - xx.

An example of a verilog PLI enhancement which prints the design hierarchy into a text file is available for licensed users at Synopsys *SolvNet* (as of 2011-01): https://solvnet.synopsys.com/retrieve/030557.html. The PLI version is the **VPI** of Std 1364-2005. The PLI function is named vpi() and is located in a file named vpi.c. It is compiled and linked into the VCS executable, along with a VCS-specific side file.

Optional readings in Palnitkar (2003):

For file I/O, see sections 9.5.1 and 9.5.5 and the Palnitkar CD example Memory.v.

Read section 14.6 on coding for logic synthesis.

Read through all of chapter 14 on logic synthesis.

Read chapter 13 on the PLI.

Chapter 25
Week 12 Class 2

25 Today's Agenda:

Lecture on deep-submicron problems and verification
Deep-submicron scaling physics.
Importance of correct logic and clocking.
Functional and timing verification.
SystemVerilog major features.
Verilog-AMS (<u>A</u>nalogue-<u>M</u>ixed <u>S</u>ignal) major features.

Continued Lab Work (Lab 23 or later)

25.1 Deep-Submicron Problems and Verification

25.2 Deep Submicron Design Problems

We restrict this discussion to electrical factors of interest to a verilog designer. Deep submicron effects related strictly to fabrication are ignored (ion implantation; details of optical proximity correction; shift to shorter, ultraviolet mask exposure wavelengths; immersion lithography; etc.).

The term "deep submicron" was adopted in the late 1990s to refer to layout pitches below about 250 nm—which is to say, below about 1/4 of a micron. At this scale, trace delays begin to overtake gate delays as primary considerations in design timing. For our purposes here, we include ***nanoscale*** technology in this deep-submicron category.

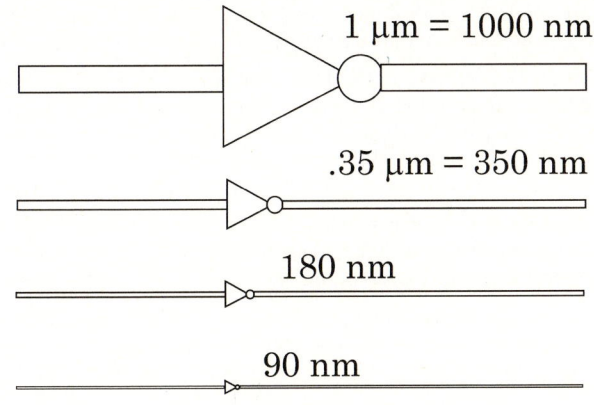

$1 \mu m = 1000$ nm

$.35 \mu m = 350$ nm

180 nm

90 nm

Fig. 25-1. Linear proportional gate schematic sizes and trace widths. Routing represented across a chip region of constant width.

Some generalities about the deep-submicron problem: Call the linear pitch size L. The major issue is that trace delays along distances L begin to contribute significantly to total delays, and anything introducing variation in trace timing becomes a major problem:

- The rate of electric charge transfer to or from a transistor is about proportional to its area, or L^2. But the delay on a trace is more or less proportional to its RC time constant. The resistance R increases linearly as trace width decreases, and capacitance C to ground decreases about the same way, so the time constant of a trace because of this reactivity remains about the same as pitch shrinks. To charge to a given voltage, the capacitance C of a trace decreases linearly as one power of L. Thus, the time to charge a trace of a given length increases about proportionally to the one remaining power of L. Individual trace lengths become important for delay calculation.

- As gate sizes shrink, distances on chip begin to increase *relative to gate size* (see figure 25-1). So, delays begin to grow because distances connecting gate pins do not shrink as rapidly as L, which is proportional to the distance between traces.

- As a greater and greater number N of gates is fabricated on a given chip area, the number of possible connections (traces) grows as something between $2N$ and N^2. Thus, on the average, as L shrinks (and chip complexity grows), the final fanout required of the average gate grows. For this reason, too, loading caused by traces begins to dominate timing estimation on the average.

In addition to propagation delay timing problems, pitches at and below 90 nm have introduced new problems related to fabrication-layer current leakage and noise. In particular, shrinking distance between traces has increased capacitance between them and therefore has increased the average cross-coupled noise.

Larger designs, lower supply voltages, and smaller pitches imply that libraries must contain a greater variety of different gates, including scan cells. Characterization must be more elaborate because of the smaller potentials and closer spacing of the gates on chip.

Library simulation and synthesis models can not simply store gate delays; instead, delays have to be calculated according to gate connectivity context, using input slew rates, output capacitance loading, current source capability, and chip operating conditions (process quality, voltage, and temperature—"PVT"). A sophisticated scalable polynomial, composite current source, or other detailed model must be used.

Also, random manufacturing process variations tend to be greater, the smaller the pitch. So, electromigration has become an issue for aluminum metal traces, requiring the more complex process of fabrication of copper traces.

As pitches have shrunk, clock speeds have increased; and, with leakage, this has put power dissipation in the forefront of all design problems. To mitigate power loss, and to reduce switching slew rates, voltage supplies on chip have dropped from 5 V or more to around 1 V.

Power loss, which goes more or less as the square of supply voltage, can be reduced by optimizing different chip areas for different operating voltages; this means that voltage level shifters have had to become commonplace. All clock and data traces generally must be level-shifted between different voltage domains. Furthermore, power consumption of many battery-operated digital devices has to be minimized by isolating different chip areas, using specialized data isolation cells, for low-speed or sleep modes of operation. Sleep modes may require retention of state by additional scan-like sequential cells, or by register duplication in always-on retention latches.

In addition to random fabrication variations, the inevitable but predictable diffraction of light at near-wavelength dimensions becomes a major factor in mask design, requiring corrective reshaping of masks (optical proximity corrections) to obtain pattern images adequate to fabricate working devices in the silicon.

Diffraction problems are equivalent to quantum-effect problems. For light, the quanta have the wavelength of the masked photons; for fabricated structures, the quanta have the wavelength of the logic-gate electrons. Low-k dielectric leakage is directly calculable in terms of quantum tunnelling of the electrons which operate the field-effect transistor components of each on-chip CMOS device.

In a different concern, huge designs have made simple full internal scan chains impractically slow. Deep submicron internal scan has to be hierarchical, with subchains nested and muxed to the TAP port, requiring complicated TAP controller logic to select subchains for the hardware tester.

At modern clock speeds, RF interactions also are becoming important, especially for serial transfer domains, such as those of ethernet, wireless antenna interfaces, and PCI Express.

All these new problems can be solved by proper physical design. While physical design is beyond the scope of this course of study, good design, especially intelligent partitioning, at the verilog source level can ease greatly the work of finding physical solutions to these, and to other, new, deep submicron problems as they arise.

Design pitches will continue to shrink until quantum uncertainty finally brakes progress in this direction. It is unclear what will be the scale at which fundamental limits prevent further reductions in digital structures, but it probably will be below 10 nm, the linear extent of 50 to 100 average atoms in the crystal structure of a solid.

25.3 The Bigger Problem

One should be aware that the deep-submicron problems definitely will increase as pitches shrink. However, based on industry surveys, even at about 130 nm, the primary causes of chip failures resulting in re-spin still are functional failure (logical design error) or clock-timing design error. According to these data, at 90 nm one would expect that perhaps only 25% of chip failures specifically will be because of deep submicron factors.

> Therefore, the verilog designer can prevent the majority of chip respin failures by entering correct functionality, with accurate assertion limits, and by preparing error-free clocking.

25.3.1 Modern Verification

Palnitkar (2003) and others group assertions with verification tools; however, the present author would prefer to view assertions as part of the design, along with timing checks and synthesis constraints.

The present author views verification as an operation independent of design entry: Verification is a practice which finds design errors. Assertions should be entered by the designer and usually are based solely on design considerations (how the design should simulate). Of course, as a result of design verification, QA assertions may be added, after the fact, to catch errors by preventing them from recurring silently.

So, we shall not discuss assertions here, with verification.

Verification of a verilog design should be understood as beyond the simple use of a compiler for a simulation or synthesis application. Compilation only checks the syntax of the language in which the design which was entered.

Functional verification usually is done during design by use of a simulator. Because the synthesized netlist is optimized by netlist library component timing, and not by verilog source code delays, simulation must be repeated, at least on a design-unit level, using a synthesized netlist and back-annotated delays.

Although a full chip, nowadays easily over five million equivalent gates, can not be debugged functionally on the typical workstation, designers may succeed in simulating a netlist of arbitrary size by using distributed computation, hardware simulation acceleration, or hardware emulation in FPGA's.

Timing verification is a major step in signoff of a physical design before tape-out for manufacturing. Any one delay out of specified limits, on a chip with twenty million traces, can cause failure and loss of millions of dollars in redesign and refabrication costs. Such failures also may cause additional, far greater, sales losses because of late entry in the product market window.

Timing verification rarely is done by simulation, because of the infeasibility of meaningfully simulating tens of millions of traces, and because of inability to evaluate simulation results even if such exhaustive simulations could be done. Instead, static timing verifiers are used which estimate delays from physical library data and back-annotated netlists. A static timing verification run can be completed on a modern design in a day.

25.3.2 Formal Verification

An important tool used for functional verification is formal verification. A formal verification tool runs statically, thus making possible the complete checking of very large designs.

There are several distinct kinds of formal verification:

- **Equivalence Checking**. This is the most popular type of formal verification; tools implementing equivalence checking of a new netlist against a "golden standard netlist" (functionally known to be correct) are mature, fast, and reliable. Typically, a verdict is reached of "equivalent", "functionally different", or "cannot determine equivalence".

  Tools which check a netlist against an RTL model are not so well perfected and often impose coding style requirements in addition to those usually demanded for synthesis. Of course, any RTL design which can be synthesized can be submitted to a netlist equivalence checker.

  There also exist equivalence checkers for system-level designs against RTL implementations; however, such tools still are experimental and are not yet widely used.

- **Model Checking**. These tools are analogous to tools which check library component characterization in *ALF* or *Liberty* against a *SPICE* model: A model is defined in a specialized language (for example, *PSL*; or, *P*roperty *S*pecification *L*anguage; IEEE Std 1850), and an implementation in an HDL such as verilog then is checked against the model. Often, such a tool can be considered a requirements checker as well as an implementation checker. Bugs can be found this way, and specific design units can be verified; but, usually, a complete verification of an entire implementation is not practical. Many of these tools can handle all possible variations of simple sequential logic.

- **Formal Proving**. This approach describes an implementation in logical statements and attempts to find inconsistencies. It can be applied fairly easily to blocks of combinational logic, but it can be more complex and difficult than an HDL when sequential logic is involved.

Some tools combine formal verification with partial simulation, allowing the designer to step through formally difficult parts by simulation and to obtain complete verification of those parts of the design most amenable to formal methods.

25.3.3 Nonlogical Factors on the Chip

We have studied only digital design in this course. Digital design requires (*a*) correct functionality and (*b*) correct timing for correct logical operation. However, independent of these two primary factors are those implied by physical interactions among logic gates, as well as by location on the chip. A working chip must have a proper distribution grid for gate power supply (VDD and VSS); and, it must be designed so that the activity of any arbitrary gate can not interfere with that of other gates.

Two important nonlogical considerations are power distribution and noise estimation:

- **Power Distribution**. As already mentioned, chips can be designed with regions operating at different voltages. The usual reason for this is (*a*) reduction of power usage, especially in battery-operated devices or (*b*) limitation of high temperatures caused by power dissipation. The adoption of IP blocks in a big chip can be facilitated if the chip can meet special power-supply requirements of the IP.

  Many devices can operate in a normal vs. a sleep mode. In the sleep mode, clocking simply may be suspended; however, the power supply also may be reduced or switched off in unused regions in order to reduce leakage. In many designs, the state of a sleepable region can be saved upon entering sleep mode and restored when normal function is resumed.

  A chip typically is bounded by its I/O pads. Just inside these pads, for chips with a single power supply, power and ground are distributed in two rings. A specific voltage region in a multivoltage chip also will be ringed by power and ground supplies. The rings provide noise isolation in addition to supplying power.

  Within any voltage region, the high and low potentials will be supplied in a grid covering the interior of the region. Each row of cells will get vias connecting its power pins to the high and the low supplies. The grid provides multiple current sources and sinks so that brief surges are averaged out, reducing the voltage variation at each individual gate.

- **Noise Estimation**. Before being taped out, a chip must be checked for noise by simulation or static verification. No possible gate activity should be capable of coupling unrelated logical states from one gate to another. Coupling simply may be capacitive; however, when changes are rapid, gate distances small, and switching transients are fast enough to cause metal traces to maintain significant potential differences along themselves, inductive coupling also must be taken into account. Switching changes also can radiate electromagnetically at trace corners, causing power loss in addition to noise.

  Coupled noise can cause changes in the logic level of nearby components; such coupling would likely be fatal to the operation of a device. More common is the problem of altered timing: A coupled potential in the direction of a changing level in a nearby trace can speed up the change; in the opposing direction, it can retard the change. Either of these effects can cause failure of the nearby operation by causing a setup or hold error.

  To prevent noise, the design must include shielding (unused power or ground traces between switching logic traces) or must allow enough distance between nearby traces that the coupling is not significant under normal operating conditions. Long parallel runs of possibly interfering logic must be rerouted. The thickness (perpendicular to chip surface) of the traces also may be varied to reduce coupling,

Digital VLSI Design with Verilog

because thick traces cross-couple capacitively better than thin ones; however, in many fabrication processes, metal or polysilicon thickness is predetermined and can not be varied locally.

In any event, the noise of contiguous logic, especially traces, must be estimated and controlled if the chip is to be fabricated reliably.

- Many of the nonlogical factors can be addressed by combining digital and analogue simulation as is done, for example, by the *Verilog-A/MS* tool which is discussed below.

25.4 System Verilog

SystemVerilog is an *Accellera* standard which has been adopted by IEEE. The 2012 standard document runs over 1300 pages, but the verilog-related subset is considerably shorter. We concentrate here on those SystemVerilog features which are related to verilog and are likely to be synthesizable by recently updated tools.

SystemVerilog consists in part of a superset of *verilog-2005* features; all of *verilog-2005* is included. This superset additionally contains many *C++* design constructs. Also included is a standalone *assertion* sublanguage.

A major goal of SystemVerilog is to make the porting of code to and from *C++* easier than it has been with *verilog-2005*. Another important goal is to make complex system-level design, as well as assertion statements, directly expressible.

A Caution

As we have seen, even the synthesizable subset of plain verilog, now almost 20 years old, has not been implemented fully (and not always correctly) by the major EDA vendors. SystemVerilog has been an IEEE standard for less than half of that time, and its synthesizable subset is more complex and varied than that of verilog.

It behooves the designer to try a small test case on any unfamiliar SystemVerilog feature before committing it to use in commercial coding, whether for simulation or synthesis. Vendors generally will be quick to fix bugs, but they have to know about them first.

25.4.1 Some Features of SystemVerilog

- User-defined types, with `typedefs` as in *C++*, and VHDL-like type-casting controls.
- Various pointer, reference, dynamic, and queue types.
- Simulation event scheduler extensions.
- Pascal-like nested `module` declarations.
- A module `final` block to run commands after a simulation `$finish`.
- VHDL-like `packages` for declarations.

- C++-like class declarations (with single inheritance, only).
- Five integer types: byte, shortint, int, integer, longint. All these except integer are unsigned types.
- A covergroup declaration for functional code coverage.
- A new PLI (Programming Language Interface) which includes the verilog VPI (Verification Procedural Interface).
- A new variable type, logic, which may be assigned procedurally or concurrently.*
- A new two-state bit type (value 0 or 1) for more efficient simulation compilation.*
- A new, unbased, unsized ' literal to express a constant value.*
- A new, timing-check-like gated event control by iff.*
- Three procedural always block variants: always_latch, always_ff, always_comb.*
- Per-module alternatives to `timescale: timeunit and timeprecision declarations.*
- Loop C-like break, continue, and return commands (no block name required).*
- A new interface type to make reuse of module I/O declarations easier.*
- A self-contained assertion language subset for assertions at module scope or in procedural code.*

The features marked "*" are likely to be useful for logic synthesis, or for simulation directly leading to synthesis, and are next discussed individually in more detail.

25.4.1.1 New Variable Type logic

In general, all variables, such as wire or reg, default to logic type if otherwise not constrained. Procedurally, logic and reg are identical, when used in reg-allowed contexts. The word logic is preferred when "reg" might be misinterpreted as referring to a hardware register. Declared logic variables are unsigned unless explicitly declared signed. Related to this, SystemVerilog includes both integer and int types; the former is the usual, signed verilog 4-state variable, and the latter is an unsigned variable.

Procedurally, an explicit logic declaration may be used thusly:

```
logic [15:0] Xin, Yin, Zin;
. . .
always@(posedge Clk)
    Xreg = {Xin, Yin};
```

Digital VLSI Design with Verilog

A typical concurrent use of these `logic` instances might be as follows:

```
fork
#2 Xin = Ain && Bin;
#3 Yin = Bin ^ Clevel;
join
```

25.4.1.2 New `bit` Type

This unsigned type is the same as a `reg`, except that it only can assume values composed of 1 and 0. If an x or z is included in a `bit` expression, it is converted to 0. For example,

```
bit[7:0] StartBit = 8'b0111_1010;
```

25.4.1.3 Unbased, Unsized ' Literals

When a ' prepends a one-bit literal with no preceding size value, and if no base unit is specification, it causes all bits to be set to the literal value given.

For example,

```
logic [7:0] a, b, c, d;
a = '0;   // Same as a = 8'b0000_0000.
b = '1;   // Same as b = 8'b1111_1111.
c = 'x;   // Same as c = 8'bxxxx_xxxx.
d = 'z;   // Same as d = 8'bzzzz_zzzz.
```

25.4.1.4 New Gated Event Control `iff`

This permits an operation to be performed if and only if a true evaluation of some specific kind succeeds. Otherwise, the operation is gated off, and it terminates with no effect.

For example, the following `always` block has no effect unless Ena is equal to 1'b1:

```
always@(ClockIn iff Ena==1'b1) . . .;
```

A change in `ClockIn` or `Ena` is what triggers evaluation of this `always` block.

25.4.1.5 Variants of the `always` Block

These process variants run within a module and differ from the simple `always` block in that any LHS variable to which they are assigned is not permitted to be assigned elsewhere in that module. The simulator will issue a warning if an assignment within the block is inconsistent with those for which the variant is designed. The `always_comb` and `always_latch` blocks do not require sensitivity lists, because all variables in the contained statements automatically are listed as sensitive.

The `always_comb` block models combinational logic. Example:

```
always_comb
  begin
  Products = B*B + B*X - 1;
  Result   = Products + xProducts;
  end
```

The `always_latch` block infers a level-sensitive latch. Example:

```
always_latch
    if (S_Z) S <= Z;
```

An `always_ff` block infers synchronous logic. Example:

```
always_ff @ (posedge Clk iff Rst == 1'b0)
    CountReg <= CountReg + 1;
```

25.4.1.6 New Timescale Alternatives

These are the `timeunit` and `timeprecision` commands, which may be used to replace `` `timescale ``. They may be included in any module, provided they are the first statements after the module header, allowing the simulation time scale to be changed module by module, if the tool permits it. Note that these new reserved words do not begin with a `` ` ``. If a `timeunit` value includes a **/** and a second argument, it becomes effectively the same as a `` `timescale `` for that module. Submodules inherit the `timeunit` and `timeprecision` values of the parent module.

Examples:

```
module MyModule (rest of header);
  timeunit 1ns;         // Time is in units of ns.
  timeprecision 100 ps; // Time is to a precision of 100 ps.
  (rest of module)
endmodule
```

```
module MyOtherModule (. . .);
  timeunit 1ns/100ps; // Same effect as "`timescale 1ns/100ps".
  . . .
endmodule
```

25.4.1.7 New Loop Commands

In addition to the verilog `disable`, SystemVerilog includes three new loop-control commands: `break`, `continue`, and `return`:

break: As in C, this causes the runtime to break out of a running loop, such as a `for` or a `while`.

Digital VLSI Design with Verilog

Example, using a modified `FindPatternBeh.v` of `Lab10`:

```
. . .
begin : Search
for ( j = 0; j <= 3; j = j ) // Might run forever.
  begin : WhileLoop
    while(1==1)   // This while thus loops forever.
      begin
        . . .
        if (CountOK==3)
          begin
            . . .
            FoundPads = 1'b1;
            break;   // Same as "disable WhileLoop;".
          end
. . .
```

continue: As in C, this triggers an immediate skip to the end of a loop statement.

Example:

```
for (i = 0; i<31; i = i+1)
   begin
   if (Reg[i] <= 8'h17)
       Reg[i] = Reg[i] + 1;
   else continue;   // Try next i.
   end
```

return: This causes an exit from a SystemVerilog `function` call or triggers an exit from a running `task`.

A SystemVerilog `function` call may be terminated by a `return`(*value*) command instead of by an assignment to the function's declared name.

For example, in our Lab11 `FnTask.v` example, instead of

```
function[7:0]   doCheckSum ( . . . );
 . . .
   doCheckSum = temp1[7:0] + temp2[7:0]^temp1[15:8] + temp2[15:8];
   end
endfunction
```

in SystemVerilog we may write,

```
function[7:0]   doCheckSum ( . . . );
 . . .
   return( temp1[7:0] + temp2[7:0]^temp1[15:8] + temp2[15:8] );
   end
endfunction
```

25.4.1.8 New interfact Type

As explained fully in chapter 25 of IEEE Std 1800, an interface is a named collection of nets or variables and is useful when a port list or other port property is repetitive among two or more modules. However, this benefit is not necessarily significant when the design process permits easy cut-and-paste repetition of port declarations.

Interfaces were briefly mentioned in *Week 8 Class 1* (section 16.1.3.3) of this Textbook. Interfaces are declared outside of modules. An interface may contain one or more of a parameter, constant, variable, function, task, modport, initial block, always block, or continuous assign, although any particular EDA tool may not implement all these features. Because the most common port declarations are of variables and nets, one would expect these to be implemented in any supporting EDA tool. An example follows, taken from the IEEE Std.

Consider this generic design fragment:

```
module memMod
   (input logic req, logic clk, logic start,
          logic[1:0] mode, logic[7:0] addr,
    output bit gnt, bit rdy,
    inout wire[7:0] data
   );
   logic avail;
   ...
endmodule
#
module cpuMod
   (input logic clk, logic gnt, logic rdy,
    output logic req, logic start, logic[7:0] addr,
           logic[1:0] mode
    inout wire[7:0] data
   );
   ...
endmodule
#
module top;
   logic req, gnt, start, rdy;
   logic clk = 0;
   logic[1:0] mode;
   logic[7:0] addr;
   wire[7:0] data;
   memMod mem(req, clk, start, mode, addr, data, gnt, rdy);
   cpuMod cpu(clk, gnt, rdy, data, req, start, addr, mode);
endmodule
```

Noticing the repetition in the module headers, it is possible to simplify those headers by means of an interface which contains their common port declarations. This is

Digital VLSI Design with Verilog

typically the simplest and most obvious use of an `interface`, to bundle a collection of variables or nets. Other, more involved uses are presented in the IEEE Std chapter referenced above.

Keeping in mind that a default `port` directionality of *inout* is assumed, the same design fragment as above may be rewritten as follows to take advantage of an `interface` declaration.

Here is the fragment rewrite:

```
interface simple_bus; // Define the interface.
   logic      req, gnt;
   logic[7:0] addr, data;
   logic[1:0] mode;
   logic      start, rdy;
endinterface: simple_bus
#
module memMod
   (simple_bus a, // Access the simple_bus interface.
    input logic clk
   );
   logic avail;
   //
   // When memMod is instantiated in module top, a.req is the
   // "logic req, gnt;" signal in the sb_intf instance of the
   // 'simple_bus' interface:
   always @(posedge clk)
      a.gnt <= a.req & avail;
   //
endmodule
#
module cpuMod(simple_bus b, input logic clk);
   ...
endmodule
#
module top;
   logic clk = 0;
   simple_bus sb_intf();        // Instantiate the interface.
   //
   memMod mem(sb_intf, clk);    // Connect the interface by position
   cpuMod cpu(.b(sb_intf), .clk(clk)); // or by name.
endmodule
```

25.4.1.9 New Assertion Language

The SystemVerilog assertion language is based on the Accellera *PSL* (Property Specification Language).

An example of the assertion language is a ***concurrent assertion*** as follows:

```
module MyModule (output . . ., input Clk1, Ena);
  . . .
  property CheckEna (Pre, Post);    // Declare the property to be asserted.
    @(posedge Pre) (Pre==1'b1) |=> ##[1:7] (Post == 1'b1);
  endproperty
  . . .
  Clk1_Ena: assert property ( CheckEna(Clk1, Ena) ) // State the assertion.
            else $display("%m [time=%d]: %s\n",
                      $time, "Ena failed to occur within 7 posedges of Clk1");
  . . .
  always@(posedge Clk1) // Violation message if no Ena within 7 clocks.
  . . .
endmodule
```

This kind of assertion is declared within a `package`, `module`, or `interface` and may be called in procedural code. Instead of the "`else $display...`" shown in the code example above, omitting the `else` action causes a default, tool-specific `$error()` system task to run if the assertion should fail. The `$error()` may be called explicitly and should accept *printf*-like formatting, the same as `$display()`.

Other messaging system tasks in SystemVerilog may be used to define severity levels; these include `$fatal()`, `$warning()`, and `$info()`. The verilog stratified event queue is extended for SystemVerilog in order to handle assertions and other enhanced language features.

Assertion messages can be verbose and thus may be used in place of comments in the code, wherever the comment would specify proper behavior of the module.

SystemVerilog assertions are simulation objects, only; they can not be synthesized and should be ignored by the synthesizer.

25.4.2 SystemVerilog Conclusion

There are, of course other SystemVerilog features useful for synthesis or synthesis-aimed simulation, but the ones discussed above are those most likely to be useful to a designer and to be implemented by an EDA vendor.

25.5 Verilog-AMS

25.5.1 Introduction

Verilog-A/MS means "Verilog Analogue / Mixed-Signal", in which "mixed" refers to mixed analogue and digital statements in the same verilog design.

Digital VLSI Design with Verilog

> **Notes on terminology:**
> We shall abbreviate *Verilog-A/MS* as **VAMS**.
> Because `analog` is a keyword in *verilog-A/MS*, the alternate spelling, "**analogue**", will be used wherever functionality is meant.

Many large ASICs include some analogue functionality. Analogue components generally require higher voltages and greater noise immunity than digital ones. VAMS may be used to simulate such an ASIC with accuracy good enough to ensure correct operation of the digital side of the design.

VAMS contains verilog (IEEE Std. 1364) as a proper subset but has no relationship with SystemVerilog. VAMS permits analogue modules as will be explained below. When a digital or analogue `module` can not itself resolve inputs of the other kind, VAMS permits a verilog source file to include declaration and instantiation of a special `connectmodule` for communication between the analogue and digital domains. Within a verilog `module`, VAMS provides `analog` blocks, similar to `always` blocks, to implement the analogue functionality of the module. Basically, VAMS allows a designer to code analogue functionality in verilog.

SPICE may be included, if necessary, in a VAMS design. A VAMS simulator is required to permit instantiation of SPICE-deck `subckt` code blocks in the same manner as verilog modules. VAMS specifies a core list of SPICE constructs required to be implemented by a conforming simulator; the precise version of SPICE otherwise is left to the simulator author.

25.5.2 Relationship to Other Languages

All VAMS analogue functionality, other than SPICE, is procedural, or is by connections among procedural blocks. However, `analog` procedural statements superpose effects; they do not define a sequence in time when they are read for execution. VAMS is closely related to verilog (IEEE Std 1364), to verilog-A (an earlier, nondigital Accellera adaptation of verilog), and to SPICE. Verilog-A is now a subset of VAMS. The analogue code in an `analog` block in VAMS actually, for the most part, uses verilog-A syntax; and, the VAMS analysis modes within a VAMS analogue domain correspond to those in SPICE.

A VAMS simulator is required to run two independent kernels, one of them computing and scheduling updates in continuous time for analogue constructs, and the other computing and scheduling in discrete time for digital constructs.

In the Synopsys tools, the *NanoSim* continuous-time simulator may be used in a VAMS program to drive *VCS* (discrete-time) and *HSPICE* (SPICE continuous-time). Mentor and Cadence also have VAMS implementations.

25.5.3 Analogue Functionality Overview

To code in VAMS, one declares verilog *wires* to be used as *nodes*, or "signals", which are connection elements which can convey current or sustain a potential (voltage). Any circuit of nodes and other elements is simulated according to Kirchoff's laws. In this context, a node between two or more points is considered a *branch* of the circuit.

Instead of the "=" and "<=" assignment operators of digital verilog, VAMS uses a contribution operator "`<+`" to associate a branch with analogue functionality. In a procedural (`analog`) block, successive contributions by default add to previous ones, representing the effect of the several connected components in the circuit. Depending on the context of a contribution, sometimes a new contribution can remove and replace a previous one. The general rule is that subsequent potentials or currents superpose on previous ones.

There are `V`(x, y) and `I`(x, y) operators to represent contributions of voltage or current, respectively. The operands x, and optionally y, define a branch by naming one or two nodes between which a potential is imposed or a current flows; when only one node is named, the other is taken to be a 0-volt, global ground.

Although in the following applications irrelevant to VLSI design, VAMS can be used to represent mechanical or hydraulic functionality such as the operation of a relay (switch) or a piston. For fluid flow, for example, this can be done by mapping the VAMS potential to hydrostatic pressure and the VAMS current to incompressible hydrodynamic flow.

VAMS can not in general be used for conditions in which the speed of light is important; it assumes immediate (procedural) effects simultaneously for all components on a node. Therefore, it can not be used for antenna design, although it can represent equilibrium states such as resonance among different nodes, as well as impedance-matching and other frequency-dependent phenomena.

25.5.4 Analogue and Digital Interaction

Analogue blocks can read outputs of digital blocks, and digital blocks can read outputs of analogue blocks, but an analogue block can not assign to a discrete-time variable, and a digital block can not assign to a continuous-time variable.

Using user-declared or predefined *natures*, different net *disciplines* may be declared for incompatible analogue and digital contexts. Nevertheless, VAMS can permit a wire to run between analogue and digital blocks usefully. This is done by `connectmodules` which define thresholds for a digital '1' or '0' in terms of voltage, permitting analogue calculation of transitions between digital levels. Mixed disciplines thus are resolved by connection rules. When disciplines of two nets are compatible, there is no further conversion required.

Synchronization of the analogue and digital simulator kernels is of some complexity.

The analogue kernel calculates all `analog` functionality in a module immediately and effectively continually, when anything changes. There is no blocking of statements or delay programming, although analogue effects which take time to occur (voltage changes on a capacitor, for example) will be simulated over time as they would in a SPICE transient analysis. The analogue kernel updates by iteration and stops when the state at that point in simulation time has been calculated to the required degree of precision. There is no minimum resolution of analogue time, except that imposed by the tool's own data formats. Like the digital kernel, the analogue kernel may calculate future events predictively, but these are placed in a queue and are not immediately made effective.

The digital kernel is the same as the verilog kernel discussed in *Week 5 Class 1*. It uses states at the current time, analogue or digital, to evaluate and modify variables according to the verilog event queue. An analogue update of a digital node is done by interpolation to the nearest digitally-resolved time.

Values of declared `real` or `integer` variables often may be calculated in either domain.

25.5.5 Example: VAMS DFF

We give here an example of a VAMS analogue model of a DFF, after Kundert and Zinke (2004; Section 6.8.3, Listing 3):

```
module DFFa (output Q, input D, Clk);
  voltage Q, D;
  localparam integer Direction = +1;   // +1 for posedge.
  integer State;
  analog
    begin
    @( cross(V(Clk) - 0.5, Direction) ) State = (V(D) > 0.5);
    V(Q) <+ transition(State? 1: 0);
    end
endmodule
```

The assumption is that we have a `VDD` of 1.0 V (digital CMOS high). The threshold-crossing operator ***cross*** permits the `State` variable assignment to occur whenever `V(Clk)` rises through 0.5V. The voltage of `Q` is assigned by the ***transition*** operator regardless of whether the preceding event control has been fulfilled or not. The *transition* operator provides a gradual, slightly sloped transition rather than an immediate digital one with infinite (vertical) slope.

25.5.6 Benefits of VAMS

The main benefit of VAMS is in relation to SPICE: It is more abstract than SPICE and simulates far faster; this makes VAMS usable for large analogue circuits.

A secondary benefit is that VAMS is convenient for large-design designers—which is to say, digital designers—because its analogue syntax is that of the familiar verilog, and its digital semantics is identical to that of verilog.

Otherwise stated, VAMS permits the easy integration of analogue functionality into large projects requiring top-down design methodologies.

25.6 Continued Lab Work (Lab 23 or Later)

25.7 Additional Study

(Optional) Read Thomas and Moorby (2002) section 11.1 for a project in which toggle-testing is used to represent power dissipation as a function of adder structure. You may wish to implement the verilog models suggested.

Optional readings in Palnitkar (2003):

Read section 15.1 on simulation, emulation, and hardware acceleration as aids to design verification. Also read the summary in section 15.4.

Read section 15.3 on formal verification.

Index

$display, **57**, 206
    example, 69, 225
$fatal, SystemVerilog, 538
$finish, 206, 245
$fullskew timing check, **366**
$hold timing check, **366**
$info, SystemVerilog, 538
$monitor, 217
$monitor, **57**
$nochange timing check, **368**
$period timing check, **368**
$recovery timing check, **367**
$recrem timing check, **368**
$removal timing check, **367**
$sdf_annotate command, 509
$setup timing check, **366**
$setuphold timing check, **367**
$skew timing check, **366**
$stop, 206, 245
    system task, 372
$strobe, **57**, 217
    example, 225
$time, **57**
$timeskew timing check, **366**
$warning, SystemVerilog, 538
$width timing check, **368**
&&&, in timing check, **371**
*, in event control, 71
`default_nettype, 233, 269
`define, **267–268**
    scope, 334
`ifdef, 267–268
    example, 49, 104
`include, example, 103
`timescale, 32, 269
–>, Event control trigger, 221
386-SX microprocessor, 476

A
Adder *vs.* counter, 137
ALF
    language, 529
    library format, 149
always block, 44, **221**, 246
    for concurrency, 253
    event control syntax, 54
    name = vars preserved, 258
    reading rationale, **217**
    scope, 335
AndOr.v, 18
Arithmetical shift, 156
Array
    addressing, 118
    multidimensional, 117
    select, 118
    verilog, **116**
Arrayed instance, **266**
Assertion
    defined, **57**, 363
    example, 85
    in serdes design, 499
assign, continuous, 18, 31
Assign-deassign, to avoid, 151
assign-deassign, to be avoided, 519
Assignment
    blocking, 54, 154, 222
    nonblocking, **54**, **55**, 154, 222
Assignment statement, 27
Asynchronous control
    flip-flop, **74**
    priority, 74
automatic function recursion, 256
automatic keyword, **256**
automatic task or function, **185**

B
Back-annotation, **507**
    SDF, 508

Backus-Naur Format (BNF), **70**
BASIC programming language, 522
Behavioral, 135, 136
    flowchart, 170
    synthesis, 177
Behavioral synch, serial clock, 169
Binary counter
    preloaded wraparound, 134
    programmed wraparound, 134
BIST. *See* Built-in self-test (BIST)
Bit select, 30
Bitwise operators, **28**, 33, **33**, 46
Block
    concurrent, **69**
    procedural, **69**
BNF. *See* Backus-Naur Format (BNF)
Boolean operators, **32**, **33**
Boundary scan, 473
Buffer, three-state, 160
`bufif1`, **158**, 232
    better cmos model, 310
    cmos model, 310
    example, 270
    switch-level model, 309
Built-in self-test (BIST), 476
    insertioncan, 477
    in isolated systems, 478
    test pattern example, 483

C
Case equality operator, 156, **250**
Case-sensitivity, verilog, 25
`case` statement, **53**
    example, 156, 158, **250**
    expression match, 249
`casex`
    to be avoided, **251**
    expression match, **251**
`casez`
    to be avoided, 253
    expression match, **252**
    wildcard match, **253**
`cell`, configuration keyword, 347
Charge strength, 309, **312**
    `trireg`, 149
    verilog, **149**

Checksum, **122**
Chip failures, causes, 527
Clock
    implementing, **55**
    serdes embedded, 288, 292, 294, 297
Clock domains, 141
    independent, 337
    indeterminate sampling, 338
    serdes, 167
    2-stage synchronizing ffs, 339
    synchronizing latches, 339
Clocked block, 73
Clock generator
    `always`, 56
    concurrent, 246
    `forever`, 56
    restartable, 247
`cmos`, switch-level primitive, **309**
Collapsing test vectors, 471
Comment
    macro regions, 43, **48**
    synthesis directive, 43
    verilog, **43**
Comment tokens, verilog, **26**
Compiler directive
    = macro, 50
    verilog list, 521, **521**
Concatenation, verilog, **120**
Concurrent block, 69
Concurrent block names, scope, 335
Conditional operator, **53**, 157
    expression match, **250**
`config`, **346**
    to be avoided, 347
    configuration keyword, 346
    scope, 335
    verilog configuration, 346
Constant, verilog, **68**
Contention, 150, 160
    in verilog, **149**
Continuous assignment, 18, **31**, 45, 73
Corner case testing, 473
Counter, **133**, 137
    behavioral, 144
    carry look-ahead, **139**
    gray code, **140**

Index

one-hot, **134**
overclocked, 142
ring, 140
ripple, **138**
synchronous, **139**
unsigned binary, 133
verilog, 100, 107
Coverage
  hardware testing, **471**
  in software, 471
  summary, 472

D

Dataflow modelling, 136
Datapath, 209
  vs. control unit, 197
DC. *See* Design compiler (DC)
Decoder
  example, 273
  tree example, 274
  verilog, **158**
Decoder example, better, 274
Decoder tree, `generate example`, **276**
Deep submicron effects, **525**
`default`, configuration keyword, 346, 347
`defparam`, to avoid, 151
`defparam`, to be avoided, **327**, 519
Delay
  avoiding procedural, 290
  blocking, 215
  conditional in `specify`, **354**
  conflict within `specify`, 356
  declared on net, **350**
  distributed, **348**
  full-path, **353**
  `ifnone` in `specify`, **355**
  intraassignment, **211**, **212**, 213
  lumped, **348**, 350
  min and max, 306
  multivalue, **213**, 214, **214**, **228**, **306**
  nonblocking, 212, 215
  not in UDP, 302
  overlap with `specify`, 356
  parallel-path, **353**
  pessimism, 214, **305**
  polarity in `specify`, 355
  procedural, 42, 213, 222
  regular, **211**
  SDF always supersedes, 357
  with strength, 305
  in synthesizable code, **204**
  timing triplet, **307**
  transport (VHDL), 212
  `trireg` to 'x,' 312
  6-value in `specify`, 354
  12-value in `specify`, 354
  vector net, 228
  to x, 214
Delay pessimism, moderated in `specify`, 374
Delay triplet, example, 350
Delay value, units, **29**
`DesDecoder`
  project synthesizable, **417**
  purpose, **291**
  redesign, **407**
Deserialization decoder, purpose, **291**
Deserializer
  generic, 289
  project schematic, 400
`Deserializer`, concurrent schematic, 415
Design compiler (DC)
  flattening logic, 87
  macro, predefined, 48
`design`, configuration keyword, 347
Design for Test (DFT), 469
  summarized, 438
Design partitioning
  rules, **336–337**
  for synthesis, 337
`design_vision`
  netlist viewer, 23
  schematic viewer, 25
D flip-flop
  from `nands`, **236**, **238**, **240**
  verilog, **61**
`disable`
  example, 169
  task or function, 186
`disable` statement, 166
D latch, verilog, **61**
`dont_touch`
  examples, 430

in verilog, 108
Drive strength, verilog, **148**
DVE gui, 21

E
ECC. *See* Error-checking and correction (ECC)
ECO, example, 293
Edge
    functional defined, **305**
    timing defined, **305**
`endconfig`, configuration keyword, 346
Equivalence checking verification, 529
Error-checking and correction (ECC), **120**, 122, **123**, 126
    finite element, **124**
    parity, **123**
    simple LFSR, 125
Error-handler, generic, 205
Error limit, pulse filter, 372
Event
    active, 212, **216**, 217
    vs. evaluation, 216
    future, **217**
    inactive (`#0`), 216, 217, **217**
    `$monitor`, 217, **217**
    nonblocking, 217, **217**
    queue example, 218
    regular, 212
`event` (keyword), 221
Event control, 44
    declared, **221**
    inline, 54
    `wait`, **221**
Event control @, **220**
Event control, –> trigger, 221
Event queue
    stratified, **215**, 216
    verilog, 151
Exponentiation, verilog, 156
Expression, defined, **71**

F
Fault simulator, 471
First-in, first-out (FIFO), 167
    bubble diagram, 198
    clock domains, 292

dataflow, 191
dual-clocking project, **416**
dual-port RAM, **416**
    introduction, **190**
    operational details, **192**
    project states, **198**
    project synthesizable, **417**
    read-write parts, 192
    schematic, 197
    state logic, 199
    transition logic, 200, 202, 203
`for`, **53**
    ease of use, 249
    examples, 248
    vs. `while`, 249
Force-release, to avoid, 151
`force-release` to be avoided, 519
`forever`, 246
    termination, 246
`forever` loop, 245
`fork-join`, 187, 212, 253, **254**
    example, 188
`for` loop, 245, 247
    in generate, 269
Formal proving verification, 529
Format specifier, **58**
    example, 58
Frame, serdes project, 98
Full-duplex serdes, 490
Full-path delay, **353**
Full scan, 475
Function
    declaration, **184**, **185**
    `Shift1`, 289
    `Shift1` improved, 290
`function`, 184, 185
    automatic, **256**
    `automatic` for recursion, 256
    example, 186
    scope, 335
    width indices, 256

G
Gate-level, 137
`generate`
    block declarations, 272

Index

conditional, **268–269**
decoder tree, **276**
downward hierarchy names, 266
looping, **269**, **271–272**
looping example, 270
loop scope quiz, **284**
`Mem1kx32gen` reqts, **282**
no nesting, 269
scope, 335
simple decoder, 273
`unrolled naming`, 272
`generate loop`, vector, **279**
`generate statement`, **267**
`genvar`
example, 270
in looping generate, **269**

H
Hard macro, defined, **108**
Header format, 1995, **42**
Header format, 2001, **42**
Header formats, contrasted, **41**
Hierarchy, in verilog, 263

I
Identifier, **69**
in ASIC library, 303
escaped, **69**
verilog, **69**
`if`, **53**, 245
expression match, 249
`iff`, example, 533
`ifnone` delay in `specify`, 355
Inertial delay, 143, **154**, 155, **372**
PATHPULSE example, 373
simulator errors, 214
`initial` block, 26, 44, 246
cautions, 56
example, 16, 27, 28
scope, 335
`inout`, not in UDP, 302
`inout` port, 234
`input` port, VHDL or SystemVerilog, 235
Instance arrays, **266**
`instance`, configuration keyword, 347
Instantiation, of module, 29

`integer`, 68
`interface`, in SystemVerilog, 335
Interface, partitioning, **335**
Internal scan, **475**
`Intro_Top`
scan chain, 85
scan I/O, 81
scan outputs, 87
`Intro_Top.v`, 15

J
JTAG port, 76, 78, 80, **88**
JTAG standard, 474

K
Keywords, lower case, 25

L
Lane, defined, **490**
Lane, PCIe, lane, 96
`large`, charge strength, 312
Latch
cts assign, 72
flip-flop preferred, 73
glitch-proof, **60**
vs. mux, **71**
synthesis, 72
synthesizable, **72**
verbose synthesis, 72
LFSR. *See* Linear feedback shift register (LFSR)
Liberty language, 529
Liberty, library format, 149
Liberty library timing checks, 364
LIFO, 190
Linear feedback shift register (LFSR), **122**, 124, 125, 478
polynomial, 123
simple, 125
Literal, 68
Literal expression, syntax, **28**
Literal syntax, **46**
Literal, syntax example, **31**
Localparam, 281
`localparam`
conditional example, 440
declaration, 324

examples, 343
sized, 428
sized state decs, 428
Logical operators, 28, **32**
Logic levels in verilog, 28, **28**

M
Macro
   example, 103
   recommended usage, **49**
Macro (compiler directive), 70
Macro, = compiler directive, 50
`medium`, charge strength, 312
Memory
   dual-port `DPMem1kx32`, **421**
   ECC, **120**
   random access, 115
   sequential access, 115
Mentor proprietary information, **3**
Messaging tasks, **56**
Model checking verification, 529
`module`, **25**, **26**
   ANSI header, **324**, 326
   for concurrency, 255
   output `reg` ports, 519
   scope, 334
   traditional header, **324**, 326
Module, example, 26
Module header, 27, **27**
   formats, **40**
`module` instance, scope, 334
MOS
   resistive strength rules, 308
   switch-level primitives, **308–309**
Mux
   schematic, 63
   switch-level model, 315
   verilog, 63

N
Named block, **165**, 166, 186
`nand`, switch-level model, 316
Net, definition, **67**
`nmos` primitive, 308, 310
Noise estimation problems, **530**
`none` implied net default, 233

`nor`, switch-level model, 316
`noshowcancelled specparam`, **375**
`not`
   example, 270
   switch-level model, 311
`notif1` example, 270
`Notifier`, 365
Notifier, in timing check, 364, **371**
Notifier `reg` example, 372, 379

O
Observability, hardware testing, **470**
Operator precedence, verilog, **157**
Operators
   bitwise vs. logical, 164
   verilog table, **156**

P
Packet serdes, 168
Packet serdes project, 98
Parallel block (`fork-join`), 187
Parallel-path delay, **353**
Parallel-serial converter, **108**
Parameter, 48
   declaration, 235
   override, 235
   vs. port declaration, 235
   `real`, 103
   `signed`, 103
`parameter`, 42, **68**, 280
   in ANSI header, 324
   generic declaration, **324**
   index range, 324
   not in literals, 107
   override by name, **325**
   override by position, **325**
   real, 324
   signed, 324, 325
Parity
   memory, **121**
   serdes data, 127
   `xor` ^ operator, 121
Partitioning, analog-digital example, 294
Part select, 30
Pass-switch primitives, **311**
Path delays, full and parallel, **353**

Index

PATHPULSE
   conflict rules, 373
   inertial delay control, **373**
   specparam, 373
PATHPULSE example, 373
PCI Express (PCIe), **96**
   analogue issues, 97
   lane, 96
Phase-locked loop (PLL), **89**
   clock-comparator, 106
   clock extraction, 168
   digital lock-in, 93
   SerialRx.v, **403**
   smooth comparator, 93
   synthesizable, **384**, **400**
   synthesizable ClockComparator, 388
   synthesizable design block diagram, **399**
   synthesizable sync function, 186
   32x, **100**
   32x blocks, 100
   32x schematic, 102
PLI, verilog, 521
PLL. *See* Phase-locked loop (PLL)
PLL 1x, **90**
   schematic, 90
   synthesizable, **389–391**
pmos primitive, 308, 310
Port connection rules, **234**
Port declaration, 27
Port map, of module, 29
Power distribution problems, **530**
Primitive
   switch-level logic, **231**
   verilog gates, **231**
primitive, 301
   scope, 334
Problems
   noise estimation, **530**
   power distribution, **530**
Procedural, 135
   assignment, **31**
   block, 69
Procedural block names, scope, 335
Property specification language (PSL), 529
PSL language, 537
pulldown gate, 232

pulldown primitive, 312
pullup gate, 232
pullup primitive, 312
Pulse filtering delay limits, 372
pulsestyle_ondetect inertia, **374**
pulsestyle_ondetect specparam, 374
pulsestyle_onevent inertia, **374**
pulsestyle_onevent specparam, 374

R

Race condition, **74**, 152, 153, 215
   defined, 152
Race, initial blocks, 154
RAM
   bidir wrapper, 131
   Mem1kx32 schematic, 129
   simple verilog, 119
   size issues, 116
rcmos primitive, 311
realtime reg type, 375
real variable, 90
Reconvergent fanout, **59**
Reduction operator, **33**, 34, 46
reg, **30**
   input port illegal, 42
   in output port, 42
   vs. trireg, 312
Register Transfer Logic (RTL), 30, 136, 169
   defined, 136
Rejection limit, pulse filter, 372
Relational expression, of 'x,' 155
Relational operator, 46, **46**
repeat, 247
repeat loop, **247**
Replication operator, 157
rnmos primitive, 308
Rounding of decimals, 103
rpmos primitive, 308
rtranif0 primitive, 311
rtranif1 primitive, 311
rtran primitive, 311

S

Scan
   boundary, **78**, 473

chain, 84
full, 475
internal, **76**, 475
Scan chain, `Intro_Top`, 85
Scheduled conflicts, 215
SDF
   annotation messages, VCS, 465
   back annotation, 508
   delay always prevails, 357
   file, **36**
   file sample, 514
   net delays, 508
   overrides delays, 508
   path delays, 508
   summary, 509
   syntax, 508, 509
   use with simulator, 509
   in verilog flow, **508**
Serdes
   class project, **98**
   clock domains, 167
   `DesDecoder`, 299
   embedded clock, 168, 288, 292
   FIFO, 97
   file organization, final, 448
   full-duplex project, 490
   packet, **111**
   project block diagram, 382, 446
`SerEncoder`, block diagram, 452
`Serializer`, project schematic, 449
Serial-parallel converter, **287**
`Set_dont_touch`, examples, 430
Shift, arithmetical, 156
Shift register, **58**
   example, 440
   parallel-load, **63**
   procedural, 64
   RTL, 64
   schematic, 59, 62, 64
   serial load, **62**
`showcancelled` inertia, **375**
`showcancelled specparam`, 374
Simulation unknowns, display values, 31
Simulators, strength spotty, 158
`small`, charge strength, 312
Soft errors, hardware, 470, 478

Source switch-level models, **312**
`specify` block, **351**
   summary, 352
   6-value delays, 354
`specify`, scope, 335
`specparam`, 351
   defined, **352**
   example, 353, 360
   with timing triplets, 353
SPICE, 310
   in verilog A/MS, 539
SR latch, 236, **237**
`SR.latch`, 19
SR latch, reset sequence, 241
`SR.v`, 19
Standard parasitic exchange format (SPEF), 508
State clock generator, synthesizable, 429
State machine
   design, **189**
   verilog, **189**
Statement, defined, **71**
Static serial clock synch, 178
Strength
   assigning, 150
   charge, **149**
   charge values, **312**
   with delay, 305
   drive, **148**
   resistive MOS rules, 309
   table, **150**
String
   verilog, 68
   verilog storage, 56
Structural, 135, 136
`supply0` net type, 233
`supply1` net type, 233
Switch level model, 149, **307**
Synopsys Design Constraint format (SDC), 520
Synopsys proprietary information, **3**
Synthesis, importance of, **37**
SystemC, can use SDC, 520
System function, 70
System task, 70
System tasks and functions, 521
   verilog list, **520**
SystemVerilog, **531**

Index

`always_comb` block, 534
`always_ff` block, 534
`always_latch` block, 534
assertion language example, 538
`break` command, 534
`continue` command, 535
`$fatal`, 538
`$info`, 538
`interface` example, **536**
`logic` type, 532–533
new assertion language, 537–538
new `bit` type, 533
new gated event control `iff`, 533
new `interface` type, 536–537
new looping commands, 534–535
new `timescale` alternatives, 534
primary features, **531–538**
`return` command, 535
structure, 531
unbased, unsigned ' literals, 533
`$warning`, 538

T
`table`
  in sequential UDP, 304
  in UDP, 302–304
TAP controller, 78, 474, 478
TAP port, 527
`task`, 184, 185
  automatic, **256**
  `automatic` vars not shared, 256
  for concurrency, 253
  concurrency example, 254
  exercise, 205
  local vars static shared, 256
  scope, 335
Task data sharing, 185
Task declaration, 184
Task, `Unload32`, 290
Testbench `Intro_Top`. example, 16
Test vectors, collapsing, 471
T flip-flop, 138
T flip-flop, 180
Three-state buffer, **157**, 160
`time reg` type, **375**

Timescale macro, **29**
Timescale specifier, 17
Timing arc
  defined, **347**
  examples, 348
Timing check, 70
  as assertion, **364**
  avoiding negative limits, **371**
  conditional event, **371**
  data event, 364
  feature summary, 364
  `$fullskew`, **366**
  `$hold`, **366**
  Liberty library, 364
  limits must be constant, 365
  negative limits, **369**, **370**
  `$nochange`, **368**
  notifier, 364, 371, 372
  `$period`, **368**
  in QuestaSim, 366
  `$recovery`, **367**
  `$recrem`, **368**
  reference event, 364, 365
  `$removal`, **367**
  `$setup`, **366**
  `$setuphold`, **367**
  `$skew`, **366**
  vs. system tasks, 363
  table of 12, **365**
  timecheck event, 365
  0 time limits, 365
  `$timeskew`, **366**
  timestamp event, 365
  `$width`, **368**
Timing paths
  and arcs, **348**
  causality, 348
Timing triplets, **307**
  example, 350
Toggle flip-flop, 138
`tranif0` primitive, 311, 315
`tranif1` primitive, 311, 315
`tran` primitive, 311, 312
Transfer-gate primitives, **311**
`triand` net type, 233

`tri` net type, 233
`tri0` net type, 233
`tri1` net type, 233
`trior` net type, 233
`trireg`
    charge strength, 149
    example, 313
    net type, 233
    pulse-filter model, 317
    switch-level net, 312
    vs. `tran` primitive, 311
TSMC proprietary information, **3**

U
UDP. *See* User-defined primitives (UDP)
`use`, configuration keyword, 347
User-defined primitives (UDP), **301**
    combinational example, 302
    sequential example, 303
    summary, 304

V
VAMS = verilog A/MS, 539
Variable-frequency oscillator (VFO)
    new comparator sampler, 383
    old comparator sampler, 383
    project `FastClock` oscillator, 385
    synthesizable, 385, 386
VCD file, **35**
VCS, schematic viewer, 23
Vector, **29**, 45
    bit significance, **32**
    example, 32
    index syntax, **30**
    logical operator, 46
    negative index, 52
    select, 119
    sign bit, 45
    type conversions, **47**
    verilog, 116
    width (type) conversion, **47**
Verification
    equivalence checking, 529
    formal, **529**
    formal proving, 529
    functional, **528**

model checking, 529
timing, **528**
Verilog
    arrayed instance, **266**
    attributes unused, 519
    clocked block, 222
    coding rules, 220, 222
    comment directives, 519
    compiler directive list, 521, **521**
    compiler directives, 518
    conditional compile, **267**
    configuration, 346
    declaration ordering, **257**
    declaration regions, 257
    hierarchical names, **264**, **265**, **333**
    hierarchy path, 263
    keywords, 518
    named block, **165**
    old ACC C routines, 523
    old TF C routines, 523
    operators, **46**
    PLI, 70, 519, **521**
    PLI VPI C routines, 523
    scope of names, **334**
    simulator file I/O, 518
    synthesizable, **75**
    synthesizable summary, 517
    system task and function list, **520**
    system tasks and functions, 518
    three wiring tricks, 519
    UDP (primitive), 301
    variable, **67**
    1995 *vs.* 2001, 517
Verilog A/MS
    `analog` block, 539
    analogue FF example, 541
    analogue functionality, **540**
    analogue/mixed signal, **538**
    applicability, 540
    benefits, **541**
    `branch`, 540
    `connectmodule`, 540
    `cross` operator, 541
    `discipline`, 540
    dual-kernel functionality, **540**
    `I` operator, 540

Index 553

nature, 540
node, 540
`operator`, 540 (please insert symbol)
SPICE allowed, 539
`transition` operator, 541
two kernels, 539
= VAMS, 539
verilog-A relationship, 539
V operator, 540
VFO. *See* Variable-frequency oscillator (VFO)
VHDL, can use SDC, 520

W
`wand` net type, 233
Watch-dog device, 260
`while`
  vs. `for`, 248
  example, 248
  header delay, 249

`while` loop, **247**
Width specifier, 17
`wire`
  implied names, **232**
  implied net, 233
  net type, 233
  other net types, 233
Wired and, **68**
Wired or, **68**
Wired-or, verilog, 144
`wor` gate, verilog, 144
`wor` net type, 233
Wrapper module methodology, **393**

X
`XorNor.v`, 20

Z
Z state, not in UDP, 302

Printed by Printforce, the Netherlands